About Island Press

Island Press is the only nonprofit organization in the United States whose principal purpose is the publication of books on environmental issues and natural resource management. We provide solutions-oriented information to professionals, public officials, business and community leaders, and concerned citizens who are shaping responses to environmental problems.

In 2003, Island Press celebrates its nineteenth anniversary as the leading provider of timely and practical books that take a multidisciplinary approach to critical environmental concerns. Our growing list of titles reflects our commitment to bringing the best of an expanding body of literature to the environmental community throughout North America and the world.

Support for Island Press is provided by The Nathan Cummings Foundation, Geraldine R. Dodge Foundation, Doris Duke Charitable Foundation, Educational Foundation of America, The Charles Engelhard Foundation, The Ford Foundation, The George Gund Foundation, The Vira I. Heinz Endowment, The William and Flora Hewlett Foundation, Henry Luce Foundation, The John D. and Catherine T. MacArthur Foundation, The Andrew W. Mellon Foundation, The Moriah Fund, The Curtis and Edith Munson Foundation, National Fish and Wildlife Foundation, The New-Land Foundation, Oak Foundation, The Overbrook Foundation, The David and Lucile Packard Foundation, The Pew Charitable Trusts, The Rockefeller Foundation, The Winslow Foundation, and other generous donors.

The opinions expressed in this book are those of the author(s) and do not necessarily reflect the views of these foundations.

About the Center for Applied Biodiversity Science

The Center for Applied Biodiversity Science (CABS) at Conservation International (CI) was launched in 1998 to strengthen the ability of CI and other institutions to identify the emerging threats to earth's biological diversity accurately and respond to them quickly. CABS brings together leading experts in science and technology to collect and interpret data about biodiversity, to forge partnerships, to plan conservation priorities, and to build strategic action plans leading to concrete conservation outcomes.

The Atlantic Forest of South America

The Center for Applied Biodiversity Science
at Conservation International

State of the Hotspots

Carlos Galindo-Leal, Scientific Editor
Philippa J. Benson, Managing Editor

The Atlantic Forest of South America: Biodiversity Status, Threats, and Outlook
edited by Carlos Galindo-Leal and Ibsen Gusmão Câmara

The Atlantic Forest of South America

Biodiversity Status, Threats, and Outlook

EDITED BY
Carlos Galindo-Leal
Ibsen de Gusmão Câmara

The Center for Applied Biodiversity Science
at Conservation International

ISLAND PRESS
Washington • Covelo • London

The Web sites that appear in this volume were each live at the time of publication. With the exception of part III, the part opening images are used by permission of Conservation International. The part III image is used by permission of Roberto Francisco.

Library of Congress Cataloging-in-Publication data.

The Atlantic Forest of South America : biodiversity status, threats, and outlook / edited by Carlos Galindo-Leal, Ibsen de Guzmão Câmara.
p. cm.—(State of the Hotspots)
Includes bibliographical references and index.
ISBN 1-55963-988-1 (cloth : alk. paper)—ISBN 1-55963-989-X (pbk.: alk. paper)
1. Forest conservation—Brazil—Mata Atlântica. 2. Biological diversity conservation—Brazil—Mata Atlântica. 3. Forest management—Brazil—Mata Atlântica. 4. Forests and forestry—South America. I. Galindo-Leal, Carlos. II. Câmara, Ibsen de Gusmão. III. Series.
SD414.B6A84 2003
333.75′16′0981—dc21

2003001903

British Cataloguing-in-Publication data available.

Design by Brighid Willson.

Printed on recycled, acid-free paper 

Manufactured in the United States of America
10 9 8 7 6 5 4 3 2 1

Contents

Foreword

The Atlantic Forest region of South America tops the world in statistics of habitat loss, with over 93 percent of the original range of the forest already gone. Approximately twelve years ago, Conservation International selected the Atlantic Forest as one of its top-priority biodiversity hotspots worldwide. This recognition drew attention and resources from international and national organizations, private institutions such as the MacArthur Foundation, and bilateral and multilateral government agencies. The reasons the Atlantic Forest deserves global attention are to be found in the wealth of information presented in this first volume of the State of the Hotspots series. Some 3,000 plant species, 35 mammals, and 104 bird species are considered threatened. Close to 70 percent of all Brazilians live within the original distribution of the Atlantic Forest, and three of the largest urban centers on the continent are also located there. In his classic work *With Broadax and Firebrand: The Destruction of the Brazilian Atlantic Forest,* Warren Dean describes in detail the history of degradation of this incredibly wealthy region. Many other treatises have also been written on its problems. The fate of the Atlantic Forest in Paraguay and Argentina is not that different.

But the history of the Atlantic Forest is beginning to be rewritten. Of all the tropical biodiversity hotspots, this region now has the best capacity to respond to the plethora of insults it has withstood over many centuries. Over 40 protected areas have been created in the last twelve years, although still less than 20 percent of the remaining area is under strict protection. Particularly in the southern and southeastern regions, but also in several other areas, conservation efforts are being carried out by a significant army of trained professionals, dozens of capable research institutions and nongovernmental organizations, and several increasingly effective government agencies, all supported by emerging democratic structures for effective governance. On-the-ground monitoring, coupled with affordable remote sensing technologies, is enabling the assessment of land-use trends, which is vital for the survival of the region's rich biodiversity. Public campaigns and environmental education programs have helped to raise awareness of the resources that are at risk of vanishing from this biome. The public is also becoming educated about the dependence of humans on vital ecosystem services, such as soil replenishment and water availability, that are rapidly dwindling throughout the extent of the forest.

The Atlantic Forest can be considered the cradle of Brazil's environmental movement and is home to its most capable universities, research centers, and nongovernmental organizations. Fundação SOS Mata Atlântica, whose principal focus is the conservation of remaining forests in the region, is the largest membership organization in Brazil. Together with Conservation International, SOS Mata Atlântica has formed the Mata Atlântica Alliance, which envisions zero deforestation. A new generation of public attorneys specializing in environmental issues is helping to ensure that progressive legislation regulating the use of natural resources and protecting biodiversity is enforced. And a growing number of park managers and rangers, responsible for hundreds of protected areas, look at their professions with a new commitment.

Several universities contribute to the training of dedicated individuals in many environmental disciplines. Citing just one example, in the graduate program in ecology, conservation, and wildlife management of the Federal University of Minas Gerais—created in 1989 with resources from Brazilian government agencies, the U.S. Fish and Wildlife Service, the MacArthur Foundation, World Wildlife Fund–U.S., Conservation International, and the Fundação Biodiversitas—more than 150 students have earned masters degrees or doctorates. Most graduates are now employed in training institutions, nongovernmental organizations, government agencies, or the private sector. Many of their theses and dissertations focused on conservation issues in the two Brazilian hotspots, the Atlantic Forest and the Cerrado. Scientists and research centers are now conducting hundreds of research projects throughout the Atlantic Forest.

The private sector has also taken up the challenge. In late 2002, four major Brazilian companies with operations in the Atlantic Forest region joined with Conservation International to form the Instituto BioAtlântica (IBIO). IBIO, a not-for-profit organization, is currently working to include the private sector in the conservation and restoration of the region and to promote sustainable development.

The results are beginning to show. The Atlantic Forest was the first major Brazilian ecosystem to be the focus of several biodiversity conservation planning exercises, culminating with a master plan for the entire biome. This work was done at the request of the Ministry of the Environment, state agencies, Conservation International, Fundaçao Biodiversitas, Fundação SOS Mata Atlântica, and Instituto de Pesquisas Ecologicas (IPE), in addition to national and international funding agencies. The master plan is now part of the National Biodiversity Strategy, Brazil's commitment under the Convention on Biological Diversity. This biodiversity blueprint is complemented by a remote sensing program coordinated by Fundação SOS Mata Atlântica and Instituto Nacional de Pesquisas Espaciais (INPE), the Brazilian space agency that is monitoring forest cover throughout the region. These exercises have spilled over into Argentina and Paraguay, creating a trinational initiative.

More than 40 new protected areas have been created in the past 10 years, including several state parks and reserves. More ambitious projects also have been proposed, such as the creation of two large-scale corridors that span most remaining vegetation; the aim is to maintain and restore connectivity between existing forest blocks. These proposals are now receiving funding from the G7 Pilot Program to Conserve the Brazilian Rain Forest (PP-G7), administered by the World Bank. The Critical Ecosystems Partnership Fund—a joint venture that includes Conservation International, the World Bank, the Global Environment Facility, the MacArthur Foundation, and the Japanese government—is ready to invest significant resources in projects conceived and implemented by nongovernmental organizations, universities, and community groups.

But much remains to be done. Most of these efforts, although providing hope for sustaining the Atlantic Forest and ensuring the protection of its biodiversity, are still in their initial stages. For these emerging initiatives to succeed, their progress must be monitored through the use of adequate indicators. A thorough examination of the Atlantic Forest, in Argentina and Paraguay as well as in Brazil, inaugurates the State of the Hotspots series, which aims to analyze the state of biodiversity in the most crucial areas worldwide. By proposing and tracking suitable indicators, the entire conservation community can invest its efforts in the most urgent activities, including but not limited to training people; monitoring species, habitats, and ecosystems; creating additional protected areas in key irreplaceable habitats; restoring forests; and educating the public. With mechanisms to track progress as well as the occasional and inevitable setbacks, we firmly believe we can make the Atlantic Forest the first success story among the global biodiversity hotspots.

—Gustavo A. B. da Fonseca, Russell A. Mittermeier, and Peter Seligmann

Preface

The Atlantic Forest of South America could be a poster child for biodiversity hotspots around the world. Less than 8 percent of the original coverage of the Atlantic forests of Brazil, Argentina, and Paraguay remains, much of that in small fragments. The rich biology of the region is hanging on by a thread as human population in the region continues its explosive growth and the aspirations of the inhabitants for a more consumptive lifestyle continue to increase demands on the environment.

Amazingly, nearly all the species known to be in the original 1 to 1.5 million-square kilometers of Atlantic Forest can still be found, though often in small isolated fragments of the original cover. Even new species of primates have been discovered there in the last decade. Biologists the world over are recognizing the richness of the Atlantic Forest and leading a charge to preserve the remnants and restore as much as possible of the former glory of the region. The more that is known about the biodiversity of the Atlantic Forest and the ongoing threats it faces, the better conservation programs can be directed.

Many scientific studies have been made in the region, but the nature of these studies usually focuses in detail on a small portion of the whole. When the authors of these studies, however, can bring the pieces together, a clear picture develops that is more powerful than the sum of the individual parts. This book brings the picture together for the first time. It summarizes much of what is known about the biological diversity of the region and focuses on the continuing threats to what is left. In its breadth and depth, this book should play a major role in informing and driving conservation programs in this very important region.

—Gordon E. Moore

Acknowledgments

We hope that this book contributes to the process of halting destruction of the Atlantic Forest hotspot and beginning its restoration. One of the first steps in this endeavor is to get people on the same page. We have been fortunate to get a variety of people from different disciplines, affiliations, regions, and countries on the same pages—literally. Each of them is intensely committed to preventing the loss of one of the greatest treasures of our planet. We are extremely grateful to all the authors for their contributions and for the trust they placed in us to adjust their chapters to fit the book requirements. As you will read in the book, they are passionate about their work and filled with optimism in the face of the alarming trends they describe.

Luiz Paulo Pinto, Maria Cecília Wey de Brito, Monica Fonseca, Paulo Gustavo, and Roberto Cavalcanti from Conservation International Brazil provided logistical support, valuable comments on several chapters, and help in recruiting an excellent team of Brazilian authors.

From the beginning of the project, Hernán Povedano and José Luis Cartes went out of their way to invite and help interview many of the contributors from Argentina and Paraguay. They continued to provide much-needed assistance throughout the development of the book.

Penny Langhammer undertook the painstaking job of creating maps to illustrate the chapters and provide the reader with a sense of place. She also made immense contributions in editing early versions of the manuscripts and in managing the entire project. Rob Waller, Greg Buppert, Lorena Bustos, and Mark Denil from Conservation International helped Penny create the maps, from obtaining map layers to editing final illustrations. Many information layers for the creation of these maps were provided by Fundação SOS Mata Atlântica, Guyra Paraguay, and Germán Palé and Guillermo Placci of Fundación Vida Silvestre Argentina.

The contributions benefited greatly from the work of the CABS publications team, led by Philippa Benson. Her relentless scrutiny and tenacious editing greatly improved the clarity and readability of the final book. Her team included the superb translators Laura Vlasman and Muriel Vasconcellos, who translated

Portuguese and Spanish versions into English in meticulous academic detail. They scrutinized the clarity of meanings, references, and consistency, often examining primary sources. Neil Lindeman and Natasha Atkins thoroughly cleaned up many chapters of the book. Glenda Fabregas contributed her ideas to early versions of the book's cover. Brighid Willson from Island Press developed the interior design and later versions of the cover. We also thank Kevin Schafer for the use of his photographs on the cover of this volume, including the background image and the muriqui (right most inset).

We were lucky to have Anthony Rylands down the hall here at CABS, with his encyclopedic knowledge of the Atlantic Forest (and his library). His careful review of parts of this book added much polish and many edges of accuracy.

Finally, Barbara Dean, Barbara Youngblood, and Laura Carrithers of Island Press took on this project enthusiastically. Of course, grateful thanks go to Silvio Olivieri, Gustavo Fonseca, and Russ Mittermeier from Conservation International, who provided guidance and much-needed support throughout the process.

Bringing together the various aspects of this book has been a complex and sometimes daunting enterprise. We hope the book gives you the same understanding and sense of urgency we have about the future of the Atlantic Forest.

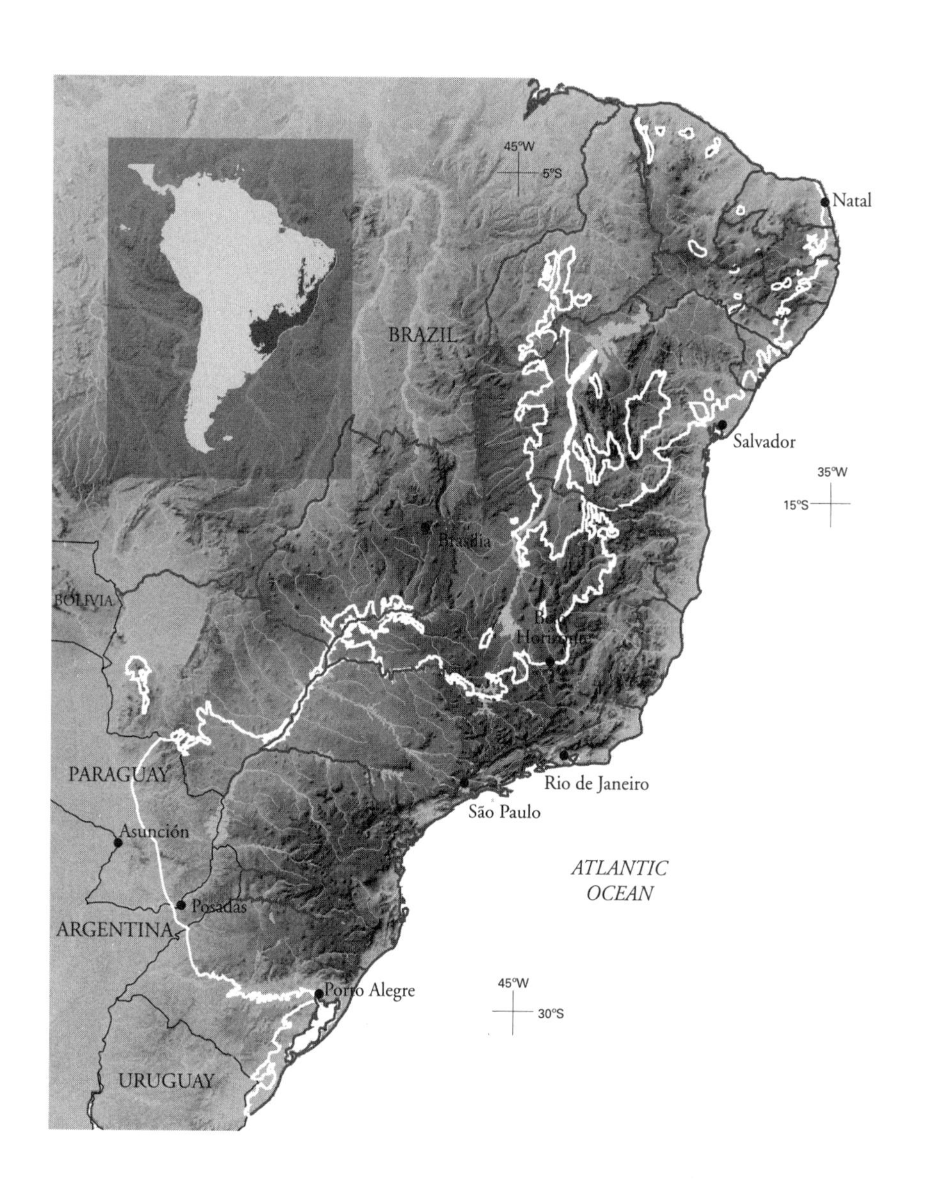

45°W
5°S
Natal
BRAZIL
Salvador
35°W
15°S
Brasília
BOLIVIA
PARAGUAY
Rio de Janeiro
São Paulo
Asunción
ATLANTIC
OCEAN
Posadas
ARGENTINA
Porto Alegre
45°W
30°S
URUGUAY

Thus the modern Atlantic Forest evolved and came finally to occupy its historical boundaries, altogether like some remote, antique empire, its origins mythical, its dynasties extending over epochs, its splendor astonishing, its inhabitants luxurious, shrewd, and conservative in their exploitation of its bounteous resources, for millennia unchallenged and unchallengeable in its perfect and total dominion, yet at its foundation utterly brittle and vulnerable.

—Warren Dean, *With Broadax and Firebrand: The Destruction of the Brazilian Atlantic Forest* (1995)

PART I

Introduction

Chapter 1

Atlantic Forest Hotspot Status: An Overview

Carlos Galindo-Leal and Ibsen de Gusmão Câmara

The Atlantic Forest is one of the world's 25 recognized biodiversity hotspots, areas where the original vegetative cover has been reduced by at least 70 percent but that together house more than 60 percent of all terrestrial species on the planet. These critical areas occupy less than 2 percent of the earth's land surface. More than 1.1 billion people live in urban and rural areas across these hotspots, and about one-quarter of them exist in dire poverty. Many of the people living in hotspots and many economies in these areas directly depend on the products of healthy ecosystems, harvesting wild plants and animals for their food, fuel, clothing, medicine, and shelter. Within the world's hotspots, then, there is a convergence of areas where millions of people live in poverty, where there is high biodiversity and endemism, and where a broad array of additional factors drives rapid habitat loss. In other words, in the world's hotspots many species, including people, share a common vulnerability and struggle for survival.

The Atlantic Forest hotspot is arguably the most devastated and most highly threatened ecosystem on the planet. It is a hotspot where the pace of change is among the fastest and, as a consequence, where the need for conservation action is most compelling. Although the Atlantic Forest is thought to have originally ranged from 1 to 1.5 million km^2, only 7–8 percent of the original forest remains (Figure 1.1).

Drivers of Biodiversity Loss

The loss of biodiversity can include the loss of ecosystems, populations, genetic variability, species, and the ecological and evolutionary processes that maintain this diversity. In the Atlantic Forest hotspot, the causes and dynamics of biodiversity loss are extraordinarily complex, fueled over time by a history of inequitable land tenure systems and local, national, and international trade relations. The specific causes of loss range from the short-term subsistence incentives of local farmers, to generalized national policies, to the global marketplace.

With its large latitudinal range, the Atlantic Forest region is remarkably

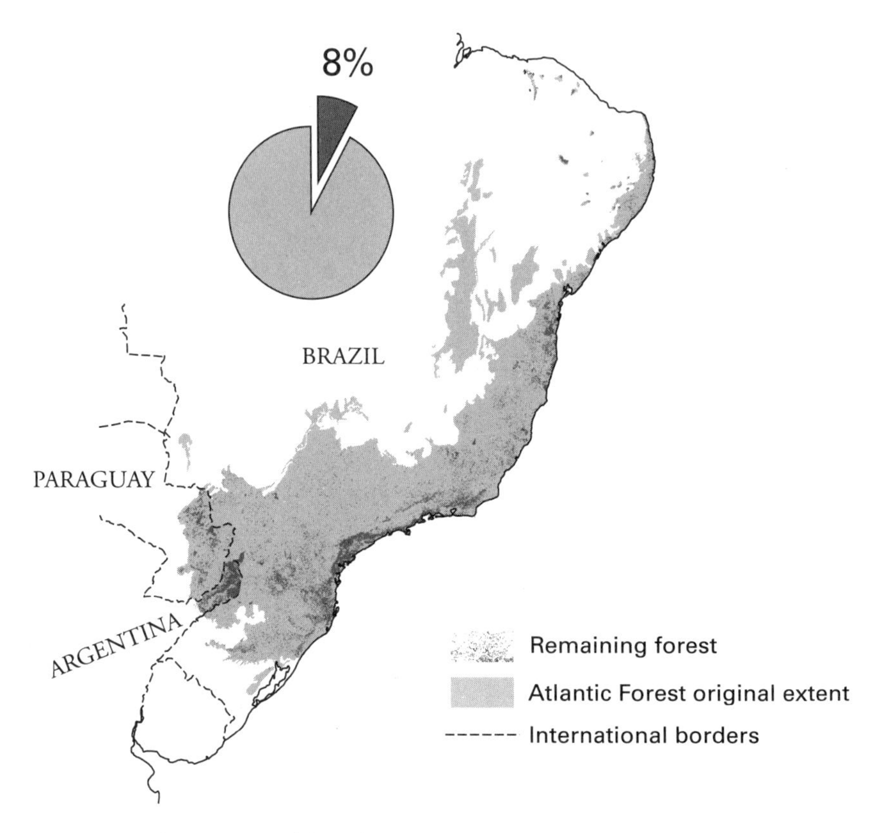

Figure 1.1. Only 8 percent of the Atlantic Forest's original extent remains.

heterogeneous. Similarly, socioeconomic conditions and pressures are diverse across the hotspot, and the status of biodiversity varies throughout because of the differential impacts of these conditions and pressures. Since colonization by the Portuguese and Spanish, the Atlantic Forest has had a long history of intensive land use for commodity exports, including cycles of exploitation of brazilwood, sugarcane, coffee, cocoa, and cattle grazing, all of which have utterly transformed the landscape. More recent drivers of biodiversity loss include intensive forms of government-subsidized soy agriculture and expanding forest plantations of pine and eucalyptus.

The fragments of the original Atlantic Forest that remain continue to deteriorate because of fuelwood harvesting, illegal logging, plant and animal poaching, and the introduction of alien species. In addition, the construction of hydropower dams has substantially contributed to habitat loss and ecological changes in the region; despite the broadly recognized ecological and social devastation rendered by dam construction, several dam projects remain under way.

Human populations are particularly high in the Atlantic Forests of Brazil, where more than 100 million people reside. In fact, populations in all three of the countries included in the Atlantic Forest hotspot (Brazil, Argentina, and Paraguay) have increased substantially over the last 50 years (Figure 1.2).

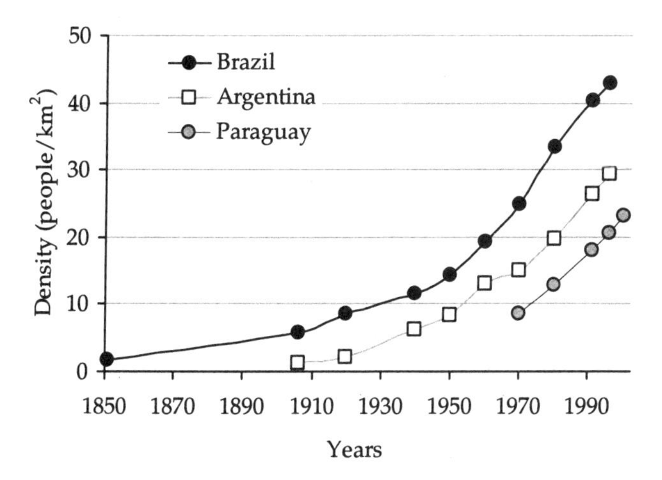

Figure 1.2. Population densities have accelerated dramatically over the past 50 years.

This growth has led to destruction of forest through uncontrolled urban expansion, industrialization, and international migration. The relationship between population growth and deforestation is unclear at best (Figure 1.3); moreover, the felling of rainforests for agricultural or urban development has not necessarily improved the quality of life for rural populations, and the expansion of tourism infrastructure has had an overall negative impact on the coastal environment.

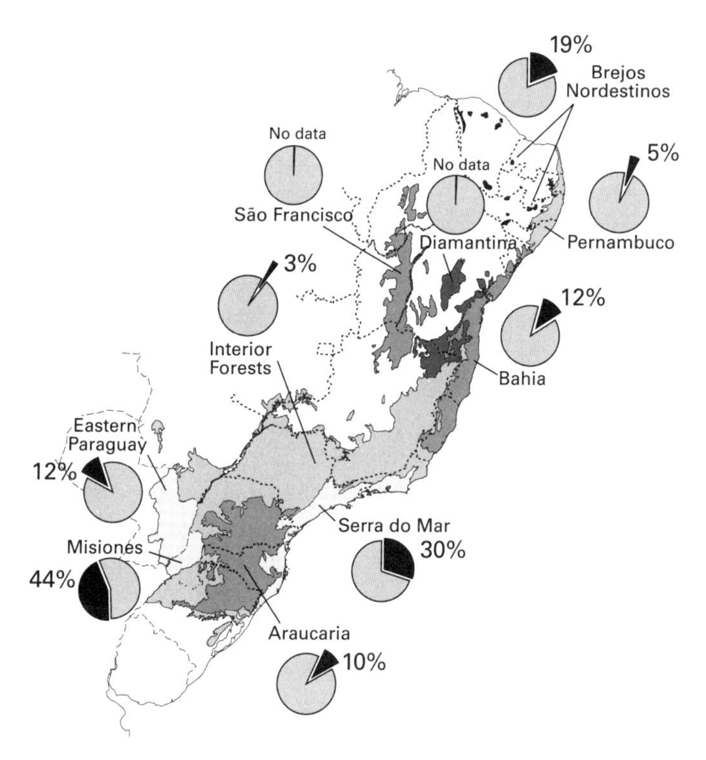

Figure 1.3. Remnant forest percentages (black) in different biogeographic areas of the Atlantic Forest hotspot.

Avoiding Extinction

Although it is possible to restore elements of biodiversity, species extinction is forever. The extensive habitat loss that has taken place in the Atlantic Forest hotspot region has endangered many scores of species. Of critical importance are endemic species, those present only in limited areas, unique because they are irreplaceable. The global Red List of Threatened Species, compiled by the World Conservation Union (IUCN), indicates that more than 110 species living in the Atlantic Forest are threatened, and of these, 29 are critically endangered (Figure 1.4).

The official threatened species list of Brazil includes more than 140 of the Atlantic Forest's terrestrial vertebrate species. In the Interior Atlantic Forest of Argentina, 22 species are officially listed as threatened, as are 35 species in Paraguay. The threats and pressures on the habitats of each of these species must be addressed without delay, with the focused goal of protecting them from immediate extinction. In some cases, intensive management, including captive breeding and translocation, are needed to restore and maintain viable populations in highly fragmented landscapes.

In addition, the vast scope of habitat loss and extreme fragmentation in the Atlantic Forest hotspot has left intact very few extensive, continuously forested ecosystems that can provide viable living space for species with large area requirements. For example, documented densities for jaguar in the southern portion of the Atlantic Forest indicate that areas larger than 10,000 km^2 would be needed to maintain long-term viability of these populations (more than 500 individuals). Also, few undisturbed areas remain that still contain complete species assemblages, where ecological and evolutionary processes are proceeding unabated. In the Atlantic Forest hotspot, only two areas reach these extents: the Serra do Mar in the São Paulo and Paraná states in Brazil and the forests that span from most of the province of Misiones in Argentina through to the Iguaçu National Park in Brazil (Figure 1.5).

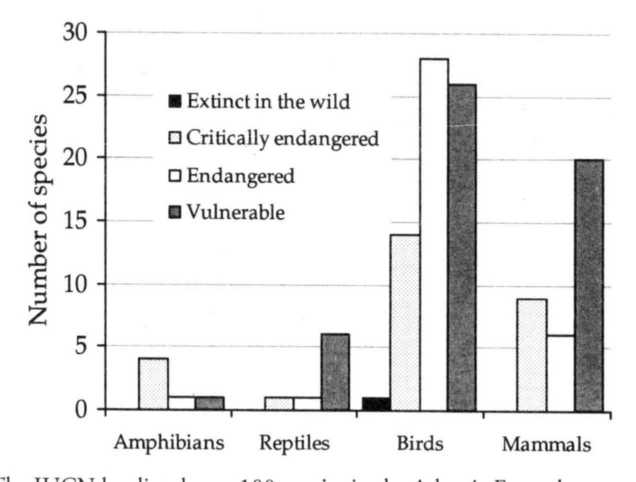

Figure 1.4. The IUCN has listed over 100 species in the Atlantic Forest hotspot as threatened.

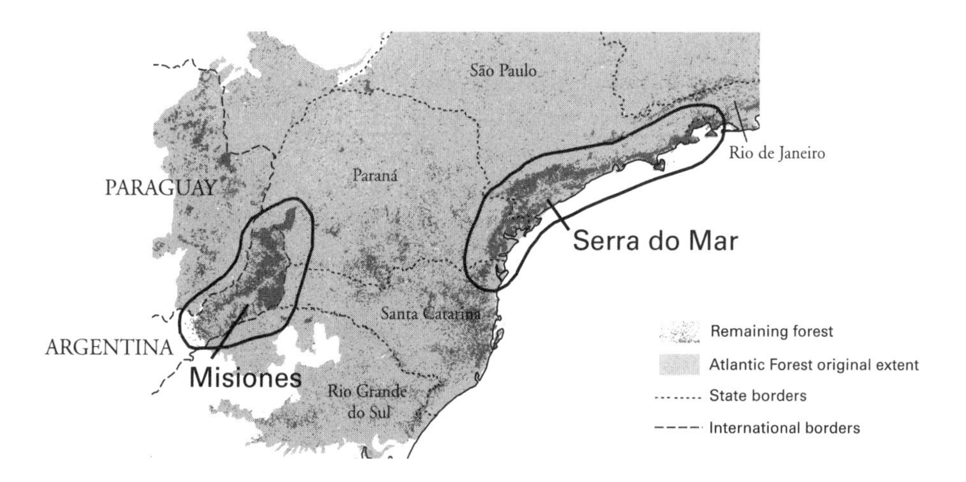

Figure 1.5. Regions with forest areas larger than 10,000 km² in the Atlantic Forest hotspot, which still contain complete species assemblages.

Marine and coastal environments have not escaped the impacts of intensive human pressures. These ecosystems are threatened by trawling and overfishing, marine traffic, industrial and domestic pollution, and the impacts of poorly planned tourism. More than 30 sites along the coast of the Atlantic Forest region have been identified as areas that need greater attention. Without such consideration, unique coastal environments and all their inhabitants will be at great risk of ruin: Coral reefs, sandy beaches, marine mammals, turtles, sharks, large and small pelagic fish, rocky coasts, mangroves, and *restingas* (coastal, sandy-soil, and scrub ecosystems) could all be lost.

Biocultural diversity has also been devastated by the unmanaged changes in the Atlantic Forest region. Vast stores of traditional knowledge of ecological systems, resource use, and natural history have been evaporating as the populations and practices of indigenous communities in the region decline. The few indigenous communities that remain are highly endangered, affected by the long-term effects of colonization and slavery and by the continuing impacts of introduced diseases, forest loss, land-use changes, and economic models that prioritize culturally foreign models of profit. Today, only approximately 134,000 indigenous people live in the Atlantic Forest hotspot.

Creating Protected Areas

Establishing protected areas has been one of the most important tools for conserving some components of biodiversity, and the number of protected areas that have been created in the Atlantic Forest has risen dramatically over the past 40 years (Figure 1.6). However, although the Atlantic Forest hotspot now contains more than 650 protected areas, most of them are small (Figure 1.7).

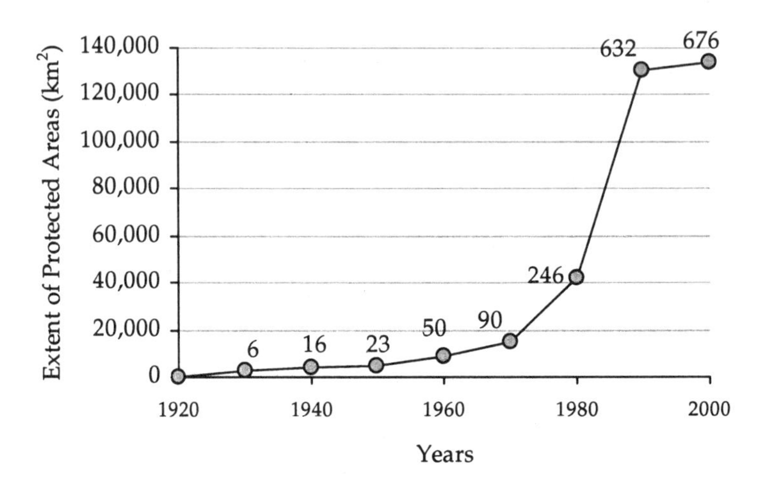

Figure 1.6. Trends in the designation of protected areas in the Atlantic Forest hotspot from 1920 to 2000.

Furthermore, it is difficult to assess the actual protection afforded by these protected areas because many of them lack the basic apparatus necessary to effectively maintain biodiversity, tools such as management plans, land tenure definition, plant and animal inventories, monitoring, and law enforcement. Although a few parks do have effective management mechanisms in place, most are only paper parks. In addition, many protected areas were created opportunistically, and their size, shape, and zoning may not be the most effective for focused conservation purposes. Less than 20 percent of remaining forest is within the strict IUCN protected area categories (Figure 1.8).

In the Atlantic Forest, some regions (e.g., Brejos Nordestinos, Pernambuco,

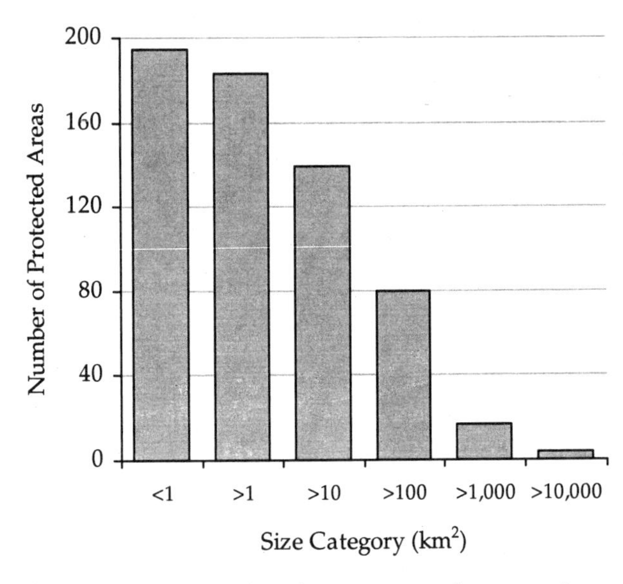

Figure 1.7. Very few protected areas in the Atlantic Forest are large enough to maintain viable populations.

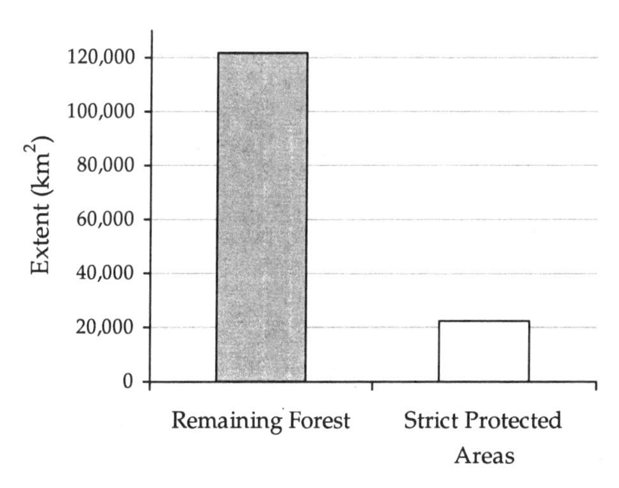

Figure 1.8. Less than 20 percent of the remaining Atlantic Forest has strict protection according to IUCN categories.

Bahia, and the Brazilian Pine Forests) need to create new protected areas, particularly areas larger than 50 km^2, sizes that can address the habitat needs of some species. Other regions need to strengthen the existing protected area systems and restore and maintain connectivity through biological corridors. Without question, financial and human resources must be enhanced to increase the effectiveness of protected area systems.

Managing Corridors

Although protected areas are necessary, in many cases they are not sufficient to maintain species with large area needs or to support broad ecological and evolutionary processes. To address these needs, conservation efforts must take place at larger, regional scales. One approach to this need involves biodiversity (or conservation) corridors, large areas that encompass both protected areas and their surrounding landscapes. The purpose of such corridors is to enhance conservation efforts by providing connectivity, the ability of landscapes and their inhabitants to remain linked through a variety of physical channels. Within corridors, many mechanisms can be used to restore and maintain the continuity of ecosystems through compatible land-uses and other conservation practices. Many individuals and groups committed to conservation in the Atlantic Forest hotspot are supporting efforts that create and promote connectivity in biodiversity corridors. At present, two regions have already been designated as biodiversity corridors in the Atlantic Forest hotspot: the Central (or Bahia) Corridor and the Sierra do Mar Corridor (Figure 1.9).

These regions are critically important because of their endemic species, most of which are threatened. Although the Bahia Corridor is highly fragmented, the Sierra do Mar Corridor contains one of the few areas in the Atlantic Forest that still has continuous tracts of forest land. As new information on biodiversity

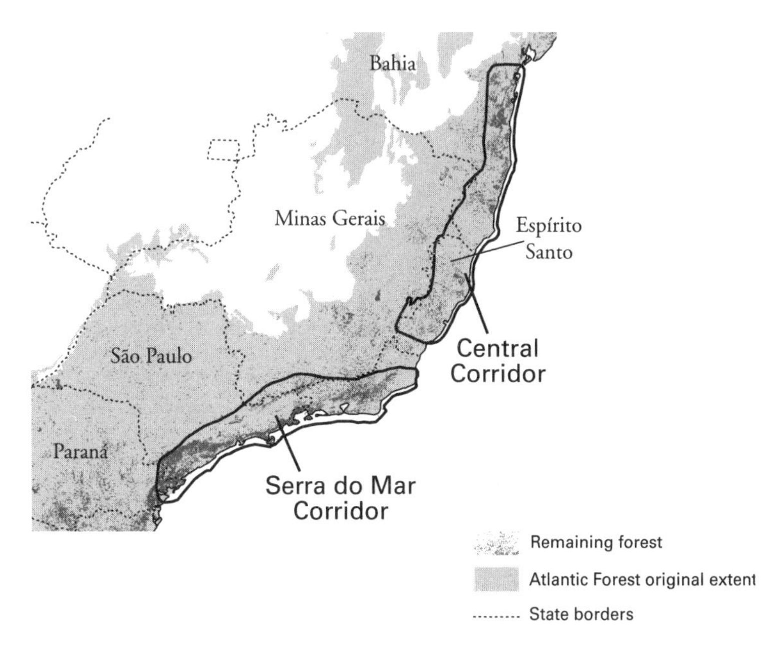

Figure 1.9. Two areas of the Atlantic Forest have been designated as biodiversity corridors.

patterns is gathered, corridors must be designed to include areas that will protect threatened species, which are now concentrated in the north of the Bahia Corridor and the Pernambuco endemic area, including Alagoas State. The region of Misiones in Argentina and Iguaçu National Park in Brazil would also benefit greatly from the establishment of a biodiversity corridor.

Conservation Capacity

Despite the many legal instruments that have been devised to protect the Atlantic Forest hotspot, inhabitants continue to engage in many illegal activities. Logging, poaching of flora and fauna, and illegal settlements all contribute to the loss and deterioration of remaining forests. In addition, lack of coordination between government agencies, both federal and state, has resulted in contradictory policies, which in turn have had severe environmental consequences. Unfortunately, the ministries of environment of the three countries at present have limited influence and are lacking in both the human and financial resources they need.

In fact, some government policies have been responsible for significant expansion of export agriculture, such as subsidies for coffee, cattle, and soy commodities, the development of which has profound ecological consequences (Figure 1.10). Recently, these subsidies have been directed toward expanding plantations and cultivating exotic monocultures, activities known to perpetuate inequalities in land ownership, income, and other anchors of socioeconomic status. These inequalities push landless peasants to continue to extend the agricultural frontier.

Nongovernment organizations have played an important role in conserva-

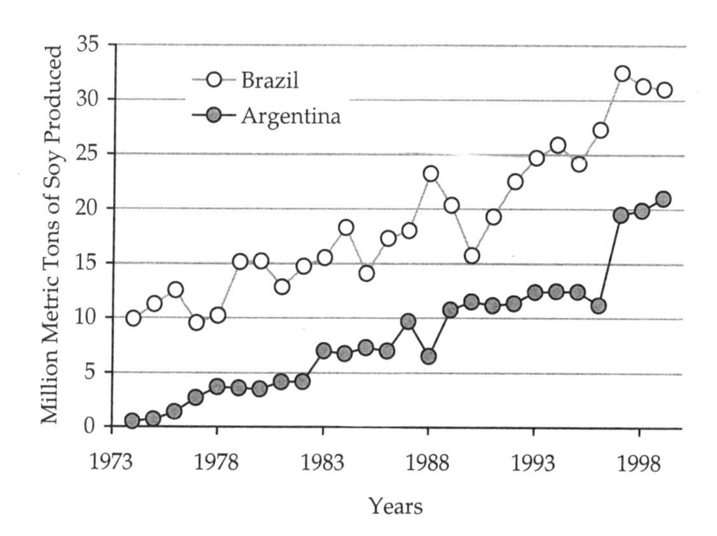

Figure 1.10. A root cause of biodiversity loss in the Atlantic Forest has been government subsidies for agricultural expansion.

tion efforts within the Atlantic Forest even though most of these organizations are not financially self-sufficient. Consequently, conservation efforts in the Atlantic Forest hotspot are highly dependent on financial resources from international institutions. In the past few years, the private sector has been playing an increasingly major role, particularly with the creation of private reserves and ecological easements.

Overall, research over the past 10 years has greatly improved the knowledge and understanding of biodiversity in the Atlantic Forest. However, as new species are discovered and described throughout the hotspot, the need for detailed and systematic inventories to be conducted across the Atlantic Forest region becomes increasingly apparent and urgent. Greater baseline knowledge about this complexity of ecosystems is essential for assessing the status of all parts of the Atlantic Forest and for building conservation actions that are strategic, transparent, and economically and environmentally sustainable.

Chapter 2

State of the Hotspots: The Dynamics of Biodiversity Loss

Carlos Galindo-Leal, Thomas R. Jacobsen, Penny F. Langhammer, and Silvio Olivieri

In the last 400 years, approximately 250 species of birds, mammals, reptiles, and amphibians have become extinct as a result of human activities. This number rises to more than 2,000 if anthropogenic extinctions before 1600 are also considered (Steadman 1995). Today, over 11,000 species of plants and animals are considered threatened, at risk of the same final fate (IUCN 2000).

This unparalleled decline of biodiversity is arguably the most critical phenomenon of our time. Within the last few decades, human impacts have shifted from local scales, via pollution and overharvesting of plant and animal populations, to global scales, most notably climate change. Volumes of evidence suggest that the planet may be facing one of the largest extinction episodes in its history (Vitousek et al. 1997; Myers et al. 2000; Hilton-Taylor 2000; Brooks et al. 2002). To design long-term solutions that will significantly slow down—or even halt—this biodiversity loss, researchers must identify the underlying sources that drive biodiversity loss and understand how they are related (Contreras-Hermosilla 2000; Stedman-Edwards 2000). Nowhere is gathering this information more important than in the biodiversity hotspots (Myers et al. 2000).

This book focuses on the state of biodiversity in the Atlantic Forest, one of the most imperiled hotspots on Earth. Each of the three countries included in the hotspot—Brazil, Argentina, and Paraguay—is discussed in its own section, with a final section covering issues that are relevant to all. Each section opens with a brief overview and is followed by chapters written by experts from a range of disciplines who analyze the status of biodiversity, the drivers of biodiversity loss, and the capacity of local institutions to provide solutions and implement effective conservation actions (Figure 2.1). The authors contributing to this book collectively draw on decades of study and personal experience regarding the Atlantic Forest, making this volume unique in its breadth, depth, and currentness of information.

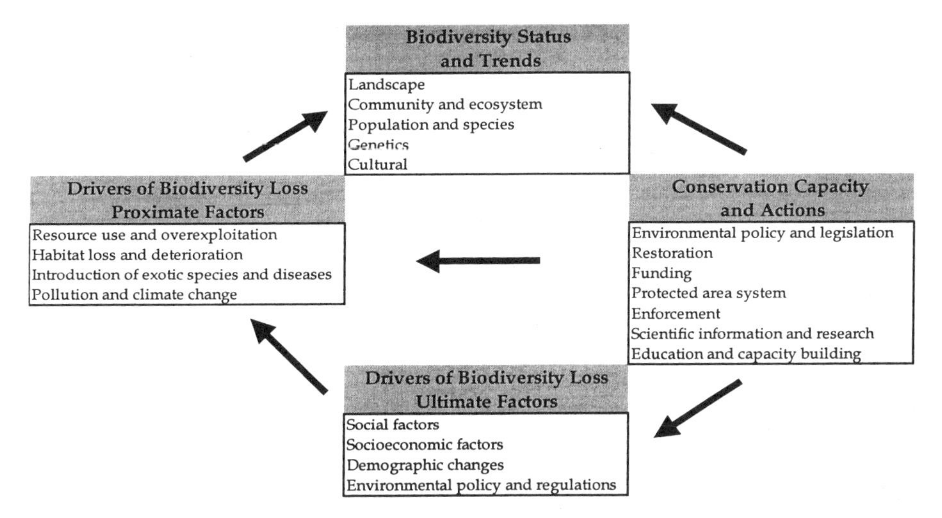

Figure 2.1. Conceptual framework for conservation issues in this book.

Consequences of Biodiversity Loss

Biodiversity loss begins with the diminishing of genetic variability and the lessening of ecological interactions and ends with the extinction of local plant and animal populations. When all the populations of a particular species are gone, that species is gone forever from the face of the planet. Species extinction is an immeasurable loss because every species contains unique genetic information, typically shaped over millions of years of evolution by complex ecological interactions.

Species extinction causes shifts in ecosystem processes, and as ecosystems become thus impoverished, their products and services decrease. These products and services include clean air and water, fertile soils, and the varieties of plants and animals that humans rely on for food, fuel, clothing, medicine, and shelter (Toledo et al. 1995; Cassis 1998; Chapter 27, this volume). In other words, the loss of biodiversity translates into loss of resources that directly affect the lives and livelihoods of human populations. Paradoxically, humans are the only species capable of realizing that the planet faces a biodiversity crisis and that we alone have the ability to build and implement solutions to this self-imposed predicament.

What Is Happening? The Status of Biodiversity

Assessing the status of biodiversity for a region is not simple. Biodiversity includes species and ecosystem dynamics at many different scales, within which ecological and evolutionary processes function and maintain genetic and population variability (Noss 1990). In turn, community assemblages interact across landscapes, responding over time to natural and anthropogenic pressures (Table 2.1).

Table 2.1. A matrix of examples of indicators of composition, structure, and function for inventories, monitoring, and evaluation on five levels of ecological organization (modified from Noss 1990).

Level	Composition	Structure	Function	Tools for Inventorying and Monitoring
Regional and landscape	Ecosystem types and extent	Frequency distribution of fragments and connectivity	Dispersal	Remote sensing
Ecosystem and community	Species identity	Water availability	Biotic interactions	Community sample
Species and population	Relative abundance	Habitat use	Demographic processes	Radio telemetry Population estimation
Genetic	Allele diversity	Polymorphism	Inbreeding effects	Electrophoresis, DNA fingerprinting
Cultural	Traditional ecological knowledge	Breadth and depth	Ecological resilience	Interviews
	Domesticated varieties		Sustainable use	

Assessing Biodiversity on Regional, Community, Population, Genetic, and Cultural Levels

At regional (also called landscape) levels, trends in the extent and fragmentation of ecosystems are important features for assessing the status of biodiversity, including the trends in connectivity (distance between fragments). Fragmentation can prompt the loss of habitat, the isolation of remaining pieces of ecosystems, the reduction of interior forest conditions, or the creation of edge effects (Chapters 5–7, 16, 25, and 31, this volume). Ecological processes are altered because fragmented landscapes favor some species and hinder others. Isolated populations of plants and animals in small patches are more vulnerable to random environmental, genetic, and demographic events, which can lead to extinction (Chapter 31, this volume). Many researchers agree (see Noss 1999; Chapter 15, this volume) that at a landscape level, the most appropriate indicators for assessing and monitoring the status of biodiversity are changes in the extent of the ecosystem, the proportion of remaining fragments of various sizes, the number of large patches, the spatial distribution of disturbances, and the degree of connectivity (Table 2.2).

Although dynamics of complete communities are also important to track in assessing biodiversity, such work often is not feasible, particularly in tropical ecosystems where too many species and too few individuals exist in each population to

Table 2.2. Examples of indicators for monitoring biodiversity on regional, community, population, genetic, and cultural levels.

Level	Indicators
Regional	
Habitat loss and fragmentation	Proportion of undisturbed remaining habitat
	Frequency distribution of remaining habitat patches
	Number of large habitat patches (categories >100 km^2, >1,000 km^2, >10,000 km^2)
	Spatial distribution of habitat loss and remnants
	Degree of connectivity
Habitat deterioration	Proportion of remaining intact habitat
	Proportion of disturbed remaining habitat
Community	Number of native species
	Number of globally extinct species
	Number of locally extinct species
	Number of threatened species
	Number of endemic species
Population	Range reduction
	Habitat extension occupied
	Frequency distribution of occupied habitat patches
	Population size trends
	Number of populations with viable size
	Number of declining populations
	Age structure
	Reproductive and survival trends
Genetic	Loss of genetically different populations
	Loss of genetic variability
	Inbreeding depression
Cultural	Loss of indigenous language
	Loss of local traditional practices
	Loss of local traditional knowledge

allow detailed evaluation of trends. A number of alternative measures have been proposed for assessing changes in community diversity and structure, among them guilds, indicator species, keystone species, and focal species (Lambeck 1997; Simberloff 1998). Similarly, population levels should also be carefully monitored, particularly those of limited-range species and threatened species, which are the primary targets for conservation because of their small or declining numbers (Caughley and Gunn 1996; Chapters 9, 16, and 17, this volume). The IUCN Red List provides a quantitative system for selecting and monitoring species that fit these criteria (Chapters 8 and 30, this volume).

Along with these very tangible components of biodiversity—landscape, communities, and populations—another element to be monitored is genetic diversity. Genetic variability may be reduced long before a species goes extinct, for as populations decrease in size, their genetic composition can be influenced by inbreeding, genetic drift, genetic bottlenecks, and founder effects (Avise and

Hamrick 1996). Studies of genetic diversity can help detect loss of variability in isolated or small populations and can help researchers understand the consequences of lack of connectivity between populations. Genetic variability can be analyzed by using DNA, proteins, and immunologic methods (Smith and Wayne 1996).

Monitoring the status of cultural diversity is another important factor because humans also influence biodiversity. Humans have domesticated and cultivated a wide variety of plants and animals to suit different environmental conditions and purposes. For example, humans have produced more than 66 varieties of corn in Mesoamerica (Sánchez-González 1994) and nearly 7,000 varieties of rice in Bangladesh (Thrupp 1997). On the other hand, contemporary large-scale agribusinesses promote monocultural production disfavoring traditional varieties of plants to the point of endangerment and increasing their vulnerability to pests and diseases (Thrupp 1997). Overall, human societies, and particularly indigenous cultures, hold rich knowledge of ecological systems, knowledge that is rapidly lost once these cultures are incorporated into capitalist production systems (Toledo et al. 1995). The alarming disappearance of native languages threatens to wipe out much of this knowledge and the insight that goes with it (Maffi 2001; Chapters 18 and 32, this volume).

Why Is This Happening? Proximate and Ultimate Causes of Biodiversity Loss

Human activities in most places now play a larger role in changing ecosystems than natural disturbances such as hurricanes, landslides, flooding, and fires. We are responsible for both direct and indirect factors that shape the landscape and hasten the rate and extent of biodiversity loss (Chapters 11 and 33, this volume).

Human actions directly cause biodiversity loss both by reducing populations of native species and by changing or eliminating their habitats. Human-induced habitat loss, which is greatly promoted by large-scale commercial agriculture (Machlis and Forester 1996; Stedman-Edwards 2000) is a leading threat to 85 percent of the 1,256 species of globally threatened plants and animals. It also plays a major role in the Atlantic Forest, as documented by authors throughout this book (Table 2.3).

Population, Consumption, and Economic Policies: Indirect Causes of Biodiversity Loss

Although the direct causes of biodiversity loss often are clear and easy to identify, they are rooted in a complex web of other social, cultural, political, and economic factors. For example, in most societies a small number of individuals consume a large proportion of resources, whereas most can barely satisfy their daily needs (FAO 2002) (Table 2.4). Inequalities of control and use of land and resources contribute to patterns of poverty and overconsumption, reinforced through his-

Table 2.3. Factors directly affecting biodiversity and their indicators.

Biodiversity Loss	Proximate Factors	Indicators
Habitat loss	Agriculture, mining, ranching, urbanization, and infrastructure development	Percentage of former ecosystem converted to large-scale agriculture, small-scale agriculture, aquaculture, mariculture, urban development, intensive agroforestry, pasture, plantations, mining and mineral exploration, infrastructure projects, irrigation, or destroyed by fire, fishing gear.
Habitat deterioration	Pollution, changes in temperature, humidity, salinity, acidity, and pH	Amount and concentration of particular groups of substances, compared with the soil, water, and air standards for such substances; agricultural pollutants (fertilizers, herbicides); industrial pollutants, acid rain.
	Removal or introduction of species	Percentage of ecosystem supporting fuel wood collection, nontimber forest products, and fishing.
Overharvesting	Harvesting for food, shelter, medicine, dyes, oils, fuel, fibers, tools, and cash	Amount of timber harvested, wildlife hunted, nontimber forest products extracted, fuelwood collected.
Species and disease introductions	Accidental	Number, ratio, relative abundance of exotic species. Ecologic and economic consequences.
	Incidental or deliberate	

tory via colonialism, international markets, and free trade (Clairmonte and Cavanagh 1988; Shiva 2000; Chapters 4, 14, 24, and 37, this volume).

Another indirect cause of biodiversity loss is the design of many government policies. In many countries, government policies are more likely to grant property rights to settlers who clear and settle forests and other biodiversity-rich areas than to those who operate under traditional land ownership rights or who practice more sustainable methods of resource use. In other words, the uncertainty and inequality of land tenure weakens incentives for land stewardship and encourages shortsighted overexploitation. Although traditional resource use patterns can be sustainable, the dynamics between poverty and inequality, higher population densities, increased reliance on the market economy, perverse subsidies, government policies, and new technologies have dramatically altered the human-nature relationship (Bawa and Dayanandan 1997; Barraclough and Ghimire 2000; Chapter 36, this volume).

Yet another cause of biodiversity loss is that of national government control over macroeconomic factors such as rates of exchange, flow of capital, levels of indebtedness, and investment, all of which create incentives for activities such as large-scale agricultural production, livestock grazing, and the harvest of forest resources (Barraclough and Ghimire 2000). In the Brazilian Amazon, for example,

Table 2.4. Drivers responsible for biodiversity loss at two ends of the spectrum.

Profit-Oriented drivers (the greedy)

		Ultimate Factors		
Biodiversity Loss	**Proximate Factors**	**Local**	**National**	**International**
Habitat loss	Industrial agriculture and fishing, aquaculture (shrimp, fish, oysters), mining, oil extraction, ranching, urbanization, and infrastructure development (dams, highways, pipelines, landfills)	Provide access and encourage immigration	National economy and debt	Financial institutions (multilateral and bilateral aid)
Habitat deterioration	Industrial pollution, changes in temperature, humidity, salinity, acidity, and pH	Population growth	National policies, laws, and incentives	Trade links
Overharvesting	Industrial fishing, forestry, wildlife trade		Structural adjustment programs	International treaties
Species and disease introductions	Agriculture, accidental		Corporate influence on governments	Demand (commodities, wildlife trade)
			National monopolies	International monopolies

Subsistence-Oriented drivers (the needy)

		Ultimate Factors		
Biodiversity Loss	**Proximate Factors**	**Local**	**National**	**International**
Habitat loss	Slash and burn agriculture, extensive ranching, destructive fishing (cyanide, dynamite)	Land use tenure and inequality	National policies, laws and incentives	
Habitat deterioration	Pollution, changes in temperature, humidity, salinity, acidity, and pH	Resource monopolies		
Overharvesting	Harvesting for food, shelter, medicine, dyes, oils, fuel, fibers, tools, and cash	Low wages		
Species and disease introductions	Agriculture, accidental	Social constraints, population growth		

most of the deforestation has its roots in government-financed subsidies that make funds easily available to investors who want to buy and clear large tracts of forest for cattle ranching (Myers and Kent 2001). Governments of developing nations use structural adjustment programs to try to strengthen economic growth and gain macroeconomic stability. Such programs promote international capital flows and privatization and often liberalize trade and exchange policies. Under such schemes, the influence of international markets and hence the foreign demand for natural resources can soar (Table 2.5) (Chapters 10, 19, and 26, this volume).

What Can We Do about It? Conservation Capacity

Conservation is a social endeavor, one that includes the actions of individuals and percolates through the activities of groups and societies as a whole. As ecosystems deteriorate and increasing numbers of species move toward the brink of extinction, effort must be focused on promoting actions that we, as individuals and as members of society, can take to protect species and the ecological processes that maintain them.

A number of specific fundamental actions are necessary to build quickly the informational and institutional capacity to support biodiversity conservation. First, conservationists need to collect current and reliable information on the status of species, communities, and ecosystems, and to understand their long-term conservation needs. Second, we need to identify and agree on specific direct and indirect causes of biodiversity loss and emerging threats. Third, we need to implement conservation strategies that benefit from past successful and unsuccessful experiences. Fourth, we must raise environmental awareness and creatively seek to implement novel approaches and incentives for conservation.

Unfortunately, the individual and institutional strengths necessary to accomplish these goals vary widely across regions within biodiversity hotspots (Chapters 12, 20, 22, and 28, this volume). Whereas academic and research institutions are most often the providers of reliable information to accomplish the first two goals (Table 2.6), government agencies usually are the entities charged with creating and enforcing the legislation and policies that can promote the third and fourth goals. However, in most of the hotspots, and particularly across the Atlantic Forest hotspot, the capacity of government institutions is seriously limited. First, the government agencies responsible for environmental issues often are given low priority in relation to other agencies; these offices are chronically understaffed and underbudgeted (Chapters 20 and 28, this volume). Second, the agencies responsible for policy development and enforcement often are rarely coordinated, a conundrum that results in contradictory policies and actions (Dudley and Stolton 1999; Chapter 28, this volume). Third, enforcement of laws and policies, if they do exist, often is weak. Last, many government initiatives are short term, alive only as long as the terms of elected officials who back them, on average a meager 4 to 6 years. The continuity of initiatives is hampered by high-level personnel changes by a newly elected administration.

An additional challenge to biodiversity conservation is rooted in the lack of

Table 2.5. Ultimate factors affecting biodiversity and examples of indicators.

Ultimate Factors	Indicators
Demographic changes	Total population
	Physiological density (number of people per unit area of arable land)
	Annual change in national population density
	Rural population density (people/hectare)
	Annual population growth rate per cultural or ethnic group
	Average annual population change in and around key habitats and protected areas
	Rate of fertility
	Infant and child mortality rate
Migration patterns	Net migration rates from urban to rural areas
	Net migration rates from rural to urban areas
	Causes of migration patterns
Economic	Rates of fluctuation in exchange rates
	Amount of debt to repay
	Gross domestic product
	Environmentally adjusted net domestic product
	Foreign aid by category
	Annual unemployment rate
	Percentage of population in poverty and poverty gap index
	Extent and revenue from mining and mineral exploration
	Revenue from key agricultural commodities
	Revenue from logging
	Current investment in infrastructure projects (roads, mines, dams, utilities)
	Planned investment in infrastructure projects (roads, mines, dams, utilities)
	Revenue from tourism, ecotourism
	Market failures in valuing the natural world
	Externalities
Policy	Subsidies: land and natural resource related (wood processing, plantations)
	Perverse incentives
	Status of land ownership (land claims) and property rights
	Percentage of national land area protected
	Number of protected areas
	Terms of timber concessions
	Programs for resettlement
	Price controls, taxes that discourage investment in conservation
	Policies related to mining and lack of environmental safeguards
Social	Percentage of high school and college graduates by gender
	Percentage adult literacy and by gender
	Percentage of population with access to sanitation
	Number of family planning centers per region
	Percentage of population with access to health care
	Male-to-female wage ratio
	Percentage of population with access to safe drinking water

Table 2.6. Contributors to and examples of indicators of local academic capacity.

Concern	Requirement	Indicator
Are there qualified professionals?	Diverse array of competent professionals	Number of qualified professionals Number of professionals with postgraduate education
Are they working on relevant issues?	Applicability to critical issues	
Is their research reliable?	Good record of published research in peer journals	Publications in first-class international journals
Is there appropriate infrastructure?	Equipment, vehicles, infrastructure, laboratories	Appropriate infrastructure
Is there access to specialized libraries, collections, and laboratories?	Access to recent research through libraries, museums	Number of specialized museums, libraries
Is there sufficient funding?	Conduct research and communicate results	Amount of funding for conservation research
Are new professionals being trained?	Specialized training for graduate and undergraduate students	Number of courses offered for new professionals
	Update training	
	Outreach	Number of publications for the general public produced by academic institutions
Is the general public informed?	Actualization	
	Outreach	

proper funding, political support, human resources, and adequate management knowledge in the government institutions that are responsible for creating and maintaining systems of protected areas (Chapters 21 and 38, this volume). Without funding, political support, or adequate management these parks functionally exist on paper only (Dudley and Stolton 1999; Galindo-Leal et al. 2000; Cifuentes et al. 2000; Bruner et al. 2001). Instead of giving attention and resources to building truly viable protected areas, governments often give priority to infrastructure development, including the construction of dams, pipelines, and highways (Chapter 35, this volume).

As threats to biodiversity continue to mount, however, civil society has been playing an increasingly active role in conservation through the formation of international, national, and regional nongovernment institutions (Margoluis and Salafsky 1998; Galindo-Leal 2000; Chapter 28, this volume). These organizations conduct a variety of activities, from identifying conservation priorities to creating and managing protected areas, all of which contribute greatly to achieving conservation goals.

The extent of the biodiversity crisis requires the participation of the government institutions, the private sector, indigenous peoples, academic institutions, and nongovernment organizations. Every sector of society must contribute if we are to stop biodiversity loss, the greatest challenge of this century.

References

Avise, J. C. and Hamrick, J. L. (eds.). 1996. *Conservation genetics: case histories from nature.* New York: Chapman and Hall.

Barraclough, S. L. and Ghimire, K. B. 2000. Social determinants of deforestation. In: Barraclough, S. L. and Ghimire, K. B. (eds.). *Agricultural expansion and tropical deforestation: poverty, international trade and land use.* pp. 1–8. London: Earthscan.

Bawa, K. and Dayanandan, S. 1997. Socioeconomic factors and tropical deforestation. *Science* 386: 562–563.

Brooks, T. M., Mittermeier, R. A., Mittermeier, C. G., da Fonseca, G. A. B., Rylands, A. B., Konstant, W. R., Flick, P., Pilgrim, J., Oldfield, S., Magin, G., and Hilton-Taylor, C. 2002. Habitat loss and extinction in the hotspots of biodiversity. *Conservation Biology* 16(4): 909–923.

Bruner, A. G., Gullison, R. E., Price, R. E., and da Fonseca, G. A. B. 2001. Effectiveness of parks in protecting tropical biodiversity. *Science* 291: 125–128.

Cassis, G. 1998. Biodiversity loss: a human health issue. *Medical Journal of Australia* 169: 568–569.

Caughley, G. and Gunn, C. 1996. *Conservation biology in theory and practice.* Oxford, U.K.: Blackwell Science.

Cifuentes, M. A., Izurieta, A., and de Faria, H. H. 2000. *Measuring protected area management effectiveness.* WWF, GTZ, IUCN. Technical Series No. 2.

Clairmonte, F. and Cavanagh, J. 1988. *Merchants of drink: transnational control of world beverages.* Penang, Malaysia: Third World Network.

Contreras-Hermosilla, A. 2000. *The underlying causes of forest decline.* Bogor, Indonesia: Center for International Forestry Research. Occasional Paper No. 30.

Dudley, N. and Stolton, S. 1999. *Threats to forest protected areas: a survey of 10 countries.* A research report from IUCN for the World Bank Alliance for Forest Conservation and Sustainable Use.

FAO (Food and Agriculture Organization of the United Nations). 2002. *Reducing poverty and hunger: the critical role of financing for food, agriculture and rural development.* International Conference of Financing for Development. Monterrey, Mexico: FAO, IFAD, WFP.

Galindo-Leal, C. 2000. Ciencia de la conservación en América Latina. *Interciencia* 25: 129–135.

Galindo-Leal, C., Weiss, S., Fay, J., and Sandler, B. 2000. Conservation priorities in the greater Calakmul Region, Mexico: correcting the consequences of a congenital illness. *Natural Areas Journal* 20(4): 376–380.

Hilton-Taylor, C. 2000. *2000 IUCN red list of threatened species.* Gland, Switzerland and Cambridge, UK: The World Conservation Union.

IUCN-SSC (The IUCN Species Survival Commission). 2000. *2000 IUCN red list of threatened species.* Online: http://www.redlist.org.

Lambeck, R. J. 1997. Focal species: a multispecies umbrella for nature conservation. *Conservation Biology* 11: 849–856.

Machlis, G. E. and Forester, D. J. 1996. The relationship between socio-economic factors and the loss of biodiversity: first efforts at theoretical and quantitative models. In: Szaro, R. C. and Johnston, D. W. (eds.). *Biodiversity in managed landscapes: theory and practice.* pp. 121–146. Oxford, England: Oxford University Press.

Maffi, L. 2001. *On biocultural diversity: linking language, knowledge, and the environment.* Washington, DC: Smithsonian Institution Press.

Margoluis, R. and Salafsky, N. 1998. *Measures of success: designing, managing, and monitoring conservation and development projects.* Washington, DC: Island Press.

Myers, N. and Kent, J. 2001. *Perverse subsidies: How misused tax dollars harm the environment and the economy.* Washington, DC: Island Press.

Myers, N., Mittermeier, R. A., Mittermeier, C. G., da Fonseca, G. A. B., and Kent, J. 2000. Biodiversity hotspots for conservation priorities. *Nature* 403: 853–858.

Noss, R. 1990. Indicators for monitoring biodiversity: a hierarchical approach. *Conservation Biology* 4: 355–364.

Noss, R. 1999. Assessing and monitoring forest biodiversity: a suggested framework and indicators. *Forest Ecology and Management* 115: 135–146.

Sánchez González, J. J. 1994. Modern variability and patterns of maize movement in Mesoamerica. In: Johannessen, S. and Hastorf, A. (eds.). *Corn and culture in the prehistoric world.* pp. 135–156. Boulder, CO: Westview Press.

Shiva, V. 2000. *Stolen harvest: the hijacking of the global food supply.* Cambridge, MA: South End Press.

Simberloff, D. 1998. Flagships, umbrellas, and keystones: is single-species management passé in the landscape era? *Biological Conservation* 83: 247–257.

Smith, T. B. and Wayne, R. K. (eds.). 1996. *Molecular genetic approaches in conservation.* New York: Oxford University Press.

Steadman, D. W. 1995. Prehistoric extinctions of Pacific island birds: biodiversity meets zooarcheology. *Science* 267: 1123–1131.

Stedman-Edwards, P. 2000. A framework for analyzing biodiversity loss. In: Wood, A., Stedman-Edwards, P., and Mang, J. (eds.). *The root causes of biodiversity loss.* pp. 11–35. London: Earthscan.

Thrupp, L. A. 1997. *Linking biodiversity and agriculture: challenges and opportunities for sustainable food security.* Washington, DC: World Resources Institute.

Toledo, V. M., Batis, A. I., Becerra, R., Martinez, E., and Ramos, C. H. 1995. The useful forest: quantitative ethnobotany of the indigenous groups of the humid tropics of Mexico. *Interciencia* 20: 177–187.

Vitousek, P. M., Mooney, H. A., Lubchenco, J., and Melillo, J. M. 1997. Human domination of earth's ecosystems. *Science* 277: 494–499.

PART II

Brazil

Chapter 3

Dynamics of Biodiversity Loss in the Brazilian Atlantic Forest: An Introduction

Luiz Paulo Pinto and Maria Cecília Wey de Brito

The Brazilian Atlantic Forest, a biodiversity mosaic, is composed of numerous vegetation types distributed along 27 degrees of south latitude, with great variations in elevation. As a true hotspot, the Brazilian Atlantic Forest has exceptional levels of biodiversity that are under enormous pressure. Unfortunately, a long history of resource extraction has eliminated most of the natural ecosystems, leaving less than 8 percent of the forest's former extent (Chapters 4 and 6).

In the last three decades, habitat fragmentation and loss have severely altered most of the Atlantic Forest hotspot, causing the local extirpation of many species. Nonetheless, the biome has proven to be extremely resilient, as the recovery of some areas and the continual discovery of new species have shown. Silva and Casteleti (Chapter 5) note that the Atlantic Forest and its associated ecosystems (*restingas* and mangroves) still shelter a significant portion of Brazil's biological diversity. Its high levels of endemism result largely from its latitudinal span and its widely varying elevation levels, which range from sea level to 2,700 m. Because of the complexity of the biodiversity and socioeconomic factors in the Brazilian Atlantic Forest, a complex mosaic is produced of biological and social situations, and researchers need to respond with conservation plans that accurately reflect the state of each region. Inland forests differ considerably from coastal ones, and the biodiversity richness and threats are not confined to terrestrial ecosystems. In Chapter 6, Hirota discusses the monitoring initiative of the Fundação SOS Mata Atlântica. The atlas developed by the initiative uses remote sensing methods to provide a periodic assessment of forest cover changes in most of the Brazilian Atlantic Forest.

Jablonski (Chapter 7) synthesizes the outcomes of a recent workshop where marine conservation priority areas were identified in the Brazilian coast. Specifically, meeting participants assessed the current state of marine biodiversity and identified threats to coral reefs, sea turtles, mammals, birds, nekton, benthic flora and fauna of the continental shelf, marine plants, and plankton.

Tabarelli and his colleagues (Chapter 8) also discuss many of the species that make up the extraordinary diversity of the Atlantic Forest to underscore the urgent need for programs and conservation actions to counter specific threats. In some cases, threats must be countered with a combination of forest conservation, captive breeding, reintroductions, and translocation, similar to the efforts undertaken by Kierulff and her colleagues (Chapter 9). Addressing threats also requires surveying the remaining endangered and endemic species populations throughout the Atlantic Forest's original distribution. Such data will allow researchers and policymakers to better assess the current status of species, to better understand the threats to their survival, and, therefore, to propose more effective conservation strategies.

Threats to the biodiversity of the Atlantic Forest are exacerbated by the fact that the region is home to approximately 70 percent of Brazil's 169 million people. Most of this population lives in megacities such as São Paulo and Rio de Janeiro. Furthermore, about 80 percent of Brazil's gross domestic product is generated in the Atlantic Forest, and the region shelters Brazil's largest industrial and silvicultural centers. As Young observes in Chapter 10, the paradox of development in the Atlantic Forest continues: although development has caused increasing threats to the biological integrity of various unique ecosystems, it has failed to improve the economic status or quality of life for local human populations.

In Chapter 11, Aguiar and colleagues extend this theme by discussing the biological importance of and socioeconomic pressures within two critical biodiversity corridors in the hotspot: the Central Corridor and the Serra do Mar Corridor. Other critical areas in the hotspot include the Pernambuco Endemism Center and the Araucaria Region.

Most of the individuals and groups committed to conservation in the Atlantic Forest recognize that the area currently covered by strictly protected areas (national parks, biological reserves, and ecological stations) is insufficient to conserve much of the forest's biodiversity. Protected areas cover less than 2 percent of the original Brazilian Forest biome, and strictly protected units currently protect only 21 percent of remnant forests (Chapter 38). Other remaining habitat fragments are highly threatened, underscoring the urgent need to expand existing protected areas and to establish new ones to connect remaining fragments. The private sector can play an important role in this effort because private protected areas can function as valuable complements to the existing system, increasing connectivity and contributing to a better representation of the priority areas included in the protected areas network.

Although the number of protected areas in the Atlantic Forest is among the highest in Brazil, the actual units of protected land are small and, in most cases, barely able to maintain viable populations of some of the species they house or to hold off the growing pressures of development. The northeast region of the Brazilian Atlantic Forest is a center of endemism in South America, for example, and contains assemblages of highly limited-range species, many of which have become critically endangered. Nevertheless, anthropogenic activities continue to expose many of the species there to pressure and threats, bringing some that are truly endangered closer to the brink of extinction. Since the protected areas in this region are small and subsistence hunting is widespread, the need is even stronger for local, regional, and national government bodies to bolster the tools they can offer to support strict conservation.

Current predictions of extinction trends in the Brazilian Atlantic Forest are based on projections of habitat loss rates and on relationships between species richness and habitat size. However, the lack of accurate information about some aspects of biodiversity makes some estimates of extinction threats somewhat speculative and highlights the need to map the geographic distribution of endangered species for all groups of fauna and flora. For example, in southern Bahia, species of vertebrates continue to be discovered and at least 12 new species of frogs and toads have recently been described. Accurate species information is essential to implementing systems for long-term monitoring in the Brazilian Atlantic Forest and to measuring the success of conservation actions.

Some positive steps are being taken in the face of these many challenges. Biome-level conservation priority assessments have been incorporated as government policy through the National Biodiversity Program (Chapter 12) and financed by the Global Environment Facility since 1997. This cycle of prioritization for the Atlantic Forest culminated in Evaluation and Priority Actions for Conservation of the Biomes of the Atlantic Forest, which was established as part of the Project on the Conservation and Sustainable Use of the Brazilian Biological Diversity of the Ministry of the Environment. This project has several objectives: to consolidate information on the biological diversity of the Atlantic Forest and identify knowledge gaps; to identify priority areas and actions based on biological importance, ecosystem integrity, and opportunity to conserve biodiversity; to identify and evaluate current and alternative uses of natural resources compatible with conservation; and to promote greater awareness and effective participation of society in conservation. The project includes more than 200 scientists and provides the best consensus-based assessment available to date of site-based biodiversity and priority conservation action for the Brazilian Atlantic Forest. The results are being published as maps, technical reports, and online databases, all of which are being broadly distributed. The Ministry of the Environment, state governments, and environmental nongovernment organizations are using project materials to define biodiversity corridors, select sites for new protected areas, assess environmental impacts, and establish institutional priorities and projects.

One organization alone could never adequately and effectively conserve the biodiversity of the Brazilian Atlantic Forest. Rather, alliances must be established that combine the efforts of government organizations, nongovernment organizations, land owners, researchers, and local communities to maximize efficiency and produce results that will be truly durable. Nongovernment organizations can play an important role in this endeavor by building a strong network of organizations that promotes the engagement of a wide range of institutions to address conservation needs across the Brazilian Atlantic Forest region.

The chapters in this section raise many important issues related to conservation in the Atlantic Forest and provide the forest's monitoring system with a preliminary demarcation. They represent the most current data and perspectives on conservation problems in this region and are intended to facilitate the work of government and nongovernment organizations, researchers, and decision makers committed to protecting the remarkable biodiversity of Brazil's Atlantic Forest.

Chapter 4

Brief History of Conservation in the Atlantic Forest

Ibsen de Gusmão Câmara

The Ecosystems of the Atlantic Forest

A colossal forest awaited the Portuguese when they arrived to colonize South America. Diversified but continuous, it extended from the Brazilian northeast to the current state of Rio Grande do Sul. Along its entire length, this vast and dense forest, the Atlantic Forest, penetrated inland to limits not well known today—in the south, certainly beyond the current borders of Argentina and Paraguay (Figure 4.1). At some point—perhaps even then—this forest was connected to the Amazon Forest. Disconnected populations of typically Amazonian animals, such as the red-handed howler monkey (*Alouatta belzebu*) and the silky anteater (*Cyclopes didactylus*), are found today in what remains of the northeastern Atlantic Forests, where they would not be unless the two forests had been linked at some time. At least 277 genera of plants have also been identified as common to both the Atlantic and the Amazon forests (Rizzini 1967). These forests may in fact have been joined at several different times during the climatic oscillations of the Pleistocene.

Atlantic Forest is a popular term with no real scientific basis. Its exact definition has been the subject of extensive discussion and widely diverging opinion. The first attempt to reach a consensus was undertaken by Fundação SOS Mata Atlântica in 1990 when it brought together 42 specialists for a workshop, including scientists, technical experts, and conservationists. At this historic meeting, which was of great importance for the conservation of the biome, participants agreed that the term *Atlantic Forest* should encompass the coastal rainforests; the mixed forests of Brazilian pine, also known as Paraná pine or araucaria (*Araucaria angustifolia*); and forests dominated by the laurel family (Lauraceae) in the south, the deciduous and semideciduous forests of the interior, and the ecosystems associated with these, including mangroves, *restingas* (coastal, sandy-soil forest and scrub), high-altitude grasslands, pockets of pastures and grasslands, and the montane *brejos* and *chãs*. These last two are forest remnants in northeast Brazil with characteristics of the Atlantic Forest. Located at high elevations in the middle of semiarid regions, they probably represent former extensions of the Atlantic Forest

Figure 4.1. Original and remaining extent of Atlantic Forest in Brazil.

maintained by local climatic conditions. This definition takes into account the original continuity of the large forest, the presence of common species, and the gradual transitions between the different vegetation formations, which make it difficult to draw strict boundaries.

After subsequent refinement, this definition was approved in 1992 by the National Council on the Environment (CONAMA), and the area was given the name "Atlantic Forest Domain." The boundaries conformed to the Map of Brazilian Vegetation published in 1988 (and revised in 1993) by the Instituto Brasileiro de Geografia e Estatística (IBGE), a federal government agency. According to the decision adopted by CONAMA and in keeping with the map, the domain comprises the dense and open ombrophilous (tolerant of wet conditions) forests (covering the montane, submontane, flatland, and *tabuleiros* [coastal low-land forest on sandy soils] rainforest of the coastal area), the mixed ombrophilous forests (corresponding to the forests of predominantly Brazilian pine and laurel in the south), and the seasonal and semideciduous forests (encompassing the subtropical forests of the southern states and the deciduous forests of the northeast), as well as the ecosystems mentioned above and the transitional zones between the forest formations, also called "areas of ecological tension."

The dense ombrophilous forests along the coast have vigorous and diverse vegetation because of the constant humidity from the ocean. In their southernmost portion, corresponding to the states of the south and southeast regions of the country, three distinct formations can be identified by their composition and structure: the forests of the coastal plain, forests in abrupt mountain areas with high humidity, and the high-altitude forests. The floristic diversity of these rainforests is greatest in the southern part of the state of Bahia and in the southeast region of the country, and it gradually diminishes farther south. In this area, the coastal plain forests are growing in Pleistocene and Holocene sediments from the nearby mountains or in former seabeds that today lie several meters above sea level. The foothill forests grow mainly on the eastern slopes of the four coastal ridges—the Serra do Mar, the Mantiqueiras, the Paranapiacabas, and the Serra Geral—in soil from the weathered layer of crystalline rock. Moisture-laden air masses blow in from the ocean, usually make for a foggy and very humid environment. Precipitation indexes reach 4,000 mm a year, the highest levels in Brazil. Toward the interior lie poorly defined areas transitioning to the seasonal forests and the forests of Brazilian pine and those dominated by the laurel family. The forests of the coastal plains are the habitat of the golden lion tamarin (*Leontopithecus rosalia*), one of the symbols of conservation in Brazil.

In the northeast region of the country the climates tend to be semiarid or to have dry winters. The mountains and low-altitude *tabuleiros* near the ocean, on which the forests thrive, do not appear until just north of Rio de Janeiro. The humid and semihumid areas rely on water vapor from the ocean, carried by trade winds, and on heavy rains caused by cold fronts from the south; rainfall is consequently dense on the tertiary *tabuleiros* near the ocean or on regoliths (fragmental and unconsolidated rock material residual or transported) of Precambrian rock more toward the interior. Between Alagoas and Paraíba the vegetation of the rainforest becomes more open. Today, after decades of uncontrolled exploitation, only a small proportion of the northeastern ombrophilous forest remains (Câmara 1991).

The mixed ombrophilous forests typically have the only two native genera of conifers, *Araucaria* and *Podocarpus,* plus various genera of laurel, such as *Ocotea* and *Nectandra,* and other families. This forest is distributed almost continuously across the southern plateau, blanketing the state of Rio Grande do Sul and the interior of Santa Catarina and Paraná, with interruptions at higher elevations of the

Table 4.1. Original extent and percentage of different forest types in the Atlantic Forest (MMA 2000b).

Forest Type	Extent (km^2)	Percentage
Dense and open ombrophilous forests	237,530	18.18
Mixed ombrophilous forests	168,916	12.93
Seasonal and semideciduous forests	635,552	48.65
Areas of ecological tension	157,747	12.07
Others	106,676	8.17
Total	1,306,421	100.00

states in São Paulo, Rio de Janeiro, and Minas Gerais. The lowest altitude at which it grows is about 500 m, and this cutoff point can be even higher in the forests' northern range. Only 2–4 percent of the original area of these forests is still reasonably well preserved (Câmara 1991).

Deciduous and semideciduous seasonal forests are located in the interior, to the west of the dense or mixed ombrophilous forests. They are composed of both evergreen and deciduous or semideciduous trees and have highly variable characteristics. Their inclusion in a single forest type is somewhat arbitrary, and although different formations could certainly be recognized, these forests have been almost totally eliminated. Originally they extended from Rio Grande do Sul to Minas Gerais, with arms reaching into the interior of Bahia and Piauí. Their soils range from extremely fertile, such as the red earth of the southern states, to very poor and sandy. These seasonal forests were poorly studied before their large-scale destruction, so we know little about their primitive floristic composition.

Between these three basic formations of the Atlantic Forest are numerous transitional areas and enclaves that are difficult to characterize; not only are their boundaries imprecise, but they also have changed along with the variations in climate over the millennia. Judging from the CONAMA definition of the Atlantic Forest Domain and the Map of Brazilian Vegetation (IBGE 1993), the forest formations described here originally covered the areas listed in Table 4.1.

The Current Status of Biodiversity

The enormous biodiversity of the Atlantic Forest results in large part from the wide range of latitude that it covers, its variations in altitude, and its diverse climatic regimes. It stretches across 27 degrees of latitude, from 3°S to 30°S. In altitude, it ranges from sea level to elevations higher than 2,700 m in the Mantiqueira and Caparaó Mountains in the states of São Paulo, Minas Gerais, Rio de Janeiro, and Espírito Santo. The current climates range from subhumid regimes and dry seasons in northeastern regions to heavy rainfall environments in parts of the coastal mountain range, the Serra do Mar.

Another factor that has influenced the floral and faunal diversity in the Brazilian Atlantic Forest is the geological and climatic history of the region as a whole. In the Cenozoic era, deep faults created sharp relief, and this topography has influenced the biomes. Also, in the Pleistocene, there were glacial periods of cold, dry climate alternated with warm and humid interglacial periods. In the coldest and driest periods, which lasted longer, the forest became fragmented into patches separated by areas of sparse *caatingas* (deciduous forests with prickly and succulent cacti), which subsequently filled in again when changes in temperature and humidity allowed. During the cold periods the Brazilian pine forests may have covered larger areas, as evidenced by the remaining small, disconnected stands of this type of forest. When climatic conditions were more favorable, the rainforests probably occupied larger areas and actually connected with the Amazon Forest. Undoubtedly, it is this great diversity of ecological conditions, coupled with the successive fragmentation and expansion of the different forest formations, that accounts for the tremendous biological diversity and the large number of endemic species that persist, despite extensive deforestation.

Published data on the number of species and the degree of endemism of the Atlantic Forest flora and fauna are variable. Mittermeier et al. (1999) estimated that 20,000 plant species, including 6,000 endemics, and 261 mammals, 620 birds, 200 reptiles, and 280 amphibians, of which 61 percent, 12 percent, 30 percent, and 90 percent, respectively, are endemic. The Ministry of the Environment (MMA 2000b), on the other hand, reported 20,000 plant species, 250 mammals, 1,020 birds, 197 reptiles, and 340 amphibians, with corresponding endemisms of 40 percent, 22 percent, 18 percent, 30 percent, and 26 percent. The differences in these estimates are due in large part to uncertainties regarding taxonomy and distributions, as well as to the use of different definitions of the exact limits of the Atlantic Forest Domain. In the case of the avifauna, the difference is exacerbated by the inclusion in the latter list of coastal species, passage migrants, and occasional visitors, which are excluded in the former. The studies that lead to these published lists (focusing particularly on the high degree of endemism of the flora, and the extent to which this rich biodiversity was threatened) prompted Conservation International to include the Atlantic Forest on its list of the world's 25 biodiversity *hotspots*, a measure that strongly highlighted the global importance of the challenge of conserving the Atlantic Forest (Mittermeier et al. 1999).

The Initial Area versus the Remaining Area

Various attempts have been made to estimate the original area of the Atlantic Forest, but there is no reliable scientific basis on which to draw precise conclusions. According to the definition of the Atlantic Forest Domain adopted by CONAMA and the 1993 Map of Brazilian Vegetation published by IBGE, the forest originally covered an area of 1,363,000 km^2, equivalent to 16 percent of the national territory. It encompassed all or part of 17 states: Piauí, Ceará, Rio Grande do Norte, Paraíba, Pernambuco, Alagoas, Sergipe, Bahia, Espírito Santo, Minas Gerais, Rio de Janeiro, São Paulo, Paraná, Santa Catarina, Rio Grande do Sul, Goiás, and Mato Grosso do Sul. By contrast, according to the most recent published data (MMA 2000a), the area remaining in 1995 was estimated at 98,878 km^2, or only 7.25 percent of the original area.

In reality, even the extremely small figure of 7.25 percent does not give a true picture of how little of the forest remains because it is likely that at the beginning of Portuguese colonization the biome probably extended over a much greater area of the northeastern states (Campos 1912; Coimbra-Filho and Câmara 1996). Moreover, the percentage of remaining area includes not only the few primary forest formations but also forests planted with exotic species and—probably what account for most of it—secondary forests in various stages of regeneration, obviously greatly impoverished from the floral standpoint. Thus, even though the current forests harbor a surprisingly high degree of biological diversity, very little remains of the authentic original forest (Dean 1996); it is reasonable to infer that many species were eliminated before they could even be described.

Moreover, it should be kept in mind that the information available on the different species of the Atlantic Forest is based on surveys of flora and fauna carried out at different times and without systematic monitoring afterward, all of which makes it difficult to arrive at a realistic assessment of the current situation. Therefore, there is little solid information about the populations in decline or about whether all the supposedly extant species are still there because some of them may have since disappeared. Meanwhile, previously unknown species are emerging, sometimes in highly unlikely places. Examples are the black-faced lion tamarin (*Leontopithecus caissara*), discovered in 1990 only a little more than 200 km from the major population center of Curitiba; the lowland tapaculo (*Stymphalornis acutirostris*), a songbird that appeared in 1995 almost within the city limits; and a previously unknown species of titi monkey (*Callicebus coimbrai*) found in the small remaining forests of the northeast and finally described in 1999.

History of the Use and Abuse of the Atlantic Forest

The destruction of the Atlantic Forest started early. Prehistoric indigenous communities, who inhabited some areas of the forest for at least 11,000 years, already practiced a rudimentary form of agriculture; however, any impact they may have had on the great forest is not perceptible today. It was shortly after the Europeans discovered Brazil in 1500 that deforestation began with the large-scale exploitation of brazilwood (*Caesalpinia echinata*), then abundant in the coastal forests from Rio de Janeiro presumably all the way to Ceará (Coimbra-Filho and Câmara 1996).

According to historical accounts, the deliberate destruction of the forest got under way when the colonists began to clear land for settlement, planting, and better defense against indigenous attacks. They burned extensive areas for these purposes and also later during military skirmishes against other Europeans. Wood was steadily consumed for all purposes; indeed, for centuries it was the only available form of fuel.

Often overlooked as an important factor in the widespread destruction of the northeast forests by early colonists is extensive cattle-raising. Large herds were driven from Bahia and Pernambuco to the interior, serving as trailblazers for subsequent human settlements. Vast areas were burned and cleared for pasture in a practice that profoundly changed the environment. Because the cattle needed access to water, the forest formations near water bodies were particularly affected. From the sixteenth to the eighteenth centuries, enormous herds occupied the flatlands along the riverbanks of the Brazilian northeast. This predatory economic activity came to be known in retrospect as the "hide cycle" (Coimbra-Filho and Câmara 1996).

In the eighteenth century, the sugarcane plantations in coastal areas from São Paulo to Rio Grande do Norte, together with mining activity in the states of São Paulo, Minas Gerais, and Goiás, further dismantled vast stretches of forest. The nineteenth century ushered in the "coffee cycle," when increasingly larger areas were turned over to cultivation in the states of Espírito Santo, Rio de Janeiro, Minas Gerais, and São Paulo, extending in the twentieth century as far as Paraná.

The devastation of the Atlantic Forest accelerated exponentially in the twentieth century. At the beginning of the 1900s the Brazilian population was barely 17 million. Fifty years later it had reached 52 million, and that number more than tripled by the end of the century. At the same time, the country became industrialized. An extensive rail network through the Atlantic Forest facilitated the opening up of new areas for cultivation, uncontrolled hunting, disorganized exploitation of timber, and the expansion of urban nuclei. But despite all this activity, at midcentury large forested areas still remained in the interior of the southeast and southern regions and along the coastal strip of Bahia and Alagoas (pers. obs.).

In the twentieth century, the timber industry eliminated nearly all the Brazilian pine forests in the southern states to provide wood for the forms needed to shape the concrete being used in the rapidly growing cities. Hardwoods were also exploited, even for firewood, with no concern for sustainable production. The Atlantic Forest was contributing nearly half of all the lumber produced in Brazil as recently as the 1970s (MMA 2000b). Today, however, the industry relies on large-scale timbering in the Amazon Forest or on plantations of exotic pine species and Australian eucalypts. The few remaining forests in the south of Bahia and the last stands of native Brazilian pine in Paraná and Santa Catarina are still exposed to predatory exploitation.

The oil crisis of the 1970s had a devastating effect on the residual forests of the northeast and the interior of São Paulo. Alcohol was used as a fuel substitute, and most of these remaining forests succumbed to extensive sugarcane planting. With the rapid expansion of the cellulose and paper industry, large areas of forest were cut and replaced with homogeneous plantings of exotic pines and eucalypts.

According to the most recent surveys (Fundação SOS Mata Atlântica, INPE, ISA 1998), between 1985 and 1995 a total of 10,368 km^2 was deforested in the southern, southeastern, and central western regions of the country, representing about 11 percent of the forest area mapped in 1985. An average area of 2.84 km^2 was cut down every day. Although protective measures had already been adopted, the destruction of the Atlantic Forest continued unabated.

Protection of the Atlantic Forest

The first measures to protect the Atlantic Forest were taken during the colonial period. In the famous Royal Letter from Portugal of 1797, the Crown dictated, apparently without any concrete results, that "every precaution should be taken to conserve the forests in the state of Brazil and prevent them from being ruined or destroyed." This concern stemmed from the need for adequate supplies of wood, then a material of strategic importance: it was needed for shipbuilding, and maintenance of maritime navigation was indispensable for both economic and military reasons. For two centuries, leading Brazilian figures have been condemning the predatory model of forest exploitation, but without taking any effective steps to curtail the process.

The first time a natural area was protected in Brazil was in 1898, when a small pocket of land was set aside in São Paulo to serve as a city park. Almost 30 years later, protected status was granted to Itatiaia National Park, and designation of Iguaçu National Park followed in 1939. Protection of natural areas then accelerated, especially after 1961. The real situation today is difficult to assess because there is no single comprehensive registry that lists, by biome, all the conservation units that have been established at various government levels. Moreover, the boundaries of some designated areas overlap, while the borders of others have never been clearly delimited (Chapter 38, this volume).

The database of conservation units (protected areas of different types) maintained by Conservation International in Brazil, which covers only federal and state units larger than 10 km^2 and omits certain others because of vegetation type or human influences, indicates that as of 2001 there were 102 conservation units of various kinds (parks, biological preserves, ecological stations, and ecological preserves) averaging 191.25 km^2. Of this number, 39 are larger than 100 km^2, and four are larger than 1,000 km^2, averaging 451.48 km^2 and 1,862.71 km^2, respectively. Conservation units larger than 10 km^2 represent 19.73 percent of the total remainder of the biome, estimated at 98,878.82 km^2. Although other data may vary from these figures, these numbers at least illustrate the situation and accurately represent the percentage of land protected. At the same time, however, the statistics are somewhat illusory given that management plans are absent, clandestine hunting and exploitation of plant resources continue unabated, and the status of various protected species remain unknown because monitoring programs are absent. A growing problem is the repeated invasion of protected areas by indigenous groups, which has created administrative conflicts between the government agencies responsible for nature conservation and those in charge of protecting the Indians (Chapter 32, this volume).

Another important step in conservation in the Atlantic Forest was the official recognition of private natural preserves. Although their total area is unknown, these areas are important for involving private land owners in the preservation effort.

Recognition of the Atlantic Forest Biosphere Reserve by the United Nations Educational, Scientific and Cultural Organization (UNESCO) in 1991–1992 was yet another key development. Covering 290,000 km^2, this area extends across 14 states, from Ceará to Rio Grande do Sul, and is administered by a national council of representatives of federal, state, and municipal governments and civil society (nongovernment organizations, the scientific community, business leaders, and people living in the reserve).

Government Legislation and Policy

A variety of legal initiatives to defend the Atlantic Forest biome have also been undertaken. Under the Forest Code adopted in 1965, the Brazilian pine may be exploited only in a rational manner and only in the forests in the southeastern, southern, and midwestern regions of the country. Every property in these areas must also preserve 20 percent of its forest cover. Strict application of this law would have kept the Atlantic Forest from shrinking to less than 8 percent of its original size.

The first specific mention of the Atlantic Forest in Brazilian legislation appears in the Federal Constitution of 1988. This constitutional provision regards the biome as national heritage and states that it should be used "according to law, under conditions that ensure preservation of the environment," specifically stipulating "insofar as the use of natural resources is concerned." The provision assumes that specific legislation will be enacted to define the area covered by the biome and to regulate its sustainable exploitation. Similar provisions were subsequently adopted by several of the states and municipalities (Table 4.2 provides a chronological list of important milestones passed since 1988).

By 1990, Congress still had not passed enabling legislation, and Brazil's president issued Decree 99.547, declaring that the ecosystems of the biome were untouchable and prohibiting the cutting down and exploitation of its vegetation. It was replaced in 1993 by Decree 750, which protected not only primary forest formations but also those in the process of natural regeneration, with certain exceptions. However, segments of society with economic interests in forest exploitation, especially rural developers and the timber industry, blocked passage of the implementing legislation and introduced a series of weakening amendments. As of 2001, eight years after introduction of the bill, the decree was still under discussion.

In 1999, CONAMA approved guidelines to promote the conservation and sustainable development of the biome, which set forth the following goals: protect the remaining biological diversity by expanding the system of conservation units; adapt the use of natural resources to ensure their conservation; recover the physiogeographic structure of the biome through initiatives to protect biological diversity, reconstitute ecological corridors, conserve the soil, and guarantee the integrity of the natural ecosystems; and align environmental with sectoral policies. For each line of action, the document listed programmatic measures for its execution and provided instruments for implementation.

To implement these actions, the Ministry of the Environment in 1998 outlined a Pilot Program for Protection of the Atlantic Forest (MMA 1998), later refined in 2000. Not only is the program compatible with the CONAMA guidelines, but for the first time, a document relating to the Atlantic Forest

Table 4.2. Chronology of important legal and regulatory milestones related to the Brazilian Atlantic Forest.

Year	Event
1988	The Federal Constitution declares the Atlantic Forest a national heritage and establishes conditions for use.
1989	Eight states include norms to protect the Atlantic Forest in their state constitutions.
1990	First assessment of the biome forest remnants using remote sensing by collaboration between Fundação SOS Mata Atlântica and Instituto Nacional de Pesquisas Espaciais.
1990	Historic Working Meeting in Atibaia, where the concept of historical coverage of the Atlantic Forest was consensually accepted and the component forest types were defined.
1990	Approval of Decree number 99.547, prohibiting logging and exploration in the Atlantic Forest.
1990	Natural Heritage Private Reserves (RPPN) are recognized officially.
1991	Publication of the first proposal for an action plan for the biome by Fundação SOS Mata Atlântica.
1991	UNESCO approved the first and second phases of the Mata Atlântica Biosphere Reserve, covering Ceará, Rio Grande do Norte, Paraíba, Pernambuco, Alagoas, Sergipe, Bahia, Santa Catarina, and Rio Grande do Sul.
1992	UNESCO approved phase 3 of the Mata Atlântica Biosphere Reserve, in agreement with the conclusions of the Atibaia Workshop.
1992	Creation of the Non-Government Organization Network for the Atlantic Forest.
1992	Presentation to Congress of the Law Initiative 3.285/92 on the Atlantic Forest regulations. In discussion in 2001.
1993	Approval of Decree 750, substituting 99.547/90, establishing norms for the protection and sustainable use of the Atlantic Forest.
1993	"Biodiversity Conservation Priorities for the Northeast Atlantic Forest" workshop in Pernambuco.
1995	Under pressure from interest groups, MMA presents to CONAMA a legal initiative to reduce the official area of the Atlantic Forest. The proposal is rejected.
1996	Workshop in Belo Horizonte, with 40 researchers, policymakers, and conservationists, to confirm the wide concept of coverage of the Atlantic Forest as approved by CONAMA and in disagreement with the initiative to reduce its size.
1998	Approval of Law 9.605, "Environmental Crime Law," not specific to the Atlantic Forest but of great importance for its conservation.
1998	Atlantic Forest, both coastal and interior, is included in the world hotspots.
1999	CONAMA approved the Policy Directions for Conservation and Sustainable Development in the Atlantic Forest.
1999	"Priority Actions for Biodiversity Conservation of the Atlantic Forest and Campos Sulinos" workshop held in Atibaia as part of the workshops promoted by MMA for the National Biodiversity Strategy.
2000	The proposal to change the Forest Code in a damaging way for the Brazilian forest, including the Atlantic Forest, was rejected by Congress as a result of strong pressure from society.
2000	Approval of Law 9.985 to create the National System of Conservation Units, improving previous legislation.

expressed concern about "cultural impoverishment" and "traditional populations," aspects with evident social significance. However, a critique of the projects approved under the program (Schuerholz 2000) by an independent consultant pointed out that socioeconomic aspects were being overemphasized to the detriment of biodiversity conservation. Among other recommendations, the consultant urged that highest priority be given to the conservation areas already established, in recognition of their critical importance for preserving biodiversity.

Conclusions

The Atlantic Forest still retains much biological diversity, despite centuries of uncontrolled exploitation, and in the last 15 years the conviction has grown both nationally and internationally that this heritage warrants urgent and effective protection. However, that sentiment conflicts with a variety of economic interests—particularly those of the agriculture and timber industries—that are lobbying to ensure continued exploitation of the land. So far, these efforts have been defeated. But if the law that has been debated in Congress since 1992 alters CONAMA's definition of which areas the Atlantic Forest comprises, only the protections afforded under the 1965 Forest Code would apply. The definitive meaning of this law will have enormous repercussions for the future of biological diversity in what remains of this once immense forest.

References

Câmara, I. G. 1991. *Plano de ação para a Mata Atlântica.* São Paulo: SOS Mata Atlântica.

Campos, G. 1912. *Mapa florestal.* Rio de Janeiro: Tipografia da Diretoria do Serviço de Estatística.

Coimbra-Filho, A. F. and Câmara, I. G. 1996. *Os limites originais do bioma Mata Atlântica na região nordeste do Brasil.* Rio de Janeiro: Fundação Brasileira para a Conservação da Natureza (FBCN).

Dean, W. 1996. *A ferro e fogo: A história e a devastação da Mata Atlântica Brasileira.* São Paulo: Companhia das Letras.

Fundação SOS Mata Atlântica, Instituto Nacional de Pesquisas Espaciais, and Instituto Socioambiental. 1998. *Atlas da evolução dos remanescentes florestais e ecossistemas associados no domínio da Mata Atlântica no período 1990–1995.* São Paulo: Fundação SOS Mata Atlântica/Instituto Nacional de Pesquisas Espaciais/Instituto Socioambiental.

IBGE (Instituto Brasileiro de Geografia e Estatística). 1993. *Mapa de vegetação do Brasil.* Rio de Janeiro: Instituto Brasileiro de Geografia e Estatística-IBGE.

Mittermeier, R. A., Myers, N., Robles Gil, P., and Mittermeier, C. G. 1999. *Hotspots.* Mexico City: Agrupación Sierra Madre, CEMEX.

MMA (Ministério do Meio-Ambiente). 1998. *Programa piloto para proteção da Mata Atlântica (PPG7/MA).* Versão 1.9. Brasília: MMA.

MMA (Ministério do Meio-Ambiente). 2000a. *Avaliação e ações prioritárias para a conservação da biodiversidade da Mata Atlântica e Campos Sulinos.* Brasília: MMA.

MMA (Ministério do Meio-Ambiente). 2000b. *Programa piloto para a proteção das florestas tropicais brasileiras: subprograma Mata Atlântica (PPG7).* Versão 1.1 setembro/2000. Brasília: MMA.

Rizzini, C. T. 1967. Delimitação, caracterização e relações da flora silvestre hiléiana. *Atas do Simpósio sobre a Biota Amazónica. Biota Amazônica* Vol. 4 (Botânica): 13–36.

Schuerholz, G. 2000. *Pilot program to conserve the Brazilian rain forest: mid-term review, Sector Mata Atlântica.* Brasília: MMA.

Chapter 5

Status of the Biodiversity of the Atlantic Forest of Brazil

José Maria Cardoso da Silva and Carlos Henrique M. Casteleti

The Atlantic Forest, one of the largest tropical forests on the planet, was the first biome to be exploited following the European colonization of Brazil (Chapter 4, this volume). Successive economic cycles and the steady expansion of the region's human population over the last five centuries have seriously compromised the ecological integrity of the Atlantic Forest's singular ecosystems. The origins of this grave environmental crisis can be traced to the history of the colonization of the region (Dean 1995; Coimbra-Filho and Câmara 1996).

Despite centuries of scientific investigation, the Atlantic Forest is still poorly known. New vertebrate species continue to be described there (Lorini and Persson 1990; Kobayashi and Langguth 1999), and many are even being found on the outskirts of some of the largest cities (Willis 1992). The fact that, despite the enormous reduction in forest cover, few known species in the region have become extinct (Chapters 8 and 30, this volume), indicates that a chance still exists for well-planned actions to conserve the biota of the Atlantic Forest. However, such actions must be based on an assessment of the current state of the biodiversity of the region and include a clearly defined set of indicators for monitoring progress.

In this chapter we offer an overview of the biodiversity of the Brazilian Atlantic Forest, emphasizing its unique biogeographic position among the tropical forests of South America, and propose a division of the Brazilian Atlantic Forest into biogeographic subregions based on the distribution of forest butterflies, mammals, and birds, the most well-known biological groups in the region. We suggest two indicators for monitoring conservation efforts: the first measures the state of biodiversity in each biogeographic subregion, and the second efforts to protect the biodiversity of the Brazilian Atlantic Forest. Based on combinations of these two indicators, we recommend actions for strategic conservation planning in each of the subregions of the Atlantic Forest.

The Biodiversity of the Atlantic Forest

The Atlantic Forest once covered an area of nearly 1,400,000 km^2 in Brazil alone. It covered a broad latitudinal strip along the Brazilian coast, from Rio Grande do Norte to Rio Grande do Sul (Rizzini 1997). More than 75 percent was forest, with enclaves of scrub, montane rupestrine (rocky) grasslands, and open countryside with low-growing vegetation (*caatinga* [open forest ecosystem consisting of thorny shrubs and stunted trees], dry xeromorphic, scrub and deciduous forest and *cerrado* [woodland–savannah ecosystems that support a unique array of drought- and fire-adapted plant species], bush savanna typical of the central plateau of Brazil), as well as coastal mangrove swamps and forests and *restingas* (coastal scrub and forest on sandy soils).

The Atlantic Forest is isolated from the two other large South American forest blocks: the Amazon Region and the Andean Forests. The Caatinga and the Cerrado, two biomes dominated by open vegetation, separate it from the Amazon Region, and the Chaco, an area of dry vegetation in the central depressions of South America, separates it from the Andean Forests. Its isolation has resulted in the evolution of a unique biota with numerous endemic species (Rizzini 1997; Myers et al. 2000). For this reason, the Atlantic Forest is consistently identified as one of South America's most distinctive biogeographic units (Müller 1973).

The evolutionary history of the Atlantic Forest is marked by periods when it was connected to other South American forest biotas, followed by periods of isolation (Prum 1988; Rizzini 1997). Its biota consequently comprises not only very old elements, which differentiated at least 3 million years ago during the Pliocene epoch, but also elements that colonized the region more recently, during the Pleistocene-Holocene transition about 10,000–20,000 years ago (Prum 1988; Hackett and Lehn 1997). The numerous related species (presumably descended from a common ancestor) now geographically dispersed throughout the region indicate that the Atlantic Forest has seen several periods of biological differentiation. For example, the four recognized species of lion tamarins (genus *Leontopithecus*) occupy distinct and isolated parts of the Atlantic Forest, in southern Bahia, Rio de Janeiro, the interior of São Paulo and the coast of Paraná. Various factors have contributed to the evolution of species with limited distributions in the Atlantic Forest, such as the formation of rivers (Silva and Straube 1996), global paleoecological changes (Haffer 1987), and regional paleoecological changes brought about by tectonic movements (Silva and Straube 1996).

The biota of the Atlantic Forest is extremely diverse (Conservation International do Brasil et al. 2000). Although our biological understanding of extensive areas is still incomplete, the region is believed to harbor 1 to 8 percent of the world's total biodiversity. Considerable environmental diversity within the Atlantic Forest biome may be the reason for the diversity of species and the high degree of endemism. Latitude is an important axis of variation: unlike most other tropical forests, the Atlantic Forest extends over 27 degrees. Latitude greatly affects the geographic distribution of lizards, for example, and only one species is found throughout the region (Vanzolini 1988). Altitude is also important: the Atlantic Forest covers terrain ranging from sea level to 2,700 m, with consequent altitudinal gradients of diversity (Holt 1928; Buzzetti 2000). Finally, there is also longitudinal variation: the forests of the interior differ significantly from those nearer to the coast (Rizzini 1997). These three factors together create a unique diversity of landscapes, and explain, at least in part, the extraordinary species diversity of the region.

Biogeographic Subdivisions

The endemic biota of the Atlantic Forest is not distributed homogeneously. Species composition varies widely, and in terms of biodiversity conservation, the entire Atlantic Forest cannot be treated as a homogeneous unit, and consideration must be given to distinct biogeographic subregions (Table 5.1).

We propose that the distribution of endemic species of forest birds, primates, and butterflies—groups that are well known in the Atlantic Forest—serve as the basis for a preliminary synthetic biogeographic classification. The biogeographic subregions of the biome that emerge fall into two types: areas of endemism and areas of transition (Figure 5.1).

Areas of endemism are subregions containing at least two endemic species, the distributions of which are superimposed. Five areas meet this criterion, including the moist forests of the northeast (Brejos Nordestinos, Pernambuco, Diamantina, and Bahia) and the coastal mountain range (Serra do Mar). Areas of transition, on the other hand, cannot be delimited on the basis of congruent distributions of endemic species. Instead, they are characterized by the presence of unique biological phenomena, such as contact zones between related species or complex mosaics of interaction between biotas of different historical evolution. There are three such areas of transition in the Atlantic Forest: São Francisco, the Interior Forests, and the Brazilian Pine (*Araucaria*) Forests. The characteristics of each of these biogeographic subregions are outlined in Table 5.1.

Indicators for Monitoring the State of Biodiversity

A set of appropriate indicators is a necessary prerequisite for any biodiversity monitoring program in the Atlantic Forest. They are particularly useful for measuring changes in the state of biodiversity, at all levels, over space and time.

Two types of indicators, in particular, are needed for monitoring the biodiversity of the Atlantic Forest: one should measure the status of biodiversity, and the other should measure society's efforts to prevent its loss. The existence of distinct biogeographic subregions, however, means that these indicators must be calculated for each one rather than for the biome as a whole.

"State of Biodiversity" Indicator

One of the best indices proposed to date for monitoring the status of a region's biodiversity is the Natural Capital Index (NCI). This index has a number of advantages: it can be applied on different spatial scales, it is easily understood, and

Table 5.1. Distribution, vegetation characteristics, and endemism of biogeographic subregions of the Atlantic Forest of Brazil.

Name of Subregion	Distribution	Characteristics	Endemism
Brejos Nordestinos	The moist forest of the northeast once extended over an estimated area of 11,960 km^2, of which only 19.4% is currently covered by forest. Today it survives as scattered enclaves in the interior of the Caatinga, an area of the Atlantic Forest with predominantly dry vegetation located in the peripheral depressions of the Brazilian northeast.	The brejos consist largely of semideciduous seasonal forest or dense ombrophilous forest growing on the humid slopes of the residual plateaus. Their biological makeup varies widely. The most extensive brejos, and those that deserve the greatest attention, are located in the following areas: the Ibiapaba mountains along the border between the states of Ceará and Piauí; the Baturité mountains and surrounding terrain along the coast of Ceará; the Araripe plateau, a transitional area that straddles the borders shared by Ceará, Pernambuco, and Piauí; and the Negra mountains in the interior of Pernambuco.	The brejos have many endemic plant and animal species (Borges 1991), including two species of amphibians, the forest frogs *Adelophryne baturitensis* and *A. maranguapensis* and the lizard *Leposoma baturitensis* (Hoogmoed et al. 1994; Rodrigues and Borges 1997), as well as a new avian species, the Araripe manakin (*Antilophia bokermanni*). The latter is known only on the humid slopes of the Araripe plateau (Coelho and Silva 1999) and was recently included on the list of threatened avian species (BirdLife International 2000).
Pernambuco	It encompasses 39,567 km^2, of which only 4.82% is still covered with forest.	It includes all the coastal Atlantic Forest north of the São Francisco River.	It is recognized as an area of endemism for plants, forest butterflies, and birds (Müller 1973; Tyler et al. 1994; Wege and Long 1995). Threatened endemic species include the Alagoas foliage-gleaner (*Phylidor novaesi*), the Alagoas (long-tailed) tyrannulet (*Phylloscartes ceciliae*), the plain spinetail (*Synallaxis infuscata*), the orange-tailed antwren (*Terenura sicki*), the Alagoas antwren (*Myrmotherula snowi*), and the seven-colored tanager (*Tangara fastuosa*). The Alagoas (razor-billed) curassow (*Mitu mitu*), currently extinct in nature, was also known only in this subregion (Sick 1997).

São Francisco	It has a 125,452 km^2 area and includes all the seasonal deciduous and semideciduous forests along São Francisco River valley in the states of Minas Gerais and Bahia. The percentage of the subregion currently covered by forests is unknown.	Preliminary descriptions of these forests indicate they hold unique vegetation and retain elements of a drier forest that was once more widely distributed in South America (Ratter et al. 1978; Prado and Gibbs 1993; Pennington et al. 2000).	This subregion is home to threatened avian species such as the Minas Gerais tyrannulet (*Phylloscartes roquettei*) (Silva 1989; Wege and Long 1995).
Diamantina	This subregion has an area measuring 82,373 km^2 and includes forests and vegetation associated with the slopes of the Diamantina plateau and surrounding tablelands. The percentage of the subregion currently covered by forest is unknown.	The landscape is composed of a mixture of small areas of stunted sparse forest, rupestrine grasslands, open pastures with low-growing vegetation, and regular forest. Seasonal deciduous and semideciduous forests predominate, with areas of dense ombrophilous forest on the slopes of some of the plateaus.	In terms of its biology, there is much that remains unknown about the biology of this subregion. It is delimited by several threatened endemic bird species: the narrow-billed antwren (*Formicivora iheringi*), the slender antbird (*Rhopornis ardesiaca*), the Bahia spinetail (*Synallaxis whitneyi*), and the Bahia tyrannulet (*Phylloscartes beckeri*). A threatened primate subspecies, the Northern Bahian blond titi (*Callicebus personatus barbarabrownae*), also appears to be endemic to this subregion.
Bahia	The Bahia subregion covers 120,954 km^2, extending from Sergipe to Espírito Santo.	About 12% of this area is still covered by forest, whereas at one time more than 83% of it was dense ombrophilous forest, with small patches of seasonal semideciduous forest, open pastures with low-growing vegetation, and open ombrophilous forest.	The subregion is recognized as an important area of endemism for terrestrial vertebrates (Müller 1973), forest butterflies (Tyler et al. 1994), and plants (Soderstrom et al. 1988). It has many threatened endemic bird and mammal species, including the pink-legged gravateiro (*Acrobatornis fonsecai*), the fringe-backed fire-eye (*Pyriglena atra*), Stresemann's bristlefront (*Merulaxis stresemanni*), the Bahia chestnut-sided tapaculo (*Scytalopus psychopompus*), the yellow-breasted capuchin (*Cebus apella xanthosternos*), and the golden-headed lion tamarin (*Leontopithecus chrysomelas*).

Continued

Table 5.1. Continued

Name of Subregion	Distribution	Characteristics	Endemism
Interior Forest	The Interior Forest extends from the northeast of Minas Gerais to Rio Grande do Sul. It is the largest biogeographic subregion of the Atlantic Forest, covering 698,344 km^2. Of this area, forests cover only 2.75% today.	The predominant phytophysiognomic type was once the seasonal semideciduous forest, which represented 62% of the area, with the rest comprised of seasonal deciduous forest, open countryside with low-growing vegetation, rupestrine grasslands, and transition areas.	The subregion could be further subdivided in terms of its vegetation (Rizzini 1997), but based on biogeographic information currently available, further subdivision is not warranted. The Interior Forest actually is a broad transition belt between the Atlantic Forest and its adjacent biomes.
Brazilian Pine Forests	This subregion is located mainly in the state of Paraná, with extensions into Santa Catarina and Rio Grande do Sul. It covers approximately 238,591 km^2, of which only 9.77% is still covered by forest.	The principal vegetation is mixed ombrophilous forest, which once covered more than 70% of the area. Small expanses of open countryside, some of it with low-growing vegetation, are found in the south. The biological characteristics of this subregion are also found in high montane enclaves of the Serra do Mar. It is a broad transition area between the Atlantic Forest and the temperate biome of South America (Leite and Klein 1990).	This subregion does not have any endemisms among its animal groups (birds, mammals, and butterflies), the main criterion used for the present biogeographic subdivision or the Atlantic Forest.
Serra do Mar	The coastal montane subregion extends from Rio de Janeiro to the northern part of Rio Grande do Sul. It covers an area of approximately 111,580 km^2, of which only 30.48% is now covered by forest.	The predominant vegetation was once dense ombrophilous forest, which accounted for more than 95% of the area, with the rest comprised of mangrove forests and salt marshes. In terms of distribution of various species (mainly birds and butterflies), the Serra do Mar might be further subdivided into northern and southern sectors, but additional in-depth study is needed.	The Serra do Mar is the subregion of the Atlantic Forest with the largest concentration of threatened endemic species of birds (Collar et al. 1997). Examples of threatened endemic species are the grey-winged cotinga (*Tijuca condita*), the kinglet calyptura (*Calyptura cristata*), Kaempfer's tody-tyrant (*Hemitriccus kaempferi*), and the restinga tyrannulet (*Phylloscartes kronei*) among the avian species, and, among mammals, the golden lion tamarin (*Leontopithecus rosalia*), and the black-faced lion tamarin (*Leontopithecus caissara*).

Figure 5.1. Biogeographic subregions of the Atlantic Forest of Brazil characterized by the presence of endemic species: Brejos Nordestinos, Pernambuco, São Francisco, Diamantina, Bahia, and Serra do Mar. Transition subregions are the Interior Forest and Brazilian Pine Forest.

has successfully passed various tests to which it has been submitted (for example, see tem Brink 2000). The NCI was developed following guidelines in the Convention on Biological Diversity, and has received the attention of the member countries of the Organization for Economic Cooperation and Development (OECD).

The NCI is the product of two measures of an ecosystem: one of quantity and one of quality. Quantity is measured as the extent of a given type of ecosystem,

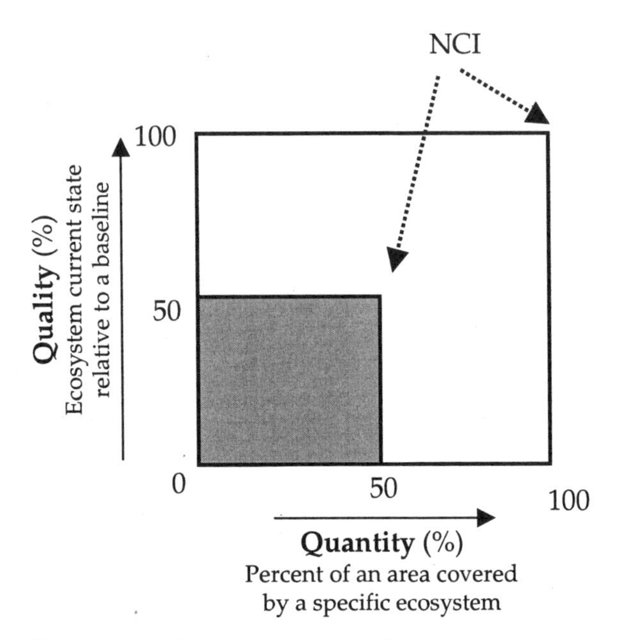

Figure 5.2. The Natural Capital Index is a combination of ecosystem quantity and quality. Quantity is measured as the percentage of a particular area covered by a given type of ecosystem, and quality is represented by a ratio that reflects an ecosystem's current state relative to a baseline state.

and quality is measured as the ecosystem's current status relative to a baseline standard (Figure 5.2). A region totally covered by unaltered natural ecosystems would have an NCI of 100 percent, while a region with quantity and quality measurements of 50 percent would have an NCI of 25 percent.

Unfortunately, the information needed to calculate these indices is available for only six of the eight biogeographic subregions. In these, most forest remnants are smaller than 2 km^2 (Figure 5.3) and, excepting the Brazilian Pine Forests and the Interior Forests, the few remaining areas that are larger than 10 km^2 represent the major part of the total forest area (Figure 5.4).

The NCIs of the six subregions vary widely (Figure 5.5). When the cutoff point for calculating forest quality is set at either 10 km^2 or 50 km^2 of the remaining forest area, the subregions rank from the largest, Serra do Mar, down through Brejos Nordestinos, Bahia, Brazilian Pine Forests, Pernambuco, and the Interior Forests. When the cutoff point is set at 100 km^2, Pernambuco and the Interior Forests change places. Of the six subregions, the Serra do Mar and the Brejos Nordestinos have retained the most natural capital, and Pernambuco and the Interior Forests have retained the least.

In general, the NCIs for the Atlantic Forest subregions show that the status of their biodiversity is critical; most of their natural wealth has been lost. The respective NCIs range from 0.86 to 22.6 when 10 km^2 is taken as the cutoff for

Figure 5.3. Distribution of forest remnants of the Atlantic Forest of Brazil. The forest remnants of the subregions São Francisco and Diamantina are not mapped, and those of the Bahia are only partially mapped.

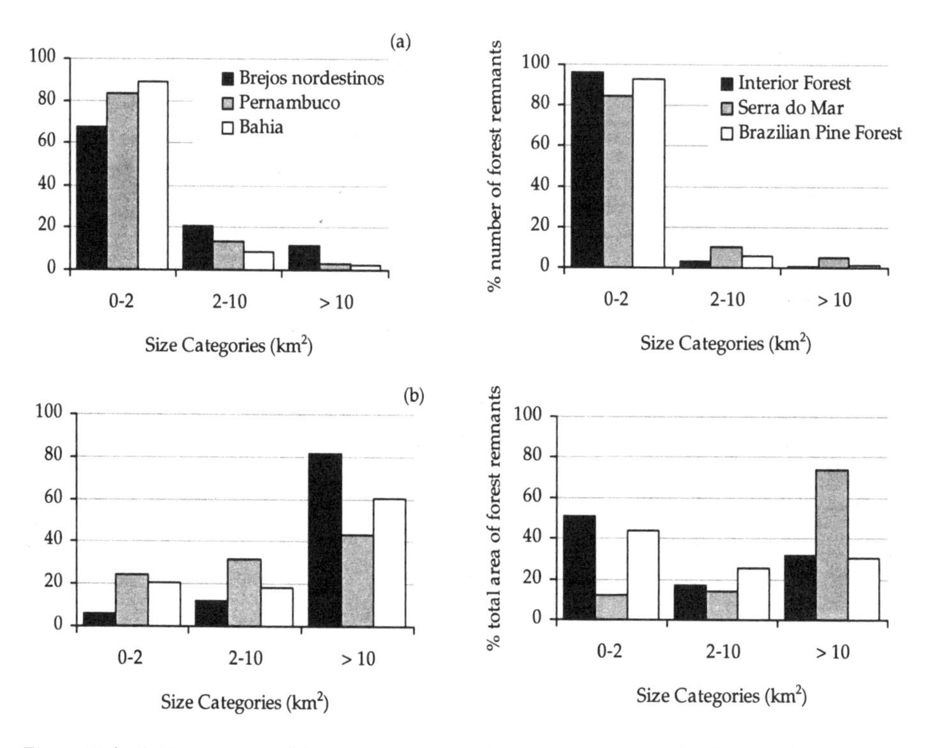

Figure 5.4. (a) Percentage of forest remnants in three size categories (km^2), and (b) percentage of total area occupied by remnants in different size categories. The vast majority of remnants are smaller than 2 km^2.

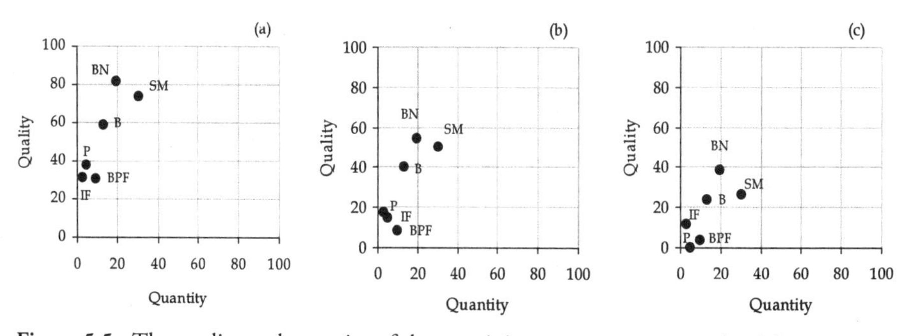

Figure 5.5. The quality and quantity of the remaining ecosystems across the Atlantic Forest regions differ dramatically. Acronyms stand for the analyzed regions (Fundação SOS Mata Atlântica 2000).

calculating quality, from 0.48 to 15.3 when the cutoff is 50 km^2, and from 0 to 8.01 when the cutoff is 100 km^2 (Figure 5.5).

"Social Response to Loss of Biodiversity" Indicator

The Biodiversity Protection Index (BPI) follows the basic model of the NCI, but assesses the quantity and quality of society's response to the loss of regional biodiversity. It is based on two measurements: the area of a region currently occupied

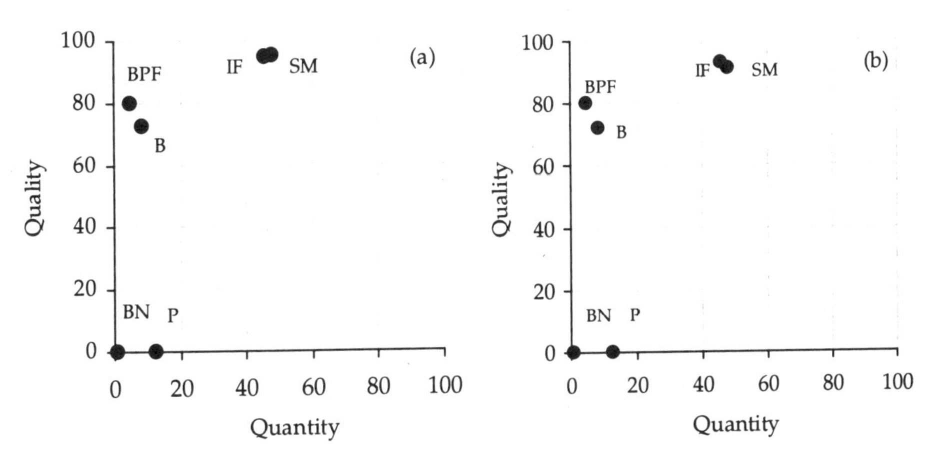

Figure 5.6. The quality and quantity of protected areas across the biogeographic regions of the Atlantic Forest differ dramatically. Acronyms stand for the analyzed regions.

by a given type of ecosystem and the area of the conservation units (CUs) under comprehensive protection (categories I and II of the World Conservation Union–IUCN) within it. The two measurements can be used to calculate the amount and quality of protection afforded to each ecosystem. Quantity is measured as the percentage of an ecosystem's current area under comprehensive protection CUs. Quality is measured as the ratio of the area of comprehensive protection CUs of at least 50 km^2 (or perhaps an even higher cutoff; see Chapter 31, this volume) to the total area of comprehensive protection CUs. Like the NCI, the index ranges from 0 to 100 percent.

The BPI focuses on comprehensive protection CUs because they are the only category of protected area capable of forming the nucleus of an effective biodiversity conservation program. Ideally, comprehensive protection CUs should be large enough and of sufficient quality to maintain populations of animal and plant species that are sensitive to anthropogenic pressures, maintain key ecological processes such as pollination and seed dispersal, serve as a source of germplasm, and serve as a control area in the assessment of environmental restoration programs. Although there is no consensus on the minimum size necessary to fulfill these functions, 50 km^2 has been adopted as a criterion for global studies on the effectiveness of protected areas (Bruner et al. 2001).

The number of comprehensive protection CUs in the six subregions ranges from three in the Brejos Nordestinos to 72 in the Serra do Mar. However, not all are of sufficient size (Figure 5.6). The amount of protection (Figure 5.7) in these terms is high in the Serra do Mar (48.08 percent) and the Interior Forests (45.75 percent), but low in the other four subregions (less than 12 percent). As for the quality of protection, regardless of whether we set the cutoff for CU size at 50 km^2 or 100 km^2, there is a striking contrast between the Serra do Mar, the Interior Forests, Bahia, and the Brazilian Pine Forests on the one hand, all with indices of at least 70 percent, and the Brejos Nordestinos and Pernambuco on the other, each with an index of 0 percent (Figure 5.6). The BPI is high for the Serra

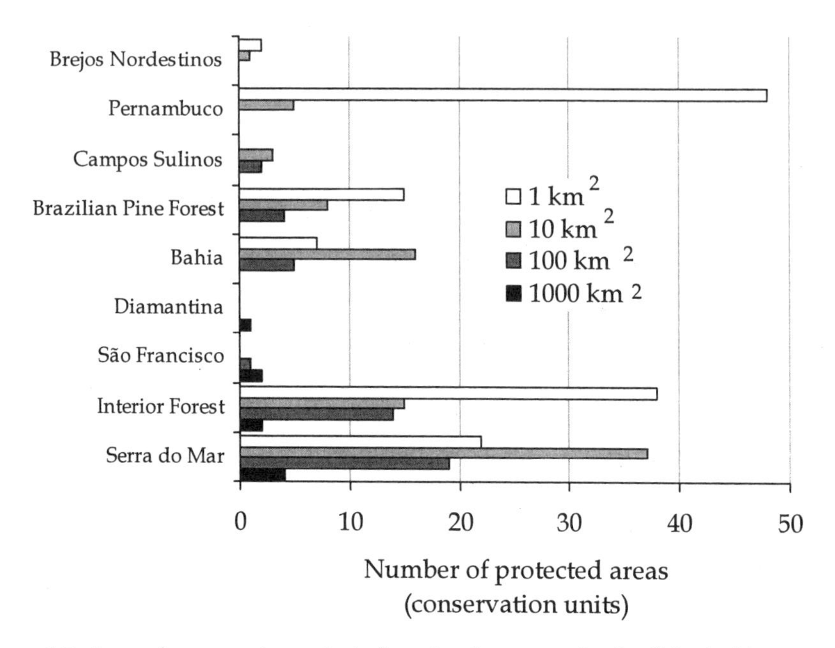

Figure 5.7. Integral conservation units in four size class categories (km^2) in the biogeographic subregions of the Atlantic Forest of Brazil.

do Mar and the Interior Forests, low for the Brazilian Pine Forests and Bahia, and very low for the Brejos Nordestinos and Pernambuco (Figure 5.6).

Evaluation of the Existing Data Sources for Calculating Indicators

Calculation of the NCI requires measures of the quantity and quality of its various ecosystems. The quantity, or size, is easily measured by calculating the percentage of each biogeographic subregion covered by the remnants of the ecosystem in question. However, measuring the quality of the ecosystem is more complicated. Ideally, an ecosystem's quality measurement should be based on a set of variables that account for the wealth and abundance of the various indicator species, as well as its structure and functioning (tem Brink 2000). However, because such information does not exist for most subregions of the Atlantic Forest, alternatives are needed. One possibility is to measure the ecosystem quality by the percentage that currently remains in areas larger than a given cutoff point. A suggested point of reference for monitoring would be a situation in which 100 percent of the remaining areas of the ecosystem are larger than the cutoff. Three cutoff points have been used: 10 km^2, 50 km^2, and 100 km^2. Evidence that the quality of an ecosystem is substantially impaired when its size is reduced justifies these cutoffs. For forests, the principal ecosystem of the Atlantic Forest, remnant areas smaller than 10 km^2 have been found to be incapable of maintaining populations of numerous species except in the forest interior (Forman 1995; Chapter 31, this volume).

Calculating the BPI also requires a measurement of quality, which can be represented by the current percentage of the biogeographic subregion covered by comprehensive protection CUs: the ratio of comprehensive protection CUs at least 50 km^2 (or 100 km^2) in size to the total area of comprehensive protection CUs in the subregion.

The best source of data for estimating the remaining area of the region's natural ecosystems is the mapping project undertaken by the SOS Atlantic Forest Foundation and the National Institute for Space Research (Instituto Nacional de Pesquisas Espaciais—INPE) (Fundação SOS Mata Atlântica 2000; Chapter 6, this volume). Maps have been drawn up using a standardized method that combines satellite images (Landsat, analogic format, scale 1:250,000) with groundtruthing in the field. All remaining areas of vegetation have been mapped and classified as forest, mangrove forest or salt marsh, permitting a quantitative analysis of the major ecosystem types. The main advantage of using this data source for calculating the NCI and the BPI is that the vegetation cover of the Atlantic Forest will be assessed every 5 years, thus offering the temporal dimension needed for an effective biodiversity monitoring program.

However, some problems exist. Most serious is that information is lacking for the remaining areas of parts of the Bahia and Diamantina subregions and for the entire São Francisco subregion (Figure 5.3). Data on the remnant forests of some of the subregions are old: data for the Pernambuco, Brejos Nordestinos, and Bahia subregions are from 1990, while information on the rest of the region is from 1995. The data for the Brejos Nordestinos and Pernambuco subregions were also collected using a different method (based on a combination of satellite images at a scale of 1:1,000,000 and aerial photographs at different scales) from that used in the more recent mapping of the southern sectors of the Atlantic Forest (Fundação SOS Mata Atlântica 2000). Calculation of the two indices is somewhat influenced by these methodological differences. The mapping of remnant forest areas completed so far by the Atlantic Forest Foundation and the INPE does not include the entire Atlantic Forest, and the NCI and the BPI can be calculated for only six of the subregions (Brejos Nordestinos, Pernambuco, Bahia, Interior Forests, Serra do Mar, and Brazilian Pine Forests).

Although the database generated by the Fundação SOS Mata Atlântica/INPE mapping project is adequate for calculating the NCI, calculating the BPI also requires a reliable digital database with information about the region's CUs. Conservation International do Brasil has a database that covers all the CUs of the Atlantic Forest, both state and federal (Chapter 38, this volume). However, although it is currently the best source of information on CUs in the region, it is still incomplete, and some reserves created at the municipal level have yet to be listed. In addition, the database does not have digital maps of all of the CUs, making more sophisticated spatial analyses not feasible. However, the Brazilian Institute for the Environment and Renewable Natural Resources (Instituto Brasileiro do Meio Ambiente e dos Recursos Naturais Renováveis—IBAMA) is now developing an information system that will include digital maps of all the

CUs in Brazil, which will be of inestimable value for the planning of conservation systems such as corridors. Unfortunately, the IBAMA information system was incomplete when the present analysis was undertaken. Many of the CUs in the Brazilian northeast and the state of Minas Gerais, have yet to be mapped at a scale suitable for spatial analysis.

A biodiversity monitoring program for the Atlantic Forest will necessarily depend on the generation of critical information at regular intervals.

- The Program for Monitoring Atlantic Forest Vegetation Cover (SOS Mata Atlântica/INPE) should be maintained and expanded to cover the entire biome. Studies of the entire region should be carried out every five years, besides studies at shorter intervals in areas considered critical.
- Urgent is the establishment of a system for monitoring the quality of forest ecosystems in all the biogeographic subregions through the generation of standardized information on population estimates of key species and species groups (especially endemic and threatened species), habitat structure, and parameters gauging ecosystem functioning.
- The IBAMA information system on conservation units should be maintained and expanded to incorporate a capacity to monitor their status through the regular analysis of satellite imagery, and in situ by resident biologists.

Defining the Destiny of the Atlantic Forest

The status of forest biodiversity, as measured by the NCI in the six subregions, is critical. Depending on the criterion used to calculate the quality of remaining forest areas, the Atlantic Forest has lost between 91 percent and 96 percent of its natural capital.

The bulk of the remaining forest areas are highly fragmented. Numerous small forests are scattered in a pattern that is fatally damaging to their long-term survival (Chapter 31, this volume). The fauna and flora of the Pernambuco and Interior Forests subregions are the most critically threatened. The destruction of the Interior Forests region implies the rupture of unique ecological and evolutionary processes characteristic of areas of transition (Silva 1998), while the loss of areas of endemism such as Pernambuco implies the complete disappearance of unique evolutionary lineages, evidenced by the large number of critically threatened terrestrial vertebrates. Species extinctions are more imminent in Pernambuco than in any other sector of the Atlantic Forest.

As measured by the BPI, society's response to the problem of biodiversity loss in the Atlantic Forest is severely inadequate, an assessment consistent with Câmara's analysis of conservation policies (Chapter 4, this volume) and the results of the Workshop on Conservation Priorities for the Brazilian Atlantic Forest (Conservation International do Brasil et al. 2000).

As previously discussed, the NCI and the BPI are independent indices: subregions with a high NCI can have a low BPI, as in the case of the Brejos Nordestinos, and areas with a low NCI can have a high BPI, as in the Interior Forests. This would seem paradoxical, but is undoubtedly the result of social, economic, and institutional differences between the subregions: local societies with serious social and economic problems responding more slowly to the loss of natural capital because the institutional capacity to meet the demand tends to be less adequate, or even lacking. Indeed, our analysis provides some support for this in that the two subregions with the lowest BPIs—namely, the Brejos Nordestinos and Pernambuco—also have the lowest economic and social indicators. Any strategy for conserving the Atlantic Forest should allow for the broad regional variation that exists both in the state of conservation (NCI) and in the efforts that are being made to protect the region (BPI). The causes of this variation should be studied through historical, geographic, cultural, and socioeconomic analyses within each of the biogeographic subregions.

Finally, a policy for the conservation of the Brazilian Atlantic Forest will not make any sense unless it is regionalized, and we recommend that plans for each of the subregions should be based on the guidelines proposed by Soulé and Terborgh (1999). By combining the NCI and the BPI, we have formulated the following recommendations to incorporate into efforts to conserve the Atlantic Forest subregions:

- Because the Serra do Mar subregion has retained much natural capital and includes the best system of protected areas, priority should be to expand this system by creating new comprehensive protection CUs in this subregion, based on detailed analyses of the region's biogeography and the efficacy of the protected areas already in place in this respect.
- In subregions with high NCI and low BPI, such as the Brejos Nordestinos, new comprehensive protection CUs should be created that are larger than 50 km^2. These new CUs should account for the environmental variation that exists in the subregion. In subregions with a low NCI and a high BPI, such as the Interior Forests, extensive areas should be restored to connect existing conservation units.
- Subregions with low NCI and low BPI, such as Pernambuco, Bahia, and the Brazilian Pine Forests, should, therefore, be given top priority for conservation action as they are the most seriously threatened. New comprehensive protection CUs larger than 50 km^2 should be created, remaining forest areas should be expanded through restoration, and ecological corridors should be established to connect them.

References

BirdLife International. 2000. *Threatened birds of the world.* Barcelona and Cambridge: Lynx Editions and BirdLife International.

Borges, D. M. 1991. *Herpetofauna do maciço de Baturité, estado do Ceará: composição, ecologia e considerações zoogeográficas.* Master's dissertation, Universidade Federal da Paraíba, João Pessoa.

Bruner, A. G., Gullison, R. E., Rice, R. E., and da Fonseca, G. A. B. 2001. Effectiveness of parks in protecting tropical biodiversity. *Science* 291: 125–128.

Buzzetti, D. R. C. 2000. Distribuição altitudinal de aves em Angra dos Reis e Parati, sul do Estado do Rio de Janeiro, Brasil. In: Alves, M. A. S., Silva, J. M. C., Van Sluys, M., Bergallo, H. G., and Rocha, C. F. D. (eds.). *A Ornitologia no Brasil: Pesquisa atual e perspectivas.* pp. 131–148. Rio de Janeiro: Editora da Universidade do Estado do Rio de Janeiro.

Coelho, G. and Silva, W. 1998. A new species of *Antilophia* (Passeriformes: Pipridae) from Chapada do Araripe, Ceará, Brazil. *Ararajuba* 6: 81–84.

Coimbra-Filho, A. and Câmara, I. G. 1996. *Os limites originais do bioma Mata Atlântica na região nordeste do Brasil.* Rio de Janeiro: Fundação Brasileira para a Conservação da Natureza.

Collar, N. J., Wege, D. C., and Long, A. J. 1997. Patterns and causes of endangerment in the New World avifauna. *Ornithological Monographs* 48: 237–260.

Conservation International do Brasil, Fundação SOS Mata Atlântica, Fundação Biodiversitas, Instituto de Pesquisas Ecológicas, Secretaria do Meio Ambiente do Estado de São Paulo, SEMAD/Instituto Estadual de Florestas–MG. 2000. *Avaliação e ações prioritárias para a conservação da biodiversidade da Floresta Atlântica e Campos Sulinos.* Brasília: MMA/SBF.

Dean, W. 1995. *With broadax and firebrand: the destruction of the Brazilian Atlantic forest.* Berkeley, CA: University of California Press.

Forman, R. T. T. 1995. *Land mosaics: the ecology of landscapes and regions.* Cambridge, England: Cambridge University Press.

Fundação SOS Mata Atlântica. 2000. Cobertura da vegetação nativa da Mata Atlântica e Campos Sulinos. In: Pinto, L. P. (ed.). *Relatório Técnico do Projeto "Avaliação e ações prioritárias para a conservação da biodiversidade da Floresta Atlântica e Campos Sulinos."* pp. 24–79. Belo Horizonte: Conservation International do Brasil.

Hackett, S. J. and Lehn, C. A. 1997. Lack of genetic divergence in a genus (*Pteroglossus*) of neotropical birds: the connection between life-history characteristics and levels of divergence. *Ornithological Monographs* 48: 267–279.

Haffer, J. 1987. Biogeography of neotropical birds. In: Whitmore, T. C. and Prance, G. T. (eds.). *Biogeography and Quaternary history in tropical America.* pp. 105–150. Oxford, England: Clarendon and Oxford University Press.

Holt, E. G. 1928. An ornithological survey of the Serra do Itatiaya, Brazil. *Bulletin of the American Museum of Natural History* 57: 251–326.

Hoogmoed, M. S., Borges, D. M., and Cascon, P. 1994. Three new species of the genus *Adelophryne* (Amphibia: Anura: Leptodactylidae) from northeastern Brazil, with remarks on the other species of the genus. *Zoologische Mededelingen Rijksmus. Natural History Leiden* 68(24): 271–300.

Kobayashi, S. and Langguth, A. 1999. A new species of titi monkey, *Callicebus* Thomas, from northeastern Brazil (Primates: Cebidae). *Revista Brasileira de Zoologia* 16: 531–551.

Leite, P. F. and Klein, R. M. 1990. Vegetação. In: Fundação Brasileiro de Geografia e Estatística, Diretoria de Geociências. *Geografia do Brasil: Região Sul.* pp. 113–150. Rio de Janeiro: IBGE.

Lorini, M. L. and Persson, V. G. 1990. Uma nova espécie de *Leontopithecus* Lesson, 1840, do sul do Brasil (Primates, Callitrichidae). *Boletim do Museu Nacional, Rio de Janeiro, nova série, Zoologia* 338: 1–14.

Müller, P. 1973. Dispersal centers of terrestrial vertebrates in the Neotropical Realm. *Biogeographica* 2: 1–244.

Myers, N., Mittermeier, R. A., Mittermeier, C. G., da Fonseca, G. A. B., and Kent, J. 2000. Biodiversity hotspots for conservation priorities. *Nature* 403: 853–858.

Pacheco, J. F. and Bauer, C. 1999. Aves. In: Pinto, L. P. (ed.). *Avaliação e Ações Prioritárias para Conservação dos Biomas Floresta Atlântica e Campos Sulinos.*

Pennington, R. T., Prado, D. E., and Pendry, C. A. 2000. Neotropical seasonally dry forests and Quaternary vegetation changes. *Journal of Biogeography* 27: 261–273.

Prado, D. E. and Gibbs, P. E. 1993. Patterns of species distributions in the dry seasonal forests of South America. *Annals of the Missouri Botanical Garden* 80: 902–927.

Prum, R. 1988. Historical relationships among avian forest areas of endemism in the neotropics. *Proceedings of the International Ornithological Congress* 19: 2562–2572.

Ratter, J. A., Askew, G. P., Montgomery, R. F., and Gifford, D. R. 1978. Observations on forests of some mesotrophic soils in central Brazil. *Revista Brasileira de Botânica* 1: 47–58.

Rizzini, C. T. 1997. *Tratado de fitogeografia do Brasil.* Rio de Janeiro: Editora Âmbito Cultural.

Rodrigues, M. T. and Borges, D. M. 1997. A new species of *Leposoma* (Squamata: Gymnophthalmidae) from a relictual forest in semiarid northeastern Brazil. *Herpetologica* 53(1): 1–6.

Sick, H. 1997. *Ornitologia Brasileira.* 2nd ed. Rio de Janeiro: Editora Nova Fronteira.

Silva, J. M. C. 1989. *Análise biogeográfica da avifauna de florestas do interflúvio Araguaia–São Francisco.* Master's dissertation, Universidade de Brasília, Brasília.

Silva, J. M. C. 1998. Birds of the Ilha de Maracá. In: Milliken, W. and Ratter, J. A. (eds.). *Maracá: the biodiversity and environment of an Amazonian rainforest.* pp. 211–229. Chichester, England: Wiley.

Silva, J. M. C. and Straube, F. 1996. Systematics and biogeography of scaled woodcreepers (Aves: Dendrocolaptidae). *Studies in Neotropical Fauna and Environment* 31: 3–10.

Soderstrom, T. R., Judziewicz, E. J., and Clark, L. G. 1988. Distribution patterns of neotropical bamboos. In: Vanzolini, P. E. and Heyer, W. R. (eds.). *Proceedings of a workshop on neotropical distribution patterns.* pp. 121–157. Rio de Janeiro: Academia Brasileira de Ciências.

Soulé, M. E. and Terborgh, J. 1999. *Continental conservation: scientific foundations of regional reserve networks.* Washington, DC: Island Press.

tem Brink, B. 2000. Biodiversity indicators for the OECD environmental outlook and strategy: a feasibility study. *Globo Report Series* 25: 1–52.

Tyler, H., Brown Jr., K. S., and Wilson, K. 1994. *Swallowtail butterflies of the Americas: a study in biological dynamics, ecological diversity, biosystematics and conservation.* Gainesville, FL: Scientific Publishers.

Vanzolini, P. E. 1988. Distributional patterns of South American lizards. In: Vanzolini, P. E. and Heyer, W. R. (eds.). *Proceedings of a workshop on neotropical distribution patterns.* pp. 317–342. Rio de Janeiro: Academia Brasileira de Ciências.

Wege, D.C. and Long, A. J. 1995. *Key areas for threatened birds in the neotropics.* Cambridge, England: BirdLife International.

Willis, E. O. and Oniki, Y. 1992. A new *Phylloscartes* (Tyrannidae) from southeastern Brazil. *Bulletin of the British Ornithologist's Club* 112: 158–165.

Chapter 6

Monitoring the Brazilian Atlantic Forest Cover

Márcia Makiko Hirota

Technical and scientific studies, research, and surveys of the Atlantic Forest invariably confirm its impressive wealth and diversity of plant and animal species, already widely recognized in national and international scientific circles (Chapters 5 and 11, this volume). Unfortunately, these investigations also reveal a grave situation of constant inroads on the Atlantic Forest and destruction of its habitats.

In recent years the SOS Atlantic Forest Foundation (Fundação SOS Mata Atlântica) and the National Institute for Space Research (Instituto Nacional de Pesquisas Espaciais, or INPE) have enlisted satellite imaging, information technology, and remote sensing to prepare the Atlas of Remaining Forest Areas and Associated Ecosystems of the Atlantic Forest (Fundação SOS Mata Atlântica et al. 1990, 1998; Fundação SOS Mata Atlântica/INPE 1992, 2001; http://www.sosmatatlantica.org.br).

The first mapping of the Atlantic Forest, undertaken with the participation of the Brazilian Institute for the Environment and Renewable Natural Resources (Instituto Brasileiro do Meio Ambiente e dos Recursos Naturais Renováveis or IBAMA), was completed in 1990. This project succeeded in identifying and mapping, at a scale of 1:1,000,000, all the remaining forest areas across the entire country. It also contributed to the definition of the original boundaries of the Atlantic Forest (Fundação SOS Mata Atlântica et al. 1990), which specify that the biome once covered an area of more than 1,363,000 km^2, or nearly 15 percent of the Brazilian national territory (Chapter 4, this volume).

Today this extensive area is home to a population of 108 million people, or more than 60 percent of the country's total population. According to the general census of 2000 (IBGE 2001), these inhabitants reside in more than 3,406 municipalities, or 62 percent of the municipalities in Brazil (Chapter 36, this volume). Data from IBGE (1997), the description of Atlantic Forest boundaries contained in Federal Decree 750/93, and the Vegetation Map of Brazil (IBGE 1993) indicate that 2,528 of these municipalities are located entirely within the biome.

The mapping project currently underway was undertaken by the SOS Atlantic

Forest Foundation and INPE to identify the remaining segments of the Atlantic Forest and associated ecosystems, such as mangrove forests and *restingas,* and to monitor them at 5-year intervals (1985, 1990, 1995, and 2000). The initial work was done at a scale of 1:250,000, but technological and methodological advances have made it possible to systematically refine this survey. For example, in the most recent phase (1995 to 2000) the scale enlarged to 1:50,000. From the outset, this project has enjoyed the active participation and contributions of many scientists, researchers, environmentalists, public agencies, and private corporations (Fundação SOS Mata Atlântica/INPE 1992–93, 2001; Fundação SOS Mata Atlântica et al. 1998).

So far, the Atlantic Forest Atlas has covered 10 states, from Bahia to Rio Grande do Sul, representing an area of 1,285,000 km^2, or 94 percent of the biome, and it is proceeding to map the remaining forest areas throughout almost this entire region with ever-increasing precision.

With regard to monitoring, the results bear out the human intervention and strong pressure on the vegetation cover, an ongoing process of uncontrolled deforestation and fragmentation of the forest. At the same time, only a small proportion of forest areas are being regenerated. These findings confirm the fragility of the biome and the extent to which its biodiversity has been compromised and continues to be threatened.

According to data gathered by the Atlas project and surveys in other regions, the Atlantic Forest has lost more than 92 percent of its original area. Today only about 100,000 km^2 of the original forest remains, most of it in fragments scattered from one end to the other, with denser concentrations in the rugged terrain of the country's southern and southeastern states (Fundação SOS Mata Atlântica et al. 1998; Conservation International do Brasil et al. 1994) (see Figure 6.1).

The causes of deforestation and overexploitation of the Atlantic Forest vary from one region to another (Chapter 11, this volume). On the basis of the Atlantic Forest Atlas and information from other sources, it can be concluded that the main problems existing in the areas surrounding the large Brazilian cities are related to development of the land for housing and tourism and selective extraction of forest resources. Real estate development is also the major factor contributing to the degradation of coastal areas, *restingas,* and mangrove forests. The cumulative effect of deforestation on a small scale exacerbates the problems for the biome as a whole. Other forms of encroachment that directly or indirectly affect the Atlantic Forests are industrial or agricultural pollution of the air, water, and soil, oil spills, mining, the construction of new roads and highways, and energy projects such as hydroelectric power plants and gas pipelines.

Deforestation of and encroachment on the Atlantic Forest compromise regions with major centers of endemism (Chapter 11, this volume), such as the state of Rio de Janeiro, which lost 305.79 km^2 between 1985 and 1990, more than 1,403.72 km^2 between 1990 and 1995, and 37.73 km^2 between 1995 and 2000, for a total of 1,747.24 km^2 in the last 15 years (see Figure 6.2; Fundação

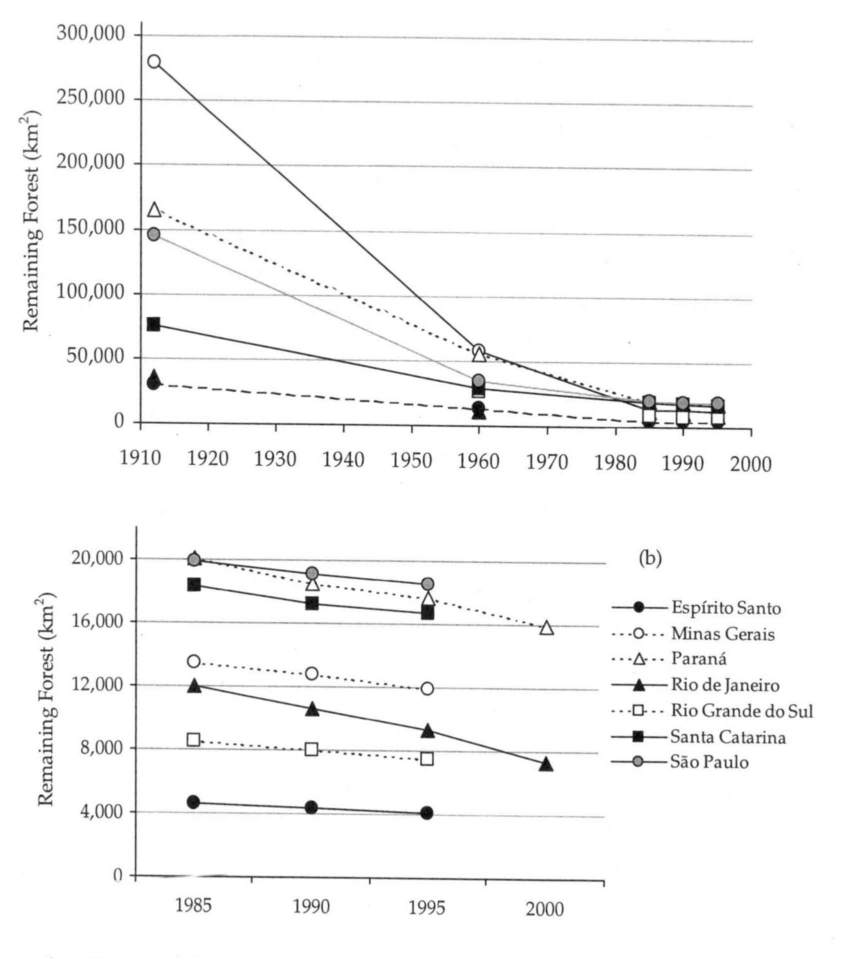

Figure 6.1. Historical (a) and current (b) trends in forest loss in several states of the Atlantic Forest of Brazil from 1910 to 2000 (from Fundação SOS Mata Atlântica et al. 1998).

SOS Mata Atlântica/INPE 2001). In the southern region of the country, the state of Santa Catarina lost 2,050 km² to deforestation between 1985 and 2000.

So far, the highest rates of deforestation have been in the state of Paraná, which in the last 15 years lost a total of 2,889.95 km²: 1,442.40 km² between 1985 and 1990, 846.09 km² between 1990 and 1995, and more than 601.46 km² between 1995 and 2000. Moreover, this state has the poorest record in terms of loss of continuous forest area.

The largest single case of Atlantic Forest deforestation since the SOS Atlantic Forest Foundation and INPE began the present monitoring exercise has been the loss of 160.86 km², appropriated for purposes of agrarian reform between 1995 and 2000 in the municipality of Rio Bonito do Iguaçu, which also lost other forest fragments amounting to 171.17 km² (Chapter 36, Figure 36.3, this volume). This area, located on the Iguaçu River in the interior of the state, had been the

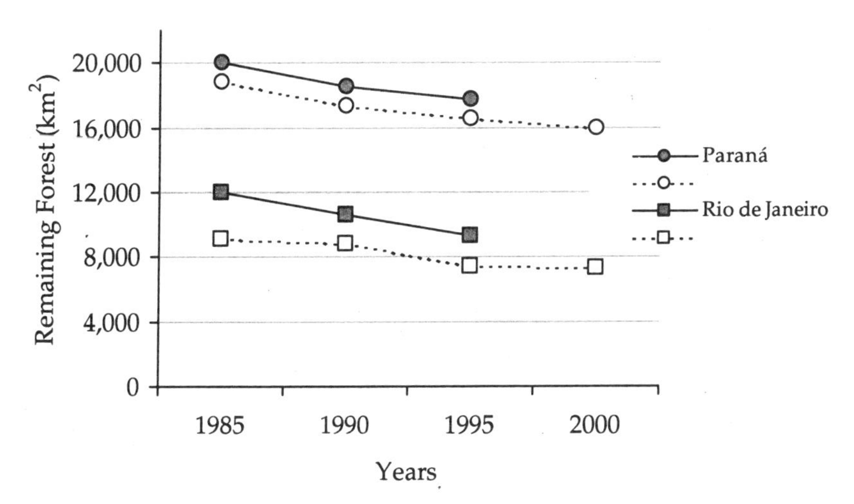

Figure 6.2. Decline in remaining forest from 1985 to 2000 in two Brazilian states. The most recent and accurate method used (open symbols) shows that remaining forest has been consistently overestimated (solid symbols).

focus of attention because damage to it was noted during the first Atlas project survey in 1985–1990. The region has been proposed as a potential area for biodiversity conservation and is considered inadequately understood, yet presumably of biological importance.

In the states of the southern region, the areas affected and still most seriously threatened are the Brazilian Pine Forests, a mixed ombrophilous forest formation that once covered an area in that region equivalent to 164,042.75 km^2. Experts estimate that the remaining forest areas in the remaining states are a mere 9 percent of their original size, and most of these areas have been greatly altered.

The Atlas project has nearly finished updating the information corresponding to the final phase (1995–2000). Preliminary data indicate that the rate of deforestation has slowed slightly, even though the situation is still very serious and human pressures on the Atlantic Forest continue to be intense. The findings so far point to a drastic reduction in the remaining areas of Atlantic Forest, amounting to more than 11,650 km^2 in the area under study over the last 15 years alone.

These examples, estimates, and brief general observations on the situation of the Atlantic Forest show that, despite achieving measures to protect it, defining priorities and strategies for its conservation, and publicizing the situation both nationally and internationally, an overall conservation policy and specific conservation actions are still needed. The designation of the Atlantic Forest as National Heritage in the Federal Constitution, the creation of hundreds of conservation units, the nomination of important areas as World Heritage sites by the United Nations, and the recognition of the Atlantic Forest as a biosphere reserve by the United Nations Educational, Scientific and Cultural Organization (UNESCO) have not been sufficient to ensure its protection (Chapter 38, this volume).

Additional problems are the lack of monitoring and efficient inspection by public agencies and the absence of immediate and positive results in the actions taken by the agencies and institutions charged with the forest's protection. More needs to be done to create awareness in the Brazilian population about the importance of the Atlantic Forest and about its direct and indirect benefits to the quality of life for all.

In addition to enlisting greater participation by civil society, stopping destruction of the Atlantic Forest will entail both finding new sustainable alternatives for economic use of the forest's natural resources and instituting viable mechanisms and incentives to encourage its preservation.

The Atlantic Forest Atlas project will continue to monitor the human impacts on the biome to provide continually refined and updated information on the changes to native vegetation in the study area.

It is important to cite the achievements of the SOS Atlantic Forest Foundation and INPE in mapping the remaining forest fragments in areas not previously studied, such as dry forests, specifically the enclaves and seasonal deciduous and semideciduous forests in the states of Piauí, Bahia, and Minas Gerais, in Serra da Bodoquena National Park and other areas of Mato Grosso do Sul, and in the extreme western and southern areas of Rio Grande do Sul. In addition, the initial and intermediate stages of regeneration are being identified and mapped. Analysis and study of these findings will undoubtedly inform new conservation efforts aimed specifically at protecting these habitats.

The challenge remaining for all is to reverse the process of devastation without further delay and to find ways to accelerate the recovery of degraded areas and expand the forest cover, thus contributing definitively to protection of the remaining forest segments and associated ecosystems of the Atlantic Forest.

References

Conservation International do Brasil, Fundação Biodiversitas, and Sociedade Nordestina de Ecologia. 1994. *Mapa de prioridades para conservação da Mata Atlântica do Nordeste.* Workshop on "áreas prioritárias para conservação da Mata Atlântica do Nordeste," 1993, Pernambuco. Belo Horizonte: Conservation International do Brasil, Fundação Biodiversitas y Sociedade Nordestina de Ecologia.

Fundação SOS Mata Atlântica/INPE. 1992–93. *Atlas da evolução dos remanescentes florestais da Mata Atlântica e ecossistemas associados no período de 1985–1990.* São Paulo: Fundação SOS Mata Atlântica/INPE.

Fundação SOS Mata Atlântica/INPE. 2001. *Atlas dos remanescentes florestais da Mata Atlântica e ecossistemas associados no período de 1995–2000.* São Paulo: Fundação SOS Mata Atlântica/INPE.

Fundação SOS Mata Atlântica/INPE/IBAMA. 1990. *Atlas dos remanescentes florestais do Domínio da Mata Atlântica.* São Paulo: Fundação SOS Mata Atlântica/INPE/IBAMA.

Fundação SOS Mata Atlântica /INPE/Instituto Socio-ambiental. 1998. *Atlas da evolução dos remanescentes florestais da Mata Atlântica e ecossistemas associados no período de 1990–1995.* São Paulo: Fundação SOS Mata Atlântica/INPE/Instituto Socio-ambiental (ISA).

IBGE (Instituto Brasileiro de Geografia e Estatística). 1993. *Mapa de vegetação do Brasil.* Rio de Janeiro: IBGE.
IBGE (Instituto Brasileiro de Geografia e Estatística). 1997. *Malha municipal.* Rio de Janeiro: IBGE.
IBGE (Instituto Brasileiro de Geografia e Estatística). 2001. *Censo populacional 2000.* Rio de Janeiro: IBGE.

Chapter 7

Conservation Priorities and Main Causes of Biodiversity Loss of Marine Ecosystems

Silvio Jablonski

The Workshop on Coastal and Marine Area Assessment and Priority Actions, held in Porto Seguro, Bahia, in October 1999, represented the culmination of the Project for the Conservation and Sustainable Utilization of Brazilian Biological Diversity (PROBIO), carried out as part of the National Biodiversity Program (PRONABIO) of the Ministry of the Environment (MMA 1999). The workshop was funded by the Global Environment Facility (GEF) through the World Bank and the Brazilian government. A total of 180 researchers representing almost all universities and research institutions with programs related to the coastal and marine areas of Brazil attended the event.

Workshop participants assessed the status of and threats to Brazil's coastal and marine habitat, focusing on a variety of groups including coral reefs, sea turtles, mammals, birds, nekton, benthic flora and fauna of the continental shelf, marine plants, and plankton. They developed a list of 31 priority conservation areas, which were chosen based on their biological, economic, and sociocultural importance as well as on the degree of anthropogenic pressure to which they are exposed. This assessment of biological importance took into account such criteria as species richness, phyletic diversity, endemisms of species and higher taxa, the number of rare or threatened species, exceptional biological phenomena such as migrations and the presence of special communities, and the ecological and functional importance of the ecosystem.

Detailed information and important references on biodiversity in the coastal and marine areas of Brazil (Table 7.1) are available in the background papers for the workshop. These texts are available on the project's Web site (MMA 1999).

Table 7.1. Groups, themes, leading experts, and institutions participating in the Workshop on Coastal and Marine Area Assessment and Priority Actions in Porto Seguro, Bahia, in October 1999.

Group	Theme	Experts	Institutions
Species preservation	Algae and marine benthic angiosperms from the coasts; diversity, exploitation, and conservation	Oliveira, E. C., Horta, P. A., Amancio, C. E., and Sant'Anna, C. L.	Instituto de Biociências da Universidade de São Paulo
Species preservation	Marine and coastal birds	Vooren, C. M. and Brusque, L. F.	Fundação Universidade Federal do Rio Grande (FURG)
Species conservation and sustainable use	Biodiversity of sharks and allies	Lessa, R., Santana, F. M., Rincón, G., Gadig, O. B. F., and El-Deir, A. C. A.	Universidade Federal Rural de Pernambuco (UFRPE)
Species conservation and sustainable use	Diagnosis of biodiversity of fish	Haimovici, M. and Klippel, S.	FURG
Species conservation and sustainable use	Diagnosis of priority actions for biodiversity of marine benthos	Belucio, L. F., Cardoso, D. N. B., Souza, M. S., Bittencourt, R. P., and Goes, E.	Universidade Federal do Pará (UFPA)
Species conservation and sustainable use	Benthic macro invertebrates in estuarine environments in Rio Grande do Sul	Bemvenuti, C. E. and Rosa-Filho, J. S.	FURG
Species conservation	Marine mammals	Zerbini, A. N. (1), Siciliano, S. (2), and Pizzorno, J. L. A. (3)	(1), Instituto de Biociências da Universidade de São Paulo; (2), Museu Nacional, Universidade Federal do Rio de Janeiro (UFRJ); (3), Universidade Federal Rural do Rio de Janeiro (UFRRJ)
Species conservation and sustainable use	Nekton; large pelagic fish	Hazin, F. H. V., Zagaglia, J. R., Hamilton, S., and Vaske Júnior, T.	UFRPE
Species conservation and sustainable use	Nekton; small pelagic fish	Cergole, M. C.	Instituto Oceanográfico da Universidade de São Paulo (IO-USP)
Species conservation and sustainable use	Plankton	Yoneda, N. T.	Universidade Federal do Paraná (UFPR)

Continued

Table 7.1. Continued

Group	Theme	Experts	Institutions
Ecosystem	Coral reefs	Castro, C. B.	Museu Nacional, UFRJ
Species conservation	Marine turtles	Sanches, T. M.	Projeto TAMAR; Instituto Brasileiro do Meio Ambiente e dos Recursos Naturais Renováveis (IBAMA)
Ecosystem	Marshes and coastal wetlands	Burger, M. I.	Museu de Ciências Naturais; Fundação Zoobotânica do Rio Grande do Sul
Ecosystem	Rocky coasts	Coutinho, R.	Instituto de Estudos do Mar Almirante Paulo Moreira (IEAPM)
Ecosystem	Diagnosis of Brazilian *restingas*	Silva, S. M.	UFPR
Ecosystem	Diagnosis of sandy beaches	Amaral, A. C. Z. (1), Amaral, E. H. M. (1), Leite, F. P. P. (1), and Gianuca, N. M. (2)	(1), Universidade Estadual de Campinas (UNICAMP); (2), FURG
Ecosystem	Mangroves, intertidal	Schaeffer-Novelli, Y.	IO-USP
Socioeconomics	Socioeconomic aspects, regional planning, and human impacts, Northeast Region (Paraíba, Pernambuco, Alagoas, Sergipe, Bahia).	Viegas, O. and Leahy, W.	Universidade Federal de Alagoas (UFAL)
Socioeconomics	Socioeconomic aspects, regional planning, and human impacts, North Region	Santos, J. U. M., Gorayeb, I. S., Bastos, M. N. C., and Costa Neto, S. V.	Museu Paraense Emilio Goeldi
Socioeconomics	Socioeconomic aspects, regional planning, and human impacts, Southeast Region	Athiê, A. A. R.	IO-USP
Socioeconomics	Socioeconomic aspects, regional planning, and human impacts, South Region	Guadagnin, D. L.	Universidade do Vale do Rio dos Sinos (UNISINOS)
Socioeconomics	Socioeconomic aspects, regional planning, and human impacts, Rio Grande do Norte, Ceará, and Piauí	Oliveira, J. E. L.	Universidade Federal do Rio Grande do Norte (UFRN)
Conservation units	Database of conservation units from coastal and marine zones in Brazil	Pereira, P. M.	Universidade Federal Fluminense (UFF)

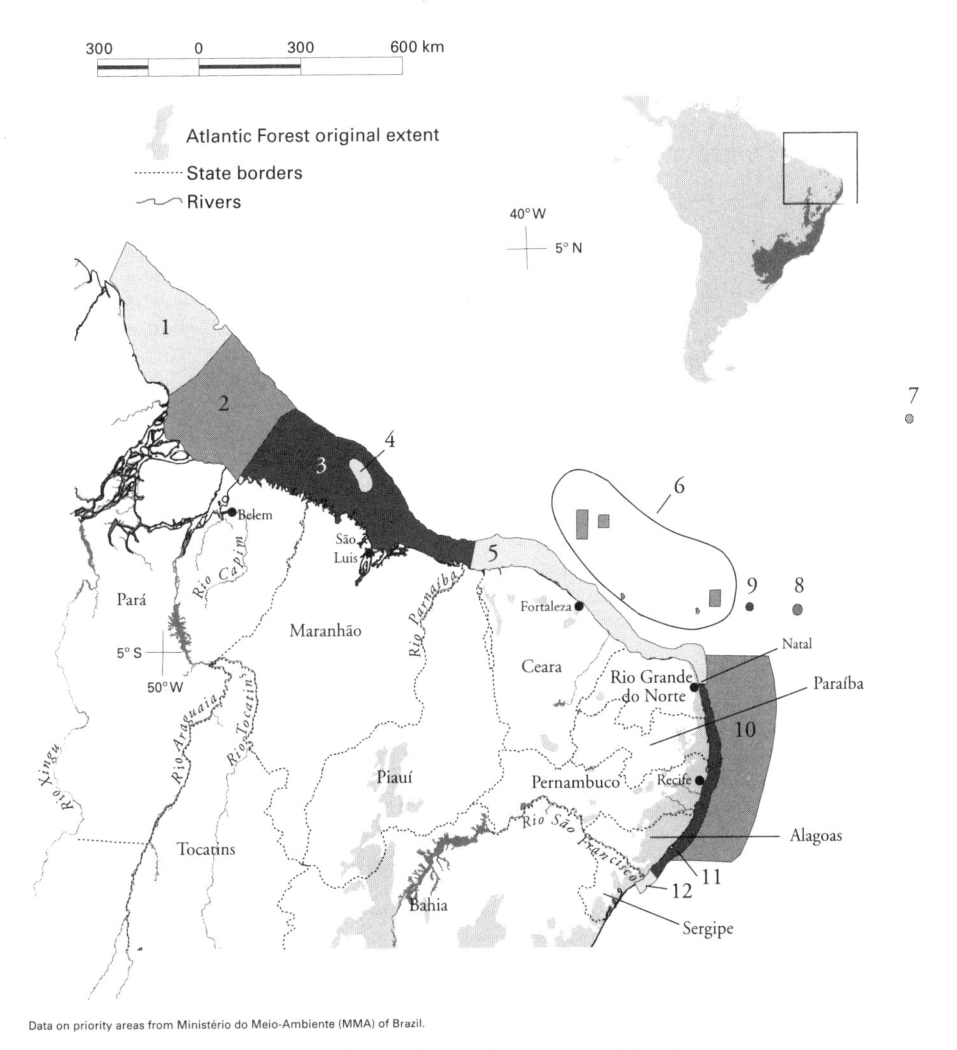

Figure 7.1. Marine priorities in the northern Atlantic coast of Brazil.

Coral Reefs

The coral reefs of Brazil, which are the only reef ecosystems of the South Atlantic, stretch for nearly 3,000 km along the northeastern coast, from southern Bahia to Maranhão. The principal coral species that make up these reefs are found only in Brazilian waters, where they contribute to the formation of structures that are without parallel elsewhere in the world.

The Abrolhos Reef Complex is important because it harbors all the true coral reef species (scleractinian corals with a calcium carbonate skeleton) that have been described to date in Brazil. A number of the reef's true coral species occur only in Brazil and, more specifically, in Bahia. For example, *Mussismilia braziliensis* and *Favia leptophylla* exist only in Bahia, and *Porites branneri* and *Siderastrea stellata* exist only in Brazilian waters. Moreover, there are two hydrocoral species that are exclusively Brazilian: *Millepora braziliensis* and *M. nitida*, the latter found only in Bahia. Among the reef-dwelling octocorals, noteworthy endemic species include *Phyllogorgia dilatata, Neospongodes atlantica, Olindagorgia gracilis, Plexaurella regia,* and *Muricea flamma,* the last three of which exist nowhere outside southern Bahia.

Because of unplanned human use over the years, several Brazilian reefs, mainly along the coast, are deteriorating rapidly. Evidence indicates that these reefs are threatened by fishing, tourism, the commercial collection of flora and fauna specimens for the souvenir industry and aquariums (Cnidaria—especially corals, hydrocorals, and reef-dwelling fish—have been the groups most affected by this potentially predatory activity), and land development along the seashore and riverbanks, as well as increased sedimentation and coastal pollution.

The workshop identified 12 large priority areas for coral reef conservation (Figures 7.1–7.3 and Tables 7.2–7.4).

Table 7.2. Areas of biological importance in the northeast zone of the Atlantic Coast of Brazil.

No.	Place	Description	Threats	Recommendations
1	Amapá Shelf	Endemism of fish species (elasmobranchs = sharks, rays, skates and chimeras). Continental shelf and oceanic adjacent areas important as nurseries for species of commercial interest.	Trawling fishing	Management, biological inventories, indirect use, and conservation unit creation needed.
2	Marajoara Gulf	From São Caetano de Odivelas, including the southern arm (Rio Pará), the Amazon River (north arm), and the Amapá shore, and the Bailique archipelago. Coastal area and continental shelf important as a nursery of several species, some of commercial interest. Important area for the distribution of marine manatee and benthic species of commercial value.	Trawling fishing	Management, biological inventories, and definition of closed fishing season needed.
3	East of Pará Shelf, Maranhão and Maranhense Gulf	Area between the southern limits of the estuary mouth to the border of Maranhão-Piauí, from the coastal zone to the 200-meter depth line. Coastal area and flat shelf with great sediment deposits, important as a nursery of benthic species (elasmobranchs and teleosts). High endemism of elasmobranchs.	Industrial and artisanal overfishing	Management, restoration, and biological inventories needed.

Table 7.2. Continued

No.	Place	Description	Threats	Recommendations
4	Parcel Manuel Luís	Great reef complex of the north region of extreme biological importance; center of endemism, high coral and reef fish diversity; biogeographic importance, Ramsar site (wetland of international importance).	Tourism, marine traffic, and fishing	Conduct management and biological inventories. Increase the conservation unit from the 50-meter limit in the direction of Álvaro bank. Finish the park management plan. Develop environmental education for fishermen and visitors. Evaluate park's carrying capacity and the potential areas for expansion.
5	Ceará to São Roque Cape	Area of the continental shelf located between the border of Maranhão with Piauí to cape São Roque, in Rio Grande do Norte. Important area because of the overlap of calcareous algae bank and the occurrence of green and red lobsters. Presence of several species of dolphins and porpoises; shelf especially rich in mesofauna.	Lobster overfishing and species captured with pelagic hooks	Management, restoration, and biological inventories needed.
6	Cadeia Norte Brasileira Banks	Area of high productivity with abundant species of high commercial value.		Management, restoration, and biological inventories needed.
7	São Pedro and São Paulo Archipelago	Region of great endemism, important as a genetic bank of marine organisms.	Industrial fishing with monoline and multiline hooks	Biological inventories needed and separation of the Environmental Protection Area (APA) Fernando de Noronha, as an independent conservation unit. Impact assessment of the research station needed.
8	Fernando de Noronha	Nesting area for 11 species of marine birds; endemism of fish species (teleosts), high phyletic diversity of marine organisms.	Tourism, lobster overfishing, recreational fishing, and domestic pollution	Management, restoration, and biological inventories needed. Assess effectiveness of the Environmental Protection Area (APA).

Continued

Table 7.2. Continued

No.	Place	Description	Threats	Recommendations
9	Rocks' Atoll	Area of high biodiversity, with endemism and nesting areas for five species of marine birds.	Illegal fishing	Management and biological inventories needed.
10	Talus of the Northeast region	Area between 05°00′ and 10°00′ S, from the continental shelf extending 100 marine miles. Area of concentration of minke and Bryde whales and other cetaceans		Biological inventories needed.
11	Continental shelf, from Cape São Roque to north of Baía de Todos os Santos, with an interrupted area in the mouth of São Francisco river.	Existence of reefs and reef communities of great importance; fishing of shrimp and fish demersals; area of concentration of humpback whale; large banks of phanerogams with manatees; priority area for marine turtles	Overfishing, industrial pollution, ports, and tourism	Management, restoration, and biological inventories needed.
12	São Francisco River Delta	Existence of species of commercial value, nursery area for pelagic fish.	Organic pesticides	Management, restoration, and biological inventories needed.

Marine Turtles

Five of the world's seven existing species of marine turtles are found in Brazilian waters, including the loggerhead sea turtle (*Caretta caretta*), the green sea turtle (*Chelonia mydas*), the leatherback sea turtle (*Dermochelys coriacea*), the hawksbill sea turtle (*Eretmochelys imbricata*), and the olive ridley sea turtle (*Lepidochelys olivacea*). Although once abundant, the populations of the species found in Brazil are now quite fragile. Human predation for the consumption of eggs and meat, mainly by coastal communities, is currently the greatest threat to marine turtles. The increased use of artificial lighting and the unrestricted use of the turtles' beach nesting grounds have also aggravated the situation. Industrial shrimp-catching and fishing gillnets and driftnets are additional mortality factors for marine turtles. Similarly, the fishing of tuna and related species in the Brazilian Exclusive Economic Zone has led to the accidental capture of many sea turtles, especially loggerheads and leatherbacks.

The workshop identified 21 priority areas for marine sea turtle conservation (see Figures 7.1–7.3 and Tables 7.2–7.4).

Marine Mammals

A total of 38 species of cetaceans (whales and dolphins) have been identified in Brazilian waters, representing approximately 49 percent of the known species in the world.

The southern right whale (*Eubalaena australis*) is on the list of endangered Brazilian fauna (Bernardes et al. 1990) and although the population is regaining lost ground, it remains at risk because its coastal habits cause it to be subject to significant anthropogenic pressure.

The humpback whale (*Megaptera novaeangliae*), also on the list of endangered species, has partially coastal habits and is found in Brazilian waters from the far south to the northeast coast. Its heaviest concentration is in the region of the Abrolhos Reef Complex (Figure 7.2), where it continues to be subject to

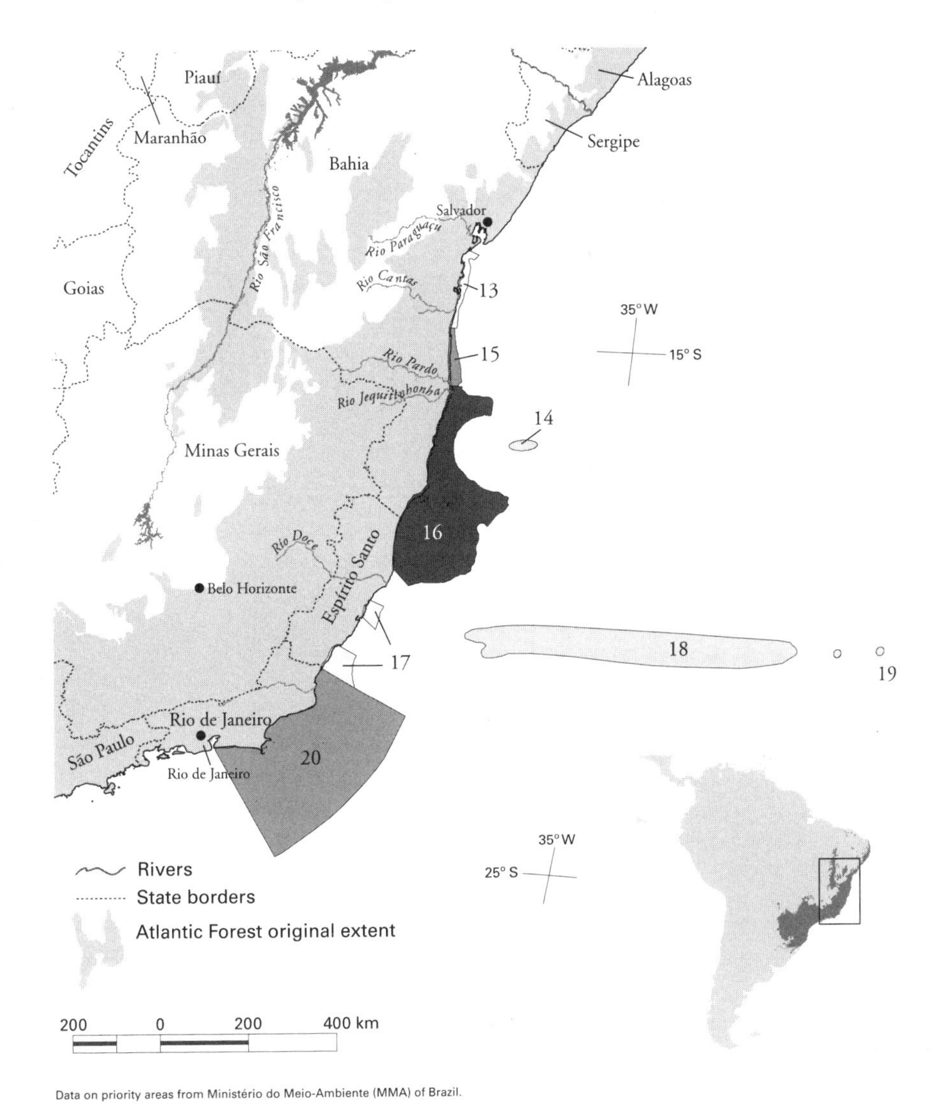

Figure 7.2. Designated marine priority areas along the central Atlantic coast of Brazil.

Table 7.3. Areas of biological importance in the central zone of the Atlantic Coast of Brazil.

No.	Place	Description	Threats	Recommendations
13	Bahia coastline	Area from the southern entrance to Todos os Santos Bay to the municipality of Ilhèus. Important area for Guiana dolphins.	Overfishing	Management, biological inventories, and restoration needed.
14	Oceanic banks of the chain north of Abrolhos	Consolidated bottom, with corals and calcareous algae.	Overfishing	Biological inventories needed.
15	Area of continental shelf, from Ilhèus until latitude 16°S, and 50 m depth			Biological inventories needed.
16	Abrolhos complex	Area of great geographic extent, formed by the Abrolhos archipelago and surrounding coral reefs and nonconsolidated bottoms. Habitat with high marine diversity, coral and fish endemism, nesting site of five marine bird species. Main reproductive site of the humpback whale in the south Atlantic.	Overfishing, tourism, marine traffic, expanding of Caravelas port, and oil exploration	Management, biological inventories, and restoration of degraded areas needed. Modification of the area of the conservation unit; urgent need of poles to tie the boats.
17	South of Espírito Santo	Area of greatest algae diversity in Brazil, with extensive colonies of calcareous and laminarious algae.	Exploration of calcareous algae banks	Biological inventories needed.
18	Oceanic banks of the chain Vitória Trindade	Underwater banks of rocky bottom with high fishing potential.	Overfishing	Biological inventories needed.

Table 7.3. Continued

No.	Place	Description	Threats	Recommendations
19	Trindade and Martin Vaz Islands	Region with unique marine communities in remote oceanic areas, with strong endemism and nesting sites for marine birds. Presence of humpback whales.	Overfishing	Management, biological inventories, restoration of terrestrial environment, and creation of conservation units of indirect use needed.
20	Cabo Frio–Campos Basin	From the border of Rio de Janeiro and Espírito Santo until Ponta de Itaipu, in Niterói, includes all the Basin of Campos. High primary productivity and high biodiversity of different groups. High algae endemism, extensive banks of laminar algae of commercial importance, nursery and feeding zone for sharks and allies, banks of phanerogams associated with the production of shrimp.	Intensive human pressure, with oil extraction and fishing	Management, biological inventories, restoration, and creation of conservation units of direct use needed.

moderate anthropogenic pressure. Since its capture by commercial whalers stopped, it has been making a comeback.

The franciscana or La Plata dolphin (*Pontoporia blainvillei*), a species endemic to northern Argentina, Uruguay, and Brazil, is also listed as endangered. Since its habitat is strictly coastal, it is subject to great anthropogenic pressure from accidental capture in fishing activities. An isolated, genetically distinct population has been found in the northern part of its area of distribution.

The tucuxi or river dolphin (*Sotalia fluviatilis*) is an exclusively coastal small cetacean found along the entire Brazilian coast, starting north of Santa Catarina Island. Its status is not well known, but in some areas it appears to be in decline. It is subject to strong anthropogenic pressure throughout most of its area of distribution.

There are only four species in the world of the order Sirenia, and two of these are found in Brazil: the West Indian manatee (*Trichechus manatus*) and the Amazonian manatee (*T. inunguis*). The latter is found only at the mouth of the

Amazon River and in the Pará River. The West Indian manatee is the most threatened marine mammal in Brazil, with discontinuous residual populations from Alagoas to Amapá totaling only a few hundred individuals at the most. In the recent past this species also inhabited the coastal area from Espírito Santo to Sergipe. Even in its critical situation, however, it continues to be hunted in several places along the northern coast.

The workshop identified 25 priority areas for marine mammal conservation (Figures 7.1–7.3 and Tables 7.2–7.4).

Coastal and Marine Birds

In establishing conservation priorities for birds, workshop participants considered a total of 111 avian species, based on their degree of association with Brazil's coastal and marine systems. Among the passerine birds, the only species included was the bicolored conebill (*Conirostrum bicolor*), which is endemic to the mangroves.

The country's northern region is the home and nesting ground of well-known endangered species such as the scarlet ibis (*Eudocimus ruber*). The region serves as a migration corridor and overwintering site for nearctic birds of the order Charadriiformes (gulls and shore birds) and also serves as a nesting ground for colonies of Ciconiiformes (wading birds). In the south and southeast, some of the coastal islands are nesting sites for terns (*Sterna* spp.), the Audubon shearwater (*Puffinus lherminieri*), the magnificent frigatebird (*Fregata magnificens*), the brown booby (*Sula leucogaster*), and the kelp gull (*Larus dominicanus*). The northern coast of Rio Grande do Sul is a resting area for migratory birds from the nearctic region and from other parts of the Southern Hemisphere.

The environmental problems that affect marine birds include the pollution of coastal waters with various forms of oil and plastic flotsam, the accidental capture of marine birds in fishing lines, and human disruption of nesting sites on coastal islands and coastal areas that birds use for resting and overwintering.

Forty priority areas were identified for conservation, 22 of them on the continent itself and 18 on various islands (Figures 7.1–7.3 and Tables 7.2–7.4).

Bottom-Dwelling and Small Open Sea Fish

Of the nearly 900 fish species that inhabit Brazil's estuarine coastal and oceanic regions, many are subject to exploitation and anthropogenic pressure, mostly by overfishing and pollution. Biodiversity is nearly uniform in large regions, and endemism is low (<5 percent) and limited to reef-dwelling species.

Traditional fishing management techniques have failed to prevent overexploitation and declining yields. As in other parts of the world, this failure is evident in most of the country's marine and estuarine environments. Creating sufficiently large marine reserves in various habitats that are coordinated with neighboring areas under exploitation to ensure recruitment and facilitate recolonization is one possible strategy for maintaining fish biodiversity in this region.

Twenty-five areas were assigned priority for conserving the biodiversity of bottom-dwelling and small open sea fish (Figures 7.1–7.3 and Tables 7.2–7.4).

Sharks and Rays

The greatest threats to sharks and rays in their natural environments are fishing, habitat destruction by human occupation of coastal lands, and various forms of pollution. Fishing in particular has a major effect on these populations. Workshop participants considered the status of the following species: the angelshark (*Squatina guggenheim*) and the hidden angelshark (*S. occulta*), the gray nurse or

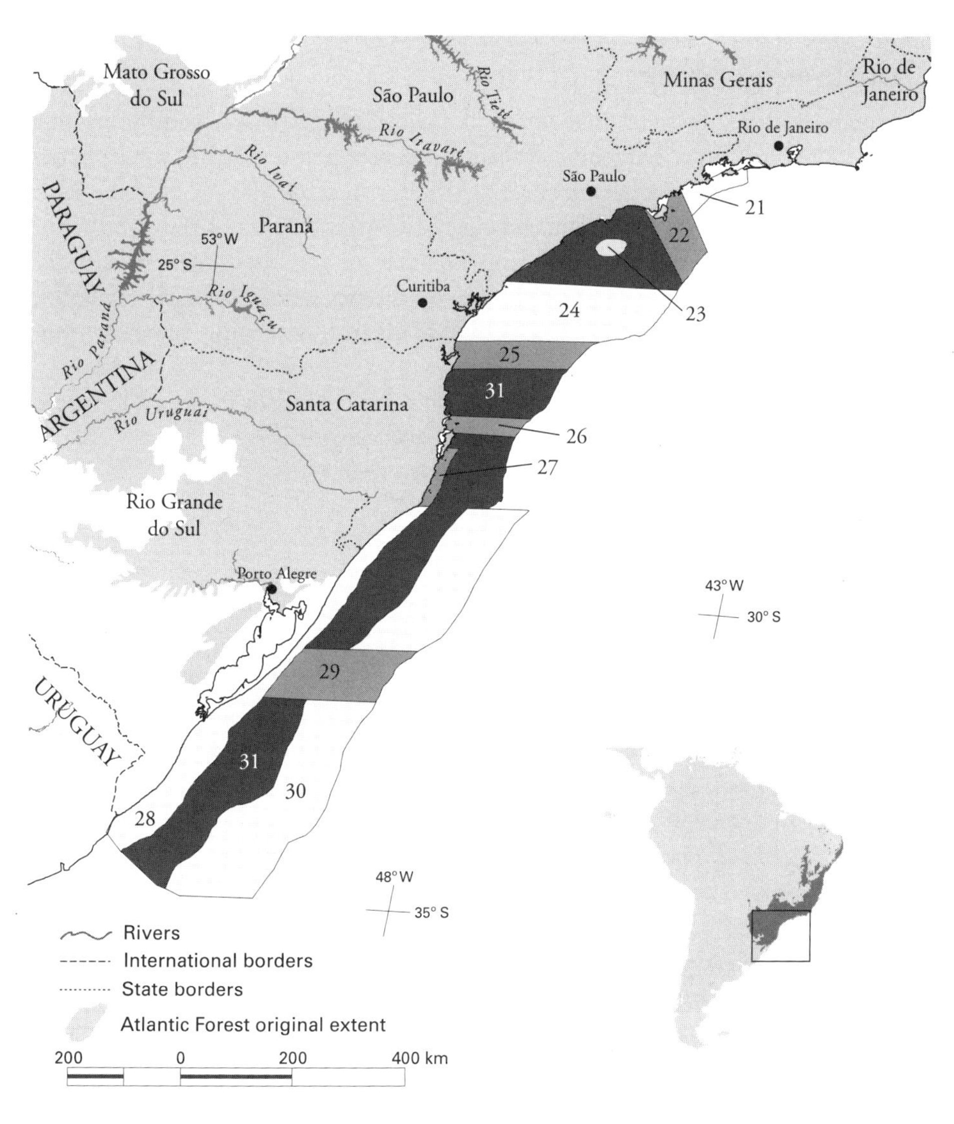

Data on priority areas from Ministério do Meio-Ambiente (MMA) of Brazil.

Figure 7.3. Designated marine priority areas along the southern Atlantic coast of Brazil.

sand tiger shark (*Carcharias taurus*), the whale shark (*Rhincodon typus*), the great white shark (*Carcharodon carcharias*), the basking shark (*Cetorhinus maximus*), the daggernose shark (*Isogomphodon oxyrhynchus*), the striped and narrownose smoothhound sharks (*Mustelus fasciatus* and *M. schmitti*), the soupfin shark (*Galeorhinus galeus*), the scalloped hammerhead shark (*Sphyrna lewini*), the Brazilian guitarfish (*Rhinobatos horkelii*), the smalltooth and largetooth sawfishes (*Pristis pectinata* and *P. perotteti*), the lesser devil rays (*Mobula hypostoma* and *M. rochebrunei*), and the manta ray (*Manta birostris*).

Fifteen broad areas were given priority for conserving the biodiversity of the elasmobranchs (sharks and rays) (Figures 7.1–7.3 and Tables 7.2–7.4).

Benthic Flora and Fauna on the Floor of the Continental Shelf

A literature survey revealed that little is known about Brazilian benthic species. With only fragmentary references available, it was therefore difficult to characterize the degree of species endemism or their status in terms of rarity. By the same token, reliable information about endangered benthic species of the continental shelf is also in short supply. Therefore, available species lists must be regarded with great caution, given the lack of studies on the space-time variation of benthic populations on the Brazilian continental shelf and systematic oceanographic operations.

Fifteen areas were recognized as biologically important for the benthic species in the areas of the Brazilian continental shelf and oceanic islands (Figures 7.1–7.3 and Tables 7.2–7.4).

Marine Plants

The marine plant category includes benthic macroalgae, blue-green algae (cyanobacteria), and angiosperms (plants with flowers) found in intertidal or permanently submerged areas, which are a fundamental link in coastal ecosystems. In addition to their role as primary producers, the fact that they fix carbon and liberate oxygen allows them to serve directly or indirectly as food sources, shelter, spawning or breeding sites, and the structural basis for a large number of animals that spend all or part of their life cycle either as epiphytes or surviving in the areas surrounding these plants. Therefore, the environments these species colonize must be protected to preserve the countless animals, such as sea turtles and manatees, that depend on them either directly or indirectly for survival.

Threats to marine plants and, by extension, the benthic ecosystems in general include the following: pollution; trawling for fish, especially in areas below the 10-m isobath and over nodules of calcareous algae (rhodoliths); highway construction near the shore, which, besides carrying soil to the sea during the construction itself, modifies drainage patterns over time and leads to uncontrolled occupation of coastal land areas in both the short and medium term; port and marina construction in areas with limited water circulation; anchor

Table 7.4. Areas of biological importance in the southeast zone of the Atlantic coast of Brazil.

No.	Place	Description	Threats	Recommendations
21	Grande Island-Ubatumirim	From point Restinga of Marambaia to Ubatumirim. Region of high biological diversity, reproductive site for several species of fish, cetaceans, and benthic organisms	Impact of trawling and other fishing techniques	Management, biological inventories, and restoration needed.
22	São Sebastião Island	Platform and talus of the region of Sebastião, to 200 m depth. High biological diversity and feeding grounds of marine turtles.	Impacts of domestic sewage, oil terminal, tourism, real estate development, and trawling	Management, biological inventories, and restoration needed.
23	Alcatrazes–Laje de Santos–Queimada Grande Islands	Islands on the continental shelf, around 12 nautical miles, in areas with nonconsolidated bottom. High level of endemism. Nesting site for marine birds.	Navy shooting range, illegal fishing, unregulated diving, tourism	Management, biological inventories, and restoration needed. Creation of a unique conservation unit including three islands with fishing management in the surrounding platform. The management area should be external to the present marine limit of the integral protection area of Laje de Santos. Also establishment of marine limits of integral protection for Alcatrazes and Queimada Grande Island.
24	Southern São Paulo and Paraná	All the continental shelf and talus, between latitude 25°S from the states of Paraná and Santa Catarina to longitude 45°W. Occurrence of important oceanographic processes for the maintenance of fishing stocks and benthic communities.	Feeding area for marine turtles.	Management, biological inventories, and restoration needed. Creation of environmental protection area in the platform.

Continued

Table 7.4. Continued

No.	Place	Description	Threats	Recommendations
25	São Francisco do Sul Shelf		Oil pipelines and fishing pressure	Management, biological inventories, and restoration needed.
26	Arvoredo Island Shelf	Continental shelf from the coastal zone to a depth of 200 m, between Porto Belo and approximately the north of Santa Catarina island. Important oceanographic processes for the biodiversity in areas adjacent to the Biological Reserve of Arvoredo Island.		Management, biological inventories, and restoration needed. Creation of environmental protection area in the platform, adjacent to the Biological Reserve Arvoredo Island.
27	"Baleia Franca" southern right whale	This region encompasses Cabo de Santa Marta Grande, the Point Andorinhas, and Santa Catarina Island, to the 50-m isobath. Area of largest reproductive concentrations of southern right whales (*Eubalena australis*) in Brazilian waters. Presence of islands, rocky coasts with high diversity of macroalgae, and southern limit of several species; coastal islands of great importance for seabirds.		Management, biological inventories, and restoration needed. Creation of environmental protection area Baleia Franca.

Table 7.4. Continued

No.	Place	Description	Threats	Recommendations
28	"Pontoporia Sul"	Coastal and marine region from Chuí to Cabo de Santa Marta, having as an external limit the isobath of 35 m. Area of distribution of a small dolphin (*Pontoporia blainvillei*), endemic to the Atlantic Coast of South America. Feeding area for five species of marine turtles, presence and breeding area of the southern right whale.	High levels of accidental capture of La Plata dolphin (*Pontoporia blainvillei*)	Management, biological inventories, and restoration needed.
29	Biodiversity Corridor Southern Shelf	Continental shelf from the coast to the isobath of 500 m, between latitudes of 30°30′S and 31°30′S, including the ocean bottom. Area of presence of marine mammals and marine turtles.	Accidental captures caused by diverse fishing activities	Management, biological inventories, and restoration needed. Creation of a biological reserve for fish biodiversity conservation.
30	Talus southern region	Continental talus from Santa Marta Cape to Chuí, between isobaths of 150 and 500 m. Presence of commercially important fish species. Intensive upwelling area.	High fishing pressure	Management and biological inventories needed.
31	South-Southeast Shelf	Continental shelf between Ubatumirim and Chuí. Reproductive area of high-commercial-value fish species. Important oceanographic processes the structure and maintenance of benthic and pelagic communities.	Overfishing and trawling	Management, biological inventories, and restoration needed. Creation of conservation units in specific sectors.

dropping in areas where the sea floor has nodules of calcareous algae or more diverse benthic communities such as algae banks and marine angiosperms; and tourist activities that involve walking on exposed reef banks at low tide or climbing on coastal rock formations. Another cause for concern is the involuntary introduction of exotic species through ballast water carried in ships or in connection with mariculture projects in the southeastern and southern regions of the country, especially in the states of São Paulo and Santa Catarina, where such organisms as the Pacific or Japanese oyster (*Crassostrea gigas*) are routinely cultivated.

Workshop participants suggested establishing reserves along the southern coast of the state of Espírito Santo to protect some of the banks of brown algae (*Laminaria* spp.) and other nodules of calcareous algae. The brown algae banks contain species of endemic algae and possibly other associated organisms that constitute a formation unparalleled anywhere else in the world. The protection of calcareous nodules formed by coralline algae is justified in light of the threats of commercial exploitation (Chapter 33, this volume).

A total of 33 areas were designated for the conservation of marine plant biodiversity (Figures 7.1–7.3 and Tables 7.2–7.4).

Plankton

Eutrophication resulting from urban and agricultural effluent has been observed at various points along the Brazilian coast (Dos Patos Lagoon, the Santos estuary/Cubatão, Sepetiba Bay, Guanabara Bay, Vitória Bay, Ilhéus, Todos os Santos Bay, and the estuary of the Jaguaribe River). The subsequent increase in biomass of opportunistic species and the decreased biodiversity of the plankton system has produced imbalances in the local ecosystems. Moreover, the accumulation of plankton cells in sediment generates an excessive organic burden for the system, resulting in anoxia in the sediment and consequent damage to the biological community. Toxin-emitting blooms of algae and bacteria have occurred frequently along the Brazilian coast, probably in association with anthropogenic impacts.

Eight areas were assigned priority for the conservation of planktonic diversity (Figures 7.1–7.3 and Tables 7.2–7.4).

Recommendations

The workshop resulted in the following general recommendations for the conservation of Brazil's marine and coastal habitats:

- Carry out further oceanographic studies, surveys of flora and fauna, studies of population dynamics, assessments of existing inventories, and studies of the dynamics of communities in areas affected by major affluent from the continent (e.g., the Marajoara Gulf and the Amapá continental shelf, the Maranhense Gulf, and the delta of the São Francisco River), areas around the oceanic banks of the northeastern and central regions of the Brazilian Exclusive Economic Zone, and certain areas of the continental shelf and slope of the northeastern, central, and southern regions.

- Conduct additional studies of fishing resources and their environments to improve mechanisms for the control of fishing through management and inspection.
- Increase studies on artificial habitats and their effects on the marine environment with a view to protecting marine biodiversity and the sustainability of resources.
- Facilitate technical and legal conditions for implementing marine conservation units designed to protect submerged oceanic banks and the migration corridors of mammals, teleostomes (fish), and elasmobranchs (sharks and rays) in continental shelf areas.
- Establish marine reserves with various degrees of restricted fishing to preserve biodiversity, ensure the recruitment of species of commercial interest in adjacent areas, and maintain a balance between sport fishing, local trade fishing, and industrial fishing.
- Conduct technical and legal studies to mitigate the impact of trawling on the sandy floor of the southern and southeastern continental shelf and the muddy floor of the continental shelf in the northern region of the Brazilian Exclusive Economic Zone.
- Intensify environmental education efforts on the subject of coastal and oceanic ecosystems, especially reefs and islands that are tourist attractions.
- Improve knowledge and control of the processes of coastal land occupation and use, which is essential to effective management of marine resources, as well as the protection and sustainable use of marine biodiversity.
- Identify new fishing resources and underexploited stocks, especially of large deep sea fish, and introduce appropriate technologies to improve selectivity and diversification of the species being caught to alleviate some of the pressure on coastal stocks and reduce the capture of accompanying fauna.
- Expand perceptions of marine resources to include not only their capacity for food production but also their biodiversity, both in terms of genetic patrimony and as a potential source of biotechnology.

Causes of Biodiversity Loss

Currently nearly one-fifth of the Brazilian population lives on the coast, representing 30 million inhabitants and a population density in the coastal area of 87 inhabitants/km^2, or five times the national average of 17/km^2. Moreover, the pattern of settlement is highly centralized, with almost 25 million inhabitants concentrated in the country's 10 largest coastal urban agglomerations. The five metropolitan areas on the coast alone account for 15 percent of the national population.

With regard to urban services, especially basic sanitation, 80 percent of Brazil's urban population lack sanitary waste disposal, and only 43 percent of urban dwellings have the benefit of septic tanks. In the coastal area the picture is aggravated by the fact that sewage is emptied into the sea (the only variable being the discharge distance).

Much of the Brazilian industrial structure is located in the coastal area, usually on the outskirts of the large urban agglomerations, including such production sectors as the chemical and petrochemical industries. Oil pollution, both chronic and acute, is a risk factor, especially in protected areas with sensitive ecosystems. Added to this problem are the activities of the ports, shipyards, and units that process cellulose and mining ore for export, all of which involve large amounts of heavy equipment with a high potential for environmental risk and impact on the coastal area.

A recent survey conducted as part of the Project for the Integrated Management of Coastal and Marine Environments under the Ministry of the Environment (MMA 2000) identified the main sources of pollution along the coast in terms of the organic burden generated (in kilograms per day) and then classified these sources according to whether they were of industrial or urban origin. The region encompassing the states of Espírito Santo, Rio de Janeiro, and São Paulo had the highest concentration of organic burden from both industrial and urban sources, especially near the Rio de Janeiro metropolitan area, where levels were found to exceed 100,000 kg/day. Large amounts of industrial runoff were also identified in northern Espírito Santo, northern Rio de Janeiro State, and the southern coast of São Paulo State. The survey confirms the presence of toxic levels of polluted runoff throughout much of the region. Large amounts of urban organic affluent were found in the estuarine and continental shelf area shared by the *municipios* surrounding the city of Belém and in the metropolitan areas of Fortaleza, Recife, and Salvador. Heavy pollution from industrial sources was identified in the area around Belém, on the northern coast of Alagoas, in the area around Maceió, and in the northernmost part of Bahia. Toxic pollutants were detected in estuaries in Ceará, southern Bahia, northern Santa Catarina, Todos os Santos Bay, and in marine areas of Alagoas and northern Bahia.

Pollution was cited as the determining factor in the pressure being exerted on coral reefs, coastal and marine birds, seabottom and small open sea fish, sharks and rays, and marine plants. An excess organic burden is responsible for the eutrophication and consequent reduction in the biodiversity of planktonic systems. Excessive sediment resulting from the human occupation and use of coastal land has had a direct effect on coral reefs and marine plants.

The destruction of fragile and complex habitats, such as mangroves, coral reefs, and estuaries, also reduces the productivity of the fishing stock. Overfishing is also a common phenomenon that has affected most of the stocks subject to commercial exploitation on the Brazilian coast, including sharks and rays, bottom-dwelling and small open sea fish, and even highly valued shrimp and lobster. Commercial trawling for shrimp and net fishing are also factors contributing to mortality in sea turtles. Marine birds and mammals are subject to accidental capture, the former by driftnets, and the mammals, especially, by gillnets.

National Priority Areas for Biodiversity Conservation

Based on a consolidation of the information in each of the subject groups described in this chapter, 31 priority areas were identified for conservation at the national level (see Figures 7.1–7.3 and Tables 7.2–7.4).

References

Bernardes, A. T., Machado, A. B. M., and Rylands, A. B. 1990. *Fauna brasileira ameaçada de extinção.* Belo Horizonte: Fundação Biodiversitas.

MMA (Ministério do Meio-Ambiente). 1999. *Avaliação e ações prioritárias para as zonas costeira e marinha,* Projeto de Conservação e Utilização Sustentável da Diversidade Biológica Brasileira (PROBIO), Programa Nacional de Biodiversidade (PRONABIO), MMA. Online: http://www/bdt.org.br/workshop/costa.

MMA (Ministério do Meio-Ambiente). 2000. Unpublished data from the inventory of sources of pollution and contamination of live marine resources in Brazil.

Chapter 8

Endangered Species and Conservation Planning

Marcelo Tabarelli, Luiz Paulo Pinto, José Maria Cardoso da Silva, and Cláudia Maria Rocha Costa

Today less than 8 percent of the original Atlantic Forest of Brazil remains (Fundação SOS Mata Atlântica et al. 1998), and this area is scattered into tens of thousands of small fragments (Ranta et al. 1998; Gascon et al. 2000; Silva and Tabarelli 2000). Despite extensive legislation designed to protect the Atlantic Forest and its biological diversity (Lima and Capobianco 1997), the loss and fragmentation of habitats, aggravated by hunting, extraction of forest products, and conversion of the forest to agricultural land, continues unabated (Almeida et al. 1995; Galetti et al. 1997; Cullen et al. 2000). Currently at least 510 species of plants, birds, mammals, reptiles, and amphibians on the Atlantic Forest register are officially threatened. Some are threatened just in the biome, some throughout Brazil, and some worldwide (MMA 2000; Tabarelli et al. 2002; Table 8.1). In addition, hundreds of other species are also threatened in specific parts of the Atlantic Forest.

Table 8.1. Total number of species, endemic, species, and threatened species for selected groups in the Atlantic Forest of Brazil.

Taxa	Total Number of Species	Endemic Species	% Endemism	Threatened Species	% Threatened	Threatened at a Regional Scale
Trees and shrubs	~20,000	~8,000	40.00	367	1.84	151
Birds	849	188	22.14	104	12.25	362
Mammals	250	55	22.00	35	14.00	113
Reptiles	197	60	30.46	3	1.52	18
Amphibians	340	90	26.47	1	0.29	16

How Many Species Are Threatened, and What Are the Principal Threats?

Because its landscape has been so radically altered, Brazil has more endangered birds, mammals, and vascular plants than almost any other country (Hilton-Taylor 2000). According to official Brazilian lists (Bernardes et al. 1990) and global lists (Hilton-Taylor 2000), at least 367 species of trees and shrubs, 104 species of birds, 35 mammals, 3 reptiles, and 1 amphibian in the Brazilian Atlantic Forest are also threatened (Table 8.1). For birds, mammals, reptiles, and amphibians, these numbers represent 10.5 percent of all the existing species in the Atlantic Forest in these groups and more than 50 percent of the threatened fauna on the official Brazilian list.

The distribution of threatened species is not homogeneous throughout the Atlantic Forest. The largest number of threatened species is in the lowland forests that bridge the states of Bahia and Espírito Santo and in the montane forests shared by the states of São Paulo and Rio de Janeiro. These regions also reported to have the greatest number of endemic species within the taxa just mentioned (Haddad and Abe 1999; Da Fonseca et al. 1999; Mendes 1999; Pacheco and Bauer 1999). Threatened endemic species also exist in many of the forests north of the São Francisco River in the "Pernambuco Center," a center of endemism distinguished for having the most endangered bird species in South America (Wegc and Long 1995).

In addition to its 510 threatened species, the Atlantic Forest of Brazil has at least 151 species of trees and shrubs, 362 species of birds, 113 mammals, 18 reptiles, and 16 amphibians that are threatened at the regional level (Table 8.1). Specifically, these species are known to be threatened in four of the southern and southeastern states, those that maintain threatened species lists. For birds, mammals, reptiles, and amphibians, these states are home to between 116 and 169 threatened species, 60 of which are regarded as probably extinct in one or more states. Although we need to know more about the Atlantic Forest in each of these states, these numbers suggest a significant threat to the biological diversity in the Atlantic Forest as a whole.

Local and regional extinctions within the Atlantic Forest usually include game species that have a broad geographic range and need good-quality habitats, such as ungulates, felines, and large frugivorous birds. These extinctions also include both restricted endemic species and endemic species with sizable ranges in this forest (Table 8.1). Although the consequences of such localized extinctions for other populations are still not known (Soulé and Terborgh 1999), Silva and Tabarelli (2000) have predicted that 33.9 percent of the species of trees and shrubs may disappear from the Pernambuco center of endemism because of the local and regional extinction of seed-dispersing vertebrates.

To identify species that are now extinct both correct identification of the species and knowledge of its original geographic range are needed. Unfortunately, the geographic ranges of most Atlantic Forest species are unknown. Scientific knowledge about some Atlantic Forest groups is good (Lewinsohn and Prado 2000), but identifying threatened species is necessarily easier for well-known taxonomic groups, such as mammals and birds. Even so, in the last decade 13 new species or subspecies of birds (Pacheco and Bauer 1999) and 3 new species of primates—certainly the most thoroughly studied group—have been scientifically described, including the black-faced lion tamarin (*Leontopithecus caissara*), discovered near one of the most populous regions of the country (Lorini and Persson 1990). Brazil's documented threatened trees and shrubs are from only 80 of the better-known families out of the more than 200 that exist in the country (Tabarelli et al. 2002), and every year dozens of new plants species are found (Prance et al. 2000). In the Pernambuco center of endemism alone, five new species of bromeliads have been described in the last two years (Leme and Siqueira Filho 2001).

Moreover, designation of a species as "threatened" depends on local or regional initiatives, such as state "red lists," undertakings that are extremely haphazard (MMA 1998). As a result, much uncertainty exists about the true number and geographic range of threatened species in the biome, especially for such groups as vascular plants. Therefore, the total number of threatened species in the Atlantic Forest at any spatial scale is most likely woefully underestimated.

The proximate causes of endangerment in the Atlantic Forest are well known. As in any tropical terrestrial ecosystem, habitat reduction is a principal threat, particularly in ecosystems with a large number of endemic species. Many Atlantic Forest endemics once had extensive geographic ranges along the Atlantic Coast and have become endangered worldwide as a result of the exploitation and reduction of their habitats throughout their areas of distribution. A notorious example is brazilwood (*Caesalpinia echinata* Lam). The first Brazilian species to become threatened, brazilwood once blanketed the Atlantic coast from Rio Grande do Norte to Rio de Janeiro, a span of 20 degrees of latitude (Melo Filho et al. 1992). In turn, many other threatened species are endemic to small areas that are now fragmented and depleted of forest cover, as in the case of the golden-faced lion tamarin in southern Bahia (Pinto and Rylands 1997). Because of the rate of fragmentation in the Pernambuco area, even the new species found here, such as the bromeliads discussed earlier, are endangered, their populations reduced to a small number of individuals scattered among the few small remaining forest areas (J. A. Siqueira Filho, pers. comm. 2002).

Loss of habitat in tropical forests is defined by much more than reduction and fragmentation of the original forest, which in the case of the Atlantic Forest has been staggering. Remaining forest habitats are also being drastically transformed by a series of often interrelated processes involving fragmentation, hunting, fire, and the extraction of forest products (Cochrane and Schulze 1999; Cochrane et al. 1999; Grelle et al. 1999; Nepstad et al. 1999; Tabarelli et al. 1999; Gascon et al. 2000; Peres 2001). Indeed, only a small fraction of the remaining 8 percent of the Atlantic Forest consists of intact forests that have undergone few anthropogenic alterations in their structure and composition (Chapter 6, this volume). Moreover, these remaining areas are unevenly distributed; in some regions less than 3 percent of the original forest cover is left (MMA 2000). Local, regional, and even global extinctions will surely soon occur in those parts of the Atlantic Forest, such as the semideciduous forests, the stands of Brazilian pine (*Araucaria*), and the Pernambuco center of endemism, now reduced to mere archipelagos of small forest fragments.

In the coming years, several factors can be expected to significantly increase the number of species counted as threatened in the Brazilian Atlantic Forest: better knowledge about the geographic distribution of taxa and the status of populations, the discovery of new species with populations limited to the few remaining forest areas, and the ongoing or increasing degradation of remaining habitats.

What Capacity Does the Conservation Unit System Have to Protect Threatened Species?

As gap analyses are undertaken to assess how much of an ecosystem's biological diversity is represented in a given system of conservation units (Primack 1995), the number of globally threatened species with protected populations in existing conservation units also must be determined. The Atlantic Forest appears to contain a large number of conservation units (Chapter 38, this volume) that may be harboring populations of many threatened species, although lack of data on the distribution of species makes this impossible to determine. At least one population of each threatened mammal species, with the exception of three small rodents, is protected in a conservation unit (Da Fonseca et al. 1994). In fact, many of the Atlantic Forest conservation units (such as the Poço das Antas and Una Biological Reserves) were created or implemented specifically to protect remaining populations of endemic mammals, especially primates (Magnanini 1978; Coimbra-Filho et al. 1993).

Given what is known about the status and distribution of the conservation units themselves, however, we can make some overall inferences, in regional and ecological terms, about whether they contain—or effectively protect—threatened species. In the Pernambuco center of endemism, for example, most of the units are small, have not yet been implemented, lack regular funding, and are surrounded by highly threatening human activities. In other words, they are nothing more than "parks on paper" (Silva and Tabarelli 2000; Uchôa Neto 2002). As a result, the threatened species in this region—at least four species of birds and three of endemic trees—lack real protection. As in the Pernambuco region, few parks, reserves, or ecological stations of more than 5 km^2 have been effectively implemented in the seasonal deciduous and semideciduous forests or in the Brazilian pine forest (Lima and Capobianco 1997; Silva and Dinoutti 1999; Chapters 5 and 38, this volume). Therefore, species whose distributions are limited to these forest types are unlikely to have populations of any notable size in the existing conservation units or to be ecologically viable over the medium and long term.

How Can Lists of Threatened Species Assist in Conservation Planning?

Unquestionably, lists are the underpinning of initiatives to protect threatened species, whether on a local, regional, or global scale. In Brazil, local priorities for designating protected areas and conservation measures are based on information about threatened endemic species (Conservation International et al. 2000). Municipal, state, and federal policies on land use and occupation are supposed to take into account the presence of threatened species (Dias 2001). Internationally, the International Convention on International Trade in Endangered Species of Wild Fauna and Flora (CITES), to which Brazil has been a party since 1975, uses lists of threatened and endangered species to curb illegal trafficking in wildlife. Lists can be powerful tools to the extent that they are used as legal instruments for action at any level.

Preparation of the lists of threatened species, or red lists, usually follows the method proposed by the World Conservation Union (IUCN), which is responsible for updating the global Red List of threatened species (Mace 1993; Lins et al. 1997; IUCN 2000). Population size, geographic distribution, and habitat availability are the basic variables used to determine the degree to which a species is threatened. In Brazil, the first list of threatened animals was published in 1973 under the imprimatur of the former Brazilian Institute of Forest Development (IBDF) and was updated in 1989 by the Brazilian Institute for the Environment and Renewable Natural Resources (IBAMA) (Bernardes et al. 1990). This official Brazilian list of threatened animal species is currently under revision. The *Livro Vermelho dos Mamíferos Brasileiros Ameaçados de Extinção*, or *Red Book of Threatened Brazilian Mammals* (Da Fonseca et al. 1994), is the first red book based on the IUCN model published in Brazil.

In 1968 the IBDF published a list of 12 threatened species of Brazilian plants and the first official list, which appeared in 1992, cited 107 threatened plant species (Melo Filho et al. 1992). However, two of IUCN's lists of globally threatened species, published in 1986–2000, identified more than 1,500 threatened species of vascular plants with seeds (Walter and Gillett 1998; Hilton-Taylor 2000), 15 times more than the number identified on the official Brazilian list. By 1990, threatened species lists reflecting the status of populations within state territories began to appear. So far, Santa Catarina (Klein 1990), Paraná (SEMA 1995a, 1995b), São Paulo (Goldenstein 1995; SEMA 1998), Minas Gerais (Machado et al. 1998; Mendonça and Lins 2000), and Rio de Janeiro (Bergallo et al. 2000) have prepared lists of threatened species. The publication of red lists at the state level appears to be a trend that will continue to grow as lists become an effective tool in developing environmental agendas and as they are recognized by states as a legal obligation. However, nine states that have territory within the Atlantic Forest domain have yet to develop lists of threatened species.

Regularly updated lists, especially when combined with basic data on the geographic range and ecology of the threatened species, can become important diagnostic tools for conservation planning in the Atlantic Forest. Particularly important indicators include the current number and past numbers of threatened species by taxonomic group, ecologic group, region, or center of endemism; the number of critical areas, the evolution of their appearance, and the disappearance of integral habitats; the number of species that have populations within conservation units or programs for their management; the number of regions, areas, or centers of endemism for which no significant conservation efforts are under way;

and the number of taxonomic groups that are poorly understood in terms of systematics or geographic range.

These indicators can then be used to reveal information about the intensity, evolution, and geography of the threats to the species; the availability of habitat and location of critical areas; the efficiency of the conservation unit system; the efficiency, gaps, and spatial coverage of conservation efforts; and the level of knowledge about regional-scale Atlantic Forest biodiversity and about the ecology and management of endangered species.

Final Considerations

The Brazilian Atlantic Forest is unquestionably one of the most threatened ecosystems not only in Brazil but in the world. Even though this is the Brazilian biome with the largest number of threatened species on all the state, national, and global lists, management of threatened species is poorly studied, and professionals with experience in this area are few (Chapter 9, this volume). The ongoing recognition of threatened species and the analysis of lists can generate important diagnostics and indicators, such as those proposed here, for planning conservation of the Atlantic Forest on any scale. Obtaining these tools will depend on establishing policies to promote the development of human resources and research with a focus on Atlantic Forest biodiversity, making available information that is currently stored in scientific collections, and creating information management centers to compile, synthesize, and disseminate regional-level information for the entire biome.

Finally, looking at extinction from the perspective of time reveals the urgency with which tools and measures must be implemented. The first list of threatened animals in Brazil identified a total of 86 species in 1972. Seventeen years later this number had jumped to 218 species. Now a newly revised list, slated for publication in 2003, is expected to contain more than 400 threatened animal species. For vascular plants, the increase is even more remarkable, jumping from 12 threatened species in 1968 to more than 1,500 today. Although efforts to identify threatened species have redoubled in the last decade, there is no question that the increasing degradation of the Brazilian ecosystems is matched by an almost exponential increase in the number of threatened species, especially in the Atlantic Forest. Attaining the goal of zero extinction in this forest will require a commitment to embark on research, threat control, habitat restoration, population management, and monitoring of the biota at a level of economic, social, political, and scientific effort never before undertaken for any ecosystem of the world.

References

Almeida, R. T., Pimentel, D. S., and Silva, E. M. C. 1995. The red-handed howler monkey in the state of Pernambuco, Northeast Brazil. *Neotropical Primates* 3: 174–176.

Bergallo, H. G., Rocha, C. F. D., Alves, M. A. S., and Sluys, M. V. (orgs.). 2000. *A Fauna ameaçada de extinção do estado do Rio de Janeiro.* Rio de Janeiro: Editora Universidade Federal do Rio de Janeiro.

Bernardes, A. T., Machado, A. B. M., and Rylands, A. B. 1990. *Fauna Brasileira ameaçada de extinção.* Belo Horizonte: Fundação Biodiversitas.

Brooks, T. and Balmford, A. 1996. Atlantic Forest extinctions. *Nature* 380: 115.

Cochrane, M. A., Alencar, A., Schulze, M. D., Souza, C. M., Nepstad, D. C., Lefebvre, P., and Davidson, E. A. 1999. Positive feedback in the fire dynamics of closed canopy tropical forests. *Science* 284: 1832–1835.

Cochrane, M. A. and Schulze, M. D. 1999. Fire as a recurrent event in a tropical forest of the eastern Amazon: effects on forest structure, biomass, and species composition. *Biotropica* 31: 2–16.

Coimbra-Filho, A. F., Dietz, L. A., Mallinson, J. J. C., and Santos, I. B. 1993. Land purchase for the Una Biological Reserve, refuge of the golden-headed lion tamarin. *Neotropical Primates* 3(1): 7–9.

Conservation International do Brasil, Fundação SOS Mata Atlântica, Fundação Biodiversitas, IPE, SMA-SP, and SEMAD-MG. 2000. *Avaliação e ações prioritárias para conservação da biodiversidade da Mata Atlântica e Campos Sulinos.* Brasília: Ministério do Meio Ambiente/SBF.

Cullen, L. Jr., Bodmer, R. E., and Padua, C. V. 2000. Effects of hunting in habitat fragments of the Atlantic Forest, Brazil. *Biological Conservation* 95: 49–56.

da Fonseca, G. A. B., Hermann, G., and Leite, Y. L. R. 1999. Macrogeography of Brazilian mammals. In: Eisenberg, J. F. and Redford, K. H. (eds.). *Mammals of the neotropics: the central tropics.* pp. 549–563. Chicago: University of Chicago Press.

da Fonseca, G. A. B., Rylands, A. B., Costa, C. M. R., Machado, R. B., and Leite, Y. L. R. 1994. *Livro vermelho dos mamíferos brasileiros ameaçados de extinção.* Belo Horizonte: Fundação Biodiversitas.

Dias, B. 2001. Demandas governamentais para o monitoramento da diversidade biológica brasileira. In: Garay, I. and Dias, B. (orgs.). *Conservação da biodiversidade em ecossistemas tropicais: avanços conceituais e revisão de novas metodologias de avaliação e monitoramento.* pp. 17–28. Rio de Janeiro: Editora Vozes.

Fundação SOS Mata Atlântica, INPE, and Instituto Socioambiental. 1998. *Atlas da evolução dos remanescentes florestais e ecossistemas associados no domínio da Mata Atlântica no período 1990–1995.* São Paulo: SOS Mata Atlântica, INPE, and ISA.

Galetti, M., Martuscelli, P., Olmos, F., and Aleixo, A. 1997. Ecology and conservation of the jacutinga (*Pipile jacutinga*) in the Atlantic Forest of Brazil. *Biological Conservation* 82: 31–39.

Gascon, C., Williamson, G. B., and da Fonseca, G. A. B. 2000. Receding forest edges and vanishing reserves. *Science* 288: 1356–1358.

Goldenstein, S. 1995. *Espécies da flora ameaçadas de extinção no estado de São Paulo: lista preliminar.* São Paulo: Secretaría de Estado do Meio Ambiente.

Grelle, C. E. V., da Fonseca, G. A. B., Fonseca, M. T., and Costa, L. P. 1999. The question of scale in threat analysis: a case study with Brazilian mammals. *Animal Conservation* 2: 149–152.

Haddad, C. F. B. and Abe, A. S. 1999. Anfíbios e répteis. In: Avaliação e ações prioritárias para a conservação da biodiversidade da Mata Atlântica e Campos Sulinos. Online: http://www.conservation.org.br.

Hilton-Taylor, C. 2000. *2000 IUCN Red List of threatened species.* Gland, Switzerland and Cambridge, UK: The World Conservation Union.

IUCN. 2000. http://www.iucn.org/themes/ssc/redlists/rlcategories2000. Updated to http://www.redlist.org/.

Klein, R. M. 1990. *Espécies raras ou ameaçadas de extinção do estado de Santa Catarina.* Rio de Janeiro: Diretoria de Geociências, Instituto Brasileiro de Geografia e Estatística.

Leme, E. M. C. and Siqueira Filho, J. A. 2001. Studies in Bromeliaceae of northeastern Brazil. *Selbyana* 22: 146–154.

Lewinsohn, T. M. and Prado, P. I. 2000. *Biodiversidade Brasileira: síntese do estado atual do conhecimento.* Unpublished technical report from the Project "Avaliação do estado de conhecimento da diversidade biológica do Brasil" (Projeto BRA97G31). Brasília: Ministério do Meio Ambiente.

Lima, A. R. and Capobianco, J. P. R. 1997. *Mata Atlântica: avanços legais e institucionais para sua conservação.* Documentos do ISA No. 4. São Paulo: Instituto Socioambiental.

Lins, L. V., Machado, A. B. M., Costa, C. M. R., and Herrmann, G. 1997. Roteiro metodológico para elaboração de listas de espécies ameaçadas de extinção. *Publicações Avulsas da Fundação Biodiversitas* 1: 1–50.

Lorini, M. L. and Persson, V. G. 1990. Nova espécie de *Leontopithecus* Lesson, 1840, do sul do Brasil (primates, Callitrichidae). *Boletim do Museu Nacional, Nova Série, Zoologia* 338: 1–13.

Mace, G. 1993. An investigation into methods for categorizing the conservation status of species. In: Edwards, P. J., May, R. M., and Webb, N. R. (eds.). *Large-scale ecology and conservation biology.* pp. 293–312. Oxford, UK: Blackwell Scientific Publications.

Machado, A. B. M., da Fonseca, G. A. B., Machado, R. B., Aguiar, L. M. S., and Lins, L. V. (eds.). 1998. *Livro vermelho das espécies ameaçadas de extinção da fauna de Minas Gerais.* Belo Horizonte: Fundação Biodiversitas.

Magnanini, A. 1978. Progress in the development of Poço das Antas Biological Reserve for *Leontophitecus rosalia rosalia* in Brazil. In: Kleiman, D. G. (ed.). *The biology and conservation of the Callitrichidae.* pp. 131–136. Washington, DC: Smithsonian Institution Press.

Melo Filho, L. E., Somner, G. V., and Peixoto, A. L. 1992. *Centuria plantarum Brasiliensium extionsionis minitata.* Rio de Janeiro: Sociedade Botânica do Brasil.

Mendes, S. L. 1999. Grupo de mamíferos: documento preliminar. In: *Avaliação e ações prioritárias para a conservação da biodiversidade da Mata Atlântica e Campos Sulinos.* Online: http://www.conservation.org.br.

Mendonça, M. P. and Lins, L. V. 2000. *Lista vermelha das espécies ameaçadas de extinção da flora de Minas Gerais.* Belo Horizonte: Fundação Biodiversitas and Fundação Zoobotânica.

MMA (Ministério do Meio Ambiente, dos Recursos Hídricos e da Amazônia Legal). 1998. *Primeiro relatório nacional para a Convenção sobre Biodiversidade Biológica.* Brasília: MMA.

MMA (Ministério do Meio Ambiente, dos Recursos Hídricos e da Amazônia Legal). 2000. *Avaliação e ações prioritárias para a conservação da biodiversidade da Mata Atlântica e Campos Sulinos.* Brasília: Conservation International do Brasil, Fundação SOS Mata Atlântica and Fundação Biodiversitas.

Nepstad, D., Veríssimo, A., Alencar, A., Nobre, C., Lima, E., Lefebvre, P., Schlesinger, P., Potter, C., Moutinho, P., Mendoza, E., Cochrane, M. A., and Brooks, V. 1999. Large-scale impoverishment of Amazon forests by logging and fire. *Nature* 398: 505–508.

Pacheco, J. F. and Bauer, C. 1999. Estado da arte da ornitologia na Mata Atlântica e Campos Sulinos. In: *Avaliação e ações prioritárias para a conservação da biodiversidade da Mata Atlântica e Campos Sulinos.* Brasília: Conservation International do Brasil, Fundação SOS Mata Atlântica and Fundação Biodiversitas.

Peres, C. A. 2001. Synergistic effects of subsistence hunting and habitat fragmentation on Amazonian forest vertebrates. *Conservation Biology* 15: 1490–1505.

Pinto, L. P. S. and Rylands, A. B. 1997. Geographic distribution of the golden-headed lion tamarin, *Leontopithecus chrysomelas:* implications for its management and conservation. *Folia Primatologica* 68: 161–180.

Prance, G. T., Beentje, H., Dransfield, J., and Johns, R. 2000. The tropical flora remains undercollected. *Annals of the Missouri Botanical Garden* 87: 67–71.

Primack, R. B. 1995. *A primer of conservation biology.* Sunderland, MA: Sinauer Associates.

Ranta, P., Blom, T., Niemela, J., Joensuu, E., and Siitonen, M. 1998. The fragmented Atlantic rain forest of Brazil: size, shape and distribution of forest fragments. *Biodiversity and Conservation* 7: 385–403.

SEMA Paraná (Secretaria de Estado do Meio Ambiente). 1995a. *Lista vermelha de plantas ameaçadas de extinção no estado do Paraná.* Curitiba: SEMA/GTZ.

SEMA Paraná (Secretaria de Estado do Meio Ambiente-PR). 1995b. *Lista vermelha de animais ameaçados de extinção do estado do Paraná.* Curitiba: SEMA/GTZ.

SEMA São Paulo (Secretaria de Estado do Meio Ambiente). 1998. *Fauna ameaçada no estado de São Paulo.* Documentos Ambientais, Série Probio.

Silva, J. M. C. and Dinnouti, A. 1999. Análise de representatividade das unidades de conservação federais de uso indireto na floresta Atlântica e campos sulinos. In: *Avaliação e ações prioritárias para a conservação da biodiversidade da Mata Atlântica e Campos Sulinos.* Online. Available: http://www.conservation.org.br.

Silva, J. M. C. and Tabarelli, M. 2000. Tree species impoverishment and the future flora of the Atlantic Forest of northeast Brazil. *Nature* 404: 72–74.

Soulé, M. E. and Terborgh, J. 1999. *Continental conservation: scientific foundations of regional reserve networks.* Washington, DC: Island Press.

Tabarelli, M., Mantovani, W., and Peres, C. A. 1999. Effects of habitat fragmentation on plant guild structure in the montane Atlantic forest of southeastern Brazil. *Biological Conservation* 91: 119–127.

Tabarelli, M., Marins, J. F., and Silva, J. M. C. 2002. La biodiversidad brasileña amenazada. *Investigación y Ciencia* 308: 42–49.

Uchôa Neto, C. A. M. 2002. *Integridade, grau de implementação e viabilidade das unidades de conservação de proteção integral na Floresta Atlântica de Pernambuco.* Master's thesis, Universidade Federal de Pernambuco, Recife.

Walter, K. S. and Gillett, H. J. (eds.). 1998. *1997 IUCN Red List of threatened plants.* Gland, Switzerland and Cambridge, UK: The World Conservation Union.

Wege, D. C. and Long, A. 1995. *Key areas for threatened birds in the tropics.* Cambridge, UK: BirdLife International.

Chapter 9

Past, Present, and Future of the Golden Lion Tamarin and Its Habitat

M. Cecília M. Kierulff, Denise M. Rambaldi, and Devra G. Kleiman

The Recent Past

The golden lion tamarin (*Leontopithecus rosalia*) is endemic to the Atlantic Forest of Brazil and is one of the most threatened primates in the world. Historically, tamarins were found throughout the coastal region of the states of Rio de Janeiro and southern Espírito Santo, but the species is now completely extinct in Espírito Santo (Coimbra-Filho 1969), and its populations have been limited to four municipalities in Rio de Janeiro: Silva Jardim, Araruama, Cabo Frio, and Saquarema (Kierulff 1993). Recently, however, reintroduction and translocation programs have expanded the range of the golden lion tamarin, bringing it into additional municipalities including Rio das Ostras, Rio Bonito, and Casimiro de Abreu (Figure 9.1).

A broad range of human activities, from deforestation for lumber extraction to agriculture, cattle ranching, and charcoal production, have reduced the habitat of the golden lion tamarin to small islands of mainly secondary vegetation (Coimbra-Filho 1969; Mittermeier et al. 1982; Kierulff 1993). Hunting has also contributed to the near complete extinction of the species in the wild. Between 1960 and 1965, about 300 golden lion tamarins were captured annually, and many were exported from Brazil for the pet market or for zoos (Coimbra-Filho and Mittermeier 1977).

Several estimates of the population of golden lion tamarins have been published over the past 40 years, all reflecting dwindling numbers. One survey found 600 individuals surviving in small patches of the 900-km^2 fragmented forest of remaining habitat (Coimbra-Filho 1969), whereas later counts found between 100 and 200 (Coimbra-Filho and Mittermeier 1977; Magnanini 1978) and up to 331 individuals (Dietz et al. 1985). A full census carried out during 1991–1992 found 560 individuals in 105 groups surviving in 105 km^2 of forest (Kierulff 1993).

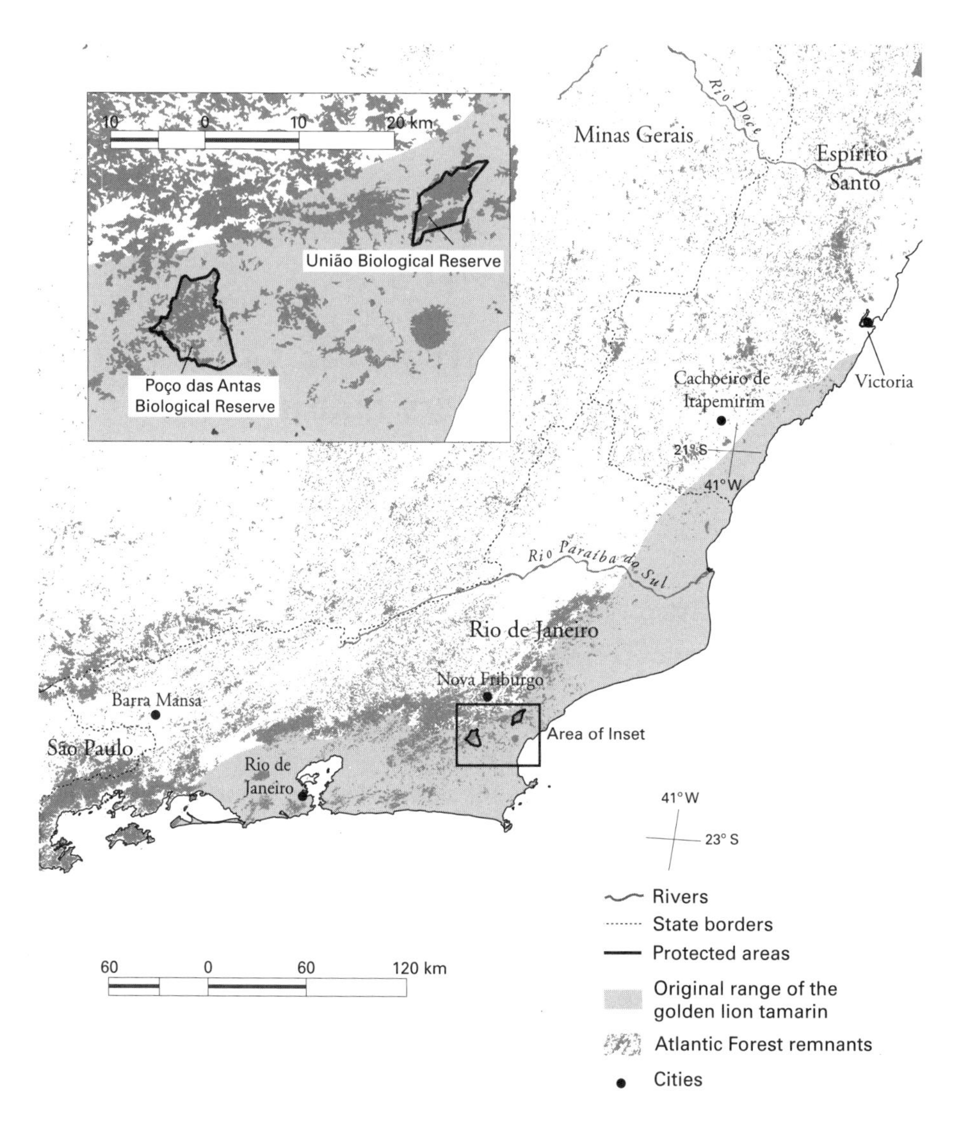

Forest cover from Fundação S.O.S. Mata Atlântica (Rio de Janeiro 2000; other states 1995). Protected areas from Instituto Brasileiro do Meio Ambiente e dos Recursos Naturais Renováveis (IBAMA). Original range of Golden Lion Tamarin from Coimbra-Filho 1969.

Figure 9.1. Original extent and remaining fragmented distribution of golden lion tamarin in Rio de Janeiro and Espírito Santo, Brazil.

The Present

At present, the largest wild population of golden lion tamarins is in the Poço das Antas Biological Reserve, which was created in 1974 specifically to protect the species (Figure 9.1). The population in the reserve has been estimated at 290 golden lion tamarins in about 50 groups (Kleiman et al. 1990; Dietz et al. 1994a). However, the forest environments tamarins need are located in only 28 km^2 of the reserve's total 55 km^2, and much of this is degraded (Kierulff 1993). During the 1991–1992 census, another 213 golden lion tamarins in 43 groups were found in four areas between 5 and 45 km^2 in size outside the reserve, and 60 individuals representing 12 groups survived in small forest fragments ranging from 0.2 to 2 km^2 (Kierulff 1993; Kierulff and Oliveira 1996).

As the numbers of tamarins declined, numerous efforts were initiated to protect the species. In 1972, a consortium of zoos and research institutions initiated the Golden Lion Tamarin Conservation Program (GLTCP) to guide the management of and research on captive populations of tamarins around the world, an effort that resulted in a viable managed zoo population. Conservation and research efforts were initiated in the field in 1983 through a program coordinated by the Smithsonian's National Zoological Park in Washington, D.C. (Kleiman et al. 1986). In 1994, the Golden Lion Tamarin Association (Associação Mico Leão Dourado, or AMLD) took over the administration of golden lion tamarin efforts in Brazil to strengthen the conservation efforts there. These efforts to conserve and study both wild and captive populations continue today.

Much research was needed to gather the knowledge necessary to build stable, managed captive populations (Kleiman et al. 1986), supported later by research on the behavioral ecology of the golden lion tamarin in the wild. Efforts to strengthen populations of golden lion tamarins in Poço das Antas, for example, have included an environmental education program focused on the local communities (Kleiman et al. 1986, 1990; Dietz 1998), a reintroduction program in which zoo-bred tamarins are released on neighboring farms next to the reserve (Beck et al. 1991), and a translocation program to rescue endangered and isolated groups (Kierulff 2000). Supporting research has also looked at other local fauna and flora (e.g., Kierulff and Oliveira 1996) and forest rehabilitation (Kleiman et al. 1990). In addition, since 1995 the AMLD has implemented a program to support the creation of Private Reserves of Natural Heritage (RPPN) to protect privately owned forest and to provide technical assistance for sustainable development through ecotourism. Eight RPPNs have been created, adding almost 8 km^2 to the available protected area (Fernandes et al. 2000).

In 1991, the Brazilian environmental agency (IBAMA) established the International Committees for the Management and Conservation of the four species of lion tamarins: golden-headed lion tamarin (*L. chrysomelas*), black-faced lion tamarin (*L. caissara*), black lion tamarin (*L. chrysopygus*) and golden lion tamarin (*L. rosalia*) (Chapter 30, this volume). At present, a single consolidated committee provides recommendations to IBAMA concerning an integrated conservation and management program for tamarins, including the management of viable captive breeding populations, the evaluation and promotion of field and captive research, and the assessment of activities that directly or indirectly affect the conservation of the four species, such as methods for managing small, fragmented populations (Kleiman and Mallinson 1998).

A wide variety of long-term studies on behavioral ecology, including analyses for population viability, have been essential for defining conservation needs and comparing the needs of wild animals with those of tamarins from reintroduction and translocation programs. Information from these kinds of studies is critical to

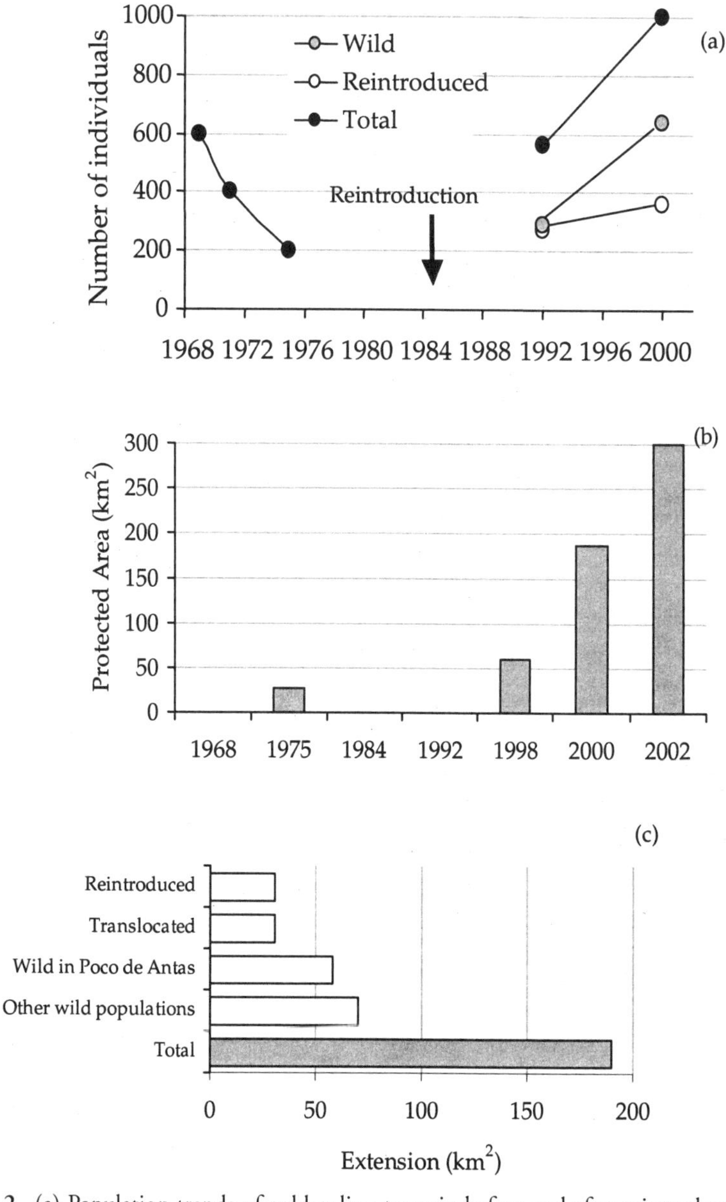

Figure 9.2. (a) Population trends of golden lion tamarin before and after reintroductions, (b) establishment of protected areas, and (c) current total extent (km^2) of Atlantic Forest in Brazil with golden lion tamarins.

effective evaluation and monitoring of species recovery activities. For example, since 1984, 153 captive-born and 7 confiscated wild-born animals have been reintroduced to the wild, mostly on privately owned ranches near the Poço das Antas Reserve. Researchers found that reintroduced tamarins need postrelease provisioning, management, and veterinary care to maximize their chances for survival and reproduction (Beck et al. 1991), but provisioning can be eventually discontinued, and the animals can become fully independent. By 2001, the reintroduced population included 360 animals in 50 groups, including surviving "founders" and their offspring, all living in forests on 21 private farms. About 360 of the estimated 1,000 golden lion tamarins currently living in the wild in the Atlantic Forest are reintroduced captive-born individuals and their descendants (Figure 9.2; Beck, pers. comm. 2002).

In 1994, the Golden Lion Tamarin Association initiated the translocation to a single forest of golden lion tamarin groups that were surviving precariously in several small forest fragments (less than 2 km^2). Groups were captured, immediately released at the new site, and thereafter monitored daily but not provisioned. To date, six tamarin groups (42 individuals) have been translocated to areas that could support their needs. By 2001, 120 golden lion tamarins including offspring, in 16 groups, have been successfully translocated, representing 12 percent of the total wild golden lion tamarin population in the world.

Initiatives to reintroduce captive-bred animals and translocate wild groups have successfully contributed to conservation of the species, both by increasing the size of the wild population and by rescuing endangered wild groups that represent valuable genetic diversity. Additionally, reintroduction and translocation together have contributed to the protection of an additional 64 km^2 of Brazilian Atlantic Forest. In this respect, the golden lion tamarin can be considered a successful flagship species (Dietz et al. 1994b). The ranches with reintroduced tamarins currently contain about 32 km^2 of forest, 19 percent of the total forest with golden lion tamarins. The translocated groups exist in the União Reserve, one of the larger and better-preserved remaining forest fragments within the original species distribution (and without a native golden lion tamarin population). The União Reserve, with a total area of 32 km^2, of which 24 km^2 is covered by forest, was transformed into a biological reserve by IBAMA in 1998, mainly to protect the translocated tamarins. União has 19 percent of the total forest containing golden lion tamarins (Figure 9.1).

The Future

Efforts to save tamarin species have made great headway in the past 30 years. In 2001, the golden lion tamarin population reached 1,000 individuals in the wild, a great step toward the 2,000 individuals that researchers have calculated as necessary to guarantee the survival of the species (Seal et al. 1990; Ballou 1990; Ballou et al. 1998). Research guided by the GLTCP over the past three decades on captive, wild, reintroduced, and translocated populations has resulted in significantly improved techniques for managing these populations and has served as a global model for an endangered species conservation program, in part by integrating zoo and field work.

Today, the major threats to the golden lion tamarin's survival—unabated forest fragmentation and the small size and isolation of the remaining populations—continue to pressure the existing populations. In 1992, the total forest area measured in the species distribution was 2,996 km^2, about 60 percent of which was spread among 8,449 fragments of less than 10 km^2 in size. In these areas, the relative densities of golden lion tamarin populations average from 4 to 9 individuals/km^2 (Kierulff 1993). Nonetheless, the 160 km^2 of public and private protected areas thought to be used by tamarins by the end of 2000 still falls far short of the approximately 300 km^2 of protected forest that is necessary to support 2,000 golden lion tamarins in the wild (Figure 9.2).

These ongoing threats are being addressed in part by continued research and conservation efforts. For example, since 1998 the AMLD has partnered with Brazilian universities to study the effects of forest fragmentation and edge creation on mammalian populations, including bats. Preliminary results have been used to design a strategy to protect the forest remnants under Brazilian federal law, to restore degraded areas and, to plant corridors between private and public forest fragments. The AMLD is also working at the watershed level to mitigate the impact of fragmentation caused by a lack of land use planning. The organization is contributing to the formulation of regional public policy, including supporting a green agrarian reform movement, to avoid the potential impact caused by the landless settlements that have been established recently in the areas surrounding the forest fragments. Finally, the AMLD is also providing assistance to local communities in the development of agroforestry systems and organic agriculture.

In the future, to avoid the problems consequent to small population size, the golden lion tamarin populations will have to be managed as a metapopulation, with thc periodic interchange of individuals between forest fragments until the tamarins can independently move along the corridors that are being implemented. The management of all golden lion tamarins as one population is necessary to prevent excessive inbreeding and the effects of unpredictable demographic and environmental changes. Such management will also bring the population closer to the size necessary to avoid the loss of genetic variability through genetic drift.

The increased number of golden lion tamarins surviving in increasing natural habitat confirms the success of the GLTCP. Currently there are 1,000 individuals in the wild surviving in 160 km^2 of forests, in two biological reserves and several RPPNs, and environmental education efforts have changed the local attitudes and behavior toward forest conservation in general. The golden lion tamarin has become a symbol for the conservation of the Atlantic Forest in Brazil as a whole (Dietz 1998). Costs for this program, in terms of surviving tamarins, decrease annually as budgets focus more on land-use planning and implementation of forest corridors. Tamarin numbers now increase mainly from natural recruitment rather than continuing reintroduction and translocation (Kleiman et al. 1991). The experience acquired and technology used to save the golden lion tamarin from extinction can be applied to other species that need similar efforts. Every day the list of threatened species around the world increases, and for the majority of these species, the causes of population decline are similar to those of the golden lion tamarin: fragmentation and small population size because of habitat destruction, degradation, and overhunting.

References

Ballou, J. D. 1990. Small population overview. In: Seal, U. S., Ballou, J. D., and Valladares-Padua, C. (eds.). *Population Viability Analysis Workshop report.* Apple Valley, MN: Captive Breeding Specialist Group (IUCN/SSC/CBSG).

Ballou, J. D., Lacy, R. C., Kleiman, D. G., Rylands, A. B., Ellis, S. (eds.). 1998. Leontopithecus *II: The second population and habitat viability assessment for lion tamarins (*Leontopithecus*).* Final Report. Apple Valley, MN: Conservation Breeding Specialist Group (IUCN/SSC).

Beck, B. B., Kleiman, D. G., Dietz, J. M., Castro, I., Carvalho, C., Martins, A., and Rettberg-Beck, B. 1991. Losses and reproduction in reintroduced golden lion tamarins *Leontopithecus rosalia. Dodo, Journal of Jersey Wildlife Preservation Trust* 27: 50–61.

Coimbra-Filho, A. F. 1969. Mico-leão, *Leontideus rosalia* (Linnaeus, 1766), situação atual da espécie no Brasil (Callitrichidae: primates). *Aéncias da Academia Brasileira de Ciências* 41 (suplemento): 29–52.

Coimbra-Filho, A. F. and Mittermeier, R. A. 1977. Conservation of the Brazilian lion tamarins *Leontopithecus rosalia.* In: Prince Rainier IV and Bourne, G. (eds.). *Primate conservation.* pp. 59–94. London: Academic Press.

Dietz, J. M., Baker, A. J., and Miglioretti, D. 1994a. Seasonal variation in reproduction, juvenile growth, and adult body mass in golden lion tamarins (*Leontopithecus rosalia*). *American Journal of Primatology* 34: 115–132.

Dietz, J. M., Coimbra-Filho, A. F., and Pessamilio, D. M. 1985. Projeto Mico-Leão 1: modelo para a conservação de espécie ameaçada de extinção. In: Thiago de Melo, M. (ed.). *Primatologia no Brasil 2.* Brasília: Sociedade Brasileira de Primatologia.

Dietz, J. M., Dietz, L. A., and Nagagata, E. Y. 1994b. The effective use of flagship species for conservation of biodiversity: the example of lion tamarins in Brazil. In: Olney, P. J. S., Mace, G. M., and Feistner, A. T. C. (eds.). *Creative conservation: interactive management of wild and captive animals.* pp. 32–49. London: Chapman and Hall.

Dietz, L. A. 1998. Community conservation education program for the golden lion tamarin in Brazil: building support for habitat conservation. In: Hoage, R. J. and Moran, K. (eds.). *Culture, the missing element in conservation and development.* Washington, DC: Kendal/Hunt Publishing Company and National Zoological Park, Smithsonian Institution.

Fernandes, R. V., Rambaldi, D. M., Bento, M. I. S., and Matsuo, P. M. 2000. The Private Reserve of Natural Heritage as a mechanism to the legal protection of the golden lion tamarin (*Leontopithecus rosalia*) habitat. *Anais do II Congresso Brasileiro de Unidades de Conservação* (Campo Grande, MS, Brasil).

Kierulff, M. C. M. 1993. *Avaliação das populações selvagens de mico-leão-dourado,* Leontopithecus rosalia, *e proposta de estratégia para sua conservação.* Master's thesis, Universidade Federal de Minas Gerais, Belo Horizonte, Brasil.

Kierulff, M. C. M. 2000. *Ecology and behaviour of translocated groups of golden lion tamarins (*Leontopithecus rosalia*).* Doctoral dissertation, University of Cambridge, UK.

Kierulff, M. C. M. and Oliveira, P. P. 1996. Re-assessing the status and conservation of the golden lion tamarin (*Leontopithecus rosalia*) in the wild. *Dodo, Journal of the Jersey Wildlife Preservation Trust* 32: 98–115.

Kleiman, D. G., Beck, B. B., Baker, A. J., Ballou, J. D., Dietz, L. A., and Dietz, J. M. 1990. Conservation program for the golden lion tamarin, *Leontopithecus rosalia. Endangered Species Update* 8(1): 18–19.

Kleiman, D. G., Beck, B. B., Dietz, J. M., and Dietz, L. A. 1991. Costs of a re-introduction and criteria for success: accounting and accountability in the Golden Lion Tamarin Conservation Program. In: Gipps, J. H. W. (ed.). *Beyond captive breeding: reintroducing endangered mammals to the wild. Symposia of the Zoological Society of London 62.* pp. 125–142. Oxford, UK: Oxford University Press.

Kleiman, D. G., Beck, B. B., Dietz, J. M., Dietz, L. A., Ballou, J. D., and Coimbra-Filho, A. F. 1986. Conservation program for the golden lion tamarins: captive research and management, ecological studies, educational strategies, and re-introduction. In: Benirschke, K. (ed.). *Primates: the road to self-sustaining populations.* pp. 959–979. New York: Springer Verlag.

Kleiman, D. G. and Mallinson, J. J. C. 1998. Recovery and management committees for lion tamarins: partnerships in conservation planning and implementation. *Conservation Biology* 12: 27–38.

Magnanini, A. 1978. Progress in the development of the Poço das Antas Biological Reserve for *Leontopithecus rosalia rosalia* in Brazil. In: Kleiman, D. G. (ed.). *The biology and conservation of the Callitrichidae.* pp. 131–136. Washington, DC: Smithsonian Institution Press.

Mittermeier, R. A., Coimbra-Filho, A. D., Constable, I. D., Rylands, A. B., and Valle, C. M. C. 1982. Conservation of primates in the Atlantic forest region of eastern Brasil. *International Zoo Yearbook* 22: 2–17.

Seal, U. S., Ballou, J. D., and Valladares-Padua, C. V. 1990. Leontopithecus: *Population Viability Analysis Workshop report.* Apple Valley, MN: Captive Breeding Specialist Group (IUCN/SSC/CBSG).

Chapter 10

Socioeconomic Causes of Deforestation in the Atlantic Forest of Brazil

Carlos Eduardo Frickmann Young

The loss of forested areas is intrinsically related to the forms of land use and production. The relationships between social and economic variables and the destruction or preservation of habitats are highly complex, and slight changes in the role of any one influence—historical, institutional, or geoenvironmental—can lead to entirely different outcomes. To understand the social and economic causes of deforestation and to identify indicators that anticipate such pressures, it is necessary to include in the analysis the heterogeneity inherent to the human processes of land occupation.

This complex interaction of factors can be seen in the socioeconomic determinants of land use in the principal Brazilian biomes. Despite similarities such as excess of rural labor and struggles for land ownership rights, the characteristics of the deforestation process differ greatly between the Amazon region, the Cerrado (central highlands), the Pantanal, and the Atlantic Forest. On one hand, for example, the commercial extraction of timber contributes more heavily to deforestation in the Amazon region than in the Cerrado, where the cultivation of grain is the main factor leading to destruction of forest. On the other hand, land speculation and predatory tourism exert greater pressure on remaining areas of the Atlantic Forest and in the Pantanal than they do elsewhere.

This chapter identifies the main socioeconomic factors that are impinging on the Atlantic Forest and discusses the difficulty in selecting and using indicators for monitoring these pressures. This discussion is not meant to be comprehensive, nor does it promote a cause-and-effect model because such models of interactions between humans and nature are oversimplified and can be counterproductive to developing effective conservation policies and actions.

The Land Occupation Cycle

Deforestation in the Atlantic Forest has been going on for a very long time, and only a few wholly undisturbed areas are left (Chapter 4 and Figures 6.1 and 6.2, this volume). In the Atlantic Forest region, causes of deforestation differ from those in regions such as the Amazon, where intense deforestation is more recent and large preserved areas remain. These differences reflect different mixes of socioeconomic pressures, although all are related to cycles of land occupation: demographic pressure and agricultural activities become less important in areas where deforestation is long-standing and more important in areas recently converted from forest to the agricultural frontier.

In general, land occupation cycles have three stages. First comes a period of population growth and rapid expansion during which agricultural activities drive deforestation. Next comes a slowing of the economic and demographic growth but with deforestation continuing in response to other pressures, such as land speculation and expansion of crop and pastureland. The final stage is a period of contracting economic and demographic pressures, as natural resources become depleted or various techniques are used to increase agricultural output while reducing the need for manual labor. In this last stage, the emergence of a class of workers who have lost their jobs or have become impoverished farmers, the "new poor," leads to exhaustion of remaining natural resources, further invasion of forest reserves, use of less suitable land for agricultural, and, ultimately, perpetuation of cycles of poverty.

In other words, this kind of land occupation cycle begins with unsustainable use of the land by populations who have a period of wealth during which people—generally those less educated—flock to take advantage of an agricultural boom. As the boom declines with overuse of once-forested lands, farmers become poor and proceed to extract maximum benefit from the scarce remaining natural resources. An example of this cycle is that of the decadent urban nuclei of the Paraíba Valley at the beginning of the twentieth century. At this time, once-abundant coffee production declined rapidly in the region, in large part because poor cultivation practices and rapid deforestation had caused the soil to erode and lose its fertility. The drop in coffee production triggered an economic and demographic exodus from the region, and its rural areas ended up being devoted mainly to raising livestock, which provides fewer jobs and less income for workers.

The following discussion of elements related to deforestation in areas of the Atlantic Forest concentrates on the states of the southeastern and southern regions, for which the *Atlas of Remaining Forest Areas* (Fundação SOS Mata Atlântica et al. 1998; Chapter 6, this volume) provides ample data on the massive decline in the Atlantic Forest in these regions.

The History and Impact of Commodity Cycles

From the sixteenth through the twentieth centuries economic cycles tied to commodities such as brazilwood, sugar, cattle, gold, and coffee all had great impact on the Brazilian economy, despite enormous differences in the modes of production and distribution of these products. The development of all these commodities caused serious damage to the Atlantic Forest, largely by failing to use sustainable approaches that would make it possible to overcome the economic and social contradictions of the colony and the empire.

Brazilwood (*Caesalpinia echinata*) is one example of rampant destruction of forest for the sake of commodity production. Brazilwood was the first product of commercial interest to emerge in the newly discovered lands, and it served as the source of the country's name. Trees were extracted at such an intensive rate (about 2 million in the first hundred years of exploitation) that by 1558, viable reserves were pushed more than 20 km in from the coast (Bueno 1998). By 1605, brazilwood had become so scarce that the Portuguese Crown began taking measures to stop the indiscriminate cutting and was sending out guards to protect the forest in the areas where extraction was most intensive (Bueno 1998). It is ironic that the species that gave Brazil its name and its reputation for natural abundance became its first natural rarity and its first victim of commodity exploitation.

Sugarcane is a very different example of the impacts of commodity production. As an introduced exotic species, sugarcane did not affect a specific natural resource but rather exerted pressure on the Atlantic Forest as a whole. Sugarcane production prompted severe deforestation in the fertile areas of the northeast coast as forests were cleared for cultivation and for firewood to fuel the sugar mills. Although the region's highly suitable soil supported sugarcane production, which continues as the basis of the economy in the region to this day, the agrarian land-owning structure around which production was organized served at the same time as an incentive for deforestation and the creation of wide social disparities. Even after the abolition of slavery, which was the engine of sugarcane cultivation and processing, the social conditions of sugarcane workers continued to be among the worst in the nation: many of the poverty pockets in Brazil are the result of five centuries of continuous planting.

The cultivation of another exotic species, coffee, followed a pattern similar to that of sugarcane, especially in its first stage, when it relied heavily on slave labor. By the early nineteenth century, coffee had become the principal source of foreign exchange for Brazil, but it brought with it land-use practices that greatly accelerated loss of forest areas in the country's southeastern region (Dean 1996). After occupying and exhausting the soil in the Paraíba Valley's "sea of hills," where unsustainable agricultural practices have been the norm up to the present day, coffee growing shifted to the interior (at first mainly in the state of São Paulo and, later, Paraná), where there is an abundance of more suitable soil. The replacement of slaves with free European immigrants as coffee growers also had a major impact on the development of coffee as a powerful commodity, which generated the excess foreign exchange that ultimately financed Brazil's industrial expansion in the twentieth century.

The growth of cattle and gold as commodities was also important in establishing the current boundaries of Brazil. However, they were at the same time primary pressures for deforestation, as hinterlands were drawn into nonsustainable cycles of use. For gold, forest areas were lost because of the techniques used for prospecting, as well as the demand for timber, firewood, and other resources by miners. The decline in the mining areas was inevitable once the reserves were exhausted (Dean 1996). For cattle, large tracts of forest were burned to clear land for new pastures. Livestock raising, which continues to be one of the principal forms of land use in Brazil, is characterized by low productivity, little demand for labor, and limited opportunities for generating income or altering the social status quo. In short, the development of both cattle and gold as commodities has led to enormous forest devastation but few social benefits.

A root problem of expanding commercial crops in Brazil is the link between the very small parcels of farmland (*minifundios*), which are not large enough for a household to subsist on, and the vast agricultural establishments that take up most of the land (*latifundios*), which are not operated on a labor-intensive basis. As commercial crops developed, large businesses transformed their workforces, moving from slave labor to sharecropping (*colonato*) to using hired labor, which became seasonal in its demands for temporary workers at peak periods. These changes resulted in the creation of an unstable and limited rural labor market that was incapable of absorbing the excess labor during nonpeak periods. Landless and underemployed workers had few options but to move to and settle in forest areas, as most preferred not to try their luck with employment in the cities.

Historically, then, the expansion of the agricultural frontier by claiming forest areas has become an escape valve to accommodate excess population. First the Atlantic Forest, then the Cerrado, and now the Amazon Forest have been cut back to address conflicts over the land. Without agrarian or economic reforms, the cycle continues, and more and more forested areas are converted to absorb migratory overflows. However, given the way the agrarian economy is organized in Brazil, there is a limit to how much excess "natural" land and nutrients can be enlisted to accommodate the nation's social problems. The crisis will come when there is no land left for the taking. Conflicts will flare up over ownership of the existing land, and workers will face the alternatives of either moving to the urban fringe, where living conditions can be precarious but chances of finding income are better, or migrating to forested areas and clearing them for settlement, thus perpetuating the cycle of deforestation.

Pressure Exerted by Agricultural Activities

These historical cycles, pushing to extend the agricultural frontier, are especially devastating for the environment in the case of conversion to pastureland. The large proportion of land dedicated to livestock is surprising in part because raising livestock is significantly less productive than raising crops (Figures 10.1 and 10.2).

Another striking imbalance in the agricultural sector in the Atlantic Forest region is the concentration of land ownership. In the southeast and northeast regions, *latifundios* (more than 1,000 ha) are more than twice the area of small land holdings (less than 50 ha). Even in the southern region, where the distribution is more balanced, holdings of 200 ha or less represent less than half the total land holdings (Figure 10.3a). When the three regions are added together, holdings of less than 50 ha occupy only 20 percent of the total area, and those of more than

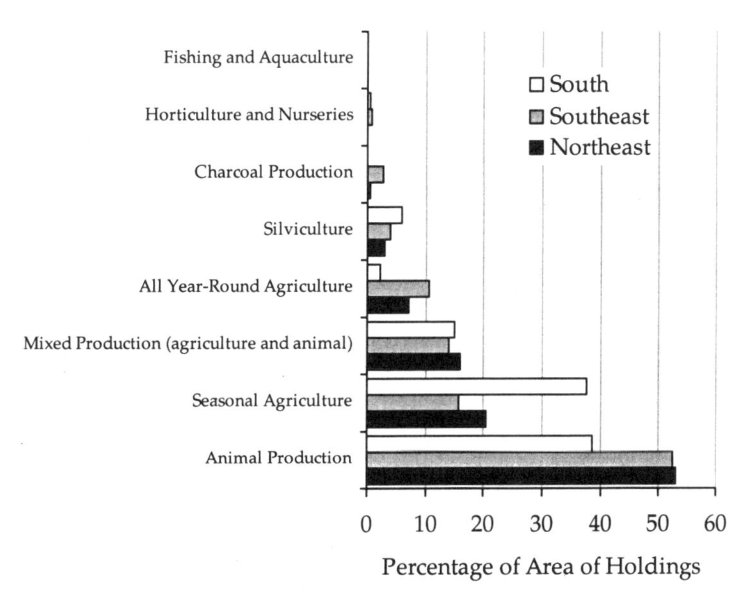

Figure 10.1. Percentage of area in three regions used by different economic activities (IBGE 1995/96).

200 ha account for 58 percent of the total. However, holdings under 50 ha account for 36 percent of the total value of production (Figure 10.3b) and also represent 76 percent of the population engaged in agriculture. By contrast, establishments larger than 1,000 ha employ only 3 percent of the agricultural labor force and generate 21 percent of the value, even though they occupy 27 percent of the total area (Figure 10.3c). Overall, then, small holdings generate much more value and many more jobs than large establishments. Therefore, in addition to the

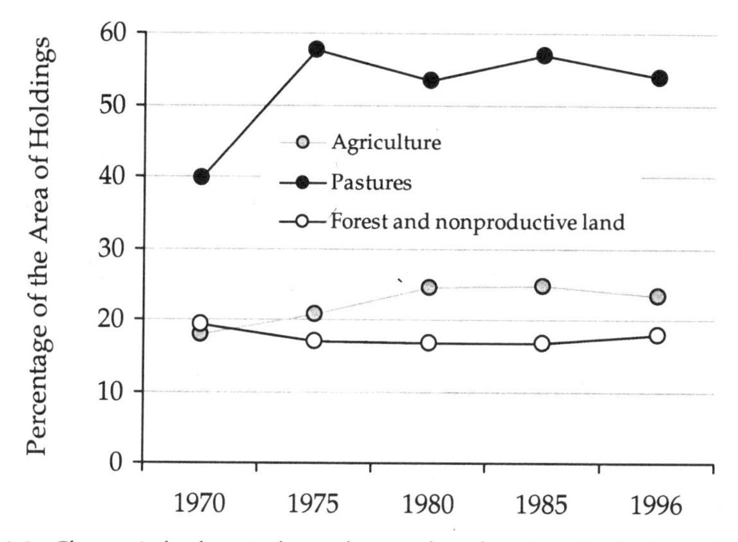

Figure 10.2. Changes in land use in the southeast and south regions (IBGE 1970, 1975, 1980, 1985, 1996).

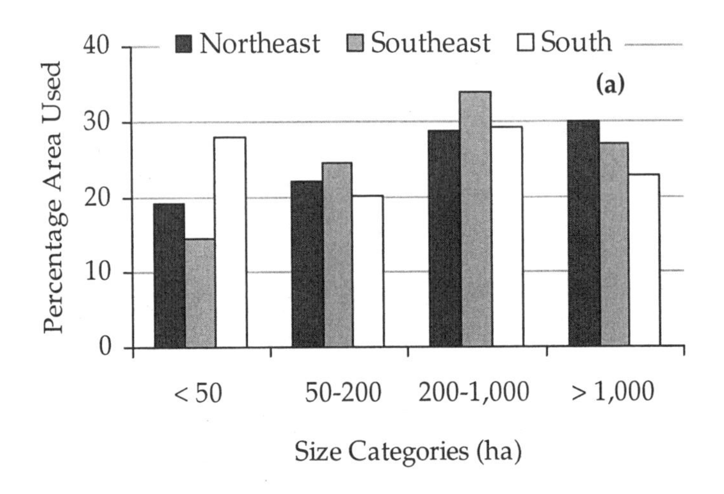

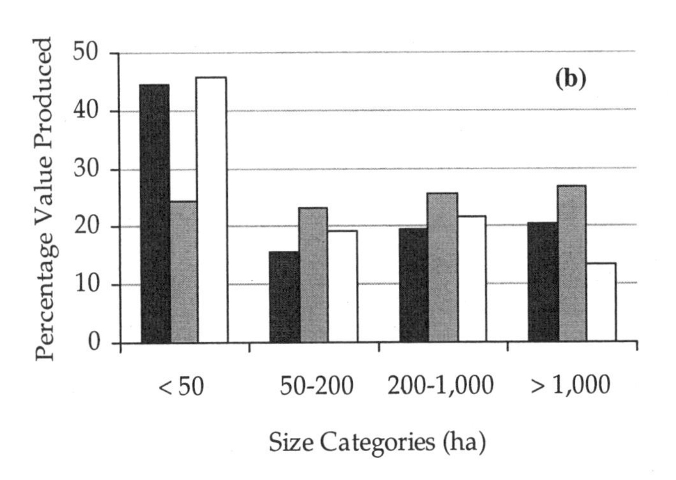

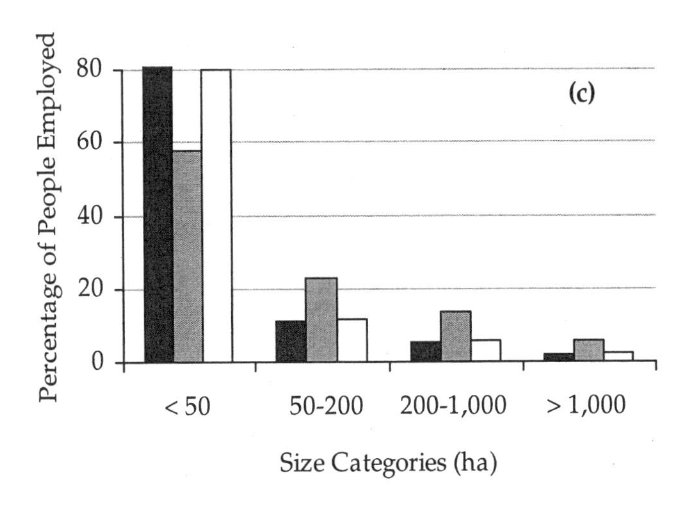

Figure 10.3. (a) Distribution of area, (b) production value, and (c) personnel employed, by size of holdings (IBGE 1996).

loss of forest areas that are being converted to agricultural use, the imbalance of land ownership and productivity creates the legacy of an unbalanced social structure, one in which large land owners benefit at the cost of rural workers, who even today have to struggle to achieve minimum conditions for survival. These disproportions also show that the expansion pressures exerted by large agricultural establishments are unrelated to demographic problems because most people work on small land holdings.

The Contribution of Public Policy

A variety of public policies have also fueled the expansion of agricultural frontiers at the expense of forested lands. For example, in the 1960s and 1970s, subsidized loans were a major incentive for rural land owners to destroy forests in order to "produce" a certain activity, usually extensive livestock raising. The fiscal crisis of the 1980s led to the gradual reduction of subsidized loans, but in their place came guaranteed minimum prices for agricultural products and supply subsidies, which also supported economically inefficient agricultural activities. The minimum price policy reduced uncertainty for farmers, especially during times of inflation, by subsidizing fuel for those with higher shipping costs in remote areas, equalizing the price of fuel throughout the country, and significantly reducing the cost of transporting produce from areas on the agricultural frontier. In the end, both these loan and price supports furthered the destruction of remaining forests.

Although some policies, such as those supporting research on sustainable forms of agriculture, have been introduced in recent years, forest areas continue to receive less favorable treatment under the rural land tax regime. Overall institutional and policy structures still heavily favor the extension of areas under cultivation and conversion of the land. For example, policies aimed at promoting exports have further fueled the expansion of cropland and pasture. Incentives for increasing Brazilian exports have had the effect of increasing demand for agricultural land and, in doing so, have triggered sharp rises in land value, especially in the southern and southeastern regions. Land speculation has encouraged the conversion of forest areas for agricultural purposes, particularly in the Cerrado and the Amazon regions, where farmers were able to buy much larger properties than they had originally owned in the south and southeast. As land is converted, the accompanying construction of highways and other infrastructure projects reduces the cost of transportation but increases the possibility of penetrating forest reserves. These development projects in turn stimulate real estate speculation and, ultimately, deforestation. Highway construction can be especially pernicious for conservation: in the case of the states in the southern and southeastern regions, the highway network grew 20 percent in just 10 years, from 843,886 km in 1985 to 882,740 km in 1990 and 1,014,114 km in 1994 (IBGE 1985, 1996). By facilitating access to previously remote regions, roads created the expectation that land prices would rise, a critical element in keeping momentum in the cycle of real estate speculation that drives deforestation.

Macroeconomic policies also have influenced deforestation through monetary policies that raise interest rates, encouraging rural producers to opt for activities that guarantee short-term results. This profiteering directly contradicts the kinds of sustainable development policies that are essential for long-term economic and employment stability. Macroeconomic policies that reduce public spending also affect forests because, despite public rhetoric to the contrary, spending cuts have made it difficult to hire enough forest rangers and to give those rangers the resources they need to manage and oversee protected areas. Policies creating circumstances that allow the invasion and degradation of forests clearly show that protecting forests is not high on the priority list of government agencies at any level.

Demographic Factors and Living Conditions

Along with public policies that have the effect of promoting deforestation, yet another contributing factor is the growth of population in rural areas. Indeed, there has been enormous demographic growth in the Atlantic Forest region at large in the last 200 years, and today it is home to more than two-thirds of the Brazilian population (Chapter 36, this volume). However, if there were an automatic correlation between population growth and deforestation, when the growth trend slows or becomes negative, one would expect to see a parallel slowdown in the loss of forest land. Quite a different picture unfolds on inspection of deforestation trends and demographic changes in rural areas in the Atlantic Forest.

Over the last four decades there has been a conspicuous exodus from the countryside for the southern and southeastern states of Brazil: in each of the southeastern states the rural population has been declining steadily since the 1960s, and in the southern states since the 1970s. Despite a cumulative population decline in rural areas of about 7.5 million, deforestation continued unabated in each of these states during this same period. Data on the remaining forest areas in the Atlantic Forest Atlas show that from 1985 to 1995 alone the cumulative loss of forest in the southern and southeastern states amounted to more than 10,000 km^2 (Fundação SOS Mata Atlântica et al. 1998; Figures 6.1 and 6.2, Chapter 6, this volume), while at the same time the rural population declined in each of these states (northeastern states have been omitted from this analysis for lack of comparable data). These figures show that demographic pressure by itself does not account for the decline of the Atlantic Forest: the rural population has been decreasing in absolute numbers for a very long time, and deforestation has continued unabated.

This decline has important implications for conservation policy because it shows that population control policies and the reduction of immigration will not necessarily solve the problem of deforestation. These population trends also underscore that much more research is needed to build a full understanding of the real socioeconomic drivers of the destruction of the Atlantic Forest.

Furthermore, the number of jobs in the agricultural sector in the southern and southeastern states fell by 2.4 million between 1985 and 1996 despite an increase in deforested area of 10,000 km^2 (Figure 10.4). In other words, the alleged correlation between deforestation and the creation of jobs again is not borne out:

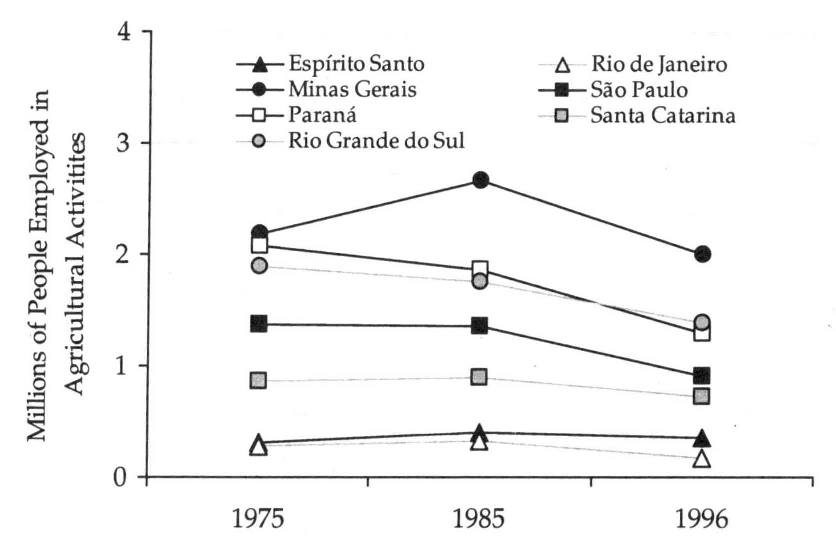

Figure 10.4. Employment trends in agricultural activities from 1975 to 1985 (IBGE 1975, 1985, 1995/96). Although there was a rise in agricultural employment in some of the states (especially Minas Gerais) during the first half of the period, there has been a downward trend in all these states since 1985.

increased deforestation has been accompanied by reduced job opportunities in the Atlantic Forest areas converted to agricultural use. A corollary to the myth that deforestation is necessary to deal with unemployment is that conversion of the forest to agricultural use is necessary for improving the living conditions of the population. If there were gains in social areas such as income, education, and housing associated with deforestation, then loss of the forest could be justified from the social standpoint even during periods of demographic exodus. However, data show that that belief is not consistent with the realities faced by the rural population today in the Atlantic Forest. The median monthly income of the rural population is very low, ranging between R$75 (US$42) in Sergipe and R$262 (US$146) in the state of São Paulo (IBGE 1999). Moreover, the income earned by half the rural population (less than R$134, or US$76) was lower than the official minimum wage at the time (R$155, or US$86; IBGE 1999). Also worth noting is the disparity in income between the rural and the urban population: except in the state of São Paulo, the median urban income is at least twice the figure earned by the rural population in each of the states in question. Thus it seems that the Atlantic Forest is being destroyed by an agricultural system that is not improving the situation of chronic poverty in rural areas.

Silviculture and Forest Product Extraction

Although the area of agricultural establishments devoted to silviculture and timber extraction is much smaller than that dedicated to raising crops and livestock, silviculture and timber activities have a significant impact in the Atlantic Forest region. Most Atlantic Forest products are harvested using nonsustainable methods, and little training on sustainable management practices is available (Young 2002). Forest development consists almost exclusively of monoculture stands of exotic wood species, especially eucalyptus and pines, a trend that is on the rise (Chapter 11, this volume).

Productive extraction is concentrated in a small number of products, with firewood and charcoal ranking as the most economically important, followed by timber extraction. Nontimber forest products represent a small proportion of the total value of vegetal extraction and include piassaba palm (*Leopoldina piassaba*), bakuri palm (*Platonia insiginis*), mangaba (*Hancornia speciosa*), the imbu tree (*Spondias tuberosa*), and the cashew nut (*Anacardium occidentale*). These goods have less economic importance in part because producers often can earn more by cultivating products that use nonsustainable extractive methods.

Policymakers must also remember that the goal of activities meant to support sustainable development is not only to protect flora and fauna but also to improve the present and future living conditions of the local populations. If the concerns of local peoples are not taken into primary consideration, sustainable development efforts may well be thwarted. For example, a valid strategy to encourage sustainable extractive activities might be to increase the prices paid to the harvester or producer through subsidies or through other economic strategies to support the product. If land conflicts with established traditional social groups can be avoided, such supports may become a stimulus for the conserving resources in reserves designed for sustainable extraction. However, if property rights are not clearly defined, price supports or subsidies can be self-defeating, encouraging predatory exploitation of public or communal lands by nonlocal producers or others. Policymakers must carefully define which products should receive subsidies, which will produce the minimum long-term damage to the environment, and which individuals or groups should be the beneficiaries of such economic strategies.

Concern for traditional communities and indigenous peoples also appears in regional differences related to nonmarketable forest products. Forest products that do not have huge market value can nonetheless be very important in traditional communities, which differ from region to region, particularly those located long distances from urban centers. If markets for these kinds of resources expand, predatory exploitation may not only destroy the capacity of the environment to replenish the resource but also interfere with traditional social forms of extraction and use (Chapters 18 and 32, this volume). The impact of this degradation on communities that depend on the local forest resources can be devastating both for the forest and for the social integrity of the populations concerned.

Indicators for Monitoring

The loss of natural habitats develops in response to a confluence of possible pressures that are specific in time and place, be they land occupation, economic cycles, agricultural pressures, public policies, demographics, silviculture, or others

(Chapter 2, this volume). This specificity makes selecting indicators for monitoring biodiversity a uniquely complicated task. The indicators must reflect up-to-date information on localities that have been carefully selected for their particular environmental importance. Data from each site should be acquired by regularly measuring variables that have been carefully selected. In priority conservation areas, it is therefore essential to set up an information network that can allow researchers and policymakers to tap into the most current and critical information for monitoring biodiversity.

Therefore, to identify the most significant pressures on biodiversity in each area, researchers should consider using simple analysis methods in tandem with synthetic indicators, looking at real-time data and statistically based predictive forecasts simultaneously, but not in the aggregate (Table 10.1). This combination of perspectives can help to create an orientation panel for decision making in the context of sustainable development (Feijó et al. 2000).

Outlook for the Future

This chapter has sought to show that the deforestation of the Atlantic Forest is the flipside of the same process of economic and social exclusion that characterized the formation of the Brazilian nation. Far from resolving social conflicts and eliminating poverty, the policies that directly or indirectly contribute to conversion of forest areas to agricultural use perpetuate a pattern in which most rural workers are confined to farming limited areas and a few large land owners control large tracts of the most productive land. The problem reflects a sad history of priorities and values in the institutional and production infrastructure of the Brazilian countryside. In all, the Atlantic Forest is being pressured by a system that drains its natural resources and is motivated almost completely by short-term thinking and an alarming absence of any value being put on long-term sustainability of land or on the interests of the poor.

But the future need not be bleak. In fact, because the present state of affairs does not hold the promise of immediate improvement in the Atlantic Forest, it is all the more important to look to the positive developments that are taking place. For example, Brazil has new laws on crimes against the environment (9605/98) and water resources (9433/97), both of which can greatly benefit conservation. The water resources legislation in particular redirects the management of water resources to river basin committees, to be created under the National Water Agency. This decentralization of water management creates the possibility that funds will be made available for the conservation and regeneration of forests in headwater areas and water body corridors because the forests render the important service of regulating the flow of water.

Positive economic experiments are under way as well. For example, new mechanisms have been adopted in some of the states to transfer some tax revenues to *municipios* according to quotas based on forest cover and other environmental criteria. Although this instrument has received some criticism, its relative success—particularly in Paraná and Minas Gerais—shows that *municipios* with large areas

Table 10.1. Pressures of biodiversity loss, biodiversity conservation, and potential indicators for the Atlantic Forest of Brazil.

Pressures of Biodiversity Loss	Pressures of Biodiversity Conservation	Potential Indicators
Unsustainable use of natural habitats	Sustainable use of natural habitats	Annual Research of Silviculture and Extraction (IBGE [Brazilian Institute of Geography and Statistics])
Nonsustainable extraction of timber, firewood, charcoal, and plant species; hunting and fishing	Sustainable extraction of plants and animals	Information of extraction from associations involved Reports on illegal trade of animals and plants (including administrative information on penalties) Price trends for illegal products in each region (e.g., timber, palmito, pet trade)
Irresponsible tourism	Ecotourism	Information on visitors and tourist capacity, National Tourism Annual Book (EMBRATUR [Brazilian Institute of Tourism]) Information on regional transportation (number of passengers, number of buses, and services) Ecotourism associations
Land uses	Protected areas	Protected areas (ministry, environmental, state, and municipal agencies)
Agriculture	Species preservation	Personnel working in environmental protection
Pastures	Recreation	Municipal agricultural census and research (IBGE) Prices of main agricultural products in the region Land prices Transportation of agricultural products
Dams	Protection of water resources	Areas included in hydroelectric projects (Eletrobrás; National Agency of Electric Energy, development plans) Supply companies

Table 10.1. Continued

Pressures of Biodiversity Loss	Pressures of Biodiversity Conservation	Potential Indicators
Urban and rural use		Housing prices Number of construction licenses Demand of rural electricity for domestic use
Public policies and institutional aspects	Community and government awareness	
Road construction	Existence of efficient conservation legislation and environmental taxes	Increase in road (DNER [National Department of Highways) and railroad networks (Ministry of Transportation); development plans
Incentives to migratory movements	Environmental education	Colonization programs; land conflicts (INCRA [National Institute for Colonization and Agrarian Reform], state colonization agencies, FETAGRI [Agricultural Workers' Federation], rural workers' unions, FUNAI [National Indian Foundation])
Indefinition and lack of respect to property rights	Protection to traditional communities	Rural credits, agriculture subsidies, and minimum prices; rural land taxes (state and municipal government agencies; CONAB [National Agricultural Supply Company])
Tax incentives and credits for agricultural conversion	Tax incentives and credits for conservation	Environmental taxes: "ICMS verde" (goods and services circulation tax) (state government agencies of finances)
Colonization and settlement projects	Payment of credits for forest environmental services (e.g., carbon conservation, water, biodiversity conservation)	

of preserved forest may receive significant financial benefits if they become actively involved in conservation.

Also, there should be some financial reward for the global benefits provided by the Atlantic Forest in terms of carbon sequestration and biodiversity preservation. Even though the resources available for such compensation may be limited in the short term, many conservationists, policymakers, and others hope that markets will be developed in the future that can contribute significantly to the preservation of forests and, in the case of carbon sequestration, to their recovery.

The main barrier to creating a global regulatory framework is the United States and its reluctance to accept its historic responsibility with regard to the problem of global warming. The recent decision of President Bush not to ratify the Kyoto Protocol, despite the advanced stage of talks on its implementation, may slow the adoption of a large-scale carbon emission market. Nevertheless, the involvement of European countries, international development agencies, and even large private corporations in this issue indicates that despite the political and diplomatic inertia, progress is being made toward establishing the International Climate Regime. The existence of vast areas of abandoned or low-productivity pasturelands, together with the accelerated carbon absorption capacity of new-growth tropical forests, offers great promise for opportunities to capture resources for reforestation of the Atlantic Forest.

Even though there are many causes for concern, there are also new opportunities to take action that will contribute to conservation. The significant progress that has been made in regulatory markets must be followed up with the implementation of economic instruments that will compensate the forest for the services it provides. Such instruments should provide for water regulation, the preservation of biodiversity, and carbon sequestration, all of which offer the possibility of generating new resources for conservation.

Policies for conserving the remaining areas of the Atlantic Forest must be part of the broader context of public policy formulation. Institutional sins of omission have contributed to environmental decline. However, significant new legislation, especially insofar as land use is concerned, includes the recognition that forest areas are productive, as seen in their reduced tax base; restrictions on the granting of subsidized credit and tax concessions; modified zoning procedures; sanctions against environmental crimes; and, most recently, reductions in rural property taxes for those who opt to preserve forest areas. Given the economic and political interests involved, however, it is not surprising that much of this institutional framework has yet to be implemented, if indeed it will ever be fully enforced. The failure to define property rights continues to create very complex social and environmental problems. As far as the social aspect is concerned, the tendency has been to assuage (but not solve) the problems by taking land away from forest areas. It is necessary to adopt a new attitude in which the forest is valued as a whole, not just for its components, to at least attempt to reverse the current trend toward destruction of our forest heritage.

References

Bueno, E. 1998. *A viagem do descobrimento: a verdadeira história da expedição de Cabral.* Rio de Janeiro: Editora Objetiva, Coleção Terra Brasilis.

Dean, W. J. 1996. *A ferro e fogo: A história e a devastação da Mata Atlântica Brasileira.* Companhia das Letras.

Feijó, C., Ramos, R. O., Young, C. E., Arrouxeles, O., and de C. Lima, F. C. G. 2000. *Contabilidade social: o novo sistema de contas nacionais do Brasil.* Rio de Janeiro: Campus.

Fundação SOS Mata Atlântica, Instituto Nacional de Pesquisas Espaciais, Instituto Socioambiental. 1998. *Atlas da evolução das remanescentes forestis e ecossistemas associados no domínio da Mata Atlântica no período 1990–1995.* São Paulo: SOS Mata Atlantica, INPE, Instituto Socioambiental.

IBGE (Instituto Brasileiro de Geografia e Estatística). *Censo Agropecuário 1970, 1975, 1980, 1985, 1996.* Rio de Janeiro: IBGE.

IBGE (Instituto Brasileiro de Geografia e Estatística). 1999. *Pesquisa nacional por amostra de domicílios.* Online: http://www.ibge.gov.br/informacoes/pnad/sint97/tabelas, January 1999.

Young, C. E. F. 2002. Economia do extrativismo em áreas de Mata Atlântica. In: Simões, L. L. and Lino, C. F. (eds.). *Sustentável Mata Atlântica: a exploração de seus recursos florestais.* pp. 173–183. São Paulo: Editora Senac.

Chapter 11

The Central and Serra do Mar Corridors in the Brazilian Atlantic Forest

Alexandre Pires Aguiar, Adriano Garcia Chiarello, Sérgio Lucena Mendes, and Eloina Neri de Matos

In the last decade, conservation priority-setting workshops in the Atlantic Forest defined biodiversity corridors (Pinto 2000, http://www.conservation.org.br/ma). These corridors represent large regional planning units comprising a mosaic of land uses and key conservation areas. Two main biodiversity corridors have been selected in the Atlantic Forest: the Central (or Bahia) and Serra do Mar Corridors (Figure 11.1), within which lie three of the four recognized centers of endemism in the Atlantic Forest (Paulista, Rio Doce, and Bahia).

Biological Importance of the Central Corridor

The Central Corridor covers about 86,000 km^2 and represents about 75 percent of the Bahia Bioregion (Chapter 5, this volume). This corridor is biologically diverse and supports many species of limited distribution, including some threatened groups. Extremely high plant species richness has been documented (454 trees species per hectare) near Una, Bahia (Thomas et al. 1998), and in central Espírito Santo (443 tree species per hectare) (Thomaz and Monteiro 1997). Similarly, high species richness of nonvolant mammals (62 species) exceeds that in other Atlantic Forest sites and most neotropical areas with inventories (Passamani et al. 2000).

The region bordering the states of Bahia and Espírito Santo is also unique because of the presence of several typical Amazonian taxa. Central Espírito Santo, in turn, harbors one of the main remnants of dense and diverse *tabuleiros* forest (Peixoto and Gentry 1990), spanning 440 km^2, which connects the Reserva Biológica de Sooretama with the Reserva Florestal de Linhares.

Southern Bahia is among the very few areas where all six Atlantic Forest primate genera occur in sympatry. Twelve primate species occur here, representing 60 percent of the primates endemic to the Atlantic Forest (Pinto 1994). Bahia is exceptionally high in bird diversity, with five new species and a new genus

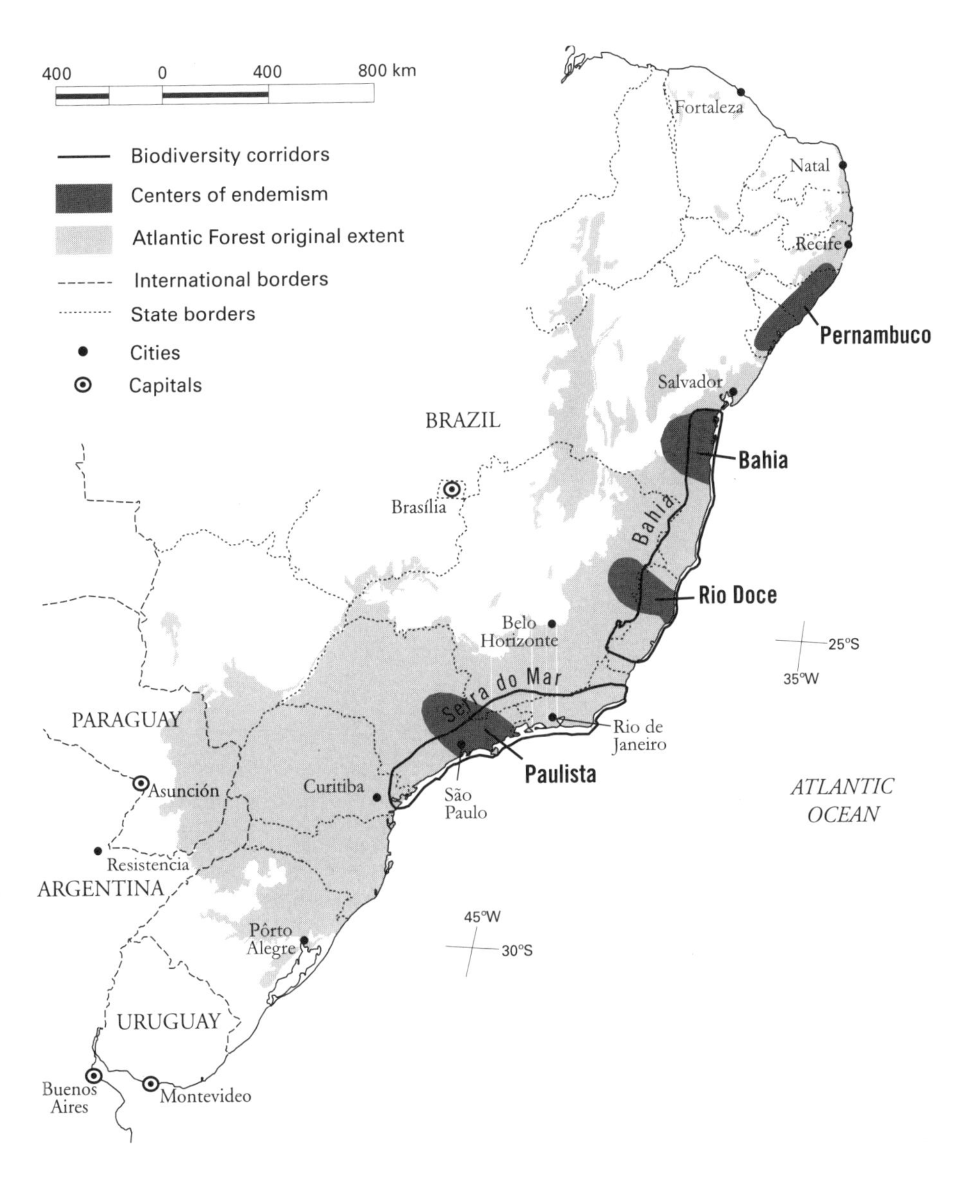

Penambuco center proposed by Muller (1973); Bahia, Rio Doce, and Paulista centers proposed by Muller (1973) and Kinzey (1982).

Figure 11.1. Location of biodiversity corridors and the four designated centers of endemism in the Atlantic Forest hotspot.

(*Acrobatornis*) recently described from the mountainous and coastal cocoa-growing regions in the south and central parts of the state. The Central Corridor contains more than 50 percent of the endemic bird species known to the Atlantic Forest (Cordeiro, pers. comm., 2001). High levels of diversity and endemism of both amphibians and reptiles also occur in the corridor. At least 12 new frog and toad species have been described recently in the Central Corridor (Silvano and Pimenta 2001).

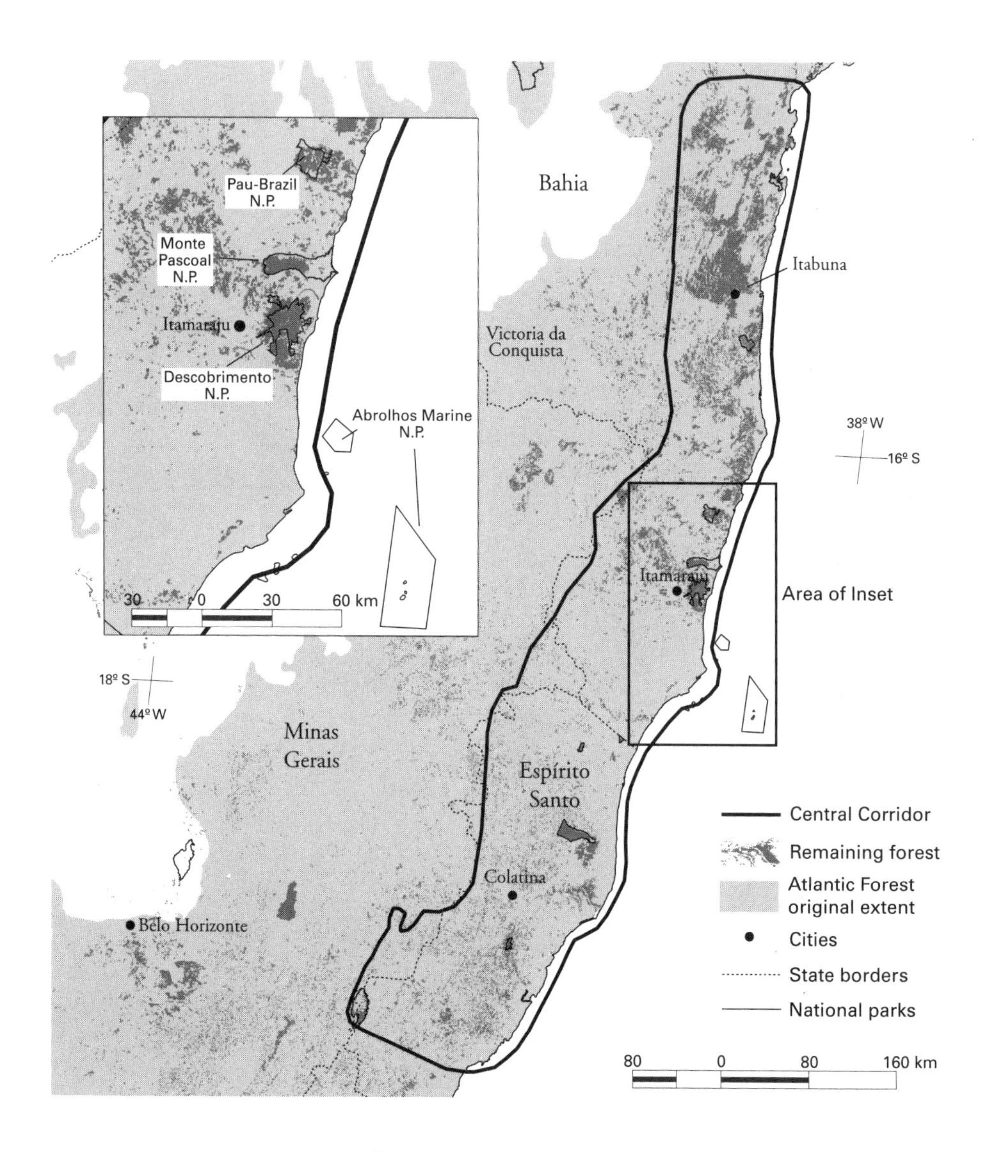

Forest cover data (1995) from Fundação S.O.S. Mata Atlântica. Protected areas data from Instituto Brasileiro do Meio Ambiente e dos Recursos Naturais Renováveis (IBAMA).

Figure 11.2. Important national parks (Descobrimento, Monte Pascoal, Pau-Brasil) in the Central Corridor, in Bahia and Espírito Santo, Brazil, protecting a total of nearly 500 km^2 of forest and Abrolhos Marine National Park.

Southern Bahia is also the region where most Brazilian cocoa (*Theobroma cacao*) is produced, in a system known locally as *cabruca* (shade cacao). About 6,500 km^2 of cocoa is cultivated in Bahia, 70 percent under the *cabruca* system (Araújo 1997). Although significantly disturbed, the *cabruca* forest is less damaging than deforestation because it supports a variety of native plants and animals (Alves 1990; Pinto 1994; Moura 1999; Pardini et al. 2000) and helps to connect protected areas. When abandoned, its biodiversity tends to increase over time, eventually approaching the level of a native forest (Alger and Caldas 1996; Sambuichi 2000).

Some of the most important blocks of forest in the Central Corridor lie in extreme southern Bahia, including national parks—Descobrimento, Monte Pascoal, Pau-Brasil—protecting a total of nearly 500 km^2 of forest. The small river basins protected by these national parks are extremely important not only to terrestrial biodiversity but also to the coral reefs and other marine ecosystems in the Abrolhos Marine National Park, the richest coral reef area in the South Atlantic (Figure 11.2).

The states of Bahia and Espírito Santo combined have about 40 protected areas, 70 percent of which are managed by the state governments. State-owned protected areas represent 69.5 percent of the total protected land, and the average size of an individual protected area in the region is 93.13 km^2. However, the high level of devastation of the Atlantic Forest in Espírito Santo and Bahia makes the enforcement of protected areas and the creation of new ones top priorities for these regions. Suitable new protected areas have already been identified, but there are not enough human and financial resources to manage even the existing ones. Crucial problems include lack of financial resources to implement management plans, insufficient technical personnel and equipment to direct and guard the units, poaching of forest products, and intentional fires (Conservation International do Brasil and IESB 2000).

Biological Importance of the Serra do Mar Corridor

The Serra do Mar Corridor is one of the richest biodiversity areas in the Atlantic Forest. The northern Serra do Mar, especially in Rio de Janeiro state, is the subregion of the Atlantic Forest with the greatest concentration of endemic species for many groups (Manne et al. 1999; Costa et al. 2000; Brown and Freitas 2000; Rocha et al. in press) and the greatest concentration of threatened bird species (Collar et al. 1997; Manne et al. 1999).

The coastal streams in the state of Rio de Janeiro have the highest level of fish endemism in the Atlantic Forest (e.g., lowland rivers and hillside streams of the São João river basin; Pinto 2000). Twelve areas in the Serra do Mar Corridor were assigned the highest priority for conservation within the Atlantic Forest, based on biodiversity and endemism. The Serra dos Órgãos stands out as a continuous forest of the montane and high-montane type, showing impressive levels of endemism, richness of invertebrates, and numbers of threatened species of mammals, amphibians, and reptiles (Bergallo et al. 2000; Pinto 2000; Rocha et al. 2000). The Itatiaia region, between Rio de Janeiro and Minas Gerais, also features high levels of endemism (Caramaschi et al. 2000; Otero et al. 2000).

The Serra da Mantiqueira mountains in the Serra do Mar also have a high diversity of plants and animals, including many endemic species of amphibians and reptiles (Costa et al. 1998) and a high diversity of small mammals (Costa et al. 2000). This region was also considered a conservation priority for the state of Minas Gerais (Figure 11.3).

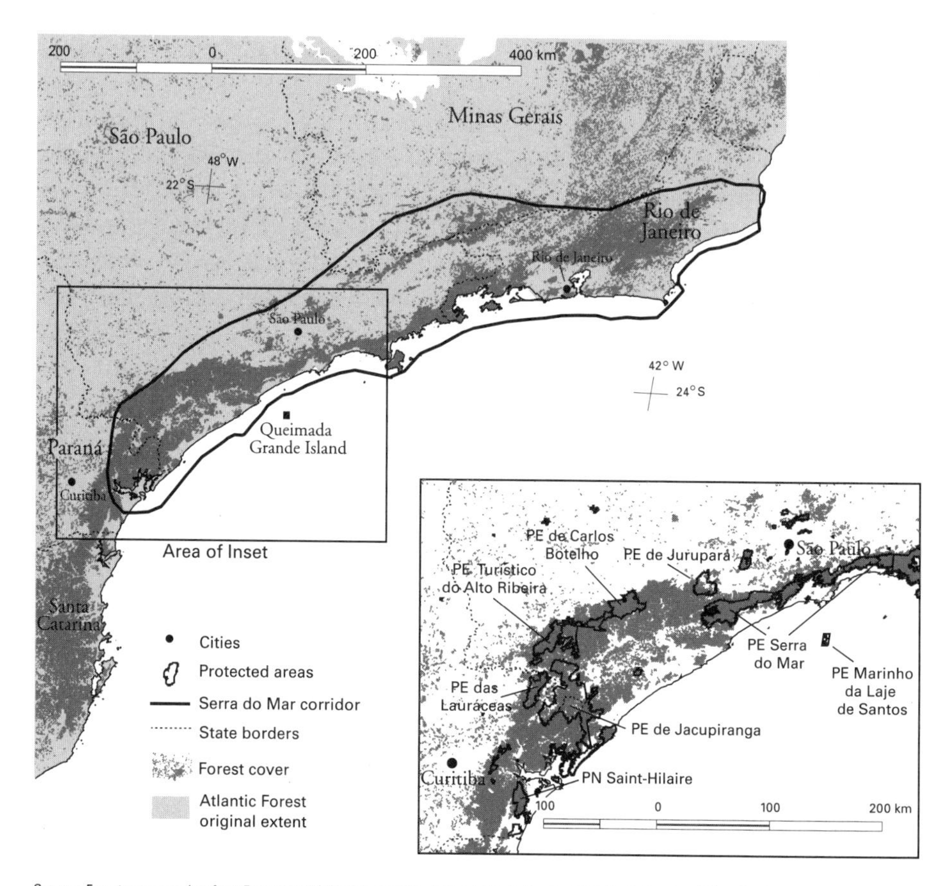

Sources: Forest remnants data from Fundação S.O.S. Mata Atlântica (Paraná, Rio de Janeiro 2000; Santa Catarina, Minas Gerais, São Paulo 1995). Protected areas data from Instituto Brasileiro do Meio Ambiente e dos Recursos Naturais Renováveis (IBAMA).

Figure 11.3. Important conservation areas within the Serra do Mar Corridor.

Important endemic invertebrates are found in specialized shoreline plant communities known as *restingas* (Platnick and Rocha 1995; Otero et al. 2000; Gonzaga and Pacheco 1990; Caramaschi et al. 2000; Rocha 2000). The Restinga of Jurubatiba, on the north coast in Rio de Janeiro State, is one of the best-preserved *restingas* of Brazil. Jurubatiba Park shows a great mosaic of well-defined ecosystems, with many rare, endemic, or threatened species, and serves as a refuge for species already extinct in other regions of Rio de Janeiro where the *restingas* are degraded or have already disappeared (Albertoni and Esteves 1999).

The Serra do Mar region includes the largest remaining block of Atlantic Forest in the slopes and tops of Serra do Mar and Serra da Mantiqueira and in adjacent flat lowlands (Figure 11.3). Although these forests are near the two largest metropolitan areas in Brazil (the cities of São Paulo and Rio de Janeiro), they remain well preserved, thanks to steep slopes that are not suitable for agriculture. Here 38 percent of the existing protected areas are owned by the federal government, and the average size of the protected areas in the Serra do Mar Corridor is more than 350 km^2 (400 km^2 for the federal reserves). The region includes some of the most important protected areas of the Atlantic Forest (e.g., Serra dos

Órgãos National Park, Serra da Bocaina National Park, and Itatiaia National Park). The prospect for long-term survival of native species is more favorable here than in any other region.

Pressures to the Central and Serra do Mar Corridors

There are many proximate and ultimate socioeconomic causes of biodiversity loss in both regions. In the following section we briefly review the most important.

Intensive Land Use

Of the 2–7 percent of the original Atlantic Forest remaining in the Central Corridor, nearly 80 percent is within areas owned by cocoa farmers. However, as a result of the cacao crisis, farmers in southern Bahia have converted up to 45 percent of their *cabruca* to pastures or other uses (SEI 1997), and farmers with large properties (averaging 14.30 km^2) have cleared about 67 percent of their land to sell the timber from the *cabruca*. Rural incentives have also contributed to deforestation. The "Pro-cacau" program, for example, has led to the devastation of 2,150 km^2 of native forest in southern Bahia (Rocha et al. in press) because credit lines are offered to farmers without adequate consideration of environmental issues (http://www.amda.org.br).

Monoculture planting of *Eucalyptus* began in Espírito Santo in the 1960s and in Bahia in the 1980s. Conditions in Bahia and northern Espírito Santo are ideal for plantations: perfect characteristics of soil and climate (http://www.veracel.com.br); a tradition of logging; low costs of land, personnel, energy, and taxes (CEPEDES and CDDH 1992); and the world's lowest production costs (Conservation International do Brasil and IESB 2000). By 1995, some 1,730 km^2 in Espírito Santo—almost 4 percent of the state's land area—had been planted with this crop, and in 20 years, *Eucalyptus* monocultures in southern Bahia have already overtaken 3,130 km^2 (Andrade 2001). The cellulose industry has been expanding operations in extreme southern Bahia and northern Espírito Santo, and the *Eucalyptus* plantations dominate part of the landscape in this portion of the Central Corridor. However, further plantations of *Eucalyptus* have recently been outlawed in Espírito Santo.

In Espírito Santo, coffee is a major source of income, and plantations are a serious threat to the forest. In the 1960s, when the coffee industry was affected by declining prices, cattle grazing emerged as an alternative, causing new and extensive deforestation in the state. Today, pastures occupy about 50 percent of the area once used for agriculture (SEAMA 2001). Proportionally, this state has been the most intensely devastated (ES-Eco 1992). Pastures, coffee, and *Eucalyptus* monoculture replace today most of the deforested area.

The human settlements resulting from the Land Reform Law in southern Bahia also have coincided, disastrously, with forested areas in the region (Rocha et al. in press). The areas deforested are small but often of great ecological importance (Alger and Caldas 1994).

Burning has been a serious and constant threat to the Atlantic Forest in Espírito Santo. In the Caparaó National Park and 10 adjacent counties, for example, 485 fires were detected by satellite imagery in September 2001 alone (Rego 2001). Nevertheless, the area authorized for controlled burning and the number of permits—mostly to clear for sugarcane plantations and cattle pastures—have increased, as has the number of fines for illegal burning.

Forest remnants in the Espírito Santo highlands are in better condition, and under better protection, than those in the lowlands, largely because the mountainous landscape makes exploitation difficult and expensive. In the lowlands, forest remnants continue to decline, particularly in the *tabuleiros* region, which covers about 25 percent of the state area. Satellite images suggest an increased rate of deforestation from 1996–2000 (Chapter 6, this volume).

Grazing is one of the most land-intensive activities in the state of Rio de Janeiro and is aggravated by the fires that are often used to clear pastures. The cattle herds in the region amount to more than 1.8 million heads—occupying 19,300 km^2, or 44.5 percent of the territory in the state (CIDE 1996)—and represent some 30 percent of the state's rural production.

Agriculture occupies 9.4 percent of the total area of Rio de Janeiro state (CIDE 1994), but the land use is far from homogeneous. Some groups closely associated with the forest remnants are subsistence farmers; among their practices are polycultivation, fallow, and *coivara,* a practice that allows the vegetation to grow back to a certain point before it is burned to increase soil fertility (Adams 2000).

The expansion of agriculture and pastures in southern Minas Gerais has also been a principal cause of environmental degradation in the region. Initially coffee cultivation spread throughout Zona da Mata and Serra da Mantiqueira, occupying foothills between mountain ranges and limiting native forests to hilltops. However, the uneven land and unsustainable cultivation techniques caused serious erosion and soil depletion. Coffee plantations then gave way to pastures, which extended to the hilltops, further fragmenting forest remnants (Brito et al. 1997).

Land use in the Paraíba Valley is very intensive and diversified, with cultivation of corn, potatoes, beans, manioc, and bananas. These are all low-yield crops, but they impede forest regeneration and involve the use of fire. Intentional fires to clear pastureland have also caused extensive damage along the frontier between Minas Gerais and Rio de Janeiro states.

The Paraíba do Sul basin was originally almost entirely covered by the Atlantic Forest, but the original vegetation now remains only in isolated patches in hilltops and other remote areas. Even so, the remaining forest is still subject to inordinate exploitation: about 20 km^2 of native vegetation was cleared in the region from 1990 to 1995 (Fundação SOS Mata Atlântica et al. 1998). The most intense deforestation in the state of Rio de Janeiro is concentrated in some municipalities of Angra dos Reis, Carmo, Santa Maria Madalena, and Campos de Goytacases. Cattle ranchers and small land owners in the Serra do Mar Corridor contribute to constant and widespread deforestation through the extraction of timber for fence stakes and subsistence agriculture.

Logging

Although logging has been practiced for five centuries in Brazil, in Bahia it has become especially intense in the past 30 years (Mesquita and Leopoldino 1999), particularly with the movement of logging companies into southern Bahia from the devastated northern Espírito Santo. In 1990, the federal government banned logging in the Atlantic Forest. However, logging companies successfully lobbied the government to be allowed to continue operating if they adopted sustainability plans, but they have not necessarily followed the recommended technical process (Mesquita in press; Rede de ONGs da Mata Atlântica 2001).

Logging companies extracted 225,000 m^3 of wood in southern Bahia in 1994, nearly 75 percent of it illegally (IESB 1997). In 1994, all logging companies with permits in this region were operating in areas supporting endangered primates (Mesquita 1997). In 2001, an expert committee evaluated 315 approved management plans, and only 32 were considered adequate. Despite legal protection, the deforestation rate in southern Bahia was greater in the early 1990s than in the 1980s (Capobianco 2001), and logging companies, legal or not, remain active in the region, showing clear expansion from 2000 to 2001 (Mesquita in press).

Fuelwood Harvesting

Much of the wood used as an energy source today is wood rejected by the cellulose industry, but this is not sufficient to meet the demand for firewood for residential heating. Therefore, the Santa Maria and Jucu rivers, for example, are still under intense deforestation pressure as a result of firewood exploitation. Low income has been one of the major factors in the use of firewood in forest regions in Rio de Janeiro. In the past, exploitation of forests for charcoal production was also a serious problem.

Plant and Animal Poaching

Wild animal trade is the third-biggest illegal trade in the world, now on the order of $10 billion per year, of which $1 billion is derived from the Brazilian market alone (http://www.renctas.org.br). The volume of illegal animal trade doubled in Brazil from 1996 to 2000, and it is estimated that 50 million animals were trapped during this period (i.e., 10 million animals per year). Wildlife trade directly affects more than 200 Brazilian species (Rocha 1995) and of these, 171 (including at least 88 endemic birds) are officially threatened (Capobianco 2001).

In Brazil, many animals exploited in local fairs are typical of the Atlantic Forest (Souza and Soares Filho 1998). At least 174 species of Brazilian fauna are being exploited commercially in Bahia alone (Freitas and Guerreiro 1998). In February 2000, an operation by the Brazilian Institute for the Environment and Renewable Natural Resources (IBAMA) in Bahia rescued 2,000 wild animals illegally held in captivity, including threatened species such as capuchin monkeys

(*Cebus xanthosternos*) and golden-headed lion tamarins (*Leontopithecus chrysomelas*).

In 1999, Espírito Santo was one of the leading states in number of penalties applied to poachers and people who collect or maintain wild animals in captivity (http://www.mma.gov.br). The following year, the environmental police of Espírito Santo rescued about 6,000 wild animals in illegal captivity, and in the first quarter of 2001 the number exceeded 2,000.

Subsistence hunting also contributes to the decline of fauna. In recent interviews, 42 percent of local residents in areas adjacent to Una Biological Reserve in Bahia admitted to hunting, and 66 percent reported that game animals are becoming less abundant in the region. Small farmers hunt more often than large farmers because they have more acute subsistence needs (Santos and Pardini 2000). Although it is not practiced on a large scale, sport hunting may affect small populations of animals such as the solitary tinamou (*Tinamus solitarius*) (Consórcio Santa Maria-Jucu 1997).

In the municipalities of Resende and Itatiaia, illegal extraction of heart palm trees (*Euterpe edulis*) is a serious problem. Organized gangs invade and camp in the forest, transport the palm hearts, and process and sell the product. In a few days, poachers can cut down thousands of palm trees, extract the heart of palm, and pack it for transportation.

Introduction of Alien Species

Many introduced species contribute to ecosystem deterioration and biodiversity loss (Chapter 33, this volume). For example, at least 16 exotic fish species are thought to be established in the rivers of the Paraíba do Sul valley (Bizerril 1998). Exotic species raised in fish hatcheries escape or are intentionally introduced into rivers, where they often occupy the niches of native species. The adaptability of ornamental plants introduced on properties near forest remnants makes them potentially threatening to native flora. Some species have proliferated outside their gardens. For example, *Impatiens balsamina,* an exotic species in the Balsaminaceae, colonizes the banks of streams and creeks, and exotic tree species such as almond trees (*Terminalia cattapa*), casuarina (*Casuarina equisetifolia*), and leucena (*Leucaena leucocephala*)—highly resistant species with great capacity for dispersal—also proliferate. Casuarinas are a particular threat to native coastal vegetation.

Urban Expansion and Industrialization

Increasing human presence near forested areas is a constant threat to biodiversity, mostly through small-scale extraction activities such as hunting, collecting ornamental and medicinal plants, capturing songbirds and ornamental birds, and poaching (Conduro and Santos 1995). Serious water pollution from untreated sewer emissions, intentional embankment of lakes, and deforestation of mangroves and *restingas* are also common effects of urban expansion (Maciel 1984; Araujo and Maciel 1998).

Coastal forests are particularly threatened by intensive and poorly planned development. The Espírito Santo coastline, for example, extends for 411 km and drains 12 river basins (SEAMA 2001). Coastal development has caused occupation or destruction of fragile ecosystems, pollution of rivers and beaches by industrial, municipal, and human waste, and deforestation. Further urban and industrial projects are still being scheduled for the *restingas* of Espírito Santo (ES-Eco 1992; SEAMA 1999).

The Rio–São Paulo region is the most industrialized in the country, and acid rain caused by the pollution often falls on forest remnants. The Paraíba do Sul basin is today one of the most industrialized areas of Brazil (the São Paulo portion of the basin, for example, has 2,730 industrial sites, which account for 10 percent of the nation's exports), and the central Paraíba region, with its high concentration of industrial sites, is also heavily polluted.

Many highways and roads cross the Serra do Mar and Serra da Mantiqueira, dividing forests and isolating many animals in fragmented habitats. Many roads, notably in southern Minas Gerais, have been built without the necessary environmental impact reports or are left unfinished, making adjacent areas vulnerable to severe erosion (e.g., Federal Highway BR-101 in southern Bahia).

Dams are also a potential threat (Chapter 35, this volume). The Companhia Energética de Minas Gerais and private investors plan to build 15 new hydroelectric dams in southern Minas Gerais, promising energy in abundance and stimulating further urbanization. Twenty thousand new businesses in this region are anticipated over the next few years.

Tourism-Related Development

The proliferation of vacation homes and tourist accommodations is a direct threat to many precious forest remnants in the state of Rio de Janeiro. Suppression of the understory forest, introduction of exotic tree species, impoundment of streams and creeks, trail construction, and feeding of wildlife disrupt the integrity of the affected forest fragments (Conduro and Santos 1995).

The northern coast of São Paulo, which extends for 161 km and encompasses 164 beaches and 17 islands, receives about 1 million visitors in the peak season (January and February). The city of Caraguatatuba—the most populous, with approximately 80,000 people—takes in some 500,000 tourists in the summer, generating about $20 million in revenue, or 25 percent of the city's annual budget. The largest areas of continuous deforestation along the northern coast coincide with real estate enterprises in the area. Untreated sewage pollutes the beaches, and the construction of vacation homes, hotels, resorts, and other amenities also creates additional pressures on the Atlantic Forest in the region.

In the state of Minas Gerais, the recent widening of the Fernão Dias highway has led to an increase in tourism in the Serra da Mantiqueira, causing various environmental problems.

Degradation of Mangroves and Restingas

Deforestation also threatens associated ecosystems of the Atlantic Forest, such as mangroves and *restingas. Restinga* consists of plant communities occurring in the littoral zone and includes beaches and dunes, and because of their sandy, loose soils, they are highly vulnerable to anthropogenic impact. A large percentage have already been cleared for mining, real estate development, and agriculture (SEAMA 2001). *Restinga* vegetation is also generally thicker and lower in height than other types of forest in the region, making it a prized source of timber and firewood for homes or small industries.

Invasion of mangrove areas, particularly by poor families looking for a place to live, is common in Espírito Santo. Wood is extracted from the mangroves to build homes, fish traps, and shrimp nurseries (MMA 1998). With the rising price of bottled gas, mangrove trees have become increasingly popular for firewood. Mangroves are also exploited for tannin, widely used in pottery and to dye and protect fishing nets. The bark of red mangrove (*Rhizophora mangle*) is the richest known source of tannin and is often removed haphazardly, causing the plant to desiccate and die.

Acknowledgments

We thank Luiz Paulo S. Pinto, André Vieira R. de Assis, Maria Cecília W. de Brito, Ivana Lamas, Luciano Petronetto, and Ana Maria da S. Ofranti for their important contributions.

References

Adams, C. 2000. *Caiçaras na Mata Atlântica: pesquisa científica versus planejamento e gestão ambiental.* São Paulo: Annablume Editora e Comunicação, FAPESP.

Albertoni, E. F. and Esteves, F. A. 1999. Jurubatiba, uma restinga peculiar. *Ciência Hoje* 25(148): 61–63.

Alger, K. and Caldas, M. 1994. The declining cocoa economy and the Atlantic Forest of southern Bahia, Brazil: conservation attitudes of cocoa planters. *The Environmentalist* 14(2): 107–119.

Alger, K. and Caldas, M. 1996. Cacau na Bahia: decadência e ameaça à Mata Atlântica. *Ciência Hoje* 117: 28–35.

Alves, M. C. 1990. *The role of cacao plantations in the conservation of the Atlantic Forest of southern Bahia, Brazil.* Master's thesis, University of Florida, Gainesville.

Andrade, M. F. 2001. Falta de informação ameaça Mata Atlântica. *A Tarde* (Salvador) 31 August 2001.

Araujo, D. S. D. and Maciel, N. C. 1998. Restingas fluminenses: biodiversidade e preservação. *Boletim FBCN* 25: 29.

Araújo, M. 1997. Conservação da Mata Atlântica na região cacaueira da Bahia. In: União dos Palmares: Conselho Nacional da Reserva da Biosfera da Mata Atlântica. *Anais do V Seminário da Reserva da Biosfera da Mata Atlântic.* pp. 11–16.

Bergallo, H. G., Rocha, C. F. D., Alves, M. A. S., and Van Sluys, M. 2000. *A fauna ameaçada de extinção do estado do Rio de Janeiro.* Rio de Janeiro: Editora UERJ.

Bizerril, C. R. S. F. 1998. *Contribuição ao conhecimento da Bacia do Rio Paraíba do Sul.* Rio de Janeiro: Aneel, CPRM.

Brito, F. R. A., Oliveira, A. M. H. C., and Junqueira, A. C. 1997. A ocupação do território e a devastação da Mata Atlântica. In: de Paula, J. A. (ed.). *Biodiversidade, população e economia: uma região de Mata Atlântica.* pp. 49–89. Belo Horizonte: UFMG/Cedeplar–ECMVS, PADCT/CIAMB.

Brown, K. S. and Freitas, A. V. L. 2000. Atlantic Forest butterflies: indicators for landscape conservation. *Biotropica* 32: 934–956.

Capobianco, J. P. R. (ed.). 2001. *Dossiê Mata Atlântica 200: Projeto Monitoramento Participativo da Mata Atlântica.* São Paulo: Rede de ONGs da Mata Atlântica, Intituto Socioambiental, Sociedade Nordestina de Ecologia.

Caramaschi, U., Carvalho e Silva, A. M. P. T., Carvalho e Silva, S. P., Gouvea, É., Izecksohn, E., Peixoto, O. L., and Pombal Jr., J. P. 2000. Anfíbios. In: Bergallo, H. G., Rocha, C. F. D., Alves, M. A. S., and Van Sluys, M. *A fauna ameaçada de extinção do estado do Rio de Janeiro.* pp. 75–78. Rio de Janeiro: Editora UERJ.

CEPEDES (Centro de Estudos e Pesquisas para o Desenvolvimento do Extremo Sul da Bahia) and CDDH (Centro de Defesa dos Direitos Humanos). 1992. *Eucalipto, uma contradição: impactos ambientais, sociais e econômicos do eucalipto e da celulose no extremo sul da Bahia.* Eunapólis: CEPEDES, CDDH.

CIDE (Centro de Informações e Dados do Rio de Janeiro). 1994. *Anuário estatístico do Rio de Janeiro.* Rio de Janeiro: Fundação Centro de Informações e Dados do Rio de Janeiro—CIDE.

CIDE (Centro de Informações e Dados do Rio de Janeiro). 1996. *Anuário estatístico do Rio de Janeiro.* Rio de Janeiro: Fundação Centro de Informações e Dados do Rio de Janeiro—CIDE.

Collar, N. J., Wege, D. C., and Long, A. J. 1997. Patterns and causes of endangerment in the New World avifauna. *Ornithological Monographs* 48: 237–260.

Conduro, L. G. S. and Santos, L. A. F. 1995. *Unidades de conservação da natureza: Conceitos básicos, definições e caracterização geral: Situação no estado do Rio de Janeiro.* Rio de Janeiro: FEEMA.

Conservation International/Center for Applied Biodiversity Science and IESB (Instituto de Estudos Socioambientais do Sul da Bahia). 2000. *Designing sustainable landscapes: the Brazilian Atlantic Forest.* Washington, DC: CI/CABS.

Consórcio Santa Maria-Jucu. 1997. *Diagnóstico e plano diretor das bacias dos rios Santa Maria e Jucu: biodiversidade, região estuarina e espaços territoriais protegidos.* Unpublished report, Vitória, Espírito Santo.

Costa, C. M. R., Herrmann, G., Martins, C. S., Lins, L. V., and Lamas, I. R. (eds.). 1998. *Biodiversidade em Minas Gerais: um atlas para sua conservação.* Belo Horizonte: Fundação Biodiversitas.

Costa, L. P., Leite, Y. L. R., da Fonseca, G. A. B., and Fonseca, M. T. 2000. Biogeography of South American forest mammals: endemism and diversity in the Atlantic Forest. *Biotropica* 32(4b): 872–881.

ES-Eco Comissão Coordenadora do Relatório Estadual sobre Meio Ambiente e Desenvolvimento. 1992. *Meio ambiente e desenvolvimento no Espírito Santo.* Relatório Final. Unpublished report, Vitória, Espírito Santo.

Freitas, M. A. and Guerreiro, W. 1998. *Registro de algumas espécies de animais silvestres no estado da Bahia.* Unpublished. Vitória da Conquista: Universidade Estadual do Sudoeste da Bahia.

Fundação SOS Mata Atlântica, INPE (Instituto Nacional de Pesquisas Espaciais), and

ISA (Instituto Socioambiental). 1998. *Atlas da evolução dos remanescentes florestais e ecossistemas associados no domínio da Mata Atlântica no período 1990–1995.* São Paulo: SOS Mata Atlântica, INPE, ISA.

Gonzaga, L. P. and Pacheco, J. F. 1990. Two new subspecies of *Formicivora serrana* (Hellmayr) from southeastern Brazil, and notes of type locality of *Formicivora deluzae.* Ménétriés. *Bulletin of the British Ornithological Club* 110(4): 187–193.

IESB (Instituto de Estudos Sócio-Ambientais do Sul da Bahia). 1997. 5 perguntas sobre a atividade madeireira no sul da Bahia. In: IESB (Instituto de Estudos Sócio-Ambientais do Sul da Bahia). *Ação das Madeireiras no Sul da Bahia.* Informes e Documentos. Unpublished.

Kinzey, W. G. 1982. Distribution of primates and forest refuges. In: Prance, G. T. (ed.). *Biological diversification in the tropics.* pp. 455–482. New York: Columbia University Press.

Maciel, N. C. 1984. Perspectivas e estratégias para uma política nacional de proteção a manguezais e estuários. *Boletim FBCN* 19: 111.

Manne, L. L., Brooks, T. M., and Pimm, S. L. 1999. Relative risk of extinction of passerine birds on continents and islands. *Nature* 399: 258–261.

Mesquita, C. A. B. 1997. Serrarias fazem festa no Sul da Bahia. *Revista Parabólicas* 33.

Mesquita, C. A. B. In press. A atividade madeireira na região cacaueira. In: Alger, K. et al. (eds). *A conservação da Mata Atlântica no Sul da Bahia.* Ilhéus: IESB (Instituto de Estudos Sócio-Ambientais do Sul da Bahia).

Mesquita, C. A. B. and Leopoldino, F. S. 1999. Por que a atividade madeireira no sul da Bahia não deve ser retomada. Unpublished. Ilhéus: IESB (Instituto de Estudos Sócio-Ambientais do Sul da Bahia).

MMA (Ministério do Meio Ambiente), dos Recursos Hídricos e da Amazônia Legal. 1998. *Primeiro relatório nacional para a Convenção sobre Diversidade Biológica.* Brasília: MMA.

Moura, R. T. M. 1999. *Análise comparativa da estrutura de comunidades de pequenos mamíferos em remanescente de Mata Atlântica e em plantio de cacau em sistema de cabruca no Sul da Bahia.* Master's thesis, Universidade Federal de Minas Gerais, Belo Horizonte.

Müller, P. 1973. *The dispersal centers of terrestrial vertebrates in the neotropical realm.* The Hague: Junk.

Otero, L. S., Brown Jr., K. S., Mielke, O. H. H., Monteiro, R. F., Costa, J., Macêdo, M. V., Maciel, N. C., Becker, J., Salgado, N. C., Santos, S. B., Moya, G. E., Almeida, J. M., and Silva, M. D. 2000. Invertebrados terrestres. In: Bergallo, H. G., Rocha, C. F. D., Alves, M. A. S., and Van Sluys, M. *A fauna ameaçada de extinção do estado do Rio de Janeiro.* pp. 53–62. Rio de Janeiro: Editora UERJ.

Pardini, R., Faria, D., and Baumgarten, J. 2000. Diretrizes biológicas para estratégias de conservação: o projeto Restauna e a Reserva Biológica de Una, sul da Bahia. In: *II Congresso Brasileiro de Unidades de Conservação.* pp. 715–721. Vol. II: *Trabalhos técnicos.* Campo Grande: Rede Nacional Pró-Unidades de Conservação, Fundação O Boticário de Proteção à Natureza.

Passamani, M., Mendes, S. L., and Chiarello, A. G. 2000. Non-volant mammals of the Estação Biológica de Santa Lúcia and adjacent areas of Santa Teresa, Espírito Santo, Brazil. *Boletim do Museu de Biologia Mello Leitão (Nova Série)* 11/12: 201–214.

Peixoto, A. L. and Gentry, A. 1990. Diversidade e composição florística da mata de tabuleiro na Reserva Florestal de Linhares (Espírito Santo, Brasil). *Revista Brasileira de Botânica* 13: 19–25.

Pinto, L. P. S. 1994. *Distribuição geográfica, população e estado de conservação do mico-leão-da-cara-dourada,* Leontopithecus chrysomelas *(Callithrichidae, Primates).* Master's thesis, Universidade Federal de Minas Gerais, Belo Horizonte.

Pinto, L. P. S. (ed.). 2000. *Avaliação e ações prioritárias para a conservação da biodiversidade da Mata Atlântica e Campos Sulinos: Relatório técnico.* Unpublished. Belo Horizonte: MMA, Conservation International do Brasil, Fundação SOS Mata Atlântica, Fundação Biodiversitas, Instituto de Pesquisas Ecológicas, Secretaria do Meio Ambiente do Estado de São Paulo, Semad/Instituto Estadual de Florestas–MG.

Platnick, N. I. and Rocha, C. F. D. 1995. On a new Brazilian spider of the genus *Trachelopachys* (Araneae, Corinnidae), with notes on misplaced taxa. *American Museum Novitates* 3153(8): 1–8.

Rede de ONGs da Mata Atlântica. 2001. *A vistoria nos planos de manejo florestal no extremo sul da Bahia, a exploração da cabruca e outras considerações.* Unpublished.

Salvador, Rego, S. C. 2001. Satélite mostra focos de incêndio no Caparaó. *A Gazeta* (Vitória) 08/09/2001.

Rocha, C. F. D. 2000. Biogeografia de répteis de restingas: distribuição, ocorrência e endemismos. In: Esteves, F. A. and Lacerda, L. D. (eds). *Ecologia de restingas e lagoas costeiras.* pp. 99–116. Macaé: Nupem, UFRJ.

Rocha, C. F. D., Van Sluys, M., Alves, M. A. S., and Bergallo, H. G. 2000. Corredores de conservação e sua importância em propostas de reflorestamento no estado do Rio de Janeiro. In: *Fundação Centro de Informações e Dados do Rio de Janeiro índice de Qualidade dos municípios-verde (IQM-verde).* Rio de Janeiro (available on CD-ROM).

Rocha, F. M. 1995. *Tráfico de animais silvestres no Brasil: documento de discussão.* Brasília: WWF, Traffic.

Rocha, R., Alger, K., Reid, J., Loureiro, W., Horlando, H., Villanueva, P. In press. Conservação através de políticas públicas. In: Alger, K. et al. (eds). *A conservação da Mata Atlântica no Sul da Bahia.* Ilhéus: IESB (Instituto de Estudos Sócio-Ambientais do Sul da Bahia).

Sambuichi, R. H. R. 2000. Levantamento fitossociológico de árvores remanescentes da Mata Atlântica em uma área de lavoura de cacau na região sul da Bahia, Brasil. In *Resumos do 51º Congresso Nacional de Botânica.* p. 241. Brasília: Sociedade Botânica do Brasil.

Santos, G. J. R. and Pardini, R. 2000. *Caracterização da caça na região do entorno da Reserva Biológica de Una-Ba. Projeto remanescentes de forestas na região de Una-BA.* Relatório final. Unpublished document. Ilhéus: Fundação Pau Brasil/IESB/UESC.

SEAMA (Secretaria do Estado do Espírito Santo para assuntos do Meio Ambiente). 1999. *Projeto Gerenciamento Costeiro/ES: proposta de trabalho para o biênio 1999/2000.* Unpublished report, Vitória.

SEAMA (Secretaria do Estado do Espírito Santo para assuntos do Meio Ambiente). 2001. *Prioridades ambientais do estado do Espírito Santo.* Unpublished report, Programa Nacional do Meio Ambiente, PNMA II.

SEI (Superintendência de Estudos Econômicos e Sociais da Bahia). 1997. Impactos da monocultura sobre o ambiente socioeconômico do Litoral Sul. In: *Série Estudos e Pesquisas 32.* Salvador: SEI.

Silvano, D. L. and Pimenta, B. V. S. 2001. *Abordagens ecológicas e instrumentos econômicos para o estabelecimento do "Corredor do Descobrimento": uma estratégia para reverter o processo de fragmentação florestal na Mata Atlântica do Sul da Bahia.* Sub-projeto Anfíbios Anuros: Síntese dos resultados preliminares. Unpublished. Ilhéus: IESB.

Souza, G. M. and Soares Filho, A. 1998. *O comércio ilegal de aves silvestres na região do*

Paraguassu e Sudoeste da Bahia. Unpublished document. Vitória da Conquista: UESB (Universidade Estadual do Sudoeste da Bahia).

Thomas, W. W., Carvalho, A. M. V., Amorim, A. M. A., Garrison, J., and Arbeláez, A. L. 1998. Plant endemism in two forests in southern Bahia, Brazil. *Biodiversity and Conservation* 7: 311–322.

Thomaz, L. D. and Monteiro, R. 1997. Composição florística da Mata Atlântica de encosta da Estação Biológica de Santa Lúcia, município de Santa Teresa, ES. *Boletim do Museu de Biologia Mello Leitão (N. Ser.)* 7: 3–48.

Chapter 12

Policy Initiatives for the Conservation of the Brazilian Atlantic Forest

José Carlos Carvalho

Brazil leads the world in megadiversity, and the Atlantic Forest is one of the five priority biodiversity hotspots. Management of this formidable biological wealth calls for urgent action based on conservation awareness and mirrored in public policies that represent the aspirations of society. To fulfill these objectives, the Brazilian government's Ministry of Environment, Water Resources, and the Legal Amazon Region has a program that calls for strengthening control and monitoring of the environment, providing incentives for adopting sustainable practices, and creating and consolidating a network of protected areas for in situ conservation of the nation's biodiversity.

By acceding to Agenda 21 and the Convention on Biological Diversity, ratified by the National Congress under Legislative Decree No. 2 dated February 3, 1994, the Brazilian government committed itself to implementing development policies that protect the environment. This commitment is consonant with the provisions in the Federal Constitution of 1988, which state that all Brazilians have the right to an ecologically balanced environment, that the environment is a blessing to be shared by all the people and is essential to a healthy quality of life, and that the state and society as a whole have the responsibility to protect and preserve it for present and future generations. The chapter in the constitution devoted to the environment specifically refers to the national heritage of the Atlantic Forest and states that its use, including the use of its natural resources, shall be governed by law under conditions that ensure its preservation.

The National Council on the Environment recently approved a set of guidelines for a policy on the conservation and sustainable development of the Atlantic Forest. The strategy was formulated by a working group of representatives from government agencies and nongovernment organizations coordinated by the Secretariat for the Formulation of Ministry of Environment Public Policies. The policy guidelines call for comprehensive actions that will promote the conservation and sustainable development of the Atlantic Forest in keeping with the principles set forth in the Federal Constitution of 1988. Following a review of the conceptual and legal bases for such a policy, the document presents guidelines and proposed program actions for ensuring compatibility between the objectives and the instruments for attaining them.

Within this context, the Brazilian government has carried out numerous activities aimed at defining a National Policy on Biodiversity based on a broad process of consultation with society related public policies, and other activities regarding the conservation of biodiversity in the Atlantic Forest and other regions of the country. Since the mid-1990s the Ministry of Environment has been carrying out the National Biodiversity Program (PRONABIO), Brazil's first government program with objectives directly related to the principles set forth in the Convention on Biological Diversity and Agenda 21. The work of PRONABIO starts from the premise that the effective protection, recovery, and sustainable use of Brazilian biological diversity must necessarily depend on measures taken by public organizations and institutions, working both in situ and ex situ as well as on private property. Its mission statement calls for the gathering, systematic compilation, and dissemination of information on biological diversity; the definition and application of instruments for the economic measurement of biological diversity; the implementation of conservation measures both in situ and ex situ; and the promotion of sustainable use of biological resources.

PRONABIO has two financial mechanisms: the Project on the Sustainable Conservation and Use of Brazilian Biological Diversity (PROBIO, also known as the National Biodiversity Project), which operates within the government, and the Brazilian Biodiversity Fund (FUNBIO), which works through private initiatives and emphasizes respect for the principles of sustainable development. PROBIO is the product of a donor agreement signed in June 1996 by the government of Brazil, the Global Environment Facility (GEF), and the World Bank, based on financial contributions equivalent to US$10 million from the National Treasury and US$10 million from GEF. It is administered by the Ministry of Environment through its Technical Secretariat (General Coordination for Biological Diversity) with support from the National Council for Scientific and Technological Development. Its main objective is to identify priority actions by fostering subprojects that will promote partnerships between the public and private sectors while generating and disseminating information about the subject.

One of PROBIO's main activities has been a series of studies conducted in all the Brazilian biomes to identify and assess priority areas for conservation. So far, 182 priority areas for biodiversity conservation have been identified in the Atlantic Forest and Campos Sulinos, 99 of which have been cited as being of extreme biological importance. These initiatives have provided an important context for the formation of institutional partnerships within organized civil society through both nongovernment and government agencies concerned with the environment as well as the principal research centers in the country. This undertaking has been gaining in importance because of the quality of the data presented and the fact that they reflect the consensus of a large number of specialists. In addition to serving as a training tool for local and regional decision making in the area of conservation, the initiative has also provided bases for the development of environmental programs.

In recent years Brazil appears to have embarked on a new period of environmental revitalization, especially in the search for solutions to conserve biodiversity. This revitalization is the result of several new initiatives that have gradually materialized, both in the form of funding—amounts that may be small but are significant nevertheless at the regional level—and in the form of national strategies to guide the most urgently needed actions. In the 1990s most resources for biodiversity conservation units and projects in Brazil came from international agencies such as the World Bank and the German Development Bank. However, new mechanisms have stimulated greater participation and commitment on the part of the federal government through the National Environment Fund and through the private sector, as with FUNBIO.

FUNBIO was created by the Ministry of Environment and the GEF through the intercession of the World Bank. It receives managerial support from the Getúlio Vargas Foundation. FUNBIO finances and mobilizes resources for programs and projects related to conservation, sustainable use of natural resources, surveys, dissemination of information, technical exchanges, and other activities related to biodiversity. Under the agreement between the Brazilian government and GEF, US$30 million was made available to create FUNBIO.

The International Pilot Program to Conserve the Brazilian Rain Forest, known as PPG7, has played an important role in the conservation and sustainable use of Amazon Forest and Atlantic Forest biodiversity. Created at the request of the Group of Seven (G7), PPG7 receives financial support from the European Union and the Netherlands and is coordinated by the World Bank. Its objective is to prevent the deforestation of Brazilian rainforests using three approaches: strengthening the public's capacity to create and execute an environmental policy, improving the management of protected areas (e.g., parks, reserves, indigenous areas, and extraction areas), and broadening the existing knowledge base on biodiversity, conservation, and sustainable use of the Brazilian rainforest. Two PPG7 projects are of special importance for the Atlantic Forest: Ecological Corridors in the Neotropical Forests and Demonstration Projects (PDA/PPG7).

A biodiversity or ecological corridor is a network of parks, reserves, and other areas of less intensive use that are administered under integrated management and aimed at guaranteeing survival of the largest possible number of species in a given region. The corridor concept has come to be known in Brazil in connection with proposals for large-scale conservation in key areas identified by the PPG7 Parks and Reserves Project. The biodiversity corridor approach is used to address environmental protection on various scales, from local to regional, and attempts to encompass various ecosystems while also maintaining or increasing levels of connectivity between the different areas (Chapter 11, this volume). The aim of the Demonstration Projects Subprogram, initiated in March 1995, is to support local initiatives that envisage conservation and sustainable use of natural resources while contributing to the knowledge base that can be applied in formulating government policies. At the end of 2001 funding in the amount of US$15.9 million was negotiated for conservation projects in the Atlantic Forest undertaken by nongovernment organizations under the aegis of the Demonstration Projects Subprogram.

The PPG7 Joint Coordination Committee recently approved a specific Atlantic Forest Subprogram, thus increasing the portfolio of projects for the biome. This new subprogram will support projects aimed at biodiversity conservation, sustainable use of natural resources, recovery of degraded areas, monitoring, and research. The proposal was developed by the Atlantic Forest Planning Advisory Unit in the Ministry of Environment.

The importance of these initiatives and studies for the implementation of national policy on biodiversity has led us to add their recommendations to the institutional goals of the Ministry of Environment and, by extension, to guide the actions of other government agencies and society as a whole aimed at securing the conservation and sustainable use of the Atlantic Forest.

PART III

Argentina

Chapter 13

Dynamics of Biodiversity Loss in the Argentinean Atlantic Forest: An Introduction

Alejandro R. Giraudo

For Argentina, a country that extends south as far as the Antarctic Ocean, the province of Misiones includes the whole of the tropical Atlantic rainforest. Misiones's forests alone contain 29 percent of Argentina's vascular plant species and 50 percent of its vertebrate species, all in only 1.1 percent of the country's total land area. Misiones still has large areas with conserved forest and is home to a notable number of rare and endangered species. The Misiones forests also include a number of species whose populations are widespread elsewhere but have become highly endangered in the Atlantic Forest, including large, charismatic species such as the tapir, the jaguar, and the giant river otter.

As a province, Misiones has been developing a system of protected areas and legal tools to support conservation and sustainable development, and there is both national and international interest in preserving biodiversity in the region. Some of the basic ingredients for effective conservation are in place, but many serious challenges remain. In the nine chapters of this part, researchers from a wide range of disciplines shed light on the origins and present status of biodiversity in Argentina. In doing so, they underscore not only that conservation must receive greater emphasis as a national social responsibility but also that a multidisciplinary approach is necessary to address the complex set of problems that underlie the impending crisis of irreparable biodiversity loss. Indeed, our major challenge in the new millennium will be to integrate the development and well-being of human populations with the conservation of biodiversity and the ecological processes that sustain both global and local integrity.

Chebez and Hilgert begin the section in Chapter 14 by describing how the present state of biodiversity in the province of Misiones has been strongly determined by the unique history of the province. In Chapter 15, Giraudo and a diverse group of experts combine rare information on the flora and fauna of Misiones. Although the extraordinary information on Misiones biodiversity presented here for the first time seems substantial, the authors emphasize that there is little information about the ecology of the Atlantic Forest, the absence of which creates a formidable roadblock to conservation action.

In an effort to address the shortage of solid data on the flora and fauna of the Atlantic Forest, in Chapter 16 Giraudo and Povedano have compiled the scanty information available on the status of large species in Misiones that are at risk, including information on population trends and area needs. The authors have also included information from observations made by distinguished naturalists and from their own conversations with local people documenting rare sightings and anecdotes. Giraudo and Povedano also discuss the contiguous and intact parcels of forest in Misiones, among the largest remaining in the Atlantic Forest, which serve as some of the last refuges for large mammals such as jaguars, pumas, and raptors. Here, these animals are increasingly threatened by habitat loss and hunting. Nonetheless, researchers estimate that the Misiones forests support as many jaguars as the entire Brazilian portion of the hotspot, an area 12 times larger.

Many other aspects of biodiversity are at great risk in Misiones. For example, three monkey species—the brown capuchin monkey and the black and the brown howler monkeys—have their southernmost range limits in Misiones and are gravely threatened, as Di Bitetti describes in Chapter 17. The culture, language, lifestyle, and indigenous knowledge of the Mbyá people of Misiones are also at risk of disappearance as the remaining communities in the province are marginalized. In Chapter 18, Sánchez and Giraudo draw on years of experience to describe and defend the Mbyá ways of life.

Designing effective conservation actions entails understanding not only the biological threats to biodiversity but also the sociological, economic, and institutional drivers of biodiversity loss. In Chapter 19, Holz and Placci outline socioeconomic pressures, ranging from those brought on by large timber companies interested in transforming the forest into plantations, to the conditions prompting poor populations to replace forest with nonsustainable agricultural crops, to the dynamics of population growth, all of which have increased pressure on Argentina's natural resources. In Chapter 20, Cinto and Bertolini outline the institutional capacities in government, nongovernment, and academic sectors, pointing to the insufficiency of both human and financial resources throughout. Likewise, in Chapter 21 Giraudo and his colleagues critically analyze the design and management of the protected area system of Misiones, underlining strengths and weaknesses.

In Chapter 22, Rey, a politician deeply committed to environmental issues, makes an urgent plea for international cooperation to create and sustain effective conservation measures throughout Argentina, particularly in the Atlantic Forest region. Rey has been instrumental in promoting the Green Corridor initiative in the Atlantic Forest region and underscores the imperative that the global community recognize the critical role that the forests of Misiones play in sustaining local biodiversity and global integrity.

Chapter 14

Brief History of Conservation in the Paraná Forest

Juan Carlos Chebez and Norma Hilgert

Early History of the Paraná Forest

The Paraná or Misiones forest covered three quarters of the Argentine province of Misiones at the time of Europeans' arrival. The forest's size has fluctuated, shrinking and growing with the cyclical changes in climate that have been witnessed by primitive local communities. The earliest recorded information on inhabitants of the Misiones territory dates back an estimated 6,000 years to the Alto Paraná culture. Since the sixteenth century, the area has been inhabited by Guaraní groups of Amazon origin (Martínez Sarasola 1992).

The subtribes mentioned by the conquistadors and missionaries of the sixteenth century have now disappeared (Métraux 1948; Chapter 32, this volume). These subtribes used to dominate the Atlantic coastal region from Barra de Cananea to Rio Grande do Sul and once spread from there along the Paraná, Uruguay, and Paraguay rivers to the delta of the Paraná River. Currently, however, the Mbyá are the only indigenous group that remain in the province (Chapter 18, this volume).

The principal elements that led to the deterioration of the forest since the Spaniards arrived in Misiones, as well as recent key activities that have contributed to its conservation, are listed chronologically in Table 14.1 (Ambrosetti 1893–94, 1894a, 1984b, 1895, 1896; Fernández Ramos 1934; Cambas 1966; Margalot 1980; Laclau 1994). We review these elements in more detail in this chapter.

The Paraná Forest

Compared with the rest of Argentina, the Misiones province is distinctive because of various unusual features that it shares in part with neighboring regions in Brazil and Paraguay (Figure 14.1). The Paraná phytogeographic province within the Amazon domain contains only two districts: Mixed Tropical Forest and Campos (Grasslands) (Cabrera 1976).

Table 14.1. History of principal contributors to forest deterioration and conservation (with contributions from Holz and Placci, Chapter 19, this volume).

Year	Event
(approx. 3000 BC–1527)	Precolonial Aboriginal populations with low population density. Intensive forest use. Low-impact agriculture, hunting, and fishing.
1527	Spanish exploration and conquest. Sebastián Gaboto travels up the Paraná River to where it meets the Paraguay River.
1609	Arrival of the Jesuits, establishment of permanent settlements under the system of *reducciones* (mission settlements) in the southern part of the Misiones territory. Socioeconomic changes over the next 150 years in stockbreeding and farming practices. Introduction of new crops on irrigated land. Practice of hunting and gathering—especially of *yerba mate*—maintained. White foreigners make increasing inroads into the forest. Land begins to be cleared for farming, hardwoods and firewood are used for home building and industries, roads are built, and mountain paths are opened to facilitate *yerba mate* harvesting. Extensive stockbreeding.
1670	The Jesuit missions grow *yerba mate* successfully. With the passage of time, the Guaraní Indian mission settlements become economically dependent on this production. At their height, the Jesuit mission settlements included more than 100,000 inhabitants (Giberti 1992).
1767	Expulsion of the Jesuits. In Misiones, the order is carried out in the middle of the following year. Abandonment of *reducciones.* Decline in the region's population and in the cultivation of *yerba mate. Yerba mate* is exploited exclusively but insufficiently from the natural populations (Giberti 1992). The abandonment of domestic stockbreeding leads to a proliferation of feral cattle.
1768	Provisional organization of the Colonial Province of Misiones, the first Rio de la Plata province, with jurisdiction over 30 communities.
1800	Dispute with Paraguay over the boundaries of the Misiones territory in the first part of the century. Exploitation of native forest and areas with *yerba mate* (starting in 1800). Concessions to private industry to exploit both native wood and *yerba mate.*
1814	Misiones territory merged with the province of Corrientes. Establishment of the first stable government. Establishment of milling industries in the southern portion of the territory. During the Corrientes government, 38 land owners gain control over most of the land in Misiones Province. Creation of large private properties from government sales. Federal land is kept mostly in the center and south of the province.
1856	The Corrientes government signs contracts for agricultural colonization on Misiones lands. Colonization promoted by the state (in federal lands). Properties have a maximum of 100 ha. They are located mostly in the center and south of the province. Colonization took place along the two main roads, determined by the province's geomorphology.
1864	Enactment of the first regulations governing the harvesting of *yerba mate* from natural populations and prohibiting the establishment of permanent settlements in that environment. Creation of reserves for cattle breeding and for the establishment of sugar- and *yerba mate*–processing facilities.

Table 14.1. Continued

Year	Event
1865	Signing of the Triple Alliance treaty, which made the Paraná River the boundary between Paraguay and Argentina, thus ending the lengthy border dispute.
1875	The Forest Pact established with the Caingangues, who lived in the hills of Misiones Province and were mounting attacks against the expeditions of *yerba mate* harvesters traveling into the Misiones territory. The agreement between Bonifacio Maydana, the *cacique,* or indigenous tribal leader (and stepson of the *cacique* Fracrán), and a *yerba mate*–harvesting expedition led by Brazilian Fructuoso Moraes Dutra provided that, from 1875 on, the banks of the Paraná River, from Corpus to Iguazú—a distance of more than 60 leagues—would be free. Only then did the *yerba mate* harvesters dare to settle along the Argentine side of the river (Peyret 1881).
1876	Plans for colonization launched with the enactment of the Federal Law on Colonization. Subdivision of the provincial territory into the northern region (forest) and the southern region (plains), with public offerings of land at different prices.
1881–1894	Misiones is declared as national territory. Demarcation of the provinces of Corrientes and Misiones. The Corrientes government gradually withdraws its public offices from Misiones. Authorization of the subdivision, sale, and deeding of public lands in Misiones. Sales of land made before federalization nullified because of lack of boundary demarcation and failure to comply with the agreed plans. Europeans who arrive from Brazil officially colonize the province. During this period, *yerba mate* harvesting and cattle breeding are the main economic activities. The province has a population of around 9,000, half of whom are Argentine citizens, whereas the rest are Brazilian or members of indigenous groups.
1895	Loss of 31,000 km^2 of land in a legal judgment won by Brazil. The area of Misiones Province is reduced to 29,801 km^2.
1897–1920	Arrival of the first Polish and Ukrainian immigrants. In Santa Ana, cultivation of *yerba mate* is restarted successfully (Giberti 1992). Before this period, the mainstay of the regional economy had been forestry, using exploited workers and a system of transport in which the logs were lashed together like a raft (known as *jangadas*) on the Paraná and Uruguay rivers.
First phase, 1897–1914	Important immigration to southern and central zone of the province influenced mostly by the state. In the Paraná river valley, the colonization was mostly by private land owners.
Second phase, 1916–1921	Colonization develops from south to north. Expansion in the number of *yerba mate* growers. The largest contribution of *yerba mate* production comes from large properties in the northwest part of the province. Also, occurrence of illegal land occupation. Large private properties in the province with native forests.
1920–1940	Expansion of farmland, which prompts a wave of colonization, especially in the Paraná corridor. The surge in immigration begins to level off only in the late 1940s. Colonization in private lands (from 1919). Land sales through businesses specialized in colonization.
1934	Creation of Iguazú National Park, first natural reserve in Misiones Province. Most of the colonization occurred in the western part of the province, with the most productive soils. The size of most properties is 25 ha.

Continued

Table 14.1. Continued

Year	Event
1950	Establishment of the first paper manufacturing plants. Intensive exploitation of native forest (approx. 1950–1980). Mechanization allowed increases in the amount of wood extraction. As a consequence, native forests were lost and degraded.
1960	Proliferation of tea plantations. Large cellulose industries established in the province. In 1960, the government initiates a system of forest incentives for plantations of exotic species, increasing pressure on the native forest.
1968–1970	Construction of a new section of National Highway 12. By 1970, the paved portion stretches to the immediate vicinity of Iguazú National Park.
1977	Enactment of Provincial Law 854 on forest management, which establishes the foundation for rational management of the forest and creates several forest and seedbed reserves; however, most are not actually implemented.
1985	Creation of the Ministry of Ecology and Renewable Natural Resources of Misiones Province. Last plans for colonization (in the 1980s). Colonization is favored near the borders to decrease immigration from neighboring countries, mostly Brazil.
1987–1989	Creation of eight provincial parks, three municipal natural parks, and two private reserves, laying the groundwork for a provincial system of protected areas.
1990	Filling of the Urugua-í dam (constructed on the Uruguay river) completed.
1991	Farming of 6,500 km^2, of which 2,600 km^2 (previously forestland) are forested with pine, araucaria, eucalyptus, and other species. The principal economic activities of the region are production of *yerba mate,* wood from native forests and plantations, citrus fruits, tea, cattle, and, to a lesser extent, tourism and farming of annual crops.
1992	Enactment of Provincial Law 2932 on Protected Areas, which creates the System of Protected Areas for the province and stipulates the categories of management and guidelines for the administration of the system.
1993	Presence in the province of 3 pulp and paper plants, 6 plywood plants, more than 1,000 sawmills, 1,400 drying facilities, 80 *yerba mate*–processing plants, meat-packing plants, and other processing plants for citrus fruits and oilseeds (Colcombet 1993).
1994	Filling of the Yacyretá dam, constructed on the Paraná River, completed.
1998	Reforesting of old *yerba mate* fields, tall native forests, and forest successions initiated (National Agricultural Technology Institute [Instituto Nacional de Tecnología Agropecuaria, INTA]). Clearing of 150 to 200 km^2 a year, of which slightly less than 50 percent is reforested areas and the rest small holdings (Colcombet 2000).
1999	Provincial Law 3631 creates the Green Corridor of Misiones with a view to safeguarding the largest continuous remnant of the Paraná Forest through its sustainable use. First attempt to legally establish a biological corridor in the country.
2000–2001	Around 9,727.49 km^2 of land in use, of which 4,077.41 km^2 are devoted to cattle farming and 2,450 km^2 to logging; some 8,000 km^2 of *capuera* (secondary forest, resulting from logging and agriculture) and 6,000 km^2 of native forest with active timber extraction under way. Productive system based, in order of importance, on cattle farming, logging in planted forests, and production of *yerba mate,* corn, tea, cassava, tobacco, black beans, citrus, and other crops (Colcombet 2000).

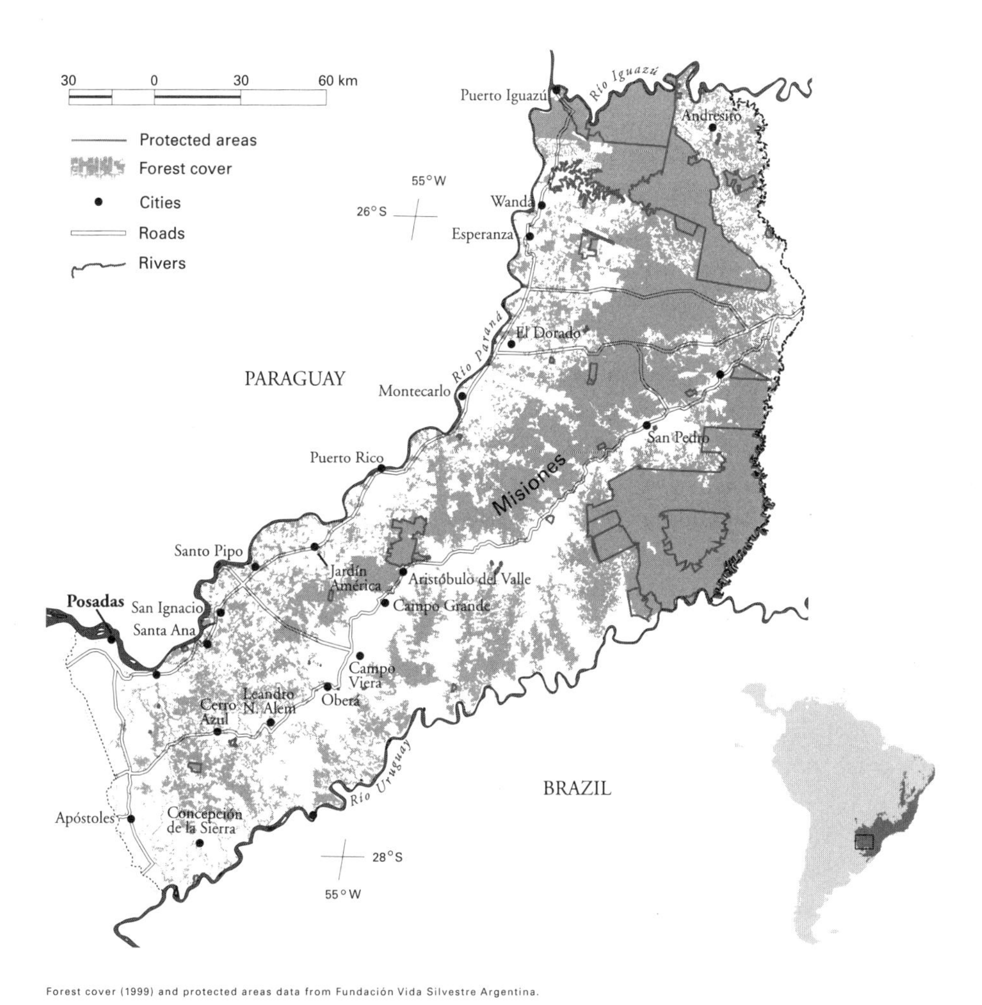

Figure 14.1. Main forested areas, access roads, and towns in the province of Misiones.

The Mixed Tropical Forest District originally occupied almost all of Misiones province and included riparian communities. It is a forest composed of two or three tree strata, one strata of bamboo and shrubs, another of herbs, and another of mosses. Adding to this complexity are various strata of epiphytes and lianas (Cabrera 1976). The upper canopy, formed by emerging trees measuring 25–42 m high, is discontinuous and primarily contains rosewood (*Aspidosperma polyneuron*), black lapacho (*Tabebuia heptaphylla*), black timbo (*Enterolobium contortisiliquum*), Paraná pine (*Araucaria angustifolia*), and cedar (*Cedrela fissilis*). In the far north of the province, rosewood is prevalent, often accompanied by heart palm (*Euterpe edulis*), while in the highlands the Paraná pine is most common (Ragonese and Castiglione 1946; Cabrera 1951, 1976; Martínez-Crovetto 1963; Chebez 1996a).

The Campos (Grasslands) District is distinguished by its great diversity of herbaceous species. Savannas of grass (*Aristida jubata*) are found in highland fields and on lateritic hillsides, red straw grass (*Andropogon lateralis*) in the lowlands, where the organic-rich subsoil rises to the surface, and balsamscale grass (*Elionurus tripsacoides* and *E. muticus*) on the edges of forest, with urunday (*Astronium balansae*) on rocky soils (Martínez-Crovetto 1963; Cabrera 1976, Chebez 1996a). Scattered throughout the grasslands are pure or mixed palm groves composed of *yatay-poñí* palm (*Butia yatay* var. *paraguariensis*) and *pindocito* palm (*Allagoptera campestris*) (Chebez 1996a, 1996b; Fontana 1998).

Current Status

Although the Paraná forest (Interior Atlantic Forest) once occupied almost a million square kilometers stretching over three countries, researchers estimate that today only 58,000 km^2 remain, less than 6 percent of the original area (Placci 2000) (Figure 14.2).

This level of profound loss is particularly evident in the Misiones forest, which was depleted by 38 to 58 percent in the twentieth century. Between 1990 and the present, three different estimates of the remaining forest in Misiones have been advanced: 11,000 km^2 (Laclau 1994), 11,300 km^2 (Chapters 15 and 19, this volume), and 16,000 km^2 (Bertonatti and Corcuera 2000). Differences in the estimates may be due to error or to different criteria and methods. However, based on estimated deforestation rates since 1997 of 150–200 km^2 per year (Colcombet, pers. comm. 2002), the amount of forest cover remaining today is approximately 9,950 km^2, a figure that is generally consistent with the sum of the current data on protected areas (4,597.65 km^2) and the current data on native forests being used for agricultural purposes (approximately 6,000 km^2) (Table 14.2). Because the protected area data include some forestland on which extractive activity is permitted, it probably overlaps somewhat with the data on forest areas allowing agricultural use, meaning that the true remaining area is probably closer to

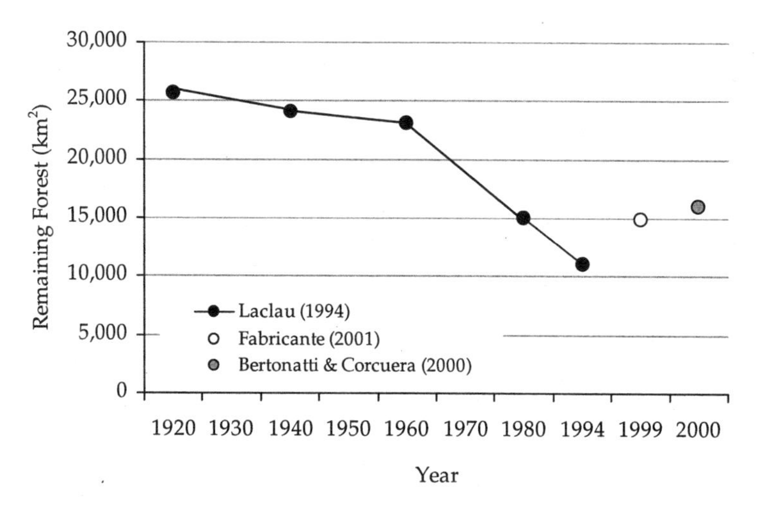

Figure 14.2. Documented estimates of forest remnants from 1920 to the present day in Misiones.

Table 14.2. Land use area for agricultural activities in Misiones, Argentina (modified from Colcombet 2000).

		Extent (km^2)	Approximate Percentage (%)
Misiones Province		29,801.00	
Protected areas		4,597.66	
Remnant forest	Approx.	10,000.00	34
Second-growth forest	Approx.	8,000.00	27
Cattle		4,077.41	14
Annual crops total		1058.29	4
	Manihot	160.00	
	Tobacco	140.00	
	Corn	425.44	
	Black beans	77.55	
	Soya	72.13	
	Vegetables	no data	
	Others	183.17	
Perennial cultures total		2,245.00	8
	Spices	s/d	
	Tea	404.87	
	Other fruits	no data	
	Sugar cane	50.66	
	Yerba mate	1,716.79	
	Citrics	72.68	
Forest management	Plantations	2,450.00	8
	Selective logging	Approx. 6,000.00	Included in remnant forest

10,000 km^2. Environments in the Misiones forest provide 6 percent of Argentina's total agricultural output (Bertonatti and Corcuera 2000), with cattle farming affecting the largest area, followed by cultivation of nonforest perennial crops and forest plantations.

Misiones province, with an area of 29,801 km^2—only 1.1 percent of the total area of Argentina—boasts the nation's greatest plant biodiversity. To date, 2,975 specific or lower taxa of vascular plants have been recorded there, 28 percent of the total number in Argentina (10,625 taxa) (Zuloaga and Belgrano 2000). The forest holds 70.79 percent of the species found in Misiones, and the grasslands account for 58.3 percent. In addition, the forest houses 41.69 percent of the species endemic to Argentina, and the grasslands house 29.2 percent. Seventy percent of faunal species native to Argentina are in Misiones, and 43 percent are found exclusively in the Misiones province. The Misiones forest ecoregion thus ranks second as a national conservation priority (Bertonatti and Corcuera 2000).

History of Use and Abuse

Historically, forest resources have been managed by the National Secretariat of Agriculture, Cattle, Fisheries, and Food (Secretaría de Agricultura, Ganadería, Pesca y Alimentación de la Nación), an agency of the Ministry of Economy, which has given higher priority to agricultural and livestock production than to conservation (Bertonatti and Corcuera 2000). In less than two centuries Argentina lost 72 to 82 percent of its indigenous forest wealth, with the Misiones forest losing 40 percent of its original area in the last century (Laclau 1994). A number of distinct uses of and influences on the forest land have contributed to this continuing decline, from timber extraction to economic recession.

Timber Extraction

Uncontrolled timber extraction has been a primary cause of forest degradation. In Misiones, logging activity was initially concentrated on "proper" timber species (e.g., cedar, *lapacho, petiribí,* and *incienso*), which were felled indiscriminately. When their populations began to decline rapidly, their extraction became unprofitable, and focus shifted to other species. Currently, permits are being issued that authorize each settler or occupant on recently settled public lands to cut 50 trees of timber or commercial value to "improve" the land (García Fernández 1998).

Agriculture and Plantations

Clearing or burning to prepare land for intensive or extensive farming, grazing, or forest plantations has also exacted a heavy environmental toll. Great tracts of land are being cleared to make way for reforesting with exotic timber species. The forestry law of Misiones permits the clearing of slopes of up to 20 percent, but it does not require any treatment to prevent erosion because it is assumed that the planted trees will protect the soil. However, this does not occur during a plantation's first year because agrochemicals prevent binding herb growth; furthermore, such protection certainly will not be available after the trees are felled. When the plant cover is eliminated, soil erodes and loses organic matter as well as the capacity to retain water.

Introduction of Exotic Species

Agricultural activities and human settlement in the forest have also resulted in the introduction of exotic species, which may become invasive (Chapter 33, this volume). In the Misiones region, however, exotic species that currently behave in an invasive manner such as the raisin tree (*Hovenia dulcis*), Chinese privet (*Ligustrum sinense*), elephant grass (*Pennisetum purpureum*), and chinaberry tree (*Melia azederach*) still do not pose a significant threat to the ecosystem (Herrera and Malmierca 1995).

The introduction of certain species may cause some of the native flora or fauna to be viewed as pests. For example, between 2001 and 2006 citrus production is expected to begin on 800 production units in the province, where the planted area currently covers 72.68 km^2 (Colcombet 2000). As a result, various bird species that have lost their native habitat due to the deforestation caused by this expanding citrus production will settle in the citrus plantations, damaging the fruit and thus becoming "pests" in their own environment (Chediack 2000).

Infrastructure Development

Another human activity that affects the Misiones forests is the construction of infrastructure, such as gas pipelines, aqueducts, roads, bridges, and dams. Examples abound of the tremendous negative impact associated with these projects. For example, the Urugua-í and Yacyretá dams have irreparably damaged the community of vertebrates of special value in the second most important locale in the region (Chebez and Casañas 2000), a tremendously sad fact given that the dam's usefulness is estimated to last only 70 years (Corcuera 1997; Bertonatti and Corcuera 2000; Chapter 35, this volume). However, a number of dams are currently planned for Misiones, including Corpus dam, Garabí dam (possibly to be located between Colonia Garabí and Garruchos, Santo Tomé department, in Corrientes province), and Roncador dam (planned for the vicinity of Puerto Panambí, Oberá department, Misiones province) (Figure 35.2, Chapter 35, this volume).

Road infrastructure also has reduced biological diversity. First, poorly constructed roads have increased erosion and landslides. Second, some high-speed roads running through protected areas (e.g., National Route 101 in Iguazú National Park and provincial roads 19 and 211, which pass through the Urugua-í Provincial Park and the Green Corridor, respectively) are heavily traveled and have no speed bumps or animal crossing points. Such roads increase edge effects, facilitate the entry of poachers and illicit palm harvesters, and increase the risk that vehicles will strike the same fauna the parks are intended to protect. In Iguazú National Park, vehicles hit an average of 200 animals every year, including tapirs (*Tapirus terrestris*) and jaguars (*Panthera onca*) (Gil 1995).

Pollution

Environmental assessment and monitoring are almost nonexistent in Argentina, so physical, chemical, and biological pollution go unabated, with problems concentrated in major cities. Water pollution may be the most serious problem, given that there are no controls on the flow of industrial wastewater, which contaminates both underground and surface rivers (Bertonatti and Corcuera 2000). In Brazil, land clearing has affected the availability of drinking water as springs have dried up or become contaminated with agrochemicals (Placci 2000). Over the past 7 years, the use of such chemicals has increased 154 percent in Argentina (Iolster and Krapovickas 1999). No information is available at present on contamination of river basins in Misiones province, but such problems are likely to arise in the future (Iolster and Krapovickas 1999). However, the enormous oil spill in the Brazilian part of the Iguazú River in July 2000 prompted the National Park Administration and a group of other official agencies to conduct biological and chemical monitoring of the Argentine portion of the river.

Population Growth

High population growth is yet another factor affecting forest conservation (García Fernández 1998). The estimated population of Misiones for 2002 is between 795,912 (http://www.misiones.arg) and 995,326 (INDEC-CELADE 1996) and is projected to grow to 1,232,201 by 2010 (Figure 14.3). The provincial population growth rate is 2.6 percent and the birth rate is 3.4 percent a year, almost twice the national rate. The estimated population density in 2000 was 28.9 inhabitants/km^2 (INDEC 2000).

Like the rest of the country, Misiones experiences a great deal of internal migration. Many inhabitants move from rural areas to urban centers, although 32 percent of residents continue to live in communities of fewer than 2,000 inhabitants (*El Territorio,* April 8, 2001). This figure is much higher than the national rural population average, which was estimated at 10–15 percent of the total population in 1990 and is projected to drop to 8 percent by 2025 (INDEC 2000). For example, it is estimated that the 2003 population of Andresito will be 303.13 percent larger than it was in 1991, whereas the population of Puerto Iguazú will grow by 171.45 percent during the same period (Wilde, *El Territorio,* May 6, 2001).

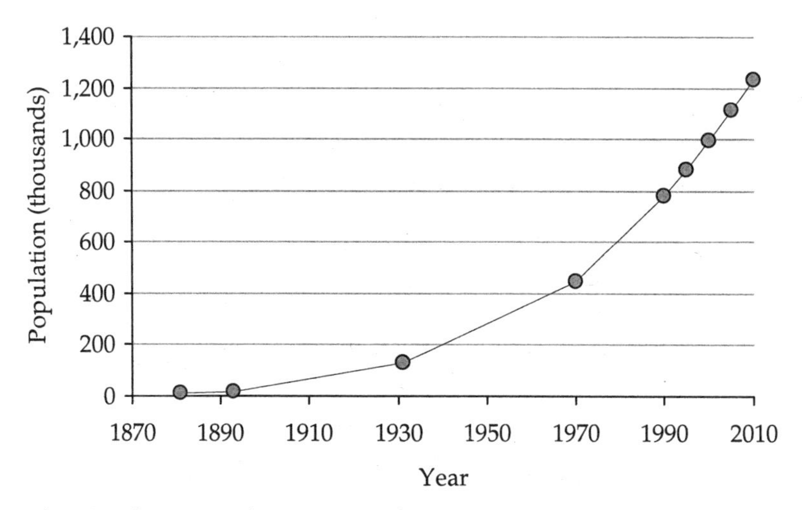

Figure 14.3. Population growth in Misiones from 1881 to 2010, with projections based on census data from 1991 (INDEC 2000).

Economic Recession

The current economic recession in Misiones has prompted many unemployed people to turn to subsistence farming, logging, and subsistence or commercial hunting and fishing. Whether legally or illegally, they engage in these activities in an unplanned and in most cases unsustainable manner (Bertonatti and Corcuera 2000). Every year new families come into the country and illegally occupy lands (public and private), many of which are natural environments, increasing local poverty in Misiones. A critical problem in the province is squatting on public lands, especially in the old forest reserves and on large private properties (García Fernández 1998). The majority of the squatters are migrant farmers from other places in the province and from Brazil (*El Territorio,* October 6, 2000), where migration is not discouraged, as squatters have received support from the Catholic prelates in the dioceses of Iguazú (*El Territorio,* November 15, 2000).

The land-use methods of squatters must not be confused with the traditional migratory farming methods of the Guaraní indigenous people (Martínez-Crovetto 1968), whose mindset regarding the use of the environment is very different from that of the squatters. The attitude of farmers who understand that they are using the land of their people is far different from that of farmers who feel they are using the land of someone who owns land to spare and is not properly putting the land to use for economic gain. The Mbyá-Guaraní people hold the former view, considering themselves part of the forest environment and therefore believing that they would cease to be Mbyá (men) without it. Therefore, the Mbyá see the environment, in all its aspects, not simply as a source of economic wealth but as the very foundation of physical, spiritual, and religious life (Chase-Sardi 1989).

Like squatting, hunting also damages the environment. Moreover, the damage caused by hunting is compounded by the fact that deforestation has already caused forest fauna to lose habitat and food sources. However, the subsistence hunting in the province that provides a source of animal protein for local small farming economies has far less impact on the forest ecosystem than many socially accepted production models, such as intensive farming, stockbreeding, and large-scale commercial farming (Giraudo and Abramson 2000), which destroy the original ecosystem and therefore are much more detrimental to biodiversity.

Subsistence hunting, which tends to be socially and culturally regulated, must be distinguished from indiscriminate sport hunting, which is a problem in the province. According to staff of the Ministry of Ecology and Renewable Natural Resources (MERNR), poaching has increased in recent years in protected areas. In addition to the already prevalent Brazilian hunters, hunters from other parts of Argentina are now coming to Misiones, tempted by the variety of species present in the forest (*El Territorio,* October 6, 2000).

With regard to commercial and sport fishing, the Argentine-Paraguayan Joint Commission on Protection of Fish Resources in 1997 agreed to keep the yearly fishing season closed longer to protect the fish population during the spawning and reproduction period of the dorado (*Salminus maxillosus,* Characidae), spotted sorubim (*Pseudoplatystoma coruscans,* Pimelodidae), pacu (*Piaractus mesopotamicus,*

Serrasalmidae), manguruyu (*Paulicea luetkeni, Pseudopimelodus zungaro zungaro,* Pimelodidae), and *bagre* (a type of catfish) (*Pimelodella* spp., *Pimelodus* spp., Pimelodidae), among other species. According to news reports, however, these populations are not being protected during their spawning period in part because the closed season is not being enforced during the crucial months (*El Territorio,* October 20, 1999; October 25, 2000).

Background to Current Protection and Conservation Efforts

Subtropical and tropical environments are characterized by a complex web of biological processes that must be maintained in a dynamic equilibrium to remain viable. Due to the interdependencies inherent to these complex environments, they are much more susceptible to functional breakdown than those located in cold or temperate latitudes, making it especially challenging to find sustainable ways of using them. Much work remains to be done in Misiones to establish a sustainable balance.

Looking back to the sixteenth century, we find that most who exploited the Misiones forest for commercial purposes were not natives of the region and did not understand the workings of the biome, insisting instead on transplanting unsuitable technologies that continue to be some of the principal causes of environmental degradation. Nevertheless, long before any protected areas were ever formally recognized in Misiones, some people became deeply committed to preserving the environment. Perhaps the first conservationist in the area was Moisés Bertoni, who arrived in Santa Ana in 1884 and later lived in Paraguay (Puerto Bertoni). He is remembered as a disseminator of indigenous knowledge and an agricultural and meteorological advisor who tried to promote "farming without burning." Another paradigmatic figure in the local history of Misiones who strove throughout his lifetime to devise production methods that would not destroy the forest was Alberto Roth, a Swiss immigrant who settled in Santo Pipó. In the mid-twentieth century he was already promoting agricultural practices that would protect the soil, cultivation of *yerba mate* (the plant from where mate drink is produced) under canopy, worm-farming, and other technologies that even now are considered cutting-edge (Peche de Bertoni and Bertoni 1984; Chebez 1996a).

If more people like Bertoni and Roth had worked in the private and public spheres in the past, conservation efforts undoubtedly would have been more successful. But trusting settlers to become guardians of the forest and depending on businesspeople to ensure the self-sustainability of their production practices now seems utopian. Therefore, as a response to the environmental degradation that has occurred since the early twentieth century, many of those who still hold environmental values in high regard have been developing a conservation strategy based on the creation of protected areas (Chebez 1996a; Rolón and Chebez 1998; Chalukian 1999).

Protected Areas

Between the time when the Iguazú National Park was created—the first protected area in Misiones—and 1990, 12 new protected areas were formed. Furthermore, in the 1980s major changes took place in the province that fostered several conservation initiatives. For example, Alberto Roth, one of those responsible for this intense activity, founded the first provincial nongovernment organization devoted to protection of the environment. Roth led a popular movement that vehemently opposed the construction of the Urugua-í dam (Chebez 1996a). Although the dam was ultimately built, the protests mounted against it led to the creation of the Ministry of Ecology and Renewable Natural Resources of Misiones (MERNR) in 1985 and the Urugua-í Provincial Park in 1988, which significantly increased the amount of land under official protection.

The Argentine Wildlife Foundation (Fundación de Vida Silvestre Argentina [FVSA]) also has created a system of six private protected areas and wildlife refuges in Misiones. In 1996, MERNR and the Argentine Ornithology Association (Aves Argentinas/Asociación Ornitológica del Plata) established the Güirá Oga Center for breeding and recovery of threatened birds in Iguazú.

In addition, MERNR has spearheaded several important conservation efforts, including the declaration of the Paraná pine (*Araucaria angustifolia*) and the rosewood (*Aspidosperma polyneuron*) as Provincial Natural Monuments (Provincial Law 2380). This law strictly prohibits the felling, sale, or destruction of any trees of these species or their by-products. Ten fauna species were also designated as natural treasures: the Brazilian merganser duck (*Mergus octosetaceus*), harpy eagle (*Harpia harpyja*), blue-winged macaw (*Ara maracana*), red-spectacled parrot (*Amazona pretrei*), giant otter (*Pteronura brasiliensis*), bush dog (*Speothos venaticus*), howler monkey (*Alouatta guariba*), jaguar (*Panthera onca*), tapir (*Tapirus terrestris*), and giant anteater (*Myrmecophaga tridactyla*). However, a regulatory framework has not been put in place to ensure enforcement of this protected legal

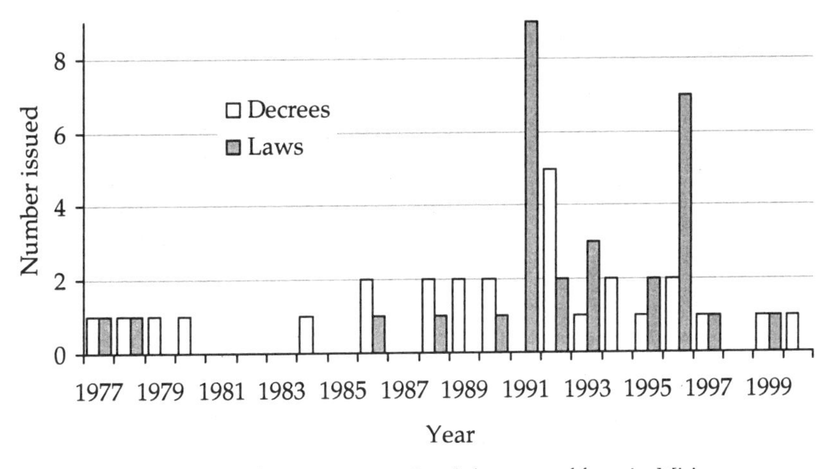

Figure 14.4. Establishment of conservation-related decrees and laws in Misiones.

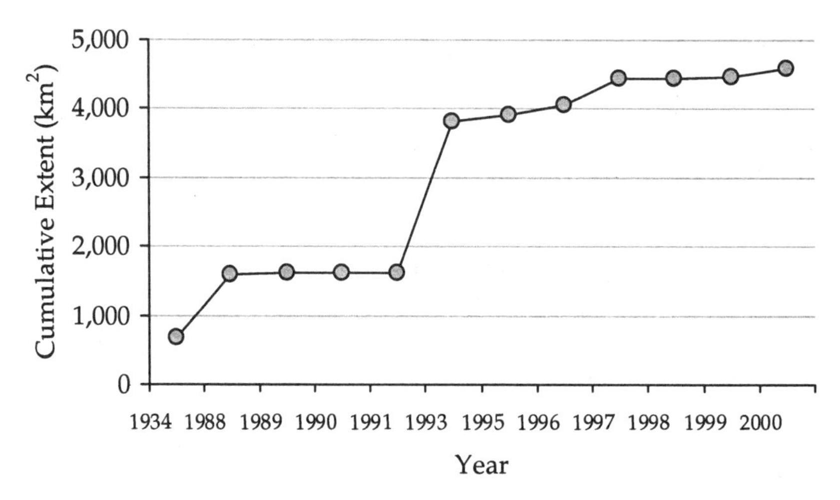

Figure 14.5. Increase in the extent of designated protected areas in Misiones since 1934.

status. The establishment of the MERNR has also streamlined other legislative aspects of conservation (Figure 14.4; Chapter 20, this volume).

Since 1934, 61 reserves have been created under various jurisdictions and domains—national government, provincial government, municipal government, public or private—with a total land area of 4,597.65 km^2. Total protected area has increased gradually over the years, with significant growth in 1934, 1988, 1993, and 1997. Between 1934 and 1989, 17 areas were created (35.35 percent of the present total), and between 1990 and 1995, 23 new areas were designated (52.8 percent of the total area protected), and an additional 18 were added by 2001 (11 percent) (Figure 14.5).

Although it is laudable that so many protected areas have been created, the economic and political realities in Argentina make adequate management of the areas a low priority. Many critical aspects of protected area management are not being given sufficient attention, including the requirements to maintain the integrity of the system within protected areas, the activities that occur in the surrounding environment, the perimeter shape and length, linkage to other sectors in the hydrographic basins involved, and even the populations or communities that need to be protected (Laclau 1994). Instead, most of the protected areas are simply public lands that were "rescued" when the opportunity arose.

In addition, Provincial Law 2932 seeks not only to protect biodiversity but also to create natural areas close to urban centers that provide city-dwellers with a place for recreation in a well-conserved natural setting, preserve the natural landscape, and promote ecotourism (Article 4, paragraphs j, k, and n)—objectives that do not necessarily demand much precision in determining the size and shape of the protected area. However, with respect to conserving biodiversity, the size, shape, and connectivity of protected areas are all very important in determining their conservation value. For instance, small fragments (less than 200 km^2) are inadequate to maintain viable populations of large mammals as well as birds, butterflies, and trees with low population densities or with naturally fragmented distribution (Chapter 31, this volume).

The Green Corridor

To prevent further fragmentation in the protected areas in the northern and eastern parts of the region, in the early 1990s conservationists began devising a strategy to consolidate the largest possible portion of this remnant to ensure its continued existence. This strategy also included protecting the upper drainage basins of major streams in the region, which supply water to large populations, and preserving an environment that affords excellent opportunities for ecotourism development (Chebez and Gil 1993; Chebez 2000; Gil and Chebez 2001).

This initiative eventually gave rise to a legislative proposal for the creation of the Green Corridor of Misiones province, which was approved by the provincial legislature as Law 3631, Decree 218/99 (Chapter 20, this volume). The law envisages sustainable management and use for 11,000 km^2, stressing the preservation of wooded areas (Figure 22.1, Chapter 22, this volume). However, several articles in both the law and in Regulatory Decree 25/01 provide state subsidies for cultivating exotic species in the conservation area, a contradiction that has hindered efforts to obtain external funding and has delayed implementation of the corridor (*El Territorio,* March 8, 2001). The current economic emergency in Argentina has also prevented funds that have been legally allocated to this project from being applied. Budgetary constraints, frequent turnover of authorities, and policies that sometimes undo conservation accomplishments oblige private environmental organizations to complement, correct, or criticize the work of the public sector when necessary (Laclau 1994; Chalukian 1999; Bertonatti and Corcuera 2000; FVSA and WWF 2000).

Many of the newly protected areas are still being organized. Most have minimal structures, need more staff, and lack operational plans, although a good proportion of the provincial parks already have approved management plans (MERNR Resolution 900/99). However, though not fully functional, the existence of numerous recently created areas reflects the government's political will to improve the regional environmental situation and ensure the survival of conservable sites.

Civil Society

Finally, ignorance has proven to be a terrible enemy of the forest. When most of the currently protected areas were designated, the prior occupants were displaced. Lacking access to information about the true value and benefit of conservation for their lives, many of these farmers understandably felt as if a source of great potential wealth was set aside without a reason (Laclau 1994). However, they are often misinformed about conservation by politicians or representatives of ecclesiastic circles and do not see protection as a benefit (*El Territorio,* September 15, 2000).

Environmental nongovernment organizations (NGOs) have joined forces in a group known as Misioneros Ecologistas (Misiones Environmentalists). NGOs also work together in the Misiones chapter of the FVSA. Additionally, a provincial ecology commission now exists to advise the MERNR, although it meets irregularly. Another organization, the Network of Ecology Associations (Red de Asociaciones Ecologistas), holds its own meetings. Finally, two prominent national NGOs with local representation in Puerto Iguazú are active in the province: the Argentine Wildlife Foundation (Fundación Vida Silvestre Argentina) and the ornithology association mentioned earlier, Aves Argentinas/Asociación Ornitológica del Plata. Nevertheless, much remains to be done to meet the conservation needs of the Paraná forest.

Conclusions

The fragile and biologically diverse Paraná forest is at a crossroads. At the provincial level, people are slowly becoming aware of the forest's importance and the need to conserve it. Misiones, which traditionally has been known for the vast hunting activity that takes place there and for the degradation and destruction of its forest (*El Territorio,* October 6, 2000), is now considered to be in the vanguard nationally in creating effective legal structures and tools for conservation. Despite its shortcomings, the province's system of protected areas is one of few worthwhile systems at the provincial level in Argentina. Indeed, the fact that Misiones is the only province with a Ministry of Ecology is clear evidence of its emerging conservationist orientation.

Still, all this is insufficient in the face of the mounting threats to the forest's survival. Recent stunning satellite images, in which the green of the Misiones forest appears in stark contrast to the cultivated and cleared lands of eastern Paraguay and southern Brazil, indicate that we still have time not only to safeguard this environment but also to explore methods of sustainable use. But we must not delude ourselves: the process of deterioration is ongoing, in large part because many people see high-impact development projects and forest plantations as an easy way to create greatly needed jobs, despite the fact that most of these jobs are temporary.

Furthermore, the province still has contradictory policies and lacks a common strategy for forest conservation. On one hand, politicians and government officials have praised the substantial increase in area planted with coniferous monocultures as a source of economic stimulus. On the other hand, they have announced a decision to create new provincial parks and have identified the development of a regulatory framework for application of the Green Corridor law as a major accomplishment. Such contradictions clearly show that political officials need to be better trained in how to coherently manage the province's natural resources.

Argentina is still not a forest-conscious country, for reasons that are beyond this chapter's scope of analysis. However, Misiones cannot be expected to act alone in confronting its environmental problems. Rather, because of this small northeastern province's critical contribution to national biodiversity, much more must be done at the national level to conserve the Misiones forest.

References

Ambrosetti, J. B. 1893–94. Viaje a las Misiones Argentinas y Brasileras en el Alto Uruguay. *Revista del Museo La Plata* III: 417–448; IV: 289–336; V: 225–250.

Ambrosetti, J. B. 1894a. Segundo viaje a Misiones por el Alto Paraná e Iguazú. *Boletín Instituto Geográfico Argentino* XV: 18–114, 247–304.

Ambrosetti, J. B. 1894b. Los Indios Cainguá del Alto Paraná (Misiones). *Boletín del Instituto Geográfico Argentino* 15: 661–744.

Ambrosetti, J. B. 1895. Los indios Kaingangues de San Pedro (Misiones) con un vocabulario. *Revista Jardín Zoológico Buenos Aires* 2: 305–387.

Ambrosetti, J. B. 1896. Tercer viaje a Misiones. *Boletín del Instituto Geográfico Argentino* XVI: 391–523.

Bertonatti, C. and Corcuera, J. 2000. *Situación ambiental Argentina 2000.* Buenos Aires: Fundación Vida Silvestre Argentina.

Cabrera, A. L. 1951. Territorios fitogeográficos de la República Argentina. *Boletín de la Sociedad Argentina de Botánica* (La Plata, Argentina) 4: 21–65.

Cabrera, A. L. 1976. Regiones fitogeográficas Argentinas. In: *Enciclopedia agricultura y jardinería.* pp. 2, 85. Buenos Aires: Ed. ACME.

Cambas, A. 1966. Cronología histórica de Misiones (período 1527–1930). In: Margalot, J. A. 1980. *Geografía de Misiones.* pp. 81–102. Buenos Aires: Edición del Autor.

Chalukian, S. C. 1999. *Cuadro de situación de las unidades de conservación de la selva Paranaense.* Unpublished report of the Fundación Vida Silvestre Argentina.

Chase-Sardi, M. 1989. El Tekoha: su organización social y los efectos negativos de la deforestación entre los Mbyá-Guaraní. *Suplemento Antropológico* (Asunción): 33–41.

Chebez, J. C. 1996a. *Fauna Misionera: catálogo sistemático y zoogeográfico de los vertebrados de la Provincia de Misiones (Argentina).* Buenos Aires: Ed. Literature of Latin America (L.O.L.A.).

Chebez, J. C. 1996b. Misiones Ñu. Campos Misioneros. Algo más que el confín de la selva. *Nuestras Aves* (Asociación Ornitológica del Plata, Buenos Aires) 34: 4–16.

Chebez, J. C. 2000. Un "Corredor Verde" para salvar a la selva. In: Bertonatti, C. and Corcuera, J. (eds.). *Situación ambiental Argentina 2000.* Buenos Aires: Fundación Vida Silvestre Argentina.

Chebez, J. C. and Casañas, H. 2000. Áreas claves para la conservación de la biodiversidad de la Provincia de Misiones, Argentina: fauna vertebrada. In: FVSA (Fundación Vida Silvestre Argentina) and WWF (World Wildlife Fund) (Coordinadores). *Memorias del taller "Visión Biológica de la Selva Atlántica."* Foz do Iguaçú: Fundación Vida Silvestre Argentina

Chebez, J. C. and Gil, G. E. 1993. Misiones Hoy: al rescate de la selva. *Nuestras Aves* (Asociación Ornitológica del Plata, Buenos Aires) XI(29): 5–9.

Chediack, A. 2000. La citricultura, los celestinos y los loros en Tucumán. In: Bertonatti, C. and Corcuera, J. (eds.). *Situación ambiental Argentina 2000.* pp. 227–229. Buenos Aires: Fundación Vida Silvestre Argentina.

Colcombet, L. 1993. *Informe Agropecuario Provincia de Misiones.* Informe Inèdito.

Colcombet, L. 2000. *Destino de las tierras ocupadas por propiedades de la Provincia de Misiones.* Unpublished report. Montecarlo, área de Extensión: INTA, EEA.

Corcuera, J. 1997. Cuenca del Plata, una integración incompleta. *Revista Vida Silvestre* (Fundación Vida Silvestre Argentina, Buenos Aires) 54: 6–15.

Fernández Ramos, R. 1934. *Misiones: a través del Primer Cincuentenario de su Federalización.* Posadas: Editado por el Autor.

Fontana, J. L. 1998. Análisis sistemático-ecológico de la flora del sur de Misiones (Argentina). *Candollea* 53: 211–300.

FVSA (Fundación Vida Silvestre Argentina) and WWF (World Wildlife Fund). 2000. *Memorias del taller "Visión Biológica de la Selva Atlántica."* Foz do Iguaçú, Brasil: G. Placci.

García Fernández, J. 1998. *Corredor Verde Misionero y Parques Provinciales Moconá, Urugua-í y Cruce Caballero.* Final report. Programa de Desarrollo Institucional Ambiental. Núcleo Provincial Misiones. Subprograma B. Sistema de Control Ambiental. Buenos Aires: Ministerio de Ecología y Recursos Naturales Renovables de Misiones.

Giberti, G. C. 1992. Yerba mate (*Ilex paraguariensis*). In: Hernández Bermejo, J. E. y L. León (eds.). *Cultivos marginados otra perspectiva de 1492.* Colección FAO, Producción Vegetal 26: 245–252 pp. Roma.

Gil, G. E. 1995. Rutas misioneras por mal camino. *Propuesta Ecológica* (Posadas, Argentina): 46–50.

Gil, G. E. and Chebez, J. C. 2001. Iguazú, el corazón del Corredor Verde: la gran selva que une tres naciones. *National Geographic, En Español* Marzo (Sección Inicial): 5–10.

Giraudo, A. R. and Abramson, R. R. 2000. Diversidad cultural y usos de la fauna silvestre por los pobladores de la selva misionera: ¿una alternativa de conservación? In: Bertonatti, C. and Corcuera, J. (eds.). *Situación ambiental Argentina 2000.* pp. 233–243. Buenos Aires: Fundación Vida Silvestre Argentina.

Herrera, J. and Malmierca, L. 1995. *Relevamiento de especies vegetales exóticas en el área Cataratas.* Unpublished report, Centro de Investigaciones Ecológicas Subtropicales.

INDEC. 2000. *Anuario estadístico de la República Argentina 2000.* INDEC Vol. XVI. Buenos Aires: Instituto Nacional de Estadística y Censos (INDEC).

INDEC-CELADE. 1996. *Serie análisis demográfico* No. 7. Buenos Aires: Instituto Nacional de Estadística y Censos (INDEC).

Iolster, P. and Krapovickas, S. 1999. *Los plaguicidas en uso en la Argentina: riesgos para las aves silvestres.* Temas de Naturalezas y Conservación No. 2, Monografías Asociación Ornitológica del Plata. Buenos Aires: BirdLife International, Aves Argentinas y Embajada Real de los Países Bajos.

Laclau, P. 1994. *La Conservación de los recursos naturales y el hombre en la Selva Paranaense.* Boletín Técnico 20. Buenos Aires: Fundación Vida Silvestre Argentina, Fondo Mundial Para la Naturaleza.

Margalot, J. A. 1980. *Geografía de Misiones.* Buenos Aires: Edición del Autor.

Martínez-Crovetto, R. 1963. Esquema fitogeográfico de Misiones (República Argentina). *Bonplandia* (Corrientes, Argentina) I(3): 171–215.

Martínez-Crovetto, R. 1968. Notas sobre la agricultura de los indios guaraníes de Misiones (República Argentina). *Etnobiológica* (Corrientes, Argentina) 10: 1–11.

Martínez Sarasola, C. 1992. *Nuestros paisanos los Indios.* Buenos Aires: Emecé.

Métraux, A. 1948. The Guaraní. In: Steward, J. H. (ed.). *Handbook of South American Indians.* Vol. 3: *The tropical forest tribes.* Bulletin 143: 69–94. Washington, DC: Bureau of American Ethnology.

Peche de Bertoni, M. A. and Bertoni, J. E. 1984. *El vigía de la selva.* Posadas: Edición del Autor.

Placci, G. 2000. El desmonte en Misiones: impactos y medidas de mitigación. In: Bertonatti, C. and Corcuera, J. (eds.). *Situación ambiental Argentina 2000.* pp. 349–354. Buenos Aires: Fundación Vida Silvestre Argentina.

Peyret, A. 1881. *Cartas sobre Misiones.* Buenos Aires: Imprenta de La Tribuna Nacional

Ragonese, A. E. and Castiglione, J. A. 1946. Los pinares de *Araucaria angustifolia* en la República Argentina. *Boletín de la Sociedad Argentina de Botánica* 1: 126–147.

Rolón, L. H. and Chebez, J. C. 1998. *Reservas naturales misioneras.* Misiones, Argentina: Editorial Universitaria, Universidad Nacional de Misiones, Ministerio de Ecología y Recursos Naturales Renovables.

Zuloaga, F. O. and Belgrano, M. 2000. Flora de Misiones. In: FVSA (Fundación Vida Silvestre Argentina) and WWF (World Wildlife Fund) (coordinadores). *Memorias del taller "Visión Biológica de la Selva Atlántica."* Foz do Iguaçú, Brasil. Fundación Vida Silvestre Argentina and World Wildlife Fund.

Chapter 15

Biodiversity Status of the Interior Atlantic Forest of Argentina

Alejandro R. Giraudo, Hernán Povedano, Manuel J. Belgrano, Ernesto Krauczuk, Ulyses Pardiñas, Amalia Miquelarena, Daniel Ligier, Diego Baldo, and Miguel Castelino

The Interior Atlantic Forest extends over nearly 825,000 km^2 in northeastern Argentina, Paraguay, and southeastern Brazil (Laclau 1994; Morello and Matteucci 1999; Chapter 5, this volume). Known in Argentina as the Paraná or Misiones Forest, the Interior Atlantic Forest is one of a group of moist tropical forests in South America and is recognized as an area of high endemism and diversity, both in neotropical regions and globally (ICPB 1992; Laclau 1994; Stotz et al. 1996). The Interior Atlantic Forest is a tropical semideciduous seasonal forest with two climatic seasons, including a tropical season with intense summer rains and a subtropical season with no dry period. Winter temperatures can fall below 15°C and can cause a physiological drought (Veloso et al. 1991). The Interior Atlantic Forest is generally considered a subdivision of the Atlantic Forest, which stretches across the eastern slopes of Serra do Mar along the coast to northeast Brazil, and receives more abundant and regular rainfall than elsewhere (Martínez Crovetto 1963; Sick 1984; Stotz et al. 1996). The two forests have a significant number of species in common (Martínez Crovetto 1963), but each also has its own distinctive features (Cabrera and Willink 1973) (Figure 5.1, Chapter 5, this volume).

The Interior Atlantic Forest has suffered profound modifications and deforestation in the past 80 years. Indeed, 94 percent of the original forested area has been lost or drastically fragmented, and the remaining 6 percent shows varying degrees of modification (Laclau 1994). This ecosystem is now believed to be one of the most threatened in the world, and the province of Misiones in Argentina retains some of the most continuous and least changed portions of the Interior Atlantic Forest (Figure 19.1, Chapter 19, this volume).

The large and unfragmented forest area of Misiones has a crucial role to play in the conservation of the Interior Atlantic Forest because it offers real opportunities to preserve extensive areas, create corridors and reserves, and promote sustainable development in nonprotected areas. However, the factors responsible for loss and fragmentation of the forest—including deforestation and population growth, with the attendant pressure on resources—have not abated (Laclau 1994; Dalmau 1995; Chapter 19, this volume). In this chapter we provide a detailed review of the status of forest ecosystems and the diversity and distribution of species in the Atlantic Forest in Argentina. However, the inventories presented here reflect only a fraction of the information that researchers and policymakers need to construct effective actions to conserve the remaining biodiversity in this valuable region. We therefore conclude by discussing the necessary but difficult task of selecting robust indicators for monitoring biodiversity and other challenges that must be overcome to protect the Interior Atlantic Forest.

Biogeographic Formations

Many researchers recognize two phytogeographic formations in Misiones: the Mixed Forest District and the Campos (Grasslands) District (Cabrera 1976). Four subdivisions lie within the Mixed Forest District, while the Campos District is not subdivided (Table 15.1 and Figure 15.1).

Overall, the Mixed Forest District includes evergreen trees ranging in height from 20 to 50 m, with strata of smaller trees and a dense undergrowth of bamboo, or arborescent ferns. Species of the families Lauraceae (*Ocotea* and *Nectandra*), Fabaceae (e.g., *Lonchocarpus, Schizolobium, Parapiptadenia*), Myrtaceae, and Meliaceae prevail, but no one species is dominant because 50 or more tree species share space in densities that vary as a result of small soil or microclimatic differences. Other common species include cedar (*Cedrella fissilis*), guatambu (*Balfourodendron riedelianum*), black laurel (*Nectandra megapotamica*), cancharana (*Cabralea canjerana*), anchico (*Parapiptadenia rigida*), pink tabebuia or trumpet tree (*Tabebuia ipe*), golden trumpet tree (*Tabebuia pulcherrima*), the queen palm (*Syagrus romanzoffiana*), and 200 others of varying sizes that make up at least three different strata. An important element in these forests is bamboo, including tacuarembó (*Chusquea*), tacuaruzú and yatevó (*Guadua*), and tacuapí (*Merostachys*). Within these general characteristics of the district, four formations with additional special characteristics can be distinguished (Table 15.1).

The rivers of Misiones, with their clear waters, stone-filled channels, rapids and falls, and the deep alluvial soils on their banks, also account for some of the particularities of the flora and fauna of the region.

Extent and Status of the Forest Cover

Originally, the Interior Atlantic Forest in Misiones covered an area of about 25,700 km^2 (Laclau 1994), or 86 percent of the province. Including the forest islets in northeastern Corrientes, the total area would have been about 26,450 km^2 (Morello and Matteucci 1999). Estimates of the remaining area of the Interior Atlantic Forest in Misiones vary depending on the criteria applied and the sources of information, ranging from 11,000 km^2 to 17,000 km^2 (Laclau

Table 15.1. Biogeographic subdivisions of the Interior Atlantic Forest in Misiones, Argentina.

Formations (Extent)	Distribution	Characteristics	Species	Conservation Status
Rosewood and Assai palm forest (2,300 km^2).	In the extreme north of Misiones on the pediplain of the Paraná and Iguazú rivers (Fig. 15.1). Its southern boundary approximately follows the Urugua-í River.	It is characterized by the former abundance of rosewood (*Aspidosperma polyneuron*) and Assai palm (*Euterpe edulis*), although now only relicts of these species are found in some areas of Iguazú National Park, the Urugua-í, and neighboring areas.	Several plant species are endemic to this type of forest, including *Begonia descoleana, Cyperus andreanus var. yguazuensis, Peperomia misionense,* and *Podostemum comatum.* There are also some animals found exclusively or more often in this formation. These include several snakes, such as *Chironius exoletus, Oxyrhopus petola, Sibynomorphus mikanii,* and *Bothrops moojeni,* and some birds, such the blue-bellied parrot (*Triclaria malachitacea*), brown-breasted bamboo tyrant (*Hemitriccus obsoletus*), and streak-capped antwren (*Terenura maculata*).	Partly protected by Iguazú National Park, but highly impacted in the eastern region.
Laurel and guatambu forest (3,200 km^2)	Located on the pediplain of the Paraná River in the central and northern portion of Misiones, merging with the rosewood and Assai palm forest in a complex transition in the north (Fig. 15.1). To the west, it is bounded by the Paraná River, by the mountains to the east, and by the Campos District to the south.	Although guatambu (*Balfourodendrum riedelianum*) and laurel (*Nectandra* spp.) are abundant, they are also common in other formations, and they might therefore more appropriately be called mixed forests (Placci and Giorgi 1993). It encompasses the gullies and flood valley of the Paraná and serves as a corridor for tropical species and also for species from open environments that inhabit the flood-prone scrublands and shrublands that form in the flood valley.	Species from the Humid Chaco and Pantanal regions have been recorded such as the snakes *Sybinomorphus turgidus* and *Eunectes notaeus* (yellow anaconda) and a couple of frogs (*Hyla raniceps, H. nana*) (Giraudo, pers. obs., 2001). Bamboo (*Guadua paraguayana*) and soapberry (*Sapindus saponaria*) are typically found along the banks of the Paraná and large tributaries (Iguazú) (J. Herrera, pers. comm. 2001). Some 30 species of birds native to the Paraná River and its tributaries have been documented (Martínez Crovetto 1963). Several species, such as the band-tailed manakin (*Pipra fasciicauda*), are found exclusively or more often in this formation.	The lowland forests are characterized by the extensive influence of the Paraná river (several kilometers inland) and the presence of deep red soils. Their conservation status is precarious, as the Paraná River was one of the first settlement routes in Misiones and suffered significant deforestation. This area has the best soils in the province for farming, and it has therefore experienced greater population pressure than other parts of Misiones.

Montane forest (15,970 km^2)	The central eastern and northeastern portions of Misiones are characterized by hills and mountains (the Central or Misiones, Imán, and Morena and Victoria mountains) and escarpments, with broken valleys and steep slopes (gradients ranging from 10 to more than 30 percent). This subunit is the largest forest formation in Misiones and also the largest remaining area of Atlantic Forest (Fig. 15.1).	The forest formations in this region are similar to the laurel and guatambu forests with some distinctive elements and an abundance of tree ferns known locally as "chachi" (*Trichipteris, Alsophyla,* and *Dicksonia*). For that reason, it has been labeled as the Tree Fern District (Martínez Crovetto 1963).	Several species appear to be exclusive or more common to this formation, including snakes (*Xenodon neuwiedi* and *Oxyrhopus clathratus*) (Giraudo 2001), birds (the rufous-tailed antthrush, *Chamaeza ruficauda*) (Castelino, pers. obs. 2001), and flora (*Nothoscordum moconense* and *Cyperus burkartii, Dyckia* spp.). Some bird species also seem to be limited to or more common on the mountain slopes along the Uruguay River. These include the red-rumped warbling finch (*Poospiza lateralis*), the thick-billed saltator (*Saltator maxillosus*), and the diademed tanager (*Stephanophorus diadematus*) (Castelino, Krauczuk, and Giraudo, pers. obs. 2000). Some fish, such as *Astyanax ojiara,* are also exclusive to the Uruguay (Azpelicueta and García 2000).	The tree ferns have been affected by their proximity to Oberá. In addition, the ferns are exploited as ornamental plants, and they are affected by widespread destructive forestry practices. At present, they are found mainly in eastern Misiones, in Yabotí Biosphere Reserve, on Guarani land, and in the Moconá and Piñalito provincial parks.
Montane Araucaria Forest (2,100 km^2)	Occupies the northeast region, which is the highest part of the province (Fig. 15.1). This formation is partly transitional between the Mixed Forests and the Paraná pine forests of the Brazilian Planalto (plateau) (Martínez Crovetto 1963; Cabrera 1976).	It is characterized by the presence of Paraná pine also known as araucaria (*Araucaria angustifolia*) and montane pine (*Podocarpus lambertii*). These are not true pines (Family Pinaceae) but are locally known by that name.	These two "pine" species and some animal species are exclusive to this formation, such the Araucaria tit-spinetail (*Leptasthenura setaria*) and the pit viper cotiara (*Bothrops cotiara*). Birds of the family Rhinocryptidae that inhabit this formation, such as the spotted bamboowren (*Psilorhamphus guttatus*) and the mouse-colored tapaculo (*Scytalopus speluncae*), are related to Andean-Patagonian lineages (Sick 1984).	Montane araucaria forests were decimated because of the economic value of the Paraná pine. Today only a few hundred hectares of highly fragmented and isolated forestland remain.

Continued

Table 15.1. Continued

Formations (Extent)	Distribution	Characteristics	Species	Conservation Status
Campos (Grasslands) District (3,800 km^2)	This district extends from southern Misiones to the northeastern part of the neighboring province of Corrientes and forms a gradual transition between the Paraná and Chaco biogeographic provinces (Cabrera 1976).	It is characterized by savannas, which alternate with woodlands and forests of urunday trees (*Astronium balansae*), a species characteristic of the humid Chaco. Urunday colonizes grasslands that are infrequently burned. The range limits of species from Atlantic Forest, the Chaco and Pampas regions converge in this district (Giraudo 2001). The forest formations take the form of islets, or "capones," and galleries along the banks of watercourses, in a vast matrix of grasslands or savannas known locally as campos that dominate in terms of area.	The highly diverse Campos District includes woodlands, Urunday forests, diverse types of grassland and scrubland, wetlands, and groves of yatay palm (*Butia yatay* subsp. *Paraguayensis* and *Allagoprtera campestris*). Some 1,074 plant species have been recorded in this district, as have some 700 species and subspecies of terrestrial vertebrates (i.e., amphibians, reptiles, birds, and mammals), which together represent approximately 40 percent of all the life forms known in Argentina, although the area of the Campos District amounts to less than 0.7 percent of the total area of Argentina (Krauczuk 1996; Fontana 1998, 2000; Giraudo 1996, 2001). Fifty-nine snake species live in the district, twice the number in several of the Mixed Forest formations, which range between 25 and 33 species (Giraudo 2001).	Some plant species exclusive to Argentina in this formation are the dwarf palms pindocito (*Allagoptera campestris*), yatay (*Butia yatay* subsp. *paraguayensis*), *Qualea cordata* (Vochysiaceae), *Agarista paraguayensis* (Ericaceae), and *Parodia schumanniana* (Cactaceae). As for fauna, some exclusive species include snakes (*Rachidelus brazili* and *Clelia quimi*), turtles (*Phrynops vanderhaegei*), and birds such as burnished-buff (rufous-crowned) tanager (*Tangara cayana*) and the least nighthawk (*Chordeiles pusillus*) (Krauczuk 2000). Recently, three new species have been documented: the snake *Apostolepis quirogai* (Giraudo and Scrocchi 1998) and two amphibians, a toad (*Melanophryniscus* sp.) (Baldo and Basso 2000) and a frog (*Scinax* sp.) (Faivovich and Baldo, pers. comm. 2001).

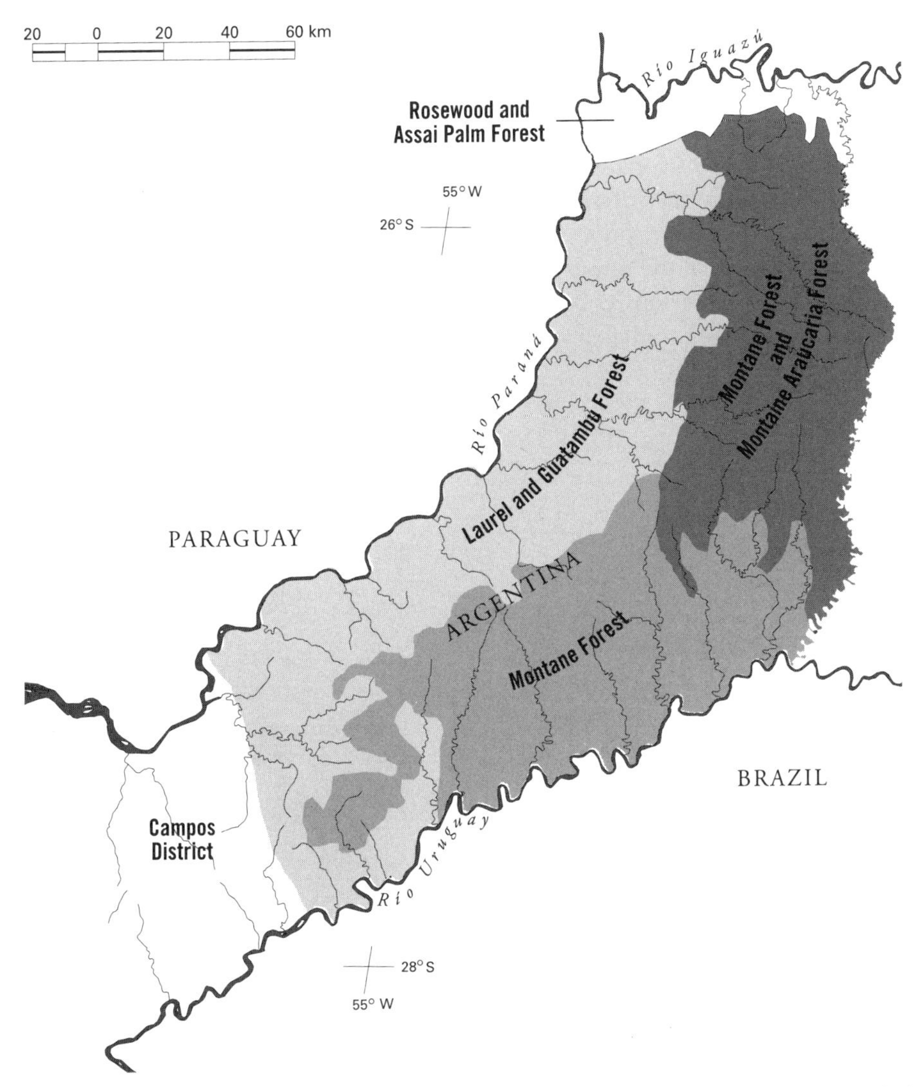

Figure 15.1. Approximate boundaries of biogeographic formations of Misiones (modified from Giraudo 2001).

1994; Morello and Matteucci 1999; Perucca and Ligier 2000). Holz and Placci (Chapter 19) recently constructed an estimate of 11,303.04 km^2, representing 57.5 percent of the original area, which is perhaps the most reliable estimate because it is based on more recent sources of information. Approximately 7,000 km^2 (between 41 and 55 percent) of the forests have been subject to selective cutting (Morello and Matteucci 1999; Perucca and Ligier 2000).

Two large forest blocks in Misiones are of key importance to the overall conservation of the Atlantic Forest, both because of their substantial size and because of the possibility of maintaining corridors between them. The first block is in the central northern sector of the province, including an area of 5,500 to 6,000 km^2 (Figure 15.2). The second forest block (the eastern block) is the region of the Yabotí

Biosphere Reserve, covering a forest area of some 3,000 km^2. These areas sustain significant populations of species with large space needs (Chapter 16, this volume).

The heterogeneity of the Atlantic Forest in Misiones results from relief, altitude, types of soils, influence of watercourses (rivers and streams), microclimates, and factors subject to a wide range of spatial variation in the diverse and irregular landscape. In addition, different management interventions, such as selective logging of some 40 timber species, have influenced the composition of the forests.

In fact, almost all the forests in protected natural areas in Misiones underwent selective logging previously. Nevertheless, they continue to hold high diversity (Protomastro 2001) and maintain populations of threatened species and rare

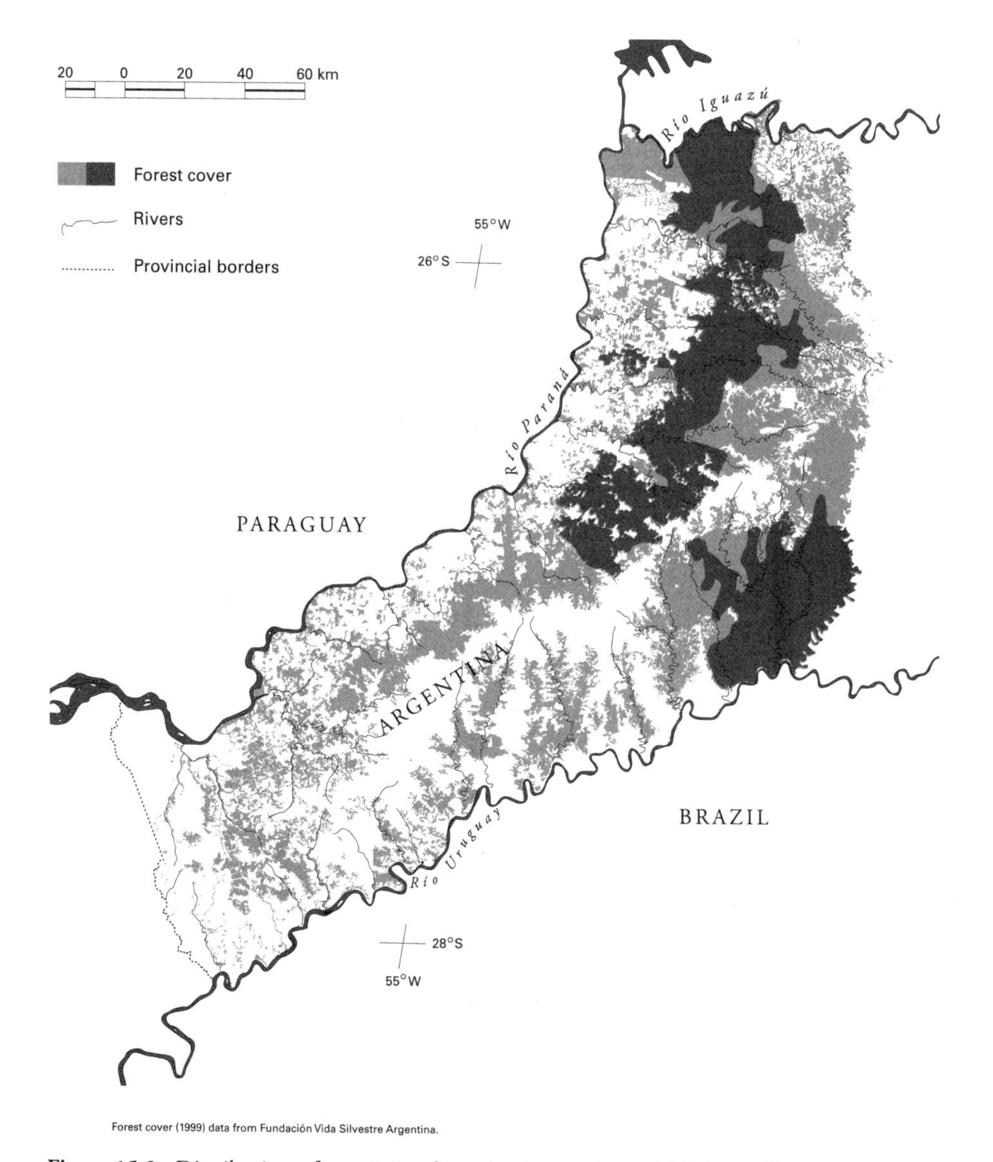

Figure 15.2. Distribution of remaining forest in the province of Misiones (from Perucca and Ligier 2000). Dark shading indicates forest in best conservation state.

wildlife. Recently, 8,940.2 km^2 (30.3 percent) of the forest cover in Misiones was classified as forest in good conservation status, distributed in blocks ranging in size from 0.09 ha to 1,321.50 km^2. The remaining 7,171.19 km^2 (24.3 percent) consists of thinned forests distributed in blocks ranging from 0.09 ha to 513.61 km^2 (Perucca and Ligier 2000). The conserved and modified forests fuse into patch mosaics. Significant fragmentation has occurred in several regions, with the laurel and guatambu, montane, and montane araucaria forests suffering the greatest destruction (Table 15.1).

Species Diversity, Endemism, and Distribution Patterns

The best-known groups in the Interior Atlantic Forest and associated grasslands are vascular plants and vertebrates, which together attest to the exceptional biodiversity in the region. Although Misiones covers only 1.1 percent of Argentina's total land area, it is the country's most diverse ecoregion, holding 3,148 taxa of vascular plants and 1,124 of vertebrates, and also accounting for 52 taxa of vascular plants and 9 of vertebrates exclusive to Argentina (Table 15.2). These figures underscore the importance of the Interior Atlantic Forest as an area of megadiversity in Argentina, and in the world as a whole.

Vascular Plants

Of the 3,148 taxa of vascular plants found in Misiones (taking into account species and infraspecific taxa), 792 taxa are found only in this province (7.33 percent of the flora in Argentina). Misiones shares a high number of

Table 15.2. Number of taxa of vascular plants and vertebrates in the Interior Atlantic Forest of Argentina and their abundance in relation to continental Argentina (Galliari et al. 1996; Zuloaga et al. 1999; Lavilla et al. 2000; Straneck and Carrizo 1992; Miquelarena, pers. comm. 2002).

Group	Interior Atlantic Forest Argentina	Argentina Total	% of Argentina Total	Endemics of Atlantic Forest	% of Endemics of Atlantic Forest	Exclusive of This Argentinean Eco-region	Endemics Exclusive to Interior Atlantic Forest of Argentina
Vascular plants	3,148	10,806	29%	No data	No data	792	52
Vertebrates	1,124	2,226	50%	No data	No data	342	13
Fish	274	450	61%	No data	No data	51	9
Terrestrial vertebrates	850	1,776	48%	190	23%	291	4
Amphibians	66	172	38%	24	38%	31	2
Reptiles	114	313	36%	41	36%	37	1
Birds	546	1,000	55%	102	19%	187	—
Mammals	124	291	43%	23	18%	36	1

Table 15.3. Exclusive and semiexclusive plant families and genera of the Interior Atlantic Forest of Misiones, Argentina.

Exclusive or Nearly Exclusive Plant Families	Best-Represented Plant Families (Number of Species)	Exclusive Plant Genera (Family)	Disjunctional Distribution
Dicksoniaceae	Poaceae (353)	*Geissomeria* (Acanthaceae)	Araucariaceae (*Araucaria*).
Dilleniaceae	Asteraceae (339)	*Xylopia* (Annonaceae)	Podocarpaceae (*Podocarpus*); *Podocarpus lambertii* is also an important element in the Yungas forests at altitudes of 1,000 to 1,700 m.
Eriocaulaceae	Fabaceae (273)	*Acisanthera, Ossaea, Pterolepis,* and *Rhynchanthera* (Melastomataceae)	Winteraceae (*Drimys*), which are also found in the sub-Antarctic forests.
Maratiaceae	Orchidaceae (134)	*Trigonia* (Trigoniaceae)	
Marantaceae	Cyperaceae (144)	*Caesaera* (Vivianiaceae)	
Ochnaceae	Euphorbiaceae (104)	*Qualea* (Vochysiaceae)	
Trigoniaceae	Rubiaceae (77)	*Websteria* (Cyperaceae)	
Triuridaceae	Verbenaceae (73)	*Sphenostigma* (Iridaceae)	
Vochysiaceae	Solanaceae (72)	*Colanthelia* and *Mesosetum* (Poaceae)	
Xyridaceae	Malvaceae (63)	*Faramea, Schenckia, Ixora,* and *Coccocypselum* (Rubiaceae) *Allagoptera* and *Euterpe* (Arecaceae) *Warrea, Wullschlaegelia, Warmingia, Vanilla, Phloeophila, Octomeria, Miltonia, Isabelia, Ligeophila,* and *Gomesa* (Orchidaceae)	

taxa with Paraguay and Brazil that are not found elsewhere in Argentina (Table 15.3).

Vertebrates

Fifty percent of all vertebrate species and subspecies found in Argentina (1,124) live in Misiones: 274 fish, 66 amphibians, 114 reptiles, 546 birds, and 124 mammals (Figure 15.3).

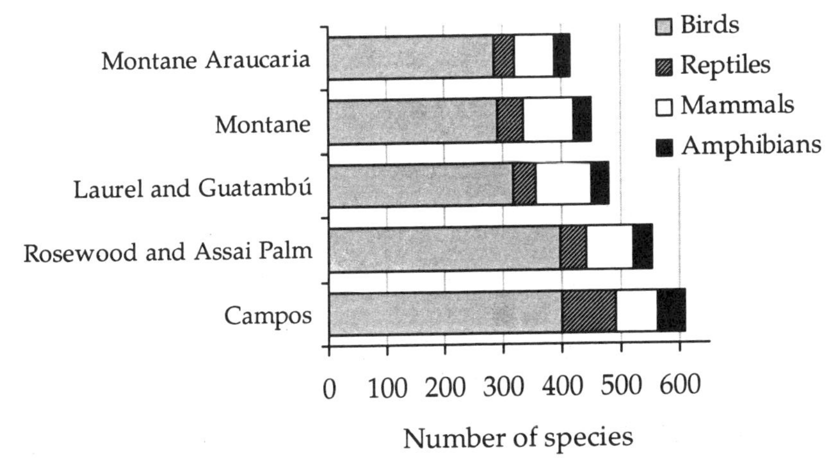

Figure 15.3. Number of vertebrate species in each biogeographic subdivision of the Interior Atlantic Forest of Misiones.

Fish

Misiones has 274 fish species: 272 native and 2 introduced species. These constitute about 61 percent of freshwater fish species in Argentina. Whereas the Paraná, Iguazú, and Uruguay rivers have been surveyed most extensively, the important interior hydrographic network of Misiones has not been sufficiently sampled, as is illustrated by the recent discovery of several new species (Table 15.4).

Although the major collector rivers such as the Paraná account for a significant part of the biodiversity of Misiones, the sectors of greatest interest with regard to fish are the upper basins of the Iguazú and Urugua-í rivers (Gómez and Chebez 1996). The high waterfalls and cascades in these river basins create conditions of isolation that favor endemism. For example, large predator species such as dorado (*Salminus maxillosus*), pirá pytá or piraputanga (*Brycon orbygnianus*), spotted sorubim (*Pseudoplatystoma coruscans*), and others are noticeably absent from upstream areas. The most prominent of these waterfalls are Iguazú Falls.

The construction of the Urugua-í dam, which took place between 1989 and 1991, made the Urugua-í River one of the most studied interior watercourses in Argentinean history and resulted in the description of several new species (Table 15.4). Unfortunately, the impact of the dam on these species has not been consistently monitored, underscoring the need for systematic and intense sampling of the important streams in Misiones as a prerequisite to accurately monitoring the status of biodiversity in the region.

Amphibians

To date, 66 species of amphibians have been recognized in Misiones and northeastern Corrientes: 38 percent of the total number of species in Argentina (Table 15.2). Of these, 24 are exclusive to the Atlantic Forest and 30 are found only in the Misiones ecoregion. Toads and frogs (*Anura*) are the best-

Table 15.4. Recently described or potentially new fish species, known in only few streams of Misiones. Type locality is the site where the species was collected for its description.

Species	Comments	Source
1. *Astyanax ojiara*	Known only from the type locality: Yabotí stream, affluent of the Uruguay River.	Azpelicueta and García 2000
2. *Astyanax* sp. A.	Cuñapirú stream, related to *A. fasciatus.*	
3. *Bryconamericus sylvicola*	Species related to *B. iheringi,* described for streams in the upper and medium region of the Urugua-í River, tributary of the Paraná River.	Braga 1999
4. *Bryconamericus* sp.	Species under description for the Urugua-í and Cuñapirú streams.	
5. *Cichlasoma tembe*	New species previously known only from the type locality: Urugua-í stream above the waterfall. New localities include the Uruguay River tributaries.	Casciotta, Gómez, and Toresani 1995
6. *Gymnogeophagus che*	Described only for the Urugua-í stream in the Paraná River basin.	Casciotta, Gómez, and Toresani 2000.
7. *Hemiancistrus* sp.	First reference for the genus in Argentina from a tributary of the Uruguay River.	
8. *Oligosarcus menezesi*	Known only from the type locality: Urugua-í stream.	Miquelarena and Protogino 1996
9. *Rhamdella* sp. nov.	From Cuñapirú, Urugua-í, and Horqueta streams	Miquelarena and Bockman 2000

represented group, with 60 species in total. In Argentina, caecilians (Gymnophiona), amphibians that resemble giant earthworms, are found only in the Interior Atlantic Forest, and three of four species (*Luetkenotyphlus brasiliensis, Siphonops paulensis,* and *S. annulatus*) have been found only in the Misiones region.

The global trend toward decline of amphibian populations again underscores the need for monitoring in the Atlantic Forest of Argentina. However, even the taxonomy of this group is little known in the region. Of the 66 known species, 5 are still being described, 3 have taxonomic problems, and at least 20 have been re-described or have undergone nomenclature changes in the last 10 years.

Reptiles

The Interior Atlantic Forest of Argentina harbors 114 species and subspecies of reptiles, representing 36 percent of the taxa known in Argentina (Table 15.2). The largest group is snakes, with 83 taxa, which translates into 73 percent of the reptiles and 84 percent of all the known snakes in Argentina. Of the reptile species and subspecies recorded, 41 (36 percent of the total) are endemic to the Interior Atlantic Forest and 36 taxa (32 percent) are exclusive to the Misiones region.

Birds

Birds are the most diverse group of vertebrates in the area: Misiones and neighboring areas of Corrientes contain 546 species (Table 15.2). The Misiones province includes more than half of the known taxa in Argentina, 103 species or subspecies (19 percent) of which are endemic to the Atlantic Forest (as a whole) and related savannas. Some birds identified as native to the Atlantic-Pampas region (Stotz et al. 1996), such as the chestnut-backed tanager (*Tangara preciosa*), red-rumped warbling finch (*Poospiza lateralis*), diademed tanager (*Stephanophorus diadematus*), and indigo (glaucous-blue) grosbeak (*Cyanoloxia glaucocaerulea*), are in fact Atlantic endemic species that migrate into areas of the Pampas following gallery forests along rivers or other favorably humid areas. In fact, several species of Atlantic birds fly for kilometers into the Cerrado through gallery forests (Silva 1996).

Mammals

A total of 124 mammal species have been recorded in the Interior Atlantic Forest of Argentina (Table 15.2), accounting for 43 percent of the mammals in continental Argentina. Of these, 46 taxa (37 percent) and 2 families (spiny rats and tree rats, Echymidae, and the paca, Agoutidae) are native to the ecoregion. The most diverse groups are bats (38 species, 31 percent) and rodents (34 taxa, 25 percent). Other well-represented orders are carnivores (18 species) and marsupials (15 species). However, more inventories of small mammals are needed, especially in forest environments and in little-studied areas such as southeastern Misiones.

The Conservation Status of Misiones Species

Of 850 species and subspecies of amphibians, reptiles, birds, and mammals in the Interior Atlantic Forest of Argentina, 20 percent (163 taxa) are threatened or near-threatened at the national level (Fraga 1996; Lavilla et al. 2000; Díaz and Ojeda 2000) (Table 15.5). Of this group of threatened or near-threatened species, 64 (39 percent) are endemic to the Atlantic Forest.

Several large mammals and birds are endangered in Misiones and Argentina in general (Chebez 1994; Chapter 16, this volume). The conservation status of many species in Argentina, and possibly in South America, cannot be considered definitive because no precise data on populations, habitat, or other critical parameters are available. Although 14 mammal species have been classified as vulnerable and 30 others as near-threatened, researchers do not have sufficient information to assign a conservation status to many other small forest mammals (Table 15.5). Among amphibians, only the frog *Hyalinobatrachyum uranoscopum,* which is found only in the *Araucaria* forests and is very rare (Lavilla et al. 2000), is classified as threatened. The need for more data describing regional fauna is further underscored by the fact that 80 species of land vertebrates are included in the "data deficient" category. Likewise, neither fish nor plants in Argentina have been fully categorized. Nevertheless, there are laws that protect some plant species in

Table 15.5. Number and percentage of threatened, near threatened, and insufficiently known species of terrestrial vertebrates of the Interior Atlantic Forest of Argentina according to national categories (Fraga 1996; Lavilla et al. 2000; Díaz and Ojeda 2000).

	Total Number of Species	Threatened and Near Threatened		Threatened			Near Threatened	Insufficiently Known		Extinct	
		Total	%	Critically Endangered	Threatened	Vulnerable	Near Threatened	Total	%	Total	%
Amphibians	66	11	17%		1	10		10	15		
Reptiles	114	34	30%	4	7	23		13	11		
Birds	546	66	12%	3	17	26	20	43	8	1	0.20%
Mammals	124	55	44%	1	9	14	31	14	11		
Totals	850	166	20%	8	34	73	51	80	9	1	0.10%

Misiones. For example, most species of orchids, one of the best-represented families in the province, are considered seriously threatened as a result of destruction or modification of their habitat and, to a lesser extent, mass extraction or collection of specimens for commercial purposes (Johnson 2001).

Loss of Unique Populations

The Interior Atlantic Forest of Argentina constitutes the southernmost extreme of the biome and the beginning of a transition to other biogeographic formations such as the Campos District. Populations of marginal areas probably differ genetically from those in central areas of their distribution as a consequence of their partial or complete isolation during the climatic changes of the Pleistocene. These "extratropical" forests, located on the fringes or at some distance from the tropics, contain species of tropical lineage with valuable genetic endowment because they have become resistant to more extreme climatic conditions (Brown et al. 1993; Lomolino and Channel 1995). Therefore, the Misiones forest at the edges of the southernmost area may be a critical refuge for many endangered species.

Selecting Indicators for Monitoring Biodiversity

Selecting indicators for monitoring biodiversity in the Interior Atlantic Forest is difficult both because the region is extremely diverse and because there is little solid information available on most species and processes. A robust set of indicators would need to include representative communities, groups of species (i.e., trophic groups), and individual species. Several researchers (Noss 1990; Mills et al. 1993; Stotz et al. 1996) have proposed four types of species that may be useful as potential biodiversity indicators:

- Keystone species, meaning species that influence a large proportion of the community (Bond 1993; Mills et al. 1993). The Assai palm (*Euterpe edulis*), for example, is a good candidate for a keystone species because it produces abundant edible fruits in winter periods of shortage (Placci et al. 1992; Chapter 34, this volume).
- Indicator species or groups, meaning species that provide information on the condition of many other species with similar needs.
- Umbrella species, meaning species with broad ecological needs that encompass those of other species.
- Flagship species, meaning charismatic species that capture public attention and promote public support of and participation in their conservation.

Selecting sound indicators for monitoring biodiversity is extremely important because they are among the most fundamental tools needed by researchers and planners to evaluate progress toward critical conservation objectives. For example, indicators can be used to assess the preservation of large and interconnected (e.g.,

Table 15.6. Potential indicators for the Interior Atlantic Forest of Misiones, Argentina.

Focus of Monitoring	Indicator
Extent of remaining ecosystems, habitats, or connectivity	Remaining ecosystems Fragment size, shape, quality, and conservation status or priority Degree of isolation Degree of connectivity by means of corridors
Communities, groups, and species	Structure and composition of communities Status and trends of focal species
Social response to biodiversity loss	Protected land (extent and percentage) Degree of implementation and effectiveness of protected areas

functional) areas that represent various habitats and their biodiversity or to assess the effectiveness of efforts to preserve rare landscape elements and associated species. Monitoring well-chosen indicators can inform efforts to protect critical habitats and threatened species and can also help limit the introduction and dispersal of exotic or nonnative species. Indicators can be used to support management practices that are sustainable and compatible with the natural potential of the area and to assess the impact of local conservation decisions in a regional con-

Table 15.7. Some characteristics of fragment quality in the Atlantic Forest related to extent.

Extent	Characteristics	Quality
1 km^2	Local extinction of large mammals and raptors, and some specialists (frugivores such as toucans and parrots). Disappearance of species susceptible to human impact. High edge effect, abundance of edge species and generalists. Decrease in primary forest species and those affected by edges.	In some cases moderate (see text)
10 km^2	Extreme dependence on protection. Low probability of presence of large mammals and raptors, and some specialized species. Disappearance of species hunted by humans. Bird communities may be very diverse.	Low to moderate according to protection
100 km^2	High dependence on the degree of management and protection. Cannot sustain viable populations of large mammals and raptors (jaguars or harpies) unless there is management or corridors. High hunting impact. A high proportion of forest birds remain.	Moderate to high according to protection
1,000 km^2	Moderate dependence on protection. May conserve most species and processes. Some species may be affected by hunting, but not heavily, especially under protection. May maintain large predators (jaguars, harpy eagles), but long-term persistence may require management or protection.	High to very high according to protection
10,000 km^2	Less dependence on protection. High likelihood of viable populations of large mammals and raptors. Maintenance of ecological patterns and processes.	Very high

text. Finally, indicators can be useful in planning for unexpected long-term changes and effects and for evaluating and engaging stakeholders in the conservation and sustainable use of biodiversity. Some potential indicators for the Interior Atlantic Forest are presented in Table 15.6.

Several indicators may prove useful in monitoring the remaining ecosystems or habitats and the degree of connectivity between them in the Atlantic Forest. First, a concrete and periodic measure of the remaining ecosystems is needed to reflect the extent, percentage, and rate of change of the land area of ecosystems or habitats that remain with respect to the original area.

Information regarding the size, frequency, shape, distribution, and degree of isolation of natural habitat fragments could also serve as monitoring indicators in a variety of ways. It is generally believed that large fragments have less edge effect, allow the conservation of larger populations of more species, and contain a greater variety of habitats (Chapter 31, this volume). Although there have been no studies on fragmentation in Misiones, observations indicate that several large mammals and raptors have indeed disappeared from small fragments (Giraudo, pers. obs. 2000; Chapter 31, this volume; Table 15.6). However, small fragments can protect certain species. For example, the small provincial parks of Cruce Caballero (5.22 km^2) and Piñalito (37.96 km^2) have conserved significant populations of brown

Table 15.8. Rank system to analyze the relative value and establish priorities of ecosystem remnants. Values are subjective and indicative (modified from Laurance et al. 1997).

	Conservation Value		
Criteria	**High**	**Medium**	**Low**
Representativeness of the region, biogeographic subregion, or habitat in protected areas	<5% in protected areas	5–30% in protected areas	>30% in protected areas
Endemic species	>1 present	1 present	None
Threatened species	>1 present	1 present	None
Disturbances	Mostly undisturbed	Modified	Degraded
Matrix type	Natural habitats	Natural and transformed habitats	Transformed (agriculture, urban)
Isolation from other similar forests or habitats	Connected or <100 m	Between 100 m and 1,000 m	>1,000 m
Connectivity	Connected for most species	Connected for some species	Unconnected for most species
Size	>100 km^2	Between 100 and 1 km^2	<1 km^2
Shape	Circular	Intermediate	Irregular
Habitat diversity	High (>3 habitats)	Moderate (between 3 and 2 habitats)	Low (1 habitat)

howler monkey (*Allouatta guariba*) (Chapter 17, this volume; Krauczuk, pers. obs. 2000) and a great abundance of tree ferns. Araucaria Provincial Park, which is extremely small (0.92 km^2), has important populations of vinaceous-breasted parrots (*Amazona vinacea*) and canebrake groundcreeper (*Clibanornis dendrocolaptoides*) (Krauczuk, pers. obs. 2000), both threatened species that are not present or are very rare in larger protected areas (Tables 15.7 and 15.8).

Other indicators must be developed to monitor changes in biological communities, groups, and species. To monitor the structure and composition of biological communities, for example, researchers and planners need to look at the state of the structure and composition of disturbed communities in relation to control communities (i.e., undisturbed communities) to assess their degree of deterioration. They also need to inspect the status, trends, and demography of species representing keystone, indicator, threatened, endemic, exploited, and introduced groups.

Finally, researchers need to identify and monitor indicators of society's response to biodiversity loss. This should include evaluating the status of protected land in each biogeographic subregion (extent and percentage) and inspecting the degree of implementation and effectiveness of protected areas through a qualitative and quantitative evaluation of the strengths, failures, and threats of protected areas (Chalukian 1999, 2000; Chapters 5 and 21, this volume).

Main Information Gaps

Research on biodiversity in Misiones has dealt mainly with the description of patterns (lists of species, taxonomic issues, species distribution, isolated data on natural history), but little headway has been made in studying the causal processes underlying such patterns. More knowledge of these causal processes is necessary, however, to predict trends in the face of growing modification of the environment and to identify the priority actions needed to conserve biodiversity. At the landscape level, a priority is to update classifications of current land-use practices and the quality of remaining natural environments to establish rational land-use practices, consistent with sustainable development strategies. Biological inventories are also needed in many areas (Chavez, pers. comm. 2001) because they can provide essential information for subsequent studies of ecology and biogeography, and, ultimately, for establishing or improving conservation and sustainable resource use plans. Furthermore, studies are needed to clarify the complex relationship between biodiversity and sociocultural and economic factors, as well as further studies on sustainable management. Finally, clear policies in the academic, government, and private spheres are needed to strengthen collaborative ties between professionals in these areas, among whom there is currently little interaction.

Conclusions and Recommendations

Given the urgency of conserving biodiversity in Argentina's Atlantic Forest, the top priorities for a monitoring program in the Interior Atlantic Forest of Argentina should include the following:

- Map and quantify the surface area of the remnants of Atlantic Forest and other important habitats (e.g., native grasslands, palm groves, wetlands).
- Assess the extent of ecosystem fragmentation.
- Assess the quality of remnants and fragments and identify priority areas for conservation or sustainable management.
- Measure the rate of habitat loss and pinpoint the main causes.
- Strengthen existing protected areas and create new areas in places in which biodiversity is not being adequately protected.
- Build a database that compiles all existing information on the Atlantic Forest, which should be accessible to all.
- Finance and facilitate studies on biodiversity, prioritizing inventories and studies of key processes in the functioning of ecosystems and including biological, cultural, and socioeconomic factors with a view to ensuring sustainable management of resources.
- Develop more educational programs aimed at involving native and rural inhabitants and society in general in the conservation and management of natural resources.
- Facilitate and promote greater implementation of all aspects of the Green Corridor (Chapters 20 and 22, this volume).
- Foster collaboration and exchange with other countries for joint action and research.
- Devise clear guidelines and processes for self-evaluation and external evaluation to modify objectives, actions, and plans as needed.
- Ensure the continuity of the monitoring program over time, irrespective of political or socioeconomic factors or changes.

These activities will lay the foundation for developing plans to manage and conserve the Atlantic Forest in Argentina, which should be built with input from political, social, and cultural sectors and should have both short- and long-term goals.

Acknowledgments

Thanks to Carlos Galindo-Leal, Igor Berkunsky, Julián Alonso, Vanesa Arzamendia, Marcelo Almirón, Karina Schiaffino, Justo Herrera, Hugo Chavez, Fernando Zuloaga, Guillermo Placci, Andrés Bortoluzzi, Marylin Cebolla Badie, Mario Di Bitteti, Jorge Protomastro, Alejandro Garello, José Benitez, Juan Pablo Cinto, Paula Bertolini, Hugo López, Norma Hilgert, and Juan Carlos Chebez for their contributions of information, publications, corrections, and suggestions.

References

Azpelicueta, M. M. and García, J. O. 2000. A new species of *Astyanax* (Characiformes, Characidae) from Uruguay river basin in Argentina, with remarks on hook presence in Characidae. *Revue suisse de Zoologie* 107: 245–257.

Baldo, D. and Basso, N. 2000. Una nueva especie del género *Melanophryniscus* Gallardo,

1961 del sur de la provincia de Misiones, Argentina (Anura, Bufonidae). *Res. XV Reunión Comunic.* pp. 13. Herpetól: Bariloche, Argentina.

Bond, W. J. 1993. Keystone species. In: Schulze, E. D. and Mooney, H. A. (eds.). *Biodiversity and ecosystem function.* Ecological Studies Series 99: 237–253.

Braga, L. 1999. Una nueva especie de *Bryconamericus* (Ostariophysi, Characidae) del Río Urugua-í, Argentina. *Revista del Museo Argentino de Ciencias Naturales Bernardino Rivadavia, Hidrobiología* 8(3): 21–29.

Brown, A. D., Placci, G. L., and Grau, N. R. 1993. Ecología y diversidad de las selvas subtropicales de la Argentina. In: Goin, F. and Goñi, R. (eds.). *Elementos de política ambiental.* pp. 215–222. Buenos Aires: Honorable Cámara de Diputados de la Provincia de Buenos Aires, La Plata.

Cabrera, A. L. 1976. Regiones fitogeográficas Argentinas. *Enciclopedia Argentina Agrícola y de Jardín* 2(1): 1–85.

Cabrera, A. L., and Willink, A. 1973. Biogeografía de América Latina. Serie de Biología, Monogr. No. 13. Washington, DC: Organización de los Estados Americanos (OEA), 117 pp.

Casciotta, J. R., Gómez, S. E., and Toresani, N. I. 1995. "*Cichlasoma*" *tembe*, a new cichlid species from the río Paraná basin, Argentina (Osteichthyes: Labroidei). *Ichthyological Exploration of Freshwaters* 6: 193–200.

Chebez, J. C. 1994. *Los que se van: especies Argentinas en peligro.* Buenos Aires: Editorial Albatros.

Chalukian, S.C. 1999. Cuadro de situación de las unidades de conservación de la Selva Paranaense. Buenos Aires: Fundación Vida Silvestre Argentina.

Chalukian, S. C. 2000. La protección de la Selva Paranaense en Argentina. In: Bertonatti, C. and Corcuera, J. (eds.). *La situación ambiental Argentina 2000.* pp. 383–386. Buenos Aires: Fundación Vida Silvestre Argentina.

Dalmau, H. H. 1995. *El país de los ríos muertos.* Buenos Aires: Editorial Erre Eme.

Díaz, G. B. and Ojeda, R. A. (eds.). 2000. *Libro rojo de los mamíferos amenazados de la Argentina 2000.* Buenos Aires: SAREM (Sociedad Argentina para el Estudio de los Mamíferos).

Fraga, R. M. 1996. Sección aves. In: García Fernández, J. J., Ojeda, R. A., Fraga, R. M., Díaz, G. B., and Baigún, R. J. (comp.). *Libro rojo de mamíferos y aves amenazados de la Argentina.* pp. 155–219. Buenos Aires: FUCEMA.

Galliari, C. A., Pardiñas, U. F. J., Goin, F. 1996. Lista comentada de los mamíferos argentinos. *Mastozoología Neotropical* 3: 39–61.

Giraudo, A. R. 2001. *La diversidad de serpientes de la Selva Paranaense y del Chaco Húmedo: Taxonomía, biogeografía y conservación.* Buenos Aires: Editorial L.O.L.A.

Giraudo, A. R. and Scrocchi, G. J. 1998. A new species of *Apostolepis* (Serpentes: Colubridae) and comments on the genus in Argentina. *Herpetológica* 54: 470–476.

Gómez, S. and Chebez, J. C. 1996. Peces de la provincia de Misiones. In: Chebez, J. C. *Fauna Misionera.* Monografía N° 5. pp. 108–179. Buenos Aires: Editorial LOLA.

ICPB. 1992. *Putting biodiversity on the map: global priorities for conservation.* Cambridge, UK: ICPB.

Johnson, A. E. 2001. *Las orquídeas del Parque Nacional Iguazú.* Buenos Aires: Editorial L.O.L.A.

Krauczuk, E. R. 2000. Presencia de *Chordeiles pusillus* como nidificante en la provincia de Misiones, Argentina. *Ornitologia Neotropical* 11: 85–86.

Laclau, P. 1994. La conservación de los recursos naturales renovables y el hombre en la selva Paranaense. *Boletín Técnico Fundación Vida Silvestre Argentina* (20): 139.

Laurence, W. F., Bierregard, R. O. Jr., Gascon, C., Didham, R. K., Smith, A. P., Lynam, A. J., Viana, V. M., Lovejoy, T. E., Sievin, K. E., Sites, J. W. Jr., Andersen M., Tocher, M. D., Kramer, E. A., Restrepo, C., and Moritz, C. 1997. Tropical forest fragmentation: Synthesis of a diverse and dynamic discipline. In: Laurence, W. F. and Bierregard, R. O. Jr. (eds.). *Tropical forest remnants: ecology, management, and Conservation of fragmented communities.* pp. 502–514. Chicago: University of Chicago Press.

Lavilla, E., Ponssa, M. L., Baldo, D., Basso, N., Bosso, A., Cespedez, J., Chebez, J. C., Faivovich, J., Ferrari, L., Lajmanovich, R., Langone, J., Peltzer, P., Uberdsa, C., Vaira, M., and Vera Candiotti, F. V. 2000. Categorización de los anfibios de Argentina. In: Lavilla, E. O., Richard, E., and Scrocchi, G. J. (eds.). *Categorización de los anfibios y reptiles de Argentina.* pp. 12–34. Tucumán: Asociación Herpetológica Argentina.

Lomolino, M. V. and Channel, R. 1995. Splendid isolation: patterns of geographic collapse in endangered mammals. *Journal of Mammalogy* 76(2): 335–347.

Martínez Crovetto, R. 1963. Esquema fitogeográfico de la provincia de Misiones (República Argentina). *Bomplandia* 1: 171–215.

Miquelarena, M. A. and Bockman, F. A. 2000. A new species of *Rhamdella* from northeastern Argentina (Siluriformes, Heptapteridae). 80th Annual Meeting American Society of Ichthyology and Herpetology, 260.

Miquelarena, A. M. and Protogino, L. C. 1996. Una nueva especie de *Oligosarcus* (Teleostei, Characidae) de la cuenca del Río Paraná, Misiones, Argentina. *Iheringia, Série Zoologia* 80: 111–116.

Mills, L. S., Soulé, M. E., and Doak, D. F. 1993. The keystone species concept in ecology and conservation. *BioScience* 43: 219–224.

Morello, J. and Matteucci, F. D. 1999. Biodiversidad y fragmentación de los bosques en la Argentina. In: Matteucci, F. D., Solbrig, O., Morello, J., and Halffter, G. (eds.). *Biodiversidad y uso de la tierra: conceptos y ejemplos en Latinoamérica.* Colección CEA 24. pp. 463–499. Buenos Aires: Eudeba.

Noss, R. 1990. Indicators for monitoring biodiversity: a hierarchical approach. *Conservation Biology* 4: 355–364.

Perucca, A. R. and Ligier, H. D. 2000. Clasificación de montes forestales nativos, mediante imágenes satelitales en la provincia de Misiones, Argentina. In: *IX Simposio Latinoamericano de Percepción Remota.* Puerto Iguazú, Misiones: Editorial, Sociedad Latinoamericano de Percepción Remota (SELPER).

Placci, G., Arditi, S. Y., Giorgis, P. A., and Wüthrich, A. A. 1992. Estructura del palmital e importancia de *Euterpe edulis* como especie clave en el Parque Nacional Iguazú, Argentina. *Yvyraretá* 3(3): 93–108.

Placci, L. G. and Giorgis, P. 1993. Estructura y diversidad de la selva del Parque Nacional Iguazú, Argentina. *Actas de las VII jornadas técnicas de ecosistemas forestales nativos: Uso, manejo y conservación.* Eldorado. pp. 253–267.

Protomastro, J. 2001. A test for preadaptation to human disturbances in the bird community of the Atlantic Forest. In: Albuquerque, J. L. B., Candido, J. F., Straube, F. C., and Roos, A. (eds.). *Ornitologia e conservação: da ciencia as estrategias.* São Paulo: Sociedade Brasileira de Ornitologia.

Sick, H. 1984. *Ornitología brasileira: uma introdução.* Vols. 1 and 2. Brasília: Universidade de Brasília.

Silva, J. M. C. 1996. Distribution of Amazonian and Atlantic birds in gallery forests of the Cerrado Region, South America. *Ornitologia Neotropical* 7(1): 1–18.

Stotz, D. F., Fitzpatrick, J. W., Parker, T. A. III, and Moskovits, D. F. 1996. *Neotropical birds: ecology and conservation.* Chicago: University Chicago Press.

Veloso, H. P., Rangel Filho, A. L. R., and Lima, J. C. A. 1991. *Classificação da vegetação Brasileira, adaptada a um sistema universal.* São Paulo: Fundação Instituto Brasileiro de Geografia e Estatística, IBGE.
Zuloaga, F. O., Morrone, O., and Rodríguez, D. 1999. Análisis de la biodiversidad en plantas vasculares de la Argentina. *Kurtziana* 27: 17–167.

Chapter 16

Threats of Extinction to Flagship Species in the Interior Atlantic Forest

Alejandro R. Giraudo and Hernán Povedano

Despite major threats to the biodiversity of the Atlantic Forest of Argentina, little information has been compiled or published on the status of many species native to the region. Species-specific information is a powerful tool for conservation and wildlife management, and one necessary to guide plans about where to channel resources, where to focus efforts in designing conservation areas, and where and how to assess environmental impacts (García Fernández et al. 1996). Threatened species are defined as those that currently face a high risk of extinction or that would face a high risk if direct pressures on them or their habitat were to continue (IUCN 2000). These species are of particular concern to both government agencies and national and international nongovernment organizations devoted to conservation; these agencies and organizations are especially concerned about charismatic, or flagship, species because they help to raise public awareness of the seriousness of environmental issues and problems in the region.

Very few detailed population studies have been conducted on species in the Interior Atlantic Forest. Nonetheless, records of naturalists and researchers during the 1900s yield general status trends for some birds and mammals. In particular, the work of Andrés Giai and William Henry Partridge, superb naturalists who made extensive field trips in Misiones between 1940 and 1960, has contributed a wealth of information to the literature.

In this chapter we examine the main flagship species that live in the Interior Atlantic Forest of Argentina, whose populations have declined markedly and are now at great risk. We describe the status of each species and their principal habitats along with the main factors that have influenced their decline. Gathering accurate information on the size, distribution, and status of the habitats of all remaining populations is critical for building plans that will protect these flagship species.

Threatened Flagship Species

Brazilian Merganser

The Brazilian merganser (*Mergus octosetaceus*) may always have been a rare species in Misiones and throughout its distribution area (Partridge 1956; Johnson and Chebez 1985). By the 1940s it was already considered one of the rarest birds in South America, and some researchers believed the species was already extinct, although at that time there were still large expanses of forest and streams in Misiones that remained untouched by human activity (Giai 1950, 1951, 1976; Partridge 1954, 1956). Records indicate that the Brazilian merganser once lived throughout most of the province, including the central and southern portions (Giai 1950, 1951, 1976; Partridge 1954, 1956; Johnson and Chebez 1985) (Table 16.1 and Figure 16.1).

The capture of a specimen in the Yacuy stream by Dr. Eduardo del Ponte in 1947 prompted the extensive campaigns carried out by Giai and Partridge, during which they remained in the forest for several years, roaming streams with expert guides and collecting data on Brazilian mergansers. These campaigns yielded important data on this almost unknown species, including the first records on its behavior, nesting, diet, courtship, and predators (Giai 1950, 1951, 1976; Partridge 1954, 1956). Giai and Partridge also captured the majority of the specimens held by museums in Argentina and the rest of the world (23 specimens between 1948 and 1954, 43 percent of the specimens on record in Argentina). To date, no one has replicated their sampling and collection efforts.

In Misiones, the rare Brazilian merganser has been sighted in three main areas (Table 16.1 and Figure 16.1): the lower and middle basin of the Urugua-í stream (before the filling of the Urugua-í dam, 1990 and 1991) (Johnson and Chebez 1985; Benstead et al. 1993) the Iguazú basin (before 1978) (Johnson and Chebez 1985), and the Piray Guazú basin. However, in 1960, a Brazilian merganser was photographed on the Tigre stream (Giraudo and Abramson 1998), and another was observed in the area in 1977 (Johnson and Chebez 1985).

Since the 1980s there have been a handful of additional sightings of the merganser (Johnson and Chebez 1985; Giraudo, Baldo, and Benitez, pers. obs. 1988; Johnson, Foerster, Chebez, and Giraudo, pers. obs. 1989) (Figure 16.1 and Table 16.1). In 1993, Benstead and other researchers navigated 376 km of streams, recording only one sighting of a merganser in the Piray Miní stream (Benstead et al. 1993). Between 1994 and 1996, Bosso navigated several streams in Misiones from which sightings had been reported but made no direct observations, confirming the merganser's presence only by means of information from people living along the Piray Miní and Piray Guazú streams (pers. comm. 2001). Recently two ornithologists observed a Brazilian merganzer in the Provincial Park Urugua-í in the Uruzú and La Playita rivers (Baldo and Arzamendia, pers. comm. 2002). This sighting confirms the presence of this rare species in the Atlantic Forest of Argentina.

Today, the Brazilian merganser is considered critically endangered, with fewer than 250 individuals worldwide (BirdLife International 2000; Chapter 30, this

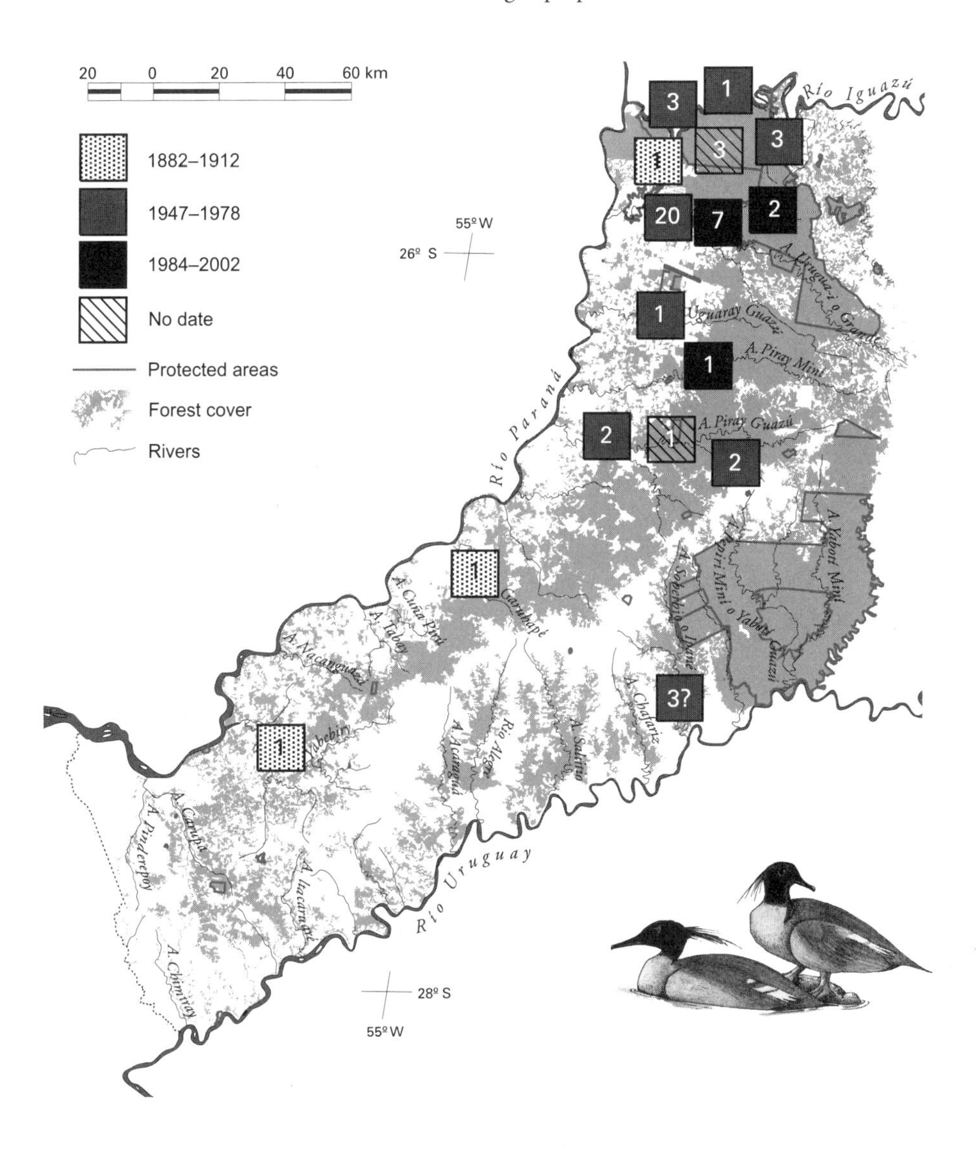

Illustration by Roberto Arreola. Forest cover (1999) and protected areas data from Fundación Vida Silvestre Argentina.

Figure 16.1. Inset squares indicate date range, location, and number of sightings of Brazilian merganser in Misiones.

volume). Changes in watercourses resulting from deforestation, pollution, hunting, and dam construction have all contributed to the declining numbers. Although the situation of the species is extremely critical, some researchers believe that sizable overall populations may still exist in the extensive water systems of Misiones. To plan realistic and effective actions to protect this rare species, researchers need to conduct thorough and extensive field studies similar to those of Giai and Partridge. A crucial factor in the conservation of the species will be the assessment and protection of the quality of the streams in which the merganser lives. Argentina has the largest number of recorded sightings of the species, so its

Table 16.1. Observations and captures of the Brazilian merganser in creeks and rivers of Misiones (Argentina) in chronological order. (Based on Giai 1950, 1951, 1976; Partridge 1954, 1956; Chebez and Johnson 1985; Benstead et al. 1993; and own data). (a) Records of Giai and Partridge in the Urugua-í River. (b) According to Johnson and Chebez (1985), Partridge and Carreras observed "several" individuals in Soberbio creek, indicated here as two individuals (see text). (c) Johnson and Chebez (1985) indicate that park warden José Gorgues observed a family in the Iguazú river (Garganta del Diablo), and several individuals between 1942 and 1978.

Creeks and Rivers	Year of Observation or Capture																						
	1882	1912	1947	1948	1949	1950	1951	1952	1953	1954	1956	1960	1969	1977	1978	1984	1988	1989	1993	2002	Without Date	Totals	%
Garuhapé	1																					1	0.02
Iguazú	1										1			2	1						3(c)	8	0.15
Yabebiry		1																				1	0.02
Yacuy			1		2																	3	0.06
Aguaray Guazú				1																		1	0.02
Urugua-í				7(a)	1(a)	2(a)	2(a)	3(a)		5(a)						5	2			1		27	0.51
Piray Guazú							1	1													3	5	0.09
Tigre												1		1								2	0.04
Uruzú																		1				1	0.02
Soberbio									2(b)				1									3	0.06
Piray Miní																			1			1	0.02
Totals	2	1	1	8	3	2	3	4	2?	5	1	1	1	3	1	5	2	1	1		6	53	1

survival there depends largely on research and conservation activities in the remnants of the Atlantic Forest of Argentina.

Blue-Winged Macaw

Over the past half century, the populations of the large blue-winged macaw (*Propyrrhura maracana*) have been drastically reduced in Argentina and other parts of its distribution area, and it is now considered vulnerable at the global level (BirdLife International 2000) and critically endangered at the national level (Fraga 1996). Although this bird was common in some areas in Argentina in the first half of the twentieth century (Eckelberry 1965; Partridge 1990), its situation is now extremely precarious, and it is very rare in Misiones (Chebez 1994). The species began to decline sharply in the 1970s, and only six sightings have been recorded since then (Chebez 1992, 1994; Saibene et al. 1996; Herrera, pers. comm. 1994). Although the reasons for the population decline are not entirely clear, deforestation may well have been a factor, particularly because the species shows a certain preference for riparian forests, a habitat that has been strongly affected by human activity in Misiones (BirdLife International 2000). The populations in Argentina are at the southernmost limit of the species' distribution area, and such marginal populations may be particularly vulnerable.

Vinaceous-Breasted Parrot

The vinaceous-breasted parrot (*Amazona vinacea*), or vinaceous Amazon, was abundant into the early 1900s in Misiones, where it was widely distributed, even in the southern part of the province (White 1882; Orfila 1938; Navas and Bó 1988). However, its populations have since declined sharply, to the point that it was erroneously declared extinct in Misiones in 1982 (Chebez 1992). The vinaceous-breasted parrot is now considered endangered at the international and national levels, with a declining population of 1,000 to 2,500 individuals (Fraga 1996; BirdLife International 2000).

The southernmost populations of the vinaceous-breasted parrot, now gone, were last recorded in the late nineteenth century in San Javier and Concepción (White 1882) and in the area of Santa Ana, Loreto, and Villa Lutetia (Menegaux 1918; Pereyra 1950; Chebez 1992; Figure 16.2). Another population was seen in Campo Viera in 1986 (Chebez 1992) and yet another in the Paraná pine forests in the northeastern part of the department of San Pedro (Giraudo et al. 1993). This area is home to a population of about 60 individuals that is seen frequently and may be the largest in the province (Krauczuk, pers. comm. 2002). The population is partially protected by the Araucaria Provincial Park, but the park is very small (0.92 km^2), and parrots fly long distances, often leaving the protected area. These parrots have also been seen in the Piñalto Provincial Park, although the birds observed may be from the Araucaria park population (Krauczuk, pers. comm. 2002).

No current data are available on the Campo Viera population, but it may well be very small—smaller still than the San Pedro population. The causes for its

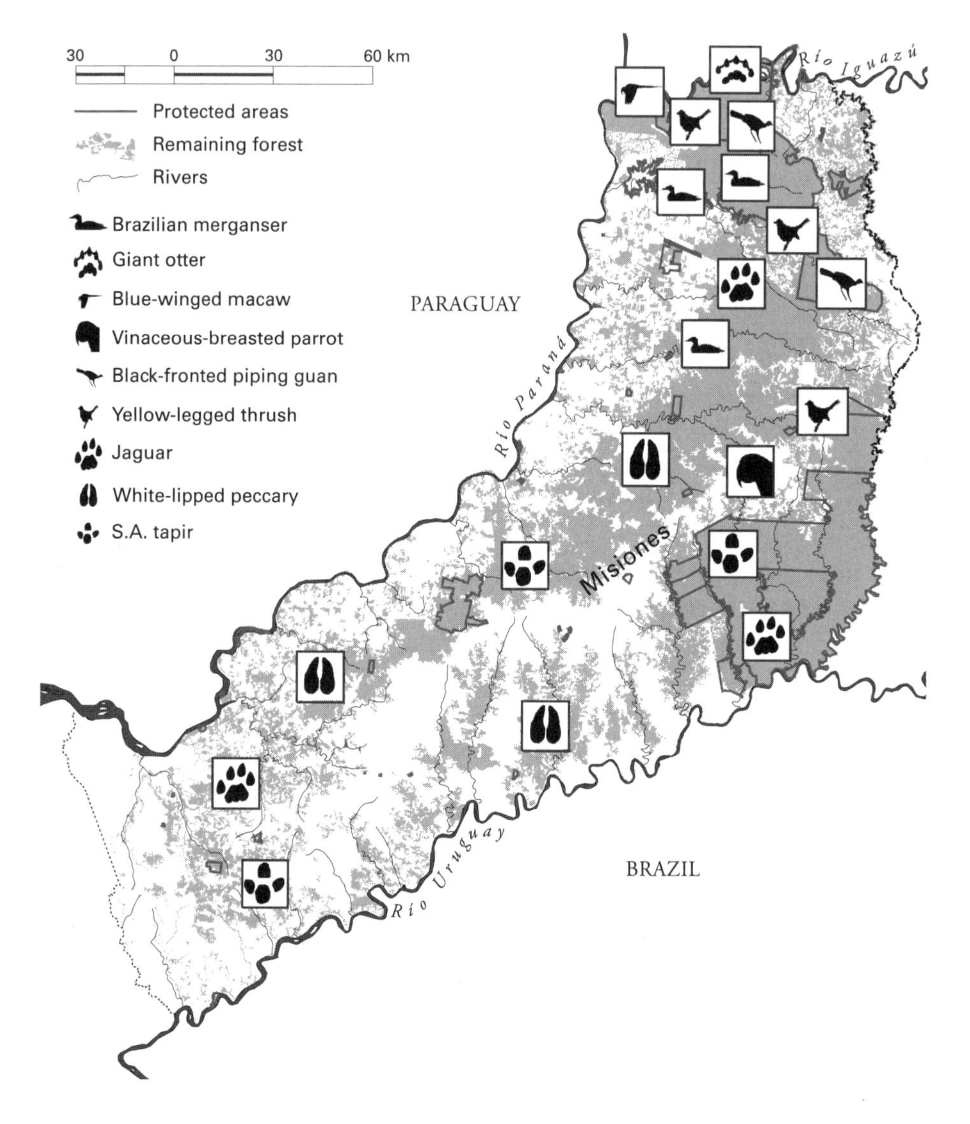

Figure 16.2. Observations of the distribution of flagship species in Misiones.

decline include reduction of the Paraná pine forests, deforestation, killing of the parrots to prevent them from attacking crops, and capture for pets (BirdLife International 2000). These factors and others have affected the Argentine populations, and they should be monitored.

Black-Fronted Piping Guan

The black-fronted piping guan (*Aburria jacutinga* or *Pipile jacutinga*) has suffered a notable reduction in its distribution, both in Brazil and in the province of Misiones, mainly because of hunting and deforestation (Sick and Texeira 1979;

Chebez 1985; Giraudo et al. 1993). The species does not seem to live in fragmented areas smaller than 100 km^2 (Giraudo and Abramson 1998), so the survival of this species in Misiones at present seems contingent on the existence of large forested areas and protected areas in which hunting is prohibited. Hunting has exerted substantial pressure on the black-fronted piping guan because it is a favorite food and much sought after by the local inhabitants. Indeed, hunting has had such an impact on this species that it has almost disappeared from large (2,000-km^2) segments of the forest (Giraudo and Abramson 1998).

In Misiones, the black-fronted piping guan lives mainly on the edges of the forest and close to streams (Giraudo, pers. obs. 1996). Two individuals were sighted on the Rincón de las Mercedes estate (September 30, 1967), near the border with Corrientes province, and a pair of cracids, possibly piping guans, were sighted in Ituzaingó in 1962. However, the identification from these sightings has not been confirmed (Short 1971). Despite intensive studies in these regions and in southern Misiones, no sightings of the black-fronted piping guans have been made for many years, and the species is therefore believed to have disappeared from the region (Giraudo 1996; Krauczuk, pers. obs. 2002). The black-fronted piping guan is considered vulnerable internationally (BirdLife International 2000) and endangered in Argentina (Fraga 1996), although protected populations do live in Iguazú National Park and Urugua-í Provincial Park.

Yellow-Legged Thrush

No recent data are available on this species. Navas and Bó (1993) reported that W. H. Partridge collected 12 specimens from the Urugua-í stream (Department of Iguazú) and in Tobuna (Department of San Pedro). Bertoni (1913) also made reference to the presence of this species in Iguazú. Although the yellow-legged thrush (*Platycichla flavipes*) is still abundant in Brazil and is easy to find elsewhere, it has not been recorded in Misiones since the 1950s, possibly because of deforestation of large areas along the border with Brazil, which may have disrupted its migratory patterns. This hypothesis might also apply to other species captured often between 1940 and 1950 but now seen only infrequently, such as the rufous-tailed attila (*Attila phoenicurus*) and the swallow-tailed cotinga (*Phibalura flavirostris*) (Partridge 1956). The yellow-legged thrush has been classified as endangered in Argentina (Fraga 1996), although it is not considered threatened internationally.

Giant Otter

The giant otter (*Pteronura brasiliensis*) has suffered a dramatic decline in the last 30 years, and it is likely to be declared extinct in Argentina soon (Chebez 1994). Although historical data mention the giant otter in various locations in northeastern Argentina in the Paraná River and streams in Misiones (Holmberg 1987; Chebez 1994), the best-documented populations are those of the Urugua-í and Aguaraí Guazú streams, which were observed during the expeditions carried out between 1948 and 1950 (Giai 1950, 1976). In the Urugua-í, the species usually was found in groups of up to nine individuals, although Giai found several giant otter burrows with two or three pups each, indicating that it was even more abundant than the more common South American or neotropical river otter (*Lontra longicaudis*) (Giai 1950, 1976; Crespo 1982).

The giant otter, also called *lobo marino* (sea wolf) in segments of the Paraná River, was recorded fairly frequently in the Paraná in the area between Itá Ibaté (Corrientes) and San Ignacio (Misiones), as confirmed by reports from local inhabitants (Giraudo, pers. obs. 1996). Noteworthy among the records obtained from surveys are the data from a fisherman from Agipé Island who indicated that during the 1960s and 1970s the species lived on Salvadora Island, located close to Agipé Island on the Paraná River between Argentina and Paraguay, where he observed otters in noisy groups and hunted them periodically. Expeditions and surveys conducted in the 1980s and 1990s yielded no sightings on the Urugua-í (Ambrosini et al. 1987; Benstead et al. 1993), Piray Guazú, Uruzú, Piray Miní (Benstead et al. 1993), Alto Iguazú (Bosso, pers. comm. 1992), or Yabotí Miní (Giraudo, pers. obs. 1991, 1995) streams. However, the South American or neotropical river otter, a species with which it coexisted, was seen frequently. The last giant otter observed in Argentina was a single individual seen swimming in the Iguazú River along the border of the national park of the same name in September 1986 (Chebez 1994; Figure 16.2).

Hunting for otter pelts appears to have been the decisive factor in the species' elimination because it has disappeared from areas that have undergone few environmental modifications. The giant otter is considered critically endangered at the national level and vulnerable at the international level (Díaz and Ojeda 2000).

Jaguar

The jaguar (*Panthera onca*), the largest American feline, has suffered more than 80 percent reduction in its distribution area within Argentina during the last 50 years and it is now found in only a handful of sectors in the provinces of Salta, Jujuy, Formosa, Chaco, Santiago del Estero (the extreme northern portion), and Misiones (Carman 1973; Ojeda and Mares 1982; Perovic and Herrán 1998; Schiaffino 2000). For that reason, the jaguar is considered endangered in Argentina (Díaz and Ojeda 2000), although internationally it is regarded as lower risk–near threatened (IUCN 2000).

In Misiones, the species has disappeared from all highly fragmented places, except those in the vicinity or adjacent to areas with more than 100 km^2 of forest cover. Generally speaking, it lives in the largest forested areas, which cover almost 10,000 km^2. Currently, the southernmost populations are found in the Arroyo Cuña Pirú valley and the area of Campo Grande (Department of Cainguás, Misiones) up to the region around the source of the Tabay stream (Giraudo and Abramson 1998). In the early 1990s, a jaguar was captured from the highlands of San José, close to Cerro Azul, in the southern part of the province (Giraudo 1996), but it is unclear whether populations of this large feline still exist in that region. In the Yacyretá area, the last known jaguars were killed by settlers between the 1950s and 1970s (Giraudo 1996; Figure 16.2).

As is the case with so many other species in the Atlantic Forest, the jaguar is most threatened by the loss of its native habitat, which is caused by deforestation and hunting. Local inhabitants hunt jaguars primarily because thcy prey on domestic animals, especially sheep and pigs (Schiaffino 2000). Jaguars attack domestic animals for a number of reasons. First, the cats have easy access to these animals because human populations are settling closer to natural areas and because the cattle are protected only by primitive management techniques. In addition, jaguars—including those that are old or sick—attack because of a scarcity of natural prey (Schiaffino 2000). Attacks on domestic stock often occur in areas adjacent to protected areas, such as Iguazú National Park and the Provincial Park of Salto Encantado del Cuñapirú (Abramson, pers. comm. 2001). The conflict between human and jaguar populations becomes more acute when authorities fail to take prompt action, although currently there is a research program aimed at reducing the occurrence of these unwanted encounters (Liva and Schiaffino 2000).

The jaguar population of Misiones is one of the largest in the Atlantic forest. According to density estimates (3.7 individuals per 100 km^2) found for Iguaçu National Park (Brazil) and Iguazú National Park (Argentina) (Crawshaw 1995), an estimated population of more than 300 jaguars could live in the forests of Misiones (10,000 km^2). The more conservative estimate put forth by Leite et al. (2002) for nonprotected areas where hunting occurs (1 jaguar per 100 km^2) would place the Argentine population around 100 individuals. These estimates suggest that the number of jaguars in the Atlantic Forest of Argentina might be equal to or up to twice the number in Brazil, where the total population has been estimated to be around 140 jaguars. Furthermore, in Brazil, jaguar populations are isolated with little possibility of contact and genetic exchange, whereas in Argentina the continuity of the forest still allows exchange between populations (Chapter 15, this volume)

South American Tapir

The South American tapir (*Tapirus terrestris*) is the largest indigenous herbivorous-frugivorous species in the region, and it is highly valued by game hunters (Giraudo and Abramson 1998). The tapir has disappeared from some regions of Misiones and has become completely extinct in the province of Corrientes (Díaz and Ojeda 2000). Its current distribution in Argentina covers less than half of its original territory (Ojeda and Mares 1982), and it is therefore considered endangered in Argentina (Díaz and Ojeda 2000). Globally, it is considered as lower risk–near threatened (IUCN 2000). The latest information on tapirs in the Yacyretá area relates to a specimen hunted on Talavera Island in 1975 (Giraudo 1996; Bosso, pers. comm. 1994). The tapir appears to have lived on the islands of Yacyretá and Talavera (Chebez 1994; Figure 16.2).

The tapir lives in the same areas as the jaguar, and in recent years it has recolonized southern portions of its distribution area, where protected areas have been created such as Valle de Cuñapirú Provincial Park, a place in which no sightings of tapirs had been recorded for 30 years (Giraudo and Abramson 1998). The biggest problem faced by this species is game hunting.

White-Lipped Peccary

Of the three peccary species present in South America, the white-lipped peccary (*Tayassu pecari*) is the most gregarious and occupies the largest territories (Terborgh 1992). In Misiones, groups of close to 200 individuals have been reported (Giai 1976) and were common in the mid-1800s, when large expanses of forest existed. Currently, these herds are seen only in areas where extensive forests remain, such as the Yabotí Biosphere Reserve (Abramson and Giraudo, pers. obs. 1991). The peccary's current distribution is roughly similar to that of the jaguar and the South American tapir (Giraudo, pers. obs. 2002; Figure 16.2).

Peccaries avoid crossing wide paved roads, which leads to even greater isolation of populations and increases their vulnerability to pursuit and hunting (Giraudo and Abramson 1998). Their gregarious behavior makes it easy for hunters to kill many animals at a time. The white-lipped peccary and the collared peccary (*Tayassu tajacu*) destroy winter crops when their food availability is low, creating conflicts with humans (Giraudo and Abramson 1998). Further research is needed to devise measures to prevent the peccaries from invading farm crops during the winter. The white-lipped peccary is considered near threatened at the national level (Díaz and Ojeda 2000), but it is not in any threatened category internationally (IUCN 2000).

How Can the Extinction of These Species Be Prevented?

The main common cause for the decline of all the species discussed here is deforestation and forest fragmentation. Nevertheless, there are biological and demographic characteristics specific to several of them that are not yet fully understood, making it difficult to determine which other factors may have contributed to their decline or disappearance. The population dynamics of some species may extend beyond the border of Argentina because they move or migrate seasonally, and the large-scale deforestation in southern Brazil could explain reduction of others. Hunting and conflicts with humans are also important causal factors for the decline of many mammals and for some birds, and the situation is exacerbated by increased access to forests via roads and highways and increasing human populations (Chapter 36, this volume). Dam construction is cited as one of the causes for the disappearance of two species, and the largest known populations of Brazilian mergansers were devastated by the Urugua-í dam (Chapter 35, this volume).

The conservation of significant forest areas through the formation of corridors would be the best way to protect the majority of these species (Chapter 31, this volume). Some species would also be afforded protection if hunting were reduced, whether through legislation, improved enforcement, or solution of the socioeconomic problems (Chapter 19, this volume) that lead many people to engage in subsistence hunting.

Monitoring of threatened species populations and research on their biology and natural history are crucial to understanding their population trends and the causes of their decline. However, academic and scientific systems in Argentina do not encourage the study of threatened species (Chehebar and Saba 1998). Peer-review systems favor the production of many scientific publications at increasingly shorter intervals. Researchers who study threatened species generally produce fewer publications and are at a disadvantage in the system and have difficulty finding funds. This is a disincentive to many researchers who might want to pursue longer-term conservation studies. The scientific system needs to devise a clear policy that will give priority to conservation and to redefining the prevailing criteria for success in the scientific community (Chehebar and Saba 1998).

Acknowledgments

Thanks to Ernesto Krauczuk and Andrés Bosso for providing unpublished information, to Karina Schiaffino and Juan Carlos Chebez for providing publications, and to Carlos Galindo-Leal for his insightful revisions and suggestions. Jorge Baldo and Yanina Arzamendia provided recent observations of the Brazilian merganser.

References

Ambrosini, S., Galliari C., and Vaccaro, O. 1987. *Proyecto de relevamiento faunístico y florístico de la Cuenca Urugua-í. Informe del grupo de mamíferos.* Unpublished report, Buenos Aires.

Benstead, P. J., Hearn, R. D., Jeffs, C. J. S., Callaghan, D. A., Calo, J., Gil, G., Johnson, A. E., and Stagi Nedelcoff, A. R. 1993. *Pato serrucho 93: an expedition to assess the current status of the Brazilian merganser,* Mergus octosetaceus, *in northeast Argentina.* Final Report, December 1993.

Bertoni, A. D. 1913. Contribución para un catálogo de aves Argentinas. *Anales de la Sociedad Científica Argentina* 75: 64–102.

BirdLife International. 2000. *Threatened birds of the world.* Barcelona and Cambridge, UK: Lynx Editions and BirdLife International.

Carman, R. L. (ed.). 1973. *De la fauna Bonaerense.* Buenos Aires: Talleres Gráficos Didot, S.C.A.

Chebez, J. C. 1985. Nuestras aves amenazadas: la yacutinga (*Aburria yacutinga*). *Nuestras Aves* 3(7): 16–17.

Chebez, J. C. 1992. Notas sobre algunas aves poco conocidas o amenazadas de Misiones (Argentina). *Aprona Boletín Científico* 21: 12–30.

Chebez, J. C. 1994. *Los que se van: especies Argentinas en peligro.* Buenos Aires: Editorial Albatros.

Chehebar, C. and Saba, S. 1998. Trampa 22: una paradoja que afecta a las especies en peligro de extinción. *Boletín de la Sociedad Biológica de Concepción, Chile* 69: 63–70.

Crawshaw, Jr., P. G. 1995. *Comparative ecology of ocelot* (Felis pardalis) *and jaguar*

(Panthera onca) *in a protected subtropical forest in Brazil and Argentina.* Doctoral dissertation, University of Florida, Gainesville.

Crespo, J. A. 1982. Ecología de la comunidad de mamíferos del Parque Nacional Iguazú, Misiones. *Revista de Ecología del Museo Argentino de Ciencias Naturales* 3(2): 43–162.

Díaz, G. B. and R. A. Ojeda (eds.) 2000. *Libro rojo de los mamíferos amenazados de la Argentina 2000.* Buenos Aires: SAREM (Sociedad Argentina para el Estudio de los Mamíferos).

Eckelberry, D. 1965. A note on the parrots of northeastern Argentina. *Wilson Bulletin* 77: 111.

Fraga, R. M. 1996. Sección III: aves. In: García Fernández, J. J., Ojeda, R. A., Fraga, R. M., Díaz, G. B., and Baigún, R. J. (comp.). *Libro rojo de mamíferos y aves amenazados de la Argentina.* pp. 155–219. Buenos Aires: FUCEMA (Fundación para la Conservación de las Especies y el Medio Ambiente).

García Fernández, J. J., Ojeda, R. A., Fraga, R. M., Diaz, G. B., and Baigun, R. J. (comp.) 1996. *Libro rojo de mamíferos y aves amenazados de la Argentina.* Buenos Aires: FUCEMA (Fundación para la Conservación de las Especies y el Medio Ambiente).

Giai, A. 1950. Notas de viajes II: por el norte de Misiones. *El Hornero* IX(2): 138–164.

Giai, A. 1951. Notas sobre la avifauna de Salta y Misiones. *El Hornero* IX: 247–276.

Giai, A. 1976. *Vida de un naturalista en Misiones.* Buenos Aires: Editorial Albatros.

Giraudo, A. R. 1996. *Impacto de la presa de yacyretá y la futura presa de garabí (Corrientes y Misiones) sobre la fauna de vertebrados tetrápodos.* Unpublished document, Santa Fe, Argentina: Final report on a fellowship for advanced studies from Consejo Nacional de Investigaciones Científicas (CONICET) (National Scientific Research Council).

Giraudo, A. R. and Abramson, R. R. 1998. *Usos de la fauna silvestre por los pobladores rurales de la Selva Paranaense de Misiones: tipos de uso, influencia de la fragmentación, posibilidades de manejo sustentable.* Boletín Técnico 42. Buenos Aires: Fundación Vida Silvestre Argentina.

Giraudo, A. R., Baldo, J. L., and Abramson, R. R. 1993. Aves observadas en el sudeste, centro y este de Misiones (Republica Argentina), con la mención de especies nuevas o poco conocidas para la provincia. *Nótulas Faunísticas* 49: 1–13.

Holmberg, E. L. 1987. Viaje a Misiones. *Boletín de la Academia Nacional de Ciencias Córdoba* 10: 1–391.

IUCN. 2000. *2000 IUCN Red List of threatened animals.* Gland, Switzerland: (IUCN) World Conservation Union.

Johnson, A. and Chebez, J. C. 1985. Sobre la situación de *Mergus octosetaceus* Vieillot (Anseriformes: Anatidae) en la Argentina. *Historia Natural, Suplemento* (1): 1–16.

Leite, M. R. P., Boulhosa, R. L. P., Galvão, F., and Cullen Jr., L. 2002. Ecology and conservation of jaguar in the Atlantic Coastal Forest, Brazil. In: Medellín, R. A., Chetkiewicz, C., Rabinowitz, A., Redford, K. H., Robinson, J. G., Sanderson, E., and Taber, A. (eds.). *Jaguares en el nuevo milenio: una evaluación de su estado, detección de prioridades y recomendaciones para la conservación del jaguar en América.* pp. 25–42. México, D.F.: UNAM (Universidad Nacional Autónoma de México), WWF (World Wildlife Fund).

Liva, A. and Schiaffino, K. 2000. *Buscando una estrategia para la conservación del yaguareté en Misiones.* 6 y 7 de Diciembre de 1999, Administración de Parques Nacionales, Delegación Regional del Nordeste Argentino, Puerto Iguazú, Misiones, Argentina. Iguazú, Misiones.

Menegaux, A. 1918. Etude d'une collection d'oiseaux faite par M. E. Wagner dans la province de Misiones (Republique Argentine). *Revue Française d'Ornithologie* 10(112–113): 288–293.

Navas, J. and Bó, N. 1988. Aves nuevas o poco conocidas de Misiones, Argentina 3. *Revista del Museo Argentino de Ciencias Naturales Bernardino Rivadavia, Ciencias Zoológicas* 15(2): 11–37.

Navas, J. R. and Bó, N. A. 1993. Aves nuevas o poco conocidas de Misiones, Argentina 5 (Adenda). *Revista del Museo Argentino de Ciencias Naturales Bernardino Rivadavia, Ciencias Zoológicas* 16: 37–50.

Ojeda, R. A. and Mares, M. A. 1982. Conservation of South American mammals: Argentina as a paradigm. In: Mares, M. A. and Genoways, H. H. (eds.) *Mammalian biology in South America.* pp. 505–539. Special Publication Series, Pymatuning Laboratory of Ecology, Vol. 6. Pittsburgh: University of Pittsburgh.

Orfila, R. 1938. Los psittaciformes argentinos. *El Hornero* 7(1): 1–21.

Partridge, W. H. 1954. Estudio preliminar sobre una colección de aves de Misiones. *Revista del Museo Argentino de Ciencias Naturales Bernardino Rivadavia, Ciencias Zoológicas* 3(2): 87–153.

Partridge, W. H. 1956. Notes on the Brazilian merganser in Argentina. *Auk* 73: 473–488.

Partridge, W. H. 1990. Aves misioneras 1. *Nuestras Aves* 8(22): 20–24.

Pereyra, J. 1950. Aves del territorio de Misiones. *Anales del Museo Nahuel Huapi "Perito Francisco Moreno"* 2: 1–40.

Perovic, P. G. and Herrán, M. 1998. Distribución del jaguar *Panthera onca* en las provincias de Jujuy y Salta, noroeste de Argentina. *Mastozoología Neotropical* 5(1): 47–52.

Saibene, C., Castelino, M., Rey, N., Calo, J., and Herrera, J. 1996. *Relevamiento de las aves del Parque Nacional Iguazú, Misiones, Argentina.* Buenos Aires: Literature of Latín América.

Schiaffino, K. 2000. Una experiencia de participación de productores rurales en un proyecto de conservación de yaguareté en Misiones: la opinión de los especialistas. Desarrollo sustentable. In: Bertonatti, C. and Corcuera, J. (eds.). *La situación ambiental Argentina 2000.* pp. 269–271. Buenos Aires: Fundación Vida Silvestre Argentina.

Short, L. L. 1971. Aves nuevas o poco comunes de Corrientes, República Argentina. *Revista del Museo Argentino de Ciencias Naturales Bernardino Rivadavia, Ciencias Zoológicas* 9: 283–309.

Sick, H. and Teixeira, D. M. 1979. Notas sobre aves brasileiras raras ou ameaçadas de extinção. *Publicações Avulsas do Museu Nacional, Rio de Janeiro* (62): 1–39.

Terborgh, J. 1992. Maintenance of diversity in tropical forest. *Biotropica* 24(2b): 283–292.

White, E. W. 1882. Notes on birds collected in the Argentine Republic. *Proccedings of the Zoological Society of London* 1882: 591–629.

Chapter 17

Outlook for Primate Conservation in Misiones

Mario S. Di Bitetti

Many species that were once plentiful in the Atlantic forests are now on the brink of extinction. In Argentina, three species of neotropical primates live in the forests of Misiones and are themselves at risk for a variety of reasons. This chapter briefly reviews the status of the tufted capuchin monkey and two species of howler monkeys, summarizes why they are at risk, and underscores why protected areas must be an essential component of any effort to conserve these animals in Argentina.

Tufted or Brown Capuchin Monkey

Although the howler monkey populations are the most intensely pressured in Misiones, the brown tufted capuchin monkey (*Cebus apella nigritus*) is also seriously at risk because of habitat destruction. The brown tufted capuchin monkey was once found in almost all forest environments in the province of Misiones, largely because of the animal's great environmental flexibility, which has enabled it to live in degraded and recovering forests as well as in mature forests. In Iguazú National Park, in the area around Iguazú Falls (Figure 17.1), the tufted capuchin currently reaches population densities of up to 16 individuals per km^2 (Di Bitetti 2001). Although there are no published estimates of the species' population density outside Iguazú National Park, groups of tufted capuchins have been found outside this protected area to some extent. The groups within the park range from 8 to 35 individuals (averaging 17, not including young of the year), with a home range of 0.81 to 2.93 km^2 (1.6 km^2 on average) (Di Bitetti 2001). The tufted capuchin is often found in the park in secondary forests and in tall forests that are abundant in rosewood (*Aspidosperma polyneuron*) and Assai palm (*Euterpe edulis*) (Di Bitetti, pers. obs. 2001).

The tufted capuchin has a generalist diet consisting mainly of fleshy fruits and arthropods, although during the season, when other foods are least available, the species eats primarily palm fruits (Terborgh 1983; Peres 1994). In Misiones, the pindo palm (*Syagrus romanzoffiana*) is a key food source for tufted capuchins (Di Bitetti 2001), although they readily eat figs (*Ficus* spp.) and vine fruits such as the

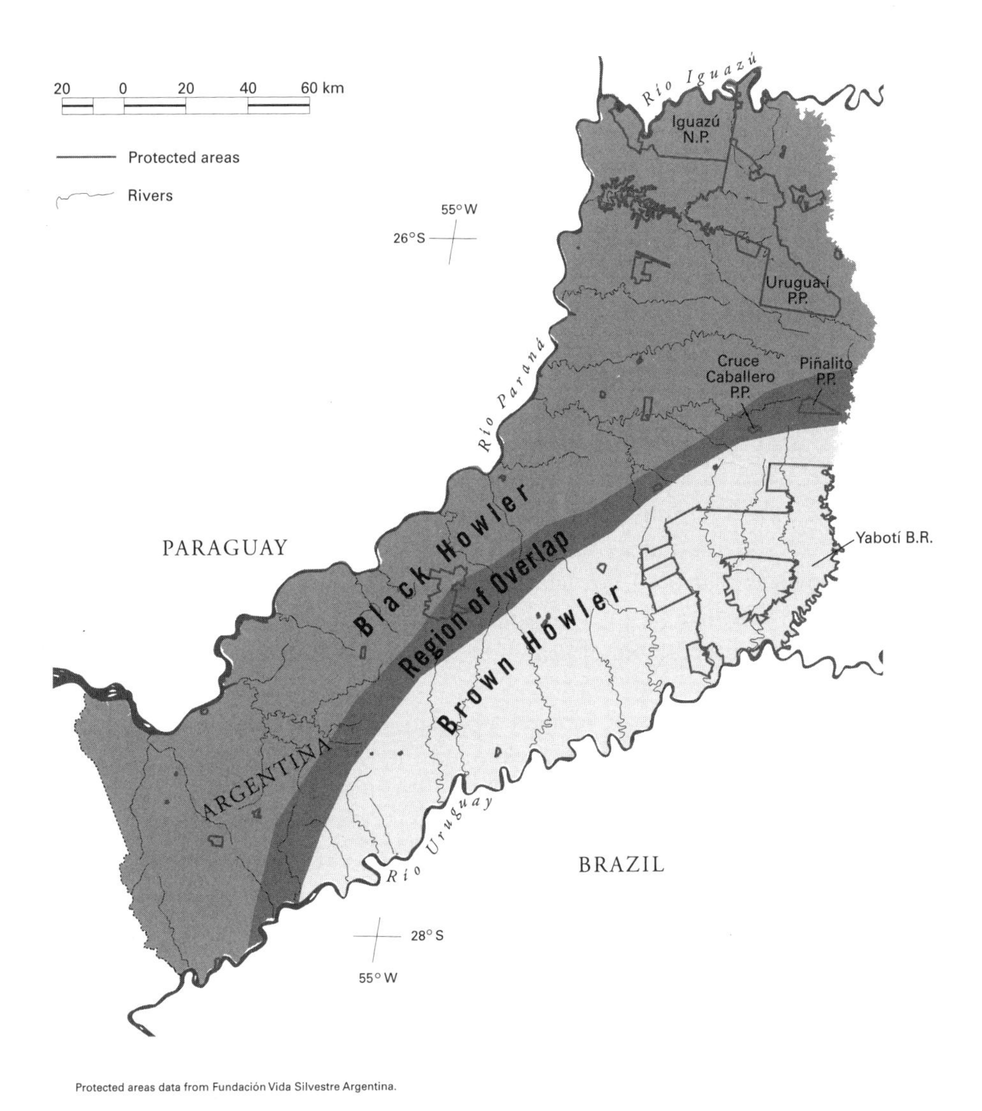

Figure 17.1. Probable distribution of several primate species in Misiones. The tufted capuchin monkey was once widespread in the province.

lemon vine (*Pereskia aculeata*) when other preferred foods are scarce. Bamboo shoots from the species *Chusquea ramosissima* are consumed all year long. The abundance of these plant species probably is largely responsible for the area's capacity to sustain tufted capuchin populations.

Howler Monkeys

Two species of howler monkey live in Misiones: the black howler monkey (*Alouatta caraya*) and the brown howler monkey (*Alouatta fusca clamitans = A. guariba clamitans*). The black howler is typical of species found in the humid

Chaco region of Argentina, especially in the flood forests along riverbanks and on the islands in the Paraná River (Rumiz 1990; Brown and Zunino 1994). In Misiones, the species is found along the banks of the Paraná River, extending east, inland, to the central mountain range that runs through the middle of the province (Figure 17.1). The brown howler monkey is a species of the Atlantic Forest, found in the central highland region of Misiones and the area east of the mountains in Argentina (Di Bitetti et al. 1994; Rylands et al. 1996).

Little is known about the current population size and distribution of either of the howler monkeys in Misiones, and no studies have been published on their diet or habitat needs in this region. In Misiones, the brown howler has been found in Paraná pine (*Araucaria angustifolia*) forests, and several authors have suggested that this tree is a key resource for the species (Crespo 1954; Cordeiro da Silva 1981; De Assis Jardim 1992). However, the brown howler monkey does not have the same distribution as the Paraná pine in most of the Atlantic Forest. What is known is that both howlers are rare in Misiones. The reasons for their low population densities are unclear, although potential causes for their declining numbers include an epidemic of yellow fever in the 1960s, habitat destruction, and hunting (Di Bitetti et al. 1994).

Although the black howler monkey population sizes are declining, they are still abundant in the humid Chaco region and therefore not considered vulnerable (Brown and Zunino 1994). However, the brown howler subspecies (*A. fusca clamitans*) in Misiones is considered vulnerable because of population decline resulting from reduction and deterioration of its habitat (IUCN 2000). In fact, the Argentine Society for the Study of Mammals (Sociedad Argentina para el Estudio de Mamíferos) considers brown howlers to be threatened with extinction in Argentina (Díaz and Ojeda 2000).

Both species of howler monkeys have been sighted recently in Piñalito Provincial Park (Baldovino and Benesovsky, pers. comm. 2001), where one species may be replacing the other (Figure 17.1). Howlers are folivorous and frugivorous, leading some researchers to believe that the availability of tender leaves with a low content of secondary compounds may be a determining factor in their population density (Peres 1997, 1999). The black howler typically is found in Chaco flood forests, which are characterized by secondary growth, whereas the brown howler is thought to need mature forest. The degradation of the Misiones forest may ultimately prove advantageous to the black howler at the expense of the brown howler (Di Bitetti et al. 1994).

Principal Threats to Misiones Primates

Without question, the primary threat to primates outside protected areas is destruction of the native forest. Forests are being cut to make room for agricultural activities, both by large lumber companies that replace the native trees with pine monocultures and by small farmers who cut forest to make needed space for their subsistence agricultural practices. As human communities encroach, epidemics of diseases and other conditions associated with humans or their domestic animals, such as yellow fever and botfly (*Dermatobia* sp.) larvae infestation, also pose a grave threat to howlers because they are especially susceptible to these maladies (Collias and Southwick 1952; Crespo 1974; Milton 1982; Timm 1994; Brown and Zunino 1994).

An additional threat to howler monkeys is hunting. Generally speaking, for example, the 54 Mbyá indigenous communities in the province of Misiones, as well as nonindigenous settlers, practice subsistence hunting in some parts of Misiones (Chapter 18, this volume). In areas with long-standing human settlements, hunting has apparently resulted in local extinction of primates, along with many other mammals and birds. Another threat is from the pet trade, for although the sale of monkeys as pets is illegal, demand for them on the local market continues to be an incentive for the capture of infants of all three primate species.

Natural predators, primarily eagles, add to the pressures on howler populations. In Misiones, primate predators included the harpy eagle (*Harpia harpyja*, which may be locally extinct) and the hawk eagles (*Spizaetus tyranus* and *Spizaetus ornatus*), which are present in Iguazú National Park. Mammal predators include the tayra (*Eira barbara*) and large felines such as the jaguar (*Panthera onca*), puma (*Felis concolor*), and ocelot (*Leopardus pardalis*) (Di Bitetti 1999).

Protection Status of Misiones Primates

Although the tufted capuchin, because of its larger home range (Rumiz 1990; Di Bitetti 2001), has greater space needs than howler monkeys, it is well protected in the various reserves in Misiones, including Iguazú National Park and Urugua-í Provincial Park. Howlers, on the other hand, are insufficiently protected because of their low population density, patchy distribution, and the small number of reserves in which the presence of these monkeys has been confirmed. The status of the brown howler is especially critical, as its presence has been confirmed in only two small reserves: Cruce Caballero Provincial Park and Piñalito Provincial Park (Table 17.1).

As discussed earlier, there is little baseline knowledge about the tufted capuchin monkey or the brown or black howler monkeys, including data on current population sizes, their habitat needs, or their diets. To design conservation programs and actions to protect these primates, therefore, this level of basic data must be collected and compiled immediately. In particular, a thorough survey of brown howler populations in Misiones is necessary to determine their demographic status and habitat needs. Information from such surveys could guide conservation plans designed to protect and connect the forest remnants that include brown howler populations and expand the two reserves in which this species lives. Without these efforts, these species may soon be forever lost.

Table 17.1. Population status and main problems for the conservation of primates in Misiones, Argentina.

Species	Tufted or Brown Capuchin (*Cebus apella*)	Black Howler (*Alouatta caraya*)	Brown Howler (*Alouatta fusca clamitans = A. guariba*)
Population status	Abundant in protected areas. Declining outside reserves.	Scarce. Small, isolated populations in its range in Misiones.	Very scarce. Small, isolated populations throughout its range.
Greatest problems facing the species in Misiones	Habitat loss to agriculture.	Habitat loss and susceptibility to disease epidemics and other maladies such as yellow fever and infestation with parasitic botfly (*Dermatobia* sp.) larvae .	Habitat loss and susceptibility to disease epidemics and other maladies such as yellow fever and infestation with parasitic botfly (*Dermatobia* sp.) larvae.
Ecological needs	Generalist species. Fruits of the pindo palm (*Syagrus romanzoffiana*) are staple of diet.	Folivorous (leaves) and frugivorous (fruits) diet. Probably more tolerant of disturbance and secondary forest than the brown howler.	Probably needs more mature forest than the black howler. Probable association with Paraná pine forests.
Conservation status of the species	Not endangered. Locally protected.	No major problems in its total distribution area, but declining. Rare in Misiones.	Considered vulnerable by the IUCN. Considered threatened with extinction by the Argentine Society for the Study of Mammals.

References

Brown, A. D. and Zunino, G. E. 1994. Hábitat, densidad y problemas de conservación de los primates de Argentina. *Vida Silvestre Neotropical* 3: 30–40.

Collias, N. and Southwick, C. W. 1952. A field study of population density and social organization in howling monkeys. *Proceedings of the American Philosophical Association* 96: 143–156.

Cordeiro da Silva, E. 1981. A preliminary survey of brown howler monkeys (*Alouatta fusca*) at the Cantareira Reserve (São Paulo, Brazil). *Revista Brasileria de Biologia* 41: 897–909.

Crespo, J. A. 1954. Presence of the reddish howling monkey (*Alouatta guariba clamitans* Cabrera) in Argentina. *Journal of Mammalogy* 35: 117–118.

Crespo, J. A. 1974. Comentarios sobre nuevas localidades de mamíferos de Argentina y de Bolivia. *Revista de Museo Argentino de Ciencias Naturales "Bernardino Rivadavia," Zoología* 11(1): 1–31.

De Assis Jardim, M. M. 1992. *Aspectos ecológicos e comportamentais de* Alouatta fusca clamitans *(Cabrera, 1940) na Estação Ecológica de Aracuri, RS, Brasil (Primates, Cebidae).* Bachelor's thesis, Universidade Federal do Rio Grande do Sul, Porto Alegre, Rio Grande do Sul, Brazil.

Díaz, G. B. and Ojeda, R. A. (eds.) 2000. *Libro rojo de mamíferos amenazados de la Argentina.* Buenos Aires: Sociedad Argentina para el Estudio de los Mamíferos.

Di Bitetti, M. S. 1999. La vida en sociedad de los primates: costos y beneficios. *Ciencia Hoy* 9(53): 32–45.

Di Bitetti, M. S. 2001. Home range use by the tufted capuchin monkey (*Cebus apella nigritus*) in a subtropical rainforest of Argentina. *Journal of Zoology, London* 253: 33–45.

Di Bitetti, M. S., Placci, G., Brown, A. D., and Rode, D. I. 1994. Conservation and population status of the brown howling monkey (*Alouatta fusca clamitans*) in Argentina. *Neotropical Primates* 2: 1–4.

IUCN. 2000. *The 2000 IUCN Red List of threatened species.* Gland, Switzerland: IUCN (World Conservation Union).

Milton, K. 1982. Dietary quality and demographic regulation in a howler monkey population. In: Leigh, E. G., Stanley Rand, A., and Windsor, D. M. (eds.). *The ecology of a tropical forest: seasonal rhythms and long-term changes.* pp. 273–289. Washington, DC: Smithsonian Institution Press.

Peres, C. A. 1994. Primate responses to phenological changes in an Amazonian terra firme forest. *Biotropica* 26: 98–112.

Peres, C. A. 1997. Effects of habitat quality and hunting pressure on arboreal folivore densities in neotropical forests: a case of study of howler monkeys (*Alouatta* spp.). *Folia Primatologica* 68: 199–222.

Peres, C. A. 1999. Effects of subsistence hunting and forest types on the structure of Amazonian primate communities. In: Fleagle, J. G., Janson, C. H., and Reed, K. E. (eds.). *Primate communities.* pp. 268–283. Cambridge, UK: Cambridge University Press.

Rumiz, D. I. 1990. *Alouatta caraya:* population density and demography in northern Argentina. *American Journal of Primatology* 21: 279–294.

Rylands, A. B., da Fonseca, G. A. B., Leite, Y. L. R., and Mittermeier, R. A. 1996. Primates of the Atlantic Forest: origin, distributions, endemism, and communities. In: Norconk, M., Garber, P., and Rosenberger, A. (eds.). *Adaptive radiations of neotropical primates.* pp. 21–51. New York: Plenum Press.

Terborgh, J. 1983. *Five New World primates: a study in comparative ecology.* Princeton, NJ: Princeton University Press.

Timm, R. M. 1994. The mammal fauna. In: McDade, L. A., Bawa, K. S., Hespenheide, H. A., and Harsthorn, G. S. (eds.). *La selva: ecology and natural history of a neotropical rainforest.* pp. 229–237. Chicago: University of Chicago Press.

Chapter 18

The Loss of Mbyá Wisdom: Disappearance of a Legacy of Sustainable Management

Angela Sánchez and Alejandro R. Giraudo

The province of Misiones is home to about 750 families, or 4,000 individuals, of the Mbyá-Guaraní indigenous group (Chapter 32, this volume). These families are distributed among some 54 communities, which vary greatly in size and composition (Table 18.1 and Figure 18.1). Historically the Iguazú River area has been the region with the greatest concentration of Guaraní population. Irrespective of the geopolitical boundaries imposed by the modern states of Argentina, Brazil, and Paraguay, the Mbyá-Guaraní generally roam freely because, for reasons linked to their cultural traditions, they consider themselves a single people.

Table 18.1. Names of Mbyá-Guaraní communities in Misiones, Argentina.

1	Santa Ana Mirí	19	Caá Cupé	37	Tekoa Yma
2	Yvoty Okara	20	Kaaguy Poty	38	Kurupayty
3	Colonia Andresito	21	Ñamandú	39	Alecrín
4	Guazurarí	22	Kaa Poty	40	Pozo Azul
5	Katupyry	23	Tajy Poty	41	Santiago de Liniers
6	Ñú Pora	24	Y Ovy	42	Yaka Pora
7	Yacutinga	25	Campo Cumplido	43	Peruti
8	El Chapa	26	Tamandua	44	El Alcazar
9	Chapai	27	Tekoa Ara Poty	45	El Doradito
10	Ojo de Agua	28	Chafariz	46	Guapoy
11	Kaaguazú	29	Arroyo Muerto	47	Fortín Mbororé
12	Takuapí	30	Guiray	48	Yryapú
13	Tabay	31	Guavira Poty	49	Sapukay
14	Leoni Poty	32	Taruma Poty	50	Arroyo Nueve
15	Marangatú	33	Fracrán	51	Paraje Mandarina
16	El Pocito	34	Caramelito	52	Kaaguy Pora
17	Yby Pytá	35	Jejy	53	Km. 278–R. Franco
18	Virgen María	36	La Internacional	54	Saracura

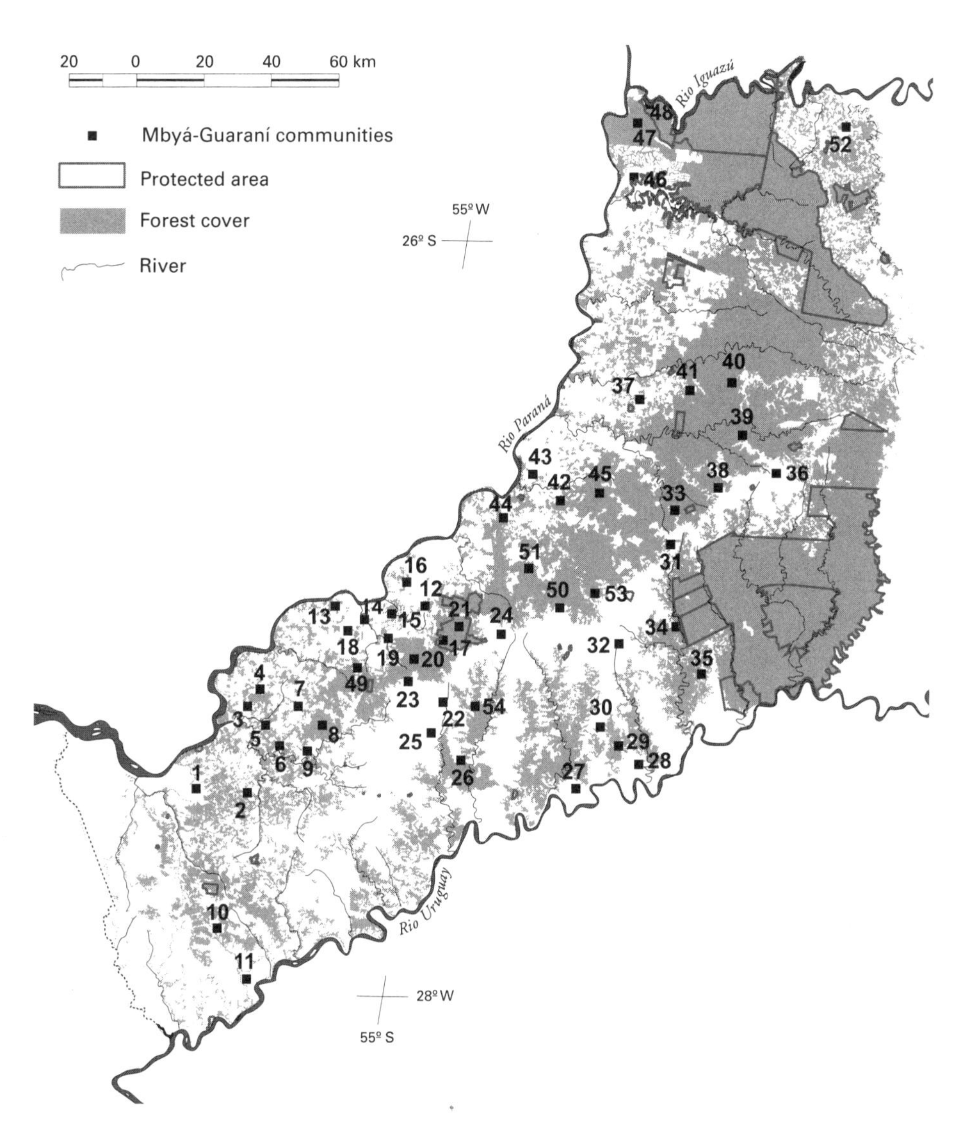

Figure 18.1. Location of Mbyá-Guaraní communities in Misiones, Argentina (Dirección de Asuntos Guaraníes de la Provincia de Misiones).

The Mbyá were originally a nomadic people, and some communities retain this lifestyle, traveling in family groups or alone among the various villages in Misiones and neighboring countries. They sometimes are called Cainguás, which means "forest dwellers" or "of the forest," a name they received because they refused to live in the Jesuit *reducciones* (mission settlements) or on the Spanish *encomiendas* (large estates granted to Spanish colonists that included rights to the use of indigenous labor) (Chapters 14 and 24, this volume). Instead, they put up fierce resistance against anyone who tried to penetrate their culture (Martínez and Crivos 1997; Cebolla Badie 2000).

The Mbyá in Misiones seem to fall into three major groups (Sánchez, pers. obs. 2002). One group is relatively sedentary, and the men of these communities hold jobs (e.g., farm labor) as the main source of their livelihood. A second group has been far less influenced by other cultures and remains much more nomadic. Although they also perform jobs like those of the more sedentary group, the basis of their economy is marketing their handicrafts, which constrains them to remain near routes where potential buyers travel. They maintain a united tribal life, preserving the rituals and other manifestations of their religion. The last group is composed of nomadic bands who practice small-scale farming in the forest. They also hunt, although to a lesser degree than in the past because the land available for hunting is limited. The Mbyá who belong to this group avoid contact with white people as much as possible. Occasionally, they come into white population centers to trade their products, especially for food and tools. Their privacy with regard to religious traditions makes them the group least influenced by other cultures.

The Mbyá's long years of contact with nonindigenous society, which has new and different cultural patterns, has led them to abandon or modify many of their customs and traditional practices. Those changes, in combination with other factors, have provoked a true crisis in many respects. Nevertheless, the Mbyá have managed to survive ethnically, preserving their language and, at least partly, their system of social and political organization. Their dialect differs from the Guaraní language currently spoken in Paraguay and northeastern Argentina, and very few researchers or other people speak it.

Land Tenure

The way in which Mbyá communities occupy or hold their lands and territories is different from that of nonindigenous people. The basic principles underpinning Guaraní thought and attitudes are territory, people, and community law, and diminishing the force of any one of them immediately weakens the others. National and provincial laws deriving from the national constitution are supposed to protect the land tenure rights of all the indigenous communities in Argentina. However, processes have occurred that make it possible to bypass the law and transfer lands to other ethnic groups. In addition to this failure to enforce the law, actions by government and nongovernment agencies have often created conflicts and divisions in traditional leadership structures. Moreover, these agencies tend to act through limited, short-term programs designed to address specific issues, whereas fundamental structural problems, such as the question of land tenure, remain unchanged.

Legal land tenure is a tool that ensures self-determination for the community and its activities and also serves as a framework for executing sustainable development programs. In the words of Cacique Dionisio Duarte, legal transfer and tenure of lands means "that we can maintain the *yvy rupa* (world), our *kaagwy rupa* (forest), and our *yvagagy rupa* (cosmos) for the full development of our culture, with respect and freedom."

Use and Knowledge of Forest Resources

Forest management is influenced by indigenous farming and hunting practices. The Mbyá live in small communities, practicing subsistence agriculture. For that purpose, they clear small plots of land in the forest, known as *milpas* or *rozados,* where they grow cassava, tobacco, corn, pumpkin, watermelon, beans, and other crops. In the surrounding forest they gather fruits and bee honey, and they hunt animals, setting ingenious traps, such as the *monde,* which they build using materials from the forest. Although these traps are effective, the Mbyá snare fewer animals than nonindigenous hunters, who use firearms and dogs. The indigenous hunting practices thus have less impact (Giraudo and Abramson 1998). They also fish, using arrows or plants such as *ychipo chimbo* (*Paullinia* sp.) and six other species, which they crush and place in small streams, where the plants release a substance that impedes the fish's breathing (Martínez Crovetto 1968). After several years, when the land loses its fertility and the animal populations in the area begin to decline, they move elsewhere, enabling the recovery of the forest segment they have been using. The Mbyá also extract two varieties of cane from the forest—*tacuarembo* (*Chusquea ramosissima*) and *tacuapi* (*Merostachys clausseni*)—which they use to weave baskets that they decorate with *güembepi* (bark of the root of the tree philodendron, or *güembe* [*Philodendron bipinnatifidum*]). They also use the extremely flexible trunk of the *ñandipá* (*Sorocea ilicifolia*) to make basket frames. Their other handicrafts include lifelike carvings of animals native to the Atlantic Forest, made of wood from the forest, which reveal a detailed knowledge of the fauna. Some Mbyá groups market their crafts, many of which are produced specifically for sale to tourists. However, objects with a religious connotation, such as the *ajaka-ete* (basket for gathering the fruits of the earth) or the *guyra-pa ete* (genuine bow and arrow), are not sold.

The Mbyá groups in Misiones have shared information on 229 species of birds (*guyra*), with details on habitat, behavior (courting, mating), nidification, and diet, including some information unrecorded in Western ornithology (Cebolla Badie 2000). They have a refined classification system that includes 38 groups or families, and they use a binomial system similar to scientific nomenclature (e.g., *Inambu guachu* = *Tinamus solitarius*), albeit with some exceptions. Birds (and the forest as a whole) form an integral part of the Mbyá culture and worldview and are considered companions and protectors. It is believed that birds give good luck to hunters and carry messages from the gods and that they bring happiness, misfortune, epidemics, and death. They are also seen as indicators of the seasons of the year and dates for planting and harvesting (Cebolla Badie 2000).

The Mbyá believe that most birds were once Mbyá men who were transformed for various reasons, and their culture preserves numerous stories about the species and their functions. They also believe in the existence of spirits who rule over the animals and will bring tragedies and bad luck if too many animals are captured. As a Mbyá proverb says, *"Guachú Ja Ete mba'eve re'ey ñanemoangeko re'ey,"* which means "the True Ruler of the Deer will not trouble us without reason" (Cadogan 1992). Thus, the Mbyá believe that if they do not hunt too many deer they will not have problems with their ruler. Similarly, they also believe that

Caaguí Yara, the spirit of the forest, will bring misfortune to anyone who hunts excessively or cuts too many trees from the forest, a belief shared by many settlers and mestizos.

The indigenous groups use the animals and plants of the forest to the fullest possible extent, consuming especially the fruits but also the seeds, sprouts, and roots of some 80 species of wild plants, the honey of 11 species of bees and wasps, and the larvae of some insects, notably the palm weevil (*Rhynchophorus palmarum*), which is extracted from the queen palm (*Arecastrum romanzoffianum*). The leaves and trunks of the queen palm are also used for roofing, and the fruit, seeds (toasted, raw, or boiled), and heart are eaten, and the trunk (the core of the stem) is ground into flour (Martínez Crovetto 1968).

The Mbyá also possess extensive knowledge of the healing properties of forest plants. They have remedies for most diseases and use the vascular plants of the surrounding forest almost completely. Many local white men have learned and used the remedies of the Mbyá and other Guaraní groups, and some even consult the *Pai Mbyá* (shaman) when they are ill. Many indigenous remedies have been passed on to other forest inhabitants and are also used by the general population. This has given rise to a small pharmaceutical industry that markets forest plants, many of which are sold in pharmacies and supermarkets in the region and throughout the country. These include *ambay* (*Cecropia pachystachya*), which is used to treat respiratory problems, and *cangorosa* or *espinheira santa* (*Maytenus ilicifolia*), which is taken as a remedy for stomach acidity. In the northern part of the province of Corrientes, a transitional area of the Interior Atlantic Forest, a total of 445 medicinal plant species have been documented (Martínez Crovetto 1981). They are used to treat almost all cases of illness among local residents and are used as antiabortives, antiallergens, sedatives, contraceptives, disinfectants, diuretics, anticancer agents, remedies for heart and sexually transmitted diseases, and myriad other purposes. Nevertheless, although partial information does exist on medicinal plants and other forest products used by the Mbyá, the exact way in which they are harvested, treated, and applied is not well known. The Mbyá shamans use medicinal plants in combination with magical and religious rituals and practices, part of their healing process that takes into account both the physical and emotional condition of the patient (López and Sánchez 1996).

Many rural residents or "colonists" of European, Paraguayan, Brazilian, or Creole origin—generally with few resources—have lived for several generations in contact with the forest. Some of them have adopted or assimilated the customs or beliefs of indigenous and other ethnic groups, and in some cases they have amassed considerable knowledge about the forest and the use of its flora and fauna. For many of these residents, hunting remains a significant economic resource, especially in the current regional economic crisis. Some of these people have a profound ecological and biological understanding of the animals they hunt. One of their hunting methods is based on precise knowledge of the fruit-eating habits of their prey, which they capture by lying in wait near plants that produce abundant fruit. Forty-four plant species used for this purpose have been documented. These people know the phenology of the plants and have developed the concept of "key fruit-bearing species" used as food sources by their prey (Giraudo and Abramson 1998).

Still, the nonindigenous population can kill more animals with firearms and dogs, and they face fewer cultural constraints, which can easily lead to overexploitation of the forest resources (Giraudo and Abramson 1998). Moreover, as their descendants migrate to towns and cities and cease to depend directly on the forest, they quickly lose their knowledge of ecosystems and begin to use resources in a less sustainable and more irresponsible way.

Causes and Consequences of the Loss of Mbyá Culture

The Mbyá peoples have undergone a process of adaptation to the region and have developed complex and efficient systems of resource management that are well suited to the environment (Chapter 32, this volume). Lack of knowledge about the contributions they are in a position to make, coupled with Western culture's rejection of and indifference toward other ways of life, have resulted in insufficient attention to the wisdom of the Mbyá.

Regrettably, because most Mbyá live in forest remnants in Misiones and in Paraguay, as the forest disappears so does the Mbyá culture. The forest is essential to full self-realization of the Mbyá (*mbya reko*). Their worldview and beliefs are closely linked to the natural environment in which they live. For example, according to Mbyá belief, the creator of the earth assumes the form of a hummingbird (*maino'i*), which, together with the golden wasp (*kavyju*), is the messenger and advisor of priests (Cadogan 1992).

The situation of the Mbyá is complicated by the fact that they generally do not have title to their lands. In many cases, they are forced out of the forest by land owners, or the lands are deforested, depriving them of their main livelihood. Western society does not welcome the integration of different cultures into its socioeconomic systems, which leads to profound poverty and identity crises in indigenous communities. As a consequence, the Mbyá suffer from severe health problems, with high mortality and malnutrition rates, especially among children. Infectious and contagious diseases occur frequently, as do other, less well-evaluated health problems, including mental disorders with social roots, such as depression and alcoholism, which have serious repercussions.

The Mbyá are aware of their situation and the serious alterations that their environment has suffered (Cebolla Badie 2000), as is evidenced by the following statements: "The forests are disappearing. There are no longer fruits, and the animals are dying off. We (the Mbyá) are surrounded, and we are near the end" (Mbyá religious leader, Cebolla Badie 2000). "We were not made to sell trees. But the men are now fighting over even little green Laurel trees. . . . And we have no more forests. If we don't buy them, we will no longer have any forests. Our father never sold them; and the forest were there for our happiness, without exception" (Pai Antonio de Fracrán, Ramos et al. 1984).

In accordance with their worldview, the Guaraní believe that causing injury to the earth is sacrilege and feel that it injures them personally. This belief is rooted in their profound respect for all that surrounds them. Many cultures of people who live in harmony with their environment and possess a wealth of resource management knowledge are disappearing. These groups, which include the Mbyá, Creoles, and other forest dwellers, are victims of an identity crisis, poverty, and unrelenting social pressures aimed at forging a cultural model that is homogeneous and therefore not adaptive to the varied biomes of the world.

A crucial component of biological ecosystems is thus disappearing: human cultural diversity, which unquestionably holds many answers to the problem of sustainable resource management that classic science is far from discovering (Giraudo and Abramson 2000).

Acknowledgments

Our thanks to Mario Di Bitteti for his valuable collaboration in drafting and editing this article and to Marylin Cebolla Badie and Carlos Galindo-Leal for their comments. We are also grateful to Ernesto Krauczuk for his help in obtaining information.

References

Cadogan, L. 1992. *Ayvu Rapyta: textos míticos de los Mbyá-Guaraníes del Guairá.* Asunción, Paraguay: Fundación León Cadogan, Centro de Estudios Antropológicos de la Universidad Católica (EADUC-CEPAG)(Biblioteca Paraguaya de Antropología, Vol. 16).

Cebolla Badie, M. 2000. El conocimiento Mbya-Guaraní de las aves: nomenclatura y clasificación. *Suplemento Antropológico* 35(2): 9–188.

Giraudo, A. R. and Abramson, R. R. 1998. Usos de la fauna silvestre por los pobladores rurales en la selva paranaense de Misiones: tipos de uso, influencia de la fragmentación y posibilidades de manejo sustentable. *Boletín Técnico de la Fundación Vida Silvestre Argentina* (47): 1–41.

Giraudo, A. R. and Abramson, R. R. 2000. Diversidad cultural y usos de la fauna silvestre por los pobladores de la selva misionera: ¿una alternativa de conservación? In: Bertonatti, C. and Corcuera J. (eds.). *La situación ambiental Argentina 2000.* pp. 233–243. Buenos Aires: Fundación Vida Silvestre Argentina.

López, N. P. and Sánchez, A. 1996. La medicina Guaraní en la era prehispánica: sus aportes. *Cirugía y Cirujanos* 64: 73–76.

Martínez M. R. and Crivos, M. 1997. Relevamiento etnográfico del Valle del Cuñapirú, Misiones. In: UNLP (Universidad Nacional de la Plata). *Relevamiento del Valle del Cuña-Piru, Aristóbulo del Valle, Provincia de Misiones.* Anexo E. La Plata, Argentina: UNLP.

Martínez Crovetto, R. 1968. La alimentación entre los indios guaraníes de Misiones (República Argentina). *Etnobiológica* 4: 1–23.

Martínez Crovetto, R. 1981. Plantas utilizadas en medicina en el NO de Corrientes. *Miscelánea* 69. Fundación Miguel Lillo, Tucumán, Argentina.

Ramos, L., Ramos B., and Martínez, A. 1984. El canto resplandeciente—Ayvu rendy vera: plegarias de *los Mbyá-Guranies de Misiones.* Carlos Martínez Gamba, compilador. Buenos Aires: Ediciones Sol (Biblioteca).

Chapter 19

Socioeconomic Roots of Biodiversity Loss in Misiones

Silvia Holz and Guillermo Placci

The tremendous biodiversity of the Interior Atlantic Forest (also known as the Paraná Forest) is being lost as a result of relentless expansion of the agricultural frontier. Although the forest once covered large portions of eastern Paraguay, northeastern Argentina, and southeastern Brazil, today only 7.81 percent (30,541.94 km^2) of the original forest cover remains, and the remnants are highly fragmented (Figure 19.1).

Deforestation has been most severe in Brazil, where only 2.72 percent (7,710.98 km^2) of the original forest remains, driven by the large-scale agricultural, urban, and industrial development. Although 13 percent (11,527.9 km^2) of the original forest still exists in Paraguay, in recent years the intensive clear-cutting to make way for agricultural development and rural settlements has moved Paraguay into first place in deforestation in Latin America. The remaining patches of forest are highly fragmented (Figure 24.1, Chapter 24, this volume).

Argentina has a larger area of continuous forest: 11,303.04 km^2 (50.9 percent of the original area), covering much of the province of Misiones with varying degrees of degradation (Figure 15.2, Chapter 15, this volume). Misiones, though poor, has a diverse economy ranging from subsistence agriculture to thriving industries, chief among them large timber companies.

A range of different socioeconomic factors have shaped the development across Interior Atlantic Forest; today the regions differ greatly in land use and tenancy patterns, conservation of forest remnants, and other factors. This chapter reviews the socioeconomic factors that have affected conservation of the Paraná Forest in Misiones, focusing on several particular processes and trends that lend insight into the situation there now.

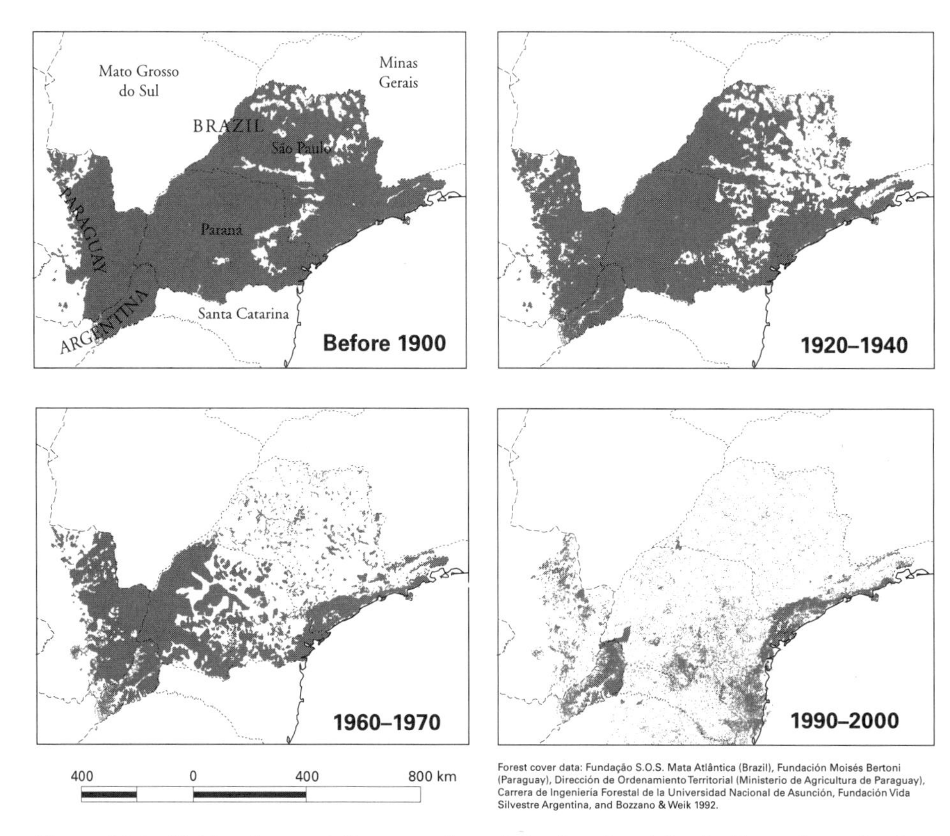

Figure 19.1. Habitat loss and fragmentation of the Interior Atlantic Forest from the late 1800s to the present. Forest covered by state mapped when available.

Impact of Human Settlements

Unlike Brazil and Paraguay, the settlement processes of Misiones took place late, and the principal objective was to populate the province and consolidate the country's borders (Schiavoni 1995). In Brazil, agricultural land was cleared across the Atlantic Forest, from the coast to the interior of the country, resulting in extensive deforestation as different crops were introduced, including cotton, sugar, and soybeans (Chapter 4, this volume). In Paraguay, development and deforestation began in Asunción (which is located at the western edge of the Paraná Forest) and progressed to the southeastern border with Argentina (Chapter 24, this volume). In Argentina, Misiones was one of the last agricultural frontiers, and its late settlement, beginning south and moving northward, was based initially on exploitation of forest resources, not on conversion of the land for agricultural production.

The Settlement Process

Misiones, like neighboring territories in Paraguay and Brazil, had been settled by humans for more than 10,000 years (Jacobus 1991). When the Spaniards arrived, Misiones was inhabited by Guaraní Indians (Chapter 32, this volume). The Spanish used the system of *encomiendas,* land grants given to conquistadors that included rights to use indigenous labor, a system of brutal repression that sparked repeated indigenous rebellions.

These uprisings, coupled with the evangelizing efforts of the Catholic Church, forced the development of a new settlement strategy know as *reducciones,* or Indian mission settlements, which were established by Jesuits and Franciscans. Over the next 150 years, this system gave rise to significant economic and social changes in ranching and farming areas, particularly through the introduction of irrigation systems, terrace farming, and crop diversification. The impact of human activities on the forest intensified as a result of the proliferation of feral cattle, clear-cutting for agriculture, use of hardwoods from the forest for housing, opening of roadways, and incursion into the forest to harvest yerba mate (Laclau 1994).

When the Jesuits were expelled from the Americas in 1767, the population of the region declined sharply, exacerbated by the wars of the time. The indigenous population, having lost its Jesuit protectors, was reduced by almost half (Báez 1926), and Jesuit territory was divided into departments, each governed by an administrator, who in turn answered to an administrator in Buenos Aires (Bartolomé 1969). Under the new priests and civil administrators, the *reducciones* entered a period of progressive decline, and thousands of Indians left the former Jesuit settlements (Poenitz and Snihur 1999).

From 1814 to 1881 Misiones was under the control of the province of Corrientes until faced with the imminent passage of the Federalization Law, when the Corrientes government sold 20,000 km^2 of the territory (Schiavoni 1995), approximately two-thirds of its current area. Twenty-nine purchasers took over the lands, and the state retained control of only a thin strip of territory in the central part of the province (Schiavoni 1995) and small areas in the south and other sectors. The impact on natural resource use patterns that developed during this period, when the ownership and control of the bulk of province's land was in the hands of a few, can still be seen today.

In 1897 the government of the Misiones territory strongly supported a policy of settlement and agricultural development that was aimed at populating the region, a step necessary to qualify as a province (Bartolomé 1969; Schiavoni 1995). By 1903 most colonizers were of foreign origin, drawn by various incentives. Settlement on private lands through "settlement companies" (private enterprises) began in 1919, when settlers from Germany, Poland, Switzerland, Japan, and England purchased land and moved into the Paraná River valley (Schiavoni 1995).

Intensification of Agricultural Development

Early in the twentieth century the provincial economy relied mainly on *mate* in the southern, central, and northeastern zones and timber extraction in the unsettled area around the Uruguay River (Schiavoni 1995). These two activities formed the base of Misiones economy. *Mate,* an indigenous drink, was adopted by the Spanish conquistadors, although the Jesuits were the first to cultivate *mate,* primarily to pay the required tax to the king. As *mate* consumption and demand increased, the natural *mate* plants were decimated (Devoto and Rotkugel 1936; Amable and Rojas 1989) and now it is rarely found in native forests. Today, *mate* tea is commonly consumed in southern Brazil, Chile, Paraguay, Uruguay, and throughout Argentina. The Argentine government provided strong incentives for *mate* production (known as "green gold") in Misiones and Corrientes, the only *mate*-producing provinces in the country.

After 1920, Misiones was settled mostly through squatting or illegal occupation of public or private lands from the descendants of previous settlers or from neighboring countries (Schiavoni 1995). Squatting continues to occur primarily in the northeastern part of the province and contributes greatly to the steady expansion of the agricultural frontier.

At this same time, tobacco farming became more widespread, especially among poor settlers. The first tea crops were also introduced, and tea production continued to expand until 1950 (Manzi 2000). Plantations of Chinese tung trees (*Aleurites* spp.), used in producing oils, were established in 1935 and expanded until 1950 (Schiavoni 1995). By the 1980s, however, the external market had disappeared, and tung tree cultivation largely ceased.

By 1940 the largest expanses of land without human settlements were located in the departments of 25 de Mayo, Guaraní, San Pedro, and General Belgrano. During this time, settlement occurred spontaneously. By 1950, the country developed an industrialization policy, and large pulp and paper companies opened in Misiones (Laclau 1994), encouraged by state credits and subsidies. By 1970, the forest area planted in exotic species in the province had quadrupled (Schiavoni 1995).

An influx of illegal Brazilians in the 1970s led to further population expansion in northeast Misiones, prompting the government to stem the inflow of immigrants (Schiavoni 1995) in part by launching settlement plans. These plans were also designed to attract well-capitalized producers and resulted in intensifying deforestation in the region.

Transformation of the Forest

Between 1810 and 1880, concessions were granted for harvesting of wild *mate* and extraction of native woods (Schiavoni 1995), an activity initially carried out by large timber companies that became the mainstay of the regional economy until 1920 (Laclau 1994). For more than half of the twentieth century, only four species were exploited commercially: cedar (*Cedrela fissilis*), black lapacho (*Tabebuia heptaphylla*), *peteribí* (*Cordia trichotoma*), and *incienso* (*Myrocarpus frondosus*). These logged forests eventually passed into the hands of the settlers, who continued to extract timber from them or level them in preparation for planting crops. The view that progress was incompatible with maintenance of the native forest (Devoto and Rotkugel 1936) was held by many and persists today in large segments of the population.

At present, the remaining 11,300 km^2 of native forest are located mainly in the northern and northeastern portions of the province, where they cover the upper river basins in areas with steep gradients and soils not suited to agriculture (Figure 15.2, Chapter 15, this volume), which are largely within the province's system of protected areas (Chapters 20 and 21, this volume).

Socioeconomic Patterns of Settlement

The settlement process resulted in tremendous heterogeneity, both in terms of cultural groups and in the spatial distribution of socioeconomic issues and consequences. Settlers of limited means populated the public lands in the *altiplano* and in the south, where the soils are less fertile than elsewhere. The arrival of large number of Brazilians seeking new livelihoods increased poverty levels and expanded the parceling of land. The lands along the banks of the Paraná River were settled by the wealthy, who purchased the best plots in the province. Between these two settlement routes remained large private properties, covered in native forests, generally held by absentee owners and exploited for timber. The Guaraní population, though greatly reduced, continued to live in isolated communities scattered through the province (Figure 18.1, Chapter 18, this volume).

Demographic Evolution

Misiones has had a unique demographic evolution. Early in the twentieth century, the province was largely uninhabited. After the settlement plans were introduced, it grew and was settled rapidly (Figure 19.2). Today, the urban population is concentrated mainly in the department capitals, with the Paraná valley the most densely populated. The province's rural population density has become the highest in the country (INDEC 1992), and the central eastern and central western regions have dense rural populations. Few native forest fragments remain in these areas, and only protected public lands and some large private properties are unsettled.

In the northeastern departments of Guaraní and San Pedro, the area with the largest remaining native forests, the rural population is also high, and most families are extremely poor. The only viable livelihood for most families is to slash and burn small areas of forest (4 to 5 ha) to grow tobacco and corn. Once productivity declines, these small parcels are abandoned, usually after 3 or 4 years. Rural dwellers also engage heavily in subsistence hunting.

Misiones is among the seven Argentine provinces with the highest annual growth rates: 28.1 percent, almost twice the estimated rate for the country as a whole. Based on projections for 1995–2000, the rate is expected to drop to 23.66 percent (INDEC 2000). The total fertility rate in Misiones is the highest in

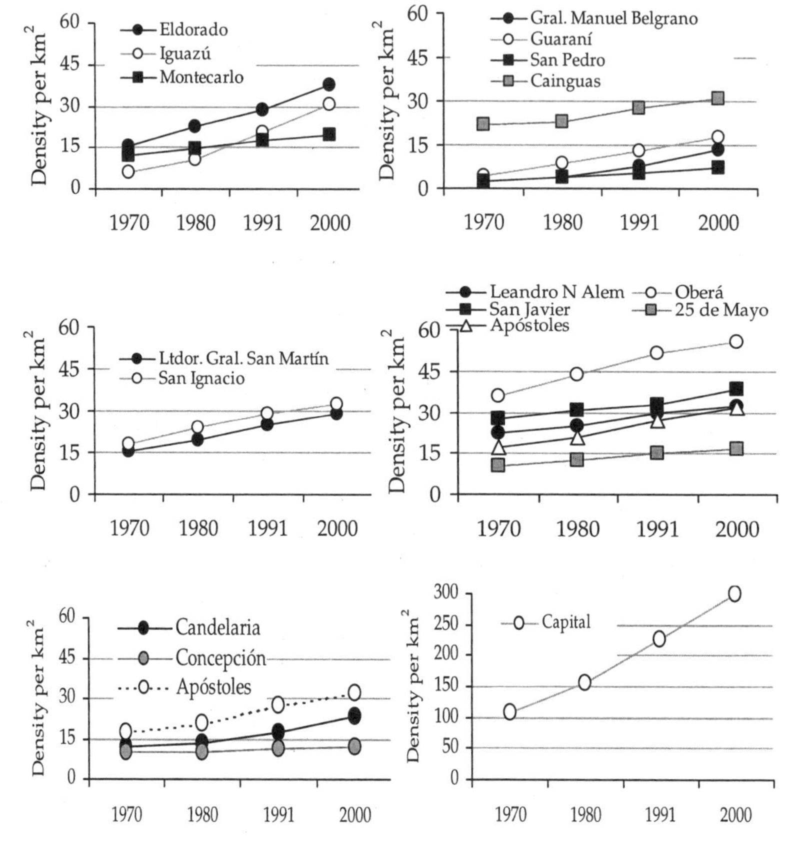

Figure 19.2. Population growth from 1970 to 2000 in different agroeconomic zones of Misiones.

Argentina (1995–2000 projections, INDEC 2000). This rapid growth, coupled with ongoing and unplanned internal migration, has caused significant changes in patterns of natural resource use.

The population growth (1980–1991) has not been uniform in all areas, however (Figure 19.2). In the northwest the urban population has grown while the rural population has decreased. In Iguazú, high rates of population growth result from internal migration prompted by job opportunities in the tourism industry and commercial activity in the border region. The northeast, the most recently settled region, continues to register rapid growth (with rates of 34 to 70 percent) because of the arrival of migrants from the south and from Brazil. The forest cover in this area has diminished markedly in recent years. The central eastern and central western regions, which were settled earlier, have experienced a shift from rural to urban population. In the southern region of the province (which comprises the capital city and its periphery), the urban population has grown as a result of migration from the rest of the province.

Current Trends

Although the latest population census was conducted in 1991, certain recent trends can be discerned. First, the economic situation of the past few years has led to a population decline in the northeast department of General Belgrano, which until the 1990s showed high growth rates as a result of the government-sponsored settlement plans. This area is strongly influenced by the Brazilian economy and by the *mate* crisis. The rural exodus, which has occurred in the northeast, northwest, and central western regions, has facilitated the expansion of new alternatives, such as the forest industry. Numerous unprofitable small farms are being bought up by timber companies at low prices, and many producers who migrate to the southern part of the province settle there to be closer to their sale markets.

The Social Arena

Development indicators (e.g., per capita output, total productivity, industrial productivity, infant mortality, literacy rates) place the province of Misiones as underdeveloped (Schiavoni 1995). The socioeconomic situation is diverse, as are the associated environmental problems. To achieve sustainable resource management, appropriate regional policies are needed.

Misiones is very much like a peninsula surrounded by neighboring countries, with a high degree of cultural exchange. It has a unique ethnic composition with predominance of groups of Polish, German, Swiss, Brazilian, and Paraguayan origin, unlike the rest of the country, most of whom trace roots to Italian or Spanish origins. The few Guaraní communities that survived the settlement process are scattered around the province and are extremely impoverished (Chapter 18, this volume). Their most critical problems are lack of land ownership, insufficient land, poor health conditions, and high rates of illiteracy (Martínez Sarasola 1992).

Basic needs are not met in about 28 percent of households in Misiones (33 percent of the provincial population) (INDEC 1993), and living conditions are extremely poor. Educational institutions (90 percent state run) lack the resources to meet the demands of this sector, where levels of formal education are extremely low: 60 percent of the population over age 13 has no more than a elementary education, and only 6 percent of adults complete secondary school (Laclau 1994). Many children must work on family farms, which often are far from schools.

In other words, the large expanses of native forest in the province of Misiones, in areas protected by the state or on private properties, are surrounded by desperately poor populations with little means of subsistence. These populations lack access to subsidies, are uncertain of selling their products (unless they produce tobacco), and have serious difficulty transporting their goods to markets; striving to make ends meet, these inhabitants exert tremendous pressure on the forest remnants. This situation makes it mandatory for any sustainable management strategy to account for social dynamics surrounding the forest remnants (Chapin and Whiteman 1998). However, current production and management models in the province do not address the specific needs and interests of poor local populations. At present, however, the church serves as a forum for discussion of socioeconomic and environmental problems and plays a key role in establishing and maintaining social relations and containing potential sources of conflict, such as alcoholism, domestic violence, and drug addiction.

The Economic Arena

The economy of the province relies basically on agricultural and forest products. Agricultural activities are strongly oriented toward production of perennial crops (principally *mate* and timber species). The main agricultural activities according to extent (percentage of the province) are stockbreeding (13.7 percent), crop farming (11.5 percent) (principally *mate*), and timber production on plantations (6.6 percent) (López et al. 2000) (Figures 19.3 and 19.4). Five primary agroeconomic zones can be distinguished (Gunther and Correa 1999).

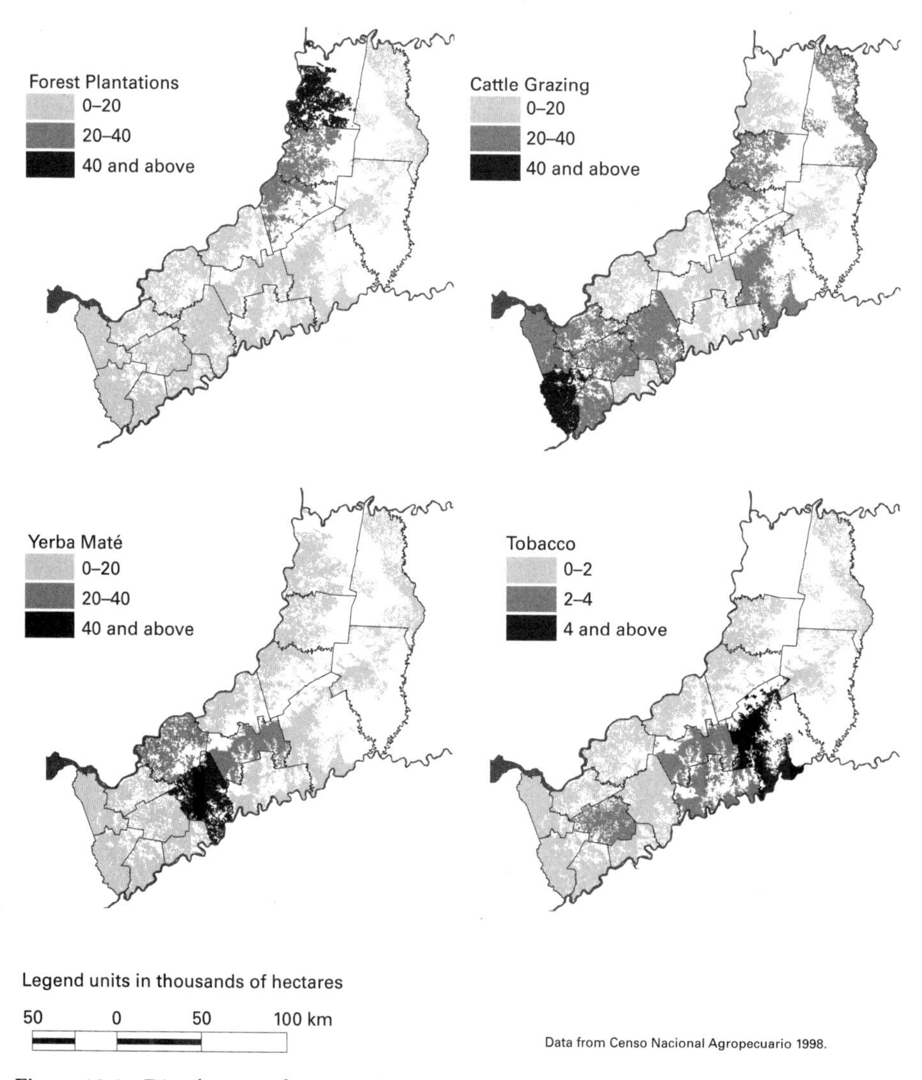

Figure 19.3. Distribution of primary land-use activities in Misiones.

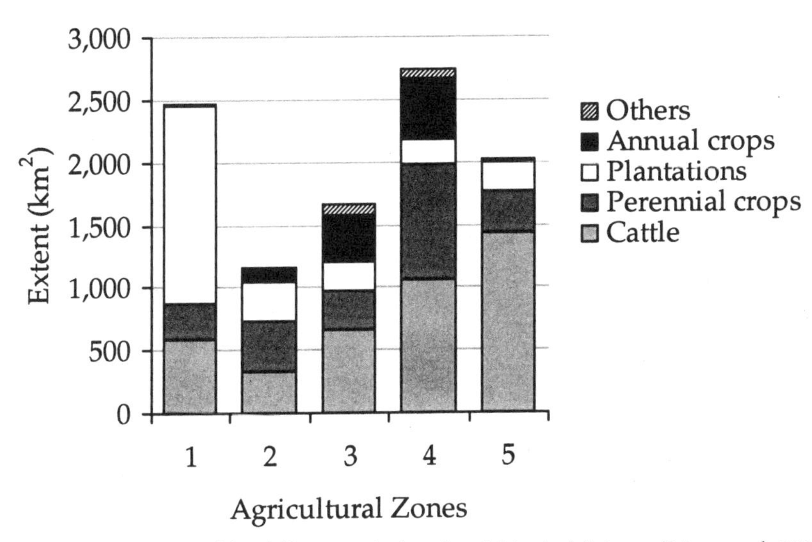

Figure 19.4. Area occupied by different agricultural activities in Misiones (López et al. 2000).

Stockbreeding

Stockbreeding occurs most often in the southern Misiones, where the cattle industry is characterized by poor animal health practices and inefficient pasture and herd management. Moreover, the producers graze their animals on planted pasturelands or in native forests that have been thinned to enhance grass productivity. However, in areas with favorable conditions for grasses and adequate cattle management, production levels are high, with annual yields of 250 to 350 kg/ha (López et al. 2000). The high productivity and earnings obtained in such places have motivated small producers to attempt stockbreeding on their farms, which in turn has become one the principal causes of the reduction in native forest cover. When pasturelands are abandoned, the forest recovers very slowly. A similar problem occurs in many other areas of Latin America and is becoming a major issue for applied research to design restoration strategies (Aide et al. 2000; Holl et al. 2000).

Farming of Annual Crops

Annual crops, particularly tobacco, are one of the few alternatives for small producers. Misiones is the chief tobacco producer in the country (Manzi 2000), and although this crop accounts for a small amount of farmland, most small-scale tobacco farmers work on small plots of land, mainly in the east of the province, which retains significant forest cover (Figures 19.3, 19.4, and 19.5).

As a result, many hectares of native forest are being replaced by tobacco farming. However, farmers have to move to neighboring forest once their productivity is depleted, usually after 3 or 4 years. Tobacco growing, a traditional activity in the province, has a tremendous social and environmental impact: it is labor-

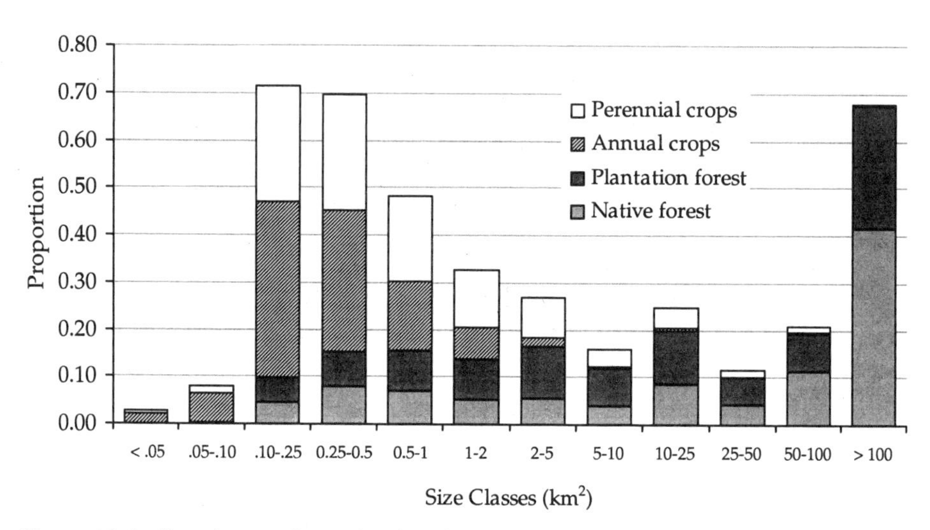

Figure 19.5. Size classes of parcels of different crop types in Misiones (Censo Nacional Agropecuario 1988; INDEC 1991).

intensive and uses large amounts of agrochemicals, which many producers are not trained to use.

Tobacco companies give producers the necessary resources for tobacco farming and then purchase their product. The social services offered by the tobacco companies are among the few such services that function adequately in the province, which explains why many small producers continue to grow tobacco even though their earnings are meager.

Perennial Crops and the Mate *Crisis*

Historically, *yerba mate* has been the most important perennial crop in Misiones. Most of the production (approximately 85 percent) is consumed in Argentina, and the rest is sold on export markets. In 1991 the production, marketing, and pricing of *mate* was deregulated, and since then its price has fallen steadily, while production has continued to increase, resulting in an oversupply. Currently, faced with another severe crisis, small and medium-sized *mate* producers are calling for new legislation to regulate activity in the sector, ensure fair prices, diversify the market, and put an end to the monopoly of the processing companies (there are approximately 100 such companies, but 78 percent of the ground *mate* is produced by only 8 large firms). In the present context, the farmers' earnings barely cover production costs.

Many *mate* farms in the province have been abandoned or sold, with severe effects on employment because approximately 25 percent of the province's workforce is employed in *mate* growing or processing (Manzi 2000). Production of *mate* tea is a multistage process that generates many jobs for plantation owners, hired labor, cooperative associations, and private companies.

Exploitation of the Native Forest

Until recent years, timber extraction has gone on in Misiones with no long-term management plans. As tree species of marketable diameter became increasingly scarce, other species began to be extracted. Currently, some 20 species are exploited commercially. Degradation of the forest remnants continues as a result of extraction of the few marketable trees still standing, coupled with inherent fragmentation consequences (Chapter 31, this volume).

Because the remaining forest fragments now have few marketable trees, the native forest has no commercial value for the owner, who must nevertheless pay taxes on the land. Furthermore, the environmental services provided—conservation of biodiversity, protection of river basins, regulation of landscape processes, scenic beauty, and others—are not valued. This situation poses a serious obstacle to conservation of the Paraná Forest because forest owners derive little benefit from maintaining the forest cover. Owners turn to a variety of activities to try to suck the last monetary value from their land. Some may extract (legally or illegally) all the marketable trees growing on their land, exacerbating the degradation of fragments, others replace the forest with more profitable activities, and still others sell their forestlands. In the current economic climate, it is increasingly difficult for producers, large and small alike, to maintain the forest remnants that exist on their lands. Providing an economic incentive for producers to conserve the native forest (through the system of "green credits," forest certification) is one of the few remaining options for halting fragmentation and degradation.

In 2001, an amendment to the Forest Law introduced a subsidy to encourage cultivation of native species on forest plantations, opening up new possibilities for production. However, for the most part producers are unaware of such laws, and even if they were, little information exists on how to manage the native forest in a sustainable manner, whether for commercial or conservation purposes. Improving the creation and dissemination of information about forest dynamics (growth, yield, and natural regeneration of native species) in a way that would allow prediction of future forest yields is critical if sustainable forestry plans are to be created (Camacho and Fenegan 1997).

No one knows how the composition and structure of forest fragments will change, nor how viable these fragment will be over time. Several studies indicate that forest fragments cannot be self-regulating because of edge effects and the constant physical and anthropogenic disturbances that impinge on them (Viana and Tabanez 1996; Viana et al. 1997; Laurance et al. 1998; Chapter 31, this volume). Exploitation authorized by government permits in otherwise protected areas of native forest continues today in areas with poor-quality soil, generally on slopes and in areas protected by law, where substitution is prohibited (Manzi 2000). The particularities of biodiversity and ecosystems in forest fragments must be much more thoroughly studied so that appropriate management practices can be developed.

Production of Exotic Forest Species

In the 1970s, Argentina developed a forest policy that provided subsidies for planting exotic species to be used for pulp and wood. These subsidies have boosted production on marginal lands in many parts of the country, restoring the productivity of highly degraded soils. In Misiones, however, the subsidies have worked largely as perverse incentives because they have reduced the costs of cutting down the native forest to plant monocultures of exotic species (principally pines, *Pinus taeda* and *P. elliotti*). For small producers, these subsidies offer tempting opportunities for quick profit. However, they often replace their crops or the native forest with species that take some 20 years to mature, during which time they earn nothing, while at the same time they must compete with large timber companies.

In sum, the system of subsidies, coupled with the impoverishment of the native forest, is transforming the forest industry in Misiones—historically based on exploitation of the native forest—into an industry based on management of exotic monocultures. These activities spawn major changes in the province, not only in terms of planted and native forest cover but also at the social level, as family enterprises are displaced by a forest industry that relies chiefly on male labor.

Political Support and Profitability

At present, strong political support exists for planting rapidly growing exotic timber species, both at the national level through the subsidy system and at the provincial level, as evidenced by the recently developed Forest Master Plan. This plan seeks to increase the forested area in Misiones within 9 years, mainly by increasing the area planted in exotic pines to 10,000 km^2, or one-third of the province's total area. However, the plan does not specify where these plantations will be located, something that should be clearly stipulated and articulated with the objectives of the Provincial Law on the Green Corridor and the Master Plan for Tourism Development in the province. In addition, the Forest Master Plan estimates only the commercial value of the native forest without taking into account the ecological role it provides.

Although the timber business is profitable, its profitability declines as logging sites move farther from road systems. Most timber activity therefore is concentrated in the Paraná River valley, in the vicinity of the pulp plants, and exotic forest species are grown primarily on large properties in the northwestern part of the province (Figure 19.3). The paper and pulp industry is represented primarily by three companies, which opened their doors in the 1970s. Although the industry grew substantially between 1991 and 1994, two of these companies (Papel Misionero and Celulosa Campana) are grappling with serious financial problems, and the third (Alto Paraná, S.A.) has recently made large investments in the modernization of its facilities to increase its productive efficiency.

Energy Generation and Infrastructure

The hydroelectric power industry is not very developed in Misiones, but the potential for expansion of this activity is one of the biggest threats to biodiversity conservation. Dam construction has many direct and indirect impacts on social and ecological systems (Chapter 35, this volume). The only significant hydroelectric generating plant is the Urugua-í dam on the Urugua-í River in northwest Misiones, the building of which flooded 88 km^2, inundating places with high concentrations of fauna (Gallardo 1986) and altering a watercourse with a large waterfall that had served as a natural barrier to biotas both upstream and downstream from it.

Although Argentina has experienced rising demand for electric power, the demands from neighboring Brazil are much greater. Misiones has great potential as a generator and supplier of hydroelectric power, but the social, economic, and environmental costs could be exceedingly high (Chapter 35, this volume). Three other major dam construction projects would also have an impact on Misiones: the Corpus Christi dam (Argentina-Paraguay) on the Paraná River (which would flood 90 km^2 of Misiones), the Garabí dam (Argentina-Brazil), also on the Uruguay River (which would flood 300 km^2 in southern Misiones), and the Roncador/Panambí (Argentina-Brazil) dam on the Uruguay River (FVSA 1996). To date, none of these dams has been built, but there is growing pressure to begin construction of the Corpus Christi dam (Figure 35.2, Chapter 35, this volume).

Tourism

Until recently, tourism was concentrated mainly in the area of Iguazú Falls and the Jesuit ruins, and little effort had been made to take advantage of other potential tourist destinations. Recently, tourism and ecotourism have increased, and the provincial government's Master Plan for Development of Tourism supports the development of different and more diverse tourist options than those offered by Brazil. In particular, the plan aims to encourage tourists to visit the forest, which would engage their interest and keep them in the province for a longer time. To ensure a viable tourist trade, the tourism industry must seek to conserve the forest and all the natural and human resources associated with it. Investment should emphasize worker training and foster long-term conservation of the Paraná Forest.

Multisectoral Participation

Until the early 1980s, agricultural activities accounted for the largest share of the gross geographic product (GGP), but industrial production has grown steadily since then as a result of the establishment of large paper manufacturing plants. In both sectors, timber and *mate* production are the most important activities. In the manufacturing industry, the principal activities are related to the pulp and paper industry (34 percent of the activity in the sector), *mate* processing (21 percent), and sawmills (10 percent) (Laclau 1994).

As for the agricultural sector, the principal activities are silviculture (48.1 percent) and production of industrial crops (43.5 percent, the primary crops being *mate* and, to a lesser extent, tea and tobacco), whereas stockbreeding accounts for only 5.8 percent (Laclau 1994). Therefore, cattle ranching, using the most land in the province, contributes only a small proportion to the GGP.

Characterization of Production Models

Productivity in Misiones is highly polarized, with large, profit-oriented enterprises on one extreme and small, labor-dependent agricultural producers on the other. The industrial production sector comprises several types of producers: family businesses that specialize in particular production methods, national companies (often headquartered in other provinces and capitalized with funds of domestic origin), and transnational corporations involved mainly in the timber sector but also in citrus and *mate* production. During the period 1988–1999 the number of small producers increased by approximately 20 percent (estimates of the Ministry of Agriculture and the National Agricultural Census). Many of the small family producers produce only for self-consumption, whereas others grow tobacco or have small *mate* or tea farms, as shown in Table 19.1. A high percent-

Table 19.1. Characteristics of the different production models in Misiones, Argentina (López et al. 2000).

Models	Small Family Production (SFP)	Capitalized Family Production (CFP)	Industrial Production (IP)
Property size range	3 to 60 ha	20 to 200 ha	200 to 5,000 ha
Cultivated area	Not more than 10 ha	13 to 50 ha	More than 150 ha
Labor	Family	Family and contracted	Contracted
Equipment	Scarce	Some machinery for agriculture	High
Capital	Lower than CFP	US$15,000–75,000	Higher than CFP
Annual crops for rent	Tobacco, corn, beans		
Subsistence crops	Manihot, yam, vegetables, orchards (mostly citrus)		
Perennial crops	*Yerba mate,* tea	*Yerba mate,* tea, orchards (mostly citrus)	*Yerba mate,* tea, citrus orchards, pine (*Pinus* spp.) plantations
Animal husbandry	Farm animals, cows (low numbers)		High numbers of cattle
Percentage of producers	70%	27%	3%
Location in agricultural zones	Distributed in the entire province but concentrated in zones 3, 4, and 5	Located in zones 1, 2, 4, and 5	Located in zones 1, 2, 4, and 5

age of these farmers do not have title to their land and therefore have no incentives to use appropriate management practices. Given the small size of their plots, these small producers are forced to farm on unsuitable soils whose productivity diminishes rapidly.

For many years, the capitalized family farming sector has experienced many financial crises as the price of their crops decline. *Mate* prices are now in a deep decline, prompting family farmers to sell their *mate* plantations to large timber companies to establish pine plantations. The timber companies generally pay very low prices for the land, but they pay cash, which is attractive to farmers.

Land Tenure System

Land distribution and the land tenure systems are cornerstones of environmental issues and solutions in Misiones because the size of properties and the way in which they are held largely determine the type of productive system that can be applied. In Misiones, smallholdings predominate (numerically), but large estates account for the greatest proportion of land, as shown in Figure 19.6.

Small properties are associated with high environmental and economic risk because the limited size of the productive area often means that resources are over-exploited and production is not diversified. Smallholders therefore are extremely vulnerable to economic and market crises. The majority of the farmland in Misiones (65 percent, which represents 67 percent of the total area of the province) is privately owned (INDEC 1991), and sharecropping and tenant farming are insignificant in terms of both number of farms and land area. Legal and illegal occupants (squatters) account for 27 percent of the farms in the province and 9 percent of its total area (INDEC 1991).

However, these statistics are from surveys done in 1988, and figures probably have changed since then. For example, it is well known that the province has seen a rise in illegal land occupancy in the past decade, and the size of large forest plantations has expanded. Many squatters occupy private lands, causing conflicts with

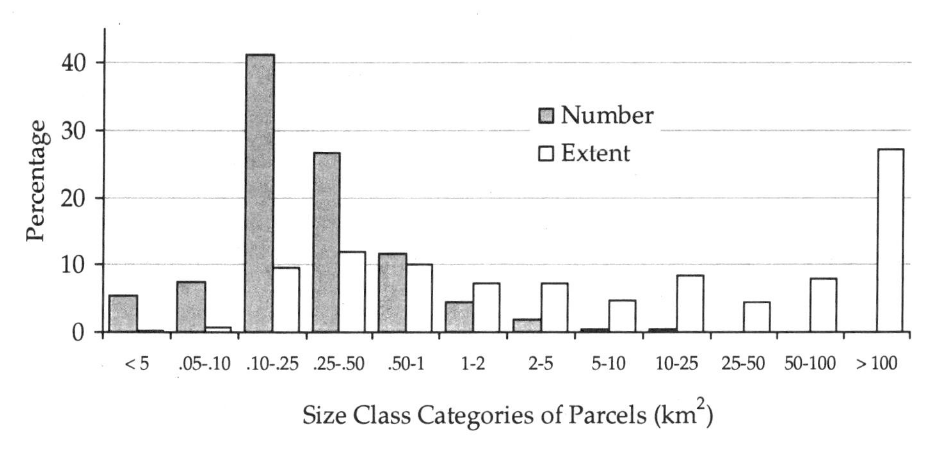

Figure 19.6. Number and area of agricultural parcels in Misiones according to class sizes (INDEC 1991).

legal owners and serious environmental degradation through poor management practices.

Regional Integration: Mercosur

In 1991 Brazil, Paraguay, Uruguay, and Argentina signed the Treaty of Asunción, establishing the Southern Cone Common Market (Mercosur), to harmonize their legislation in key areas and better integrate their trade (Abinzano 1998; Chapter 37, this volume). Despite the limitations, unresolved issues, and weaknesses of the Mercosur system, there is broad agreement in political circles and in society as a whole on the need to develop a strategy to increase the region's participation as a block in the world economy as it becomes increasingly globalized (Leichner 2000).

At present, trade between the Mercosur member countries is unbalanced (Abinzano 2000). Numerous Brazilian products come into Argentina and are sold at prices so low that local producers are unable to compete. The Argentine government has taken some measures to facilitate the export of goods to Brazil, but it has not addressed the high production costs that prevent Argentine producers from competing with their Brazilian counterparts (Abinzano 2000).

The Mercosur agreement recognizes the links between the environment, trade, economic growth, and integration and the concepts of sustainable management and internalization of externalities. Mercosur thus provides a regional framework for ecosystem conservation (Leichner 2000). In February 2001 the Mercosur countries signed a framework agreement on the environment (Mercosur/CMC/Dec. 2/01), which has the objective of preserving the region's natural heritage, encouraging sustainable use of natural resources, increasing and strengthening cooperation between member countries, promoting policies and practices that do not lead to environmental degradation, establishing common strategies, ensuring equitable conditions for competition, promoting the creation of specific funds as financial instruments for repairing damage and restoring the environment, and establishing mechanisms to prevent such damage.

Overall, then, myriad socioeconomic forces are threatening the Paraná Forest on many levels. With a backdrop of weak regulatory capacity in some government institutions, the forests are subject to the degrading influences of unplanned migration, squatting, and hunting by individuals, along with expanding forest plantations of exotic species, increased land area devoted to cattle ranching and agriculture, inappropriate uses of soil and agrochemicals, and a range of additional unsustainable practices in native forests.

Possibilities for Conservation and Regulatory Frameworks

The province of Misiones has enacted a number of laws to protect natural resources (Chapter 20, this volume). Despite enforcement problems and some cases of noncompliance, these laws are generally followed, leading to the conservation of 50 percent of the original forest cover. Nevertheless, this legislation will not, in itself, ensure linkage of the main protected forest areas. To promote connection of the forest fragments in the province with remnants in neighboring countries, another regulatory measure was devised: the Law on the Green Corridor of the Province of Misiones.

Both the Green Corridor Law and the National Forest Law (which created the system of subsidies for cultivation of native forest species) afford strong support for efforts to promote conservation and sustainable resource use. Nevertheless, they have not been widely disseminated, and a large proportion of small and medium-sized producers and municipal authorities are unaware or poorly informed of their provisions.

As described earlier, the sources of biodiversity loss in the region are complex. Some simple, economically and technically viable methods could be used to evaluate major trends in the province's environmental situation (Table 19.2). The proposed indicators would assist in the identification of areas where social or economic processes represent a conservation threat.

Conclusions

Historically, the presence of the native forest has been seen as a hindrance to development. This perception remains prevalent among many who value the forest only for its timber, and not for the ecological services the forest offers. This failure to value the forest has several primary roots. For example, many people continue to consider it necessary to clear the forest (which entails a cost) to engage in other productive activities. In addition, the forest remnants generally are degraded and lack marketable tree varieties, but those who own forestland must still pay taxes on it, making forest clearing seem attractive. The province has not yet developed the necessary technical tools and economic strategies (e.g., subsidies, carbon trading, forest certification) to ensure that the native forest remnants have value for small-, medium-, and large-scale producers. These are the types of tools and strategies that must be designed in the framework of the Green Corridor.

The socioeconomic issues relating to the use of natural resources are regionally different, and specific sustainable development strategies are needed to address the particular environmental problems of each zone. The current political and economic orientation in the province emphasizes the development of forest-based industries and ecotourism. The demand for wood from developed countries is expected to rise because they have restricted the use of their own forests while new markets have emerged in Asia. For these reasons pressure on native forests and on natural resources in general will increase. Development policies at the local and regional levels should take account of the ecological limits of natural systems so that these systems can continue providing environmental services to society.

Economic and political planning should provide for sustainable development programs for small and medium-sized producers, who constitute a major group in the province and are especially vulnerable to the current economic crisis in Argentina. Given the economic policy currently envisaged for Misiones, coupled with the growth and internal migration of the province's population, a land use

Table 19.2. Indicators to monitor resource-use variables, methods, availability, and reliability in Misiones, Argentina.

Threat for Forest Conservation	Indicator	Description and Methods	Availability and Reliability
Expansion of exotic species plantations	Native forest area. Plantation area.	Annual estimates of land-use changes through remote sensing.	Good capacity in the province to acquire and analyze imagery.
Increase in area devoted to cattle and agriculture	New areas devoted to cattle and agriculture. Area with abandoned fields (*capueras*).		
Unplanned migrations	Migration rate.	Population and agriculture censuses in areas with land-use changes to identify main local causes.	
Illegal land occupation	Percentage of different types of land tenure (owners, leasers, and squatters).		Low financial resources in the province to support these studies.
Land use	Average size of agricultural land units. Average income of producers.		
Inadequate land use and agrochemical use	Water quality (sediments, derived nitrogen compounds, coliform bacteria, pesticide traces).	Monitoring network in key watershed sites. Previously trained local people may conduct sampling. Assessment would function as an early warning system to identify areas where more intensive efforts are needed.	Low financial resources in the province to support these studies. Data obtained from these assessments are reliable.
Wildlife hunting	Population status of indicator animal species.	Rapid assessments of abundance of game animals (peccaries, felids, tapirs, deer).	Low financial resources in the province to support these studies. Data reliability depends on sampling design and execution.
State regulatory weakness	Degree of implementation of protected areas (e.g., land tenure, human resources, equipment, permanent funding, planning, illegal activities in the vicinity).	Assessment by independent organizations.	At present Fundación Vida Silvestre Argentina conducts this type of assessment in the province-protected areas and in neighboring Paraguay and Brazil (Chalukian 2000).

plan is urgently needed, one that focuses on sustainable development and biodiversity preservation. Finally, any provincial-level planning for the various sectors must be based on reliable, up-to-date information. It is therefore crucial to update current population, agricultural, and economic census data and to disseminate this information and all information related to sustainable resource management.

References

Abinzano, R. 1998. *Mercosur, un modelo de integración.* Posadas (Misiones), Argentina: Editorial Universitaria, Universidad Nacional de Misiones.

Aide, T. M., Zimmerman, J. K., Pascarella, J. B., Rivera, L., and Marcano-Vega, H. 2000. Forest regeneration in a chronosequence of tropical abandoned pastures: implications for restoration ecology. *Restoration Ecology* 8(4): 328–338.

Amable, M. A. and Rojas, L. M. 1989. *Historia de la Yerba Mate en Misiones.* Posadas (Misiones), Argentina: Editorial Montoya.

Báez, C. 1926. *Historia colonial del Paraguay y Río de La Plata.* Asunción.

Bartolomé, M. A. 1969. La situación de los Guaraníes (Mbyá) de Misiones (Argentina). *Suplemento Antropológico de la Revista del Ateneo Paraguayo* 4(2): 161–184.

Bozzano, B. E. and Weik, J. H. 1992. *El avance de la deforestación y el impacto económico: Proyecto de Planificación del Manejo de los Recursos Naturales (MAG/GT-GTZ).* Asunción: Icono.

Camacho, M. and Fenegan, B. 1997. *Efectos del aprovechamiento forestal y el tratamiento silvicultural en un bosque húmedo del NE de Costa Rica.* Serie Técnica: Informe Técnico No. 295. Turrialba, Costa Rica: CATIE (Centro Agronómico Tropical de Investigación y Enseñanza).

Censo Nacional Agropecuario 1988. Resultados Generales. Provincia de Misiones 19. Buenos Aires: Instituto Nacional De Estadística y Censos, 46p.

Chalukian, S. C. 2000. La protección de la Selva Paranaense en Argentina. In: Fundación Vida Silvestre Argentina. *Situación ambiental Argentina.* pp. 383–386. Buenos Aires: FVSA (Fundación Vida Silvestre Argentina).

Chapin, F. and Whiteman, G. 1998. Sustainable development of the boreal forest: interaction of ecological, social and business feedbacks. *Conservation Ecology* 2(2): 12.

Devoto, F. and Rotkugel, M. 1936. *Informe sobre los bosques del Parque Nacional Iguazú.* Buenos Aires: Ministerio de Agricultura de la Nación.

FVSA (Fundación Vida Silvestre Argentina). 1996. *La represa Corpus Christi y otras obras de la Cuenca del Plata.* Buenos Aires: FVSA (Fundación Vida Silvestre Argentina).

Gallardo, J. M. 1986. *Plan de relevamiento faunístico de la Cuenca del Arroyo Urugua-í (Misiones).* Posadas, Argentina: Ministerio de Ecología y Recursos Renovables de la Provincia de Misiones, EMSA (Electricidad de Misiones), Museo de Ciencias Naturales "Bernardino Rivadavia."

Gunther, D. F. and Correa, G. M. 1999. *Zonas agroeconómicas homogéneas y sistemas de producción predominantes para productores que integran "cambio rural" en la Provincia de Misiones.* Posadas, Argentina: INTA (Instituto Nacional de Tecnología Agropecuaria), Centro Regional Misiones, Estación Experimental Agropecuaria Cerro Azul.

Holl, K. D., Loik, M. E., Lin, E. H., and Samuels, I. A. 2000. Tropical montane forest restoration in Costa Rica: overcoming barriers to dispersal and establishment. *Restoration Ecology* 8(4): 339–349.

INDEC (Instituto Nacional de Estadísticas y Censos). 1991. *Censo nacional agropecuario 1988, resultados generales.* Buenos Aires: Instituto Nacional de Estadísticas y Censos.

INDEC (Instituto Nacional de Estadísticas y Censos). 1992. *Censo nacional de población y vivienda 1991.* Buenos Aires: Instituto Nacional de Estadísticas y Censos.

INDEC (Instituto Nacional de Estadísticas y Censos). 1993. *Anuario estadístico de la República Argentina.* Buenos Aires: Instituto Nacional de Estadísticas y Censos.

INDEC (Instituto Nacional de Estadísticas y Censos). 2000. *Anuario estadístico de la República Argentina 2000.* Vol. XVI. Buenos Aires: Instituto Nacional de Estadísticas y Censos.

Jacobus, A. L. 1991. Utilização de animais e vegetais na prehistoria do Rio Grande do Sul. In: Kern, A. *Arqueologia pré-histórica do Rio Grande do Sul.* pp. 63–87. Porto Alegre: Mercado Aberto.

Laclau, P. 1994. *La conservación de los recursos naturales y el hombre en la Selva Paranaense.* Boletín Técnico No. 2. Buenos Aires: Fundación Vida Silvestre Argentina.

Laurance, W. F., Ferreira, L. V., Rankin-De Merona, J. M., and Laurance, S. G. 1998. Rain forest fragmentation and the dynamics of Amazonian tree communities. *Ecology* 79 (6): 2032–2040.

Leichner Reynal, M. 2000. *Mercosur, su dimensión ambiental: comercio y prioridades políticas de inversión.* Washington, DC: WWF (World Wildlife Fund).

López, M., Noceda, C., Colcombet, L., and Gauto, J. 2000. *El sector agrario de la Provincia de Misiones.* INTA (Instituto Nacional de Tecnología Agropecuaria), Estación Experimental Agropecuaria Montecarlo/Facultad de Ciencias Forestales, Universidad Nacional de Misiones, Eldorado. Report prepared for FVSA (Fundación Vida Silvestre Argentina), Buenos Aires.

Manzi, G. A. 2000. *Desarrollo industrial de Misiones. Una óptica Crítica para la discusión con miras a la integración territorial.* Posadas (Misiones), Argentina: Editorial Universitaria, Universidad Nacional de Misiones.

Martínez Sarasola, C. 1992. *Nuestros paisanos los Indios.* Buenos Aires: Editorial Emecé.

Poenitz, A. and Snihur, E. 1999. *La herencia Misionera: identidad cultural de una región Americana.* Posadas (Misiones), Argentina: Ediciones El Territorio.

Schiavoni, G. 1995. *Colonos y ocupantes: parentesco, reciprocidad y diferenciación social en la frontera agraria de Misiones.* Posadas (Misiones), Argentina: Editorial Universitaria, Universidad Nacional de Misiones.

Viana, V. M. and Tabanez, A. A. 1996. Biology and conservation of forest fragments in the Brazilian Atlantic Moist Forest. In: Schelhas, J. and Greenberg, R. (eds.). *Forest patches in tropical landscapes.* pp. 151–167. Washington, DC: Island Press.

Viana, V. M., Tabanez, A. A., and Batista, J. L. 1997. Dynamics and restoration of forest fragments in the Brazilian Atlantic moist forest. In: Laurance, W. F. and Bierregaard Jr., R. O. F. (eds.). *Tropical forest remnants: ecology, management and conservation of fragmented communities.* pp. 351–365. Chicago: University of Chicago Press.

Chapter 20

Conservation Capacity in the Paraná Forest

Juan Pablo Cinto and María Paula Bertolini

The Paraná Forest, Argentina's portion of the Atlantic Forest, lies entirely within the province of Misiones in the extreme northeast of the country. It is bordered by the Brazilian states of Rio Grande do Sul, Santa Catarina, and Paraná and the Paraguayan departments of Itapúa and Alta Paraná (Chapters 14 and 15, this volume).

The political and land management history of Misiones, like that of the rest of Argentina, has been tumultuous. The region has been segmented and controlled in different ways under different authorities, from the reign of the Jesuits in the 1700s through the restoration of democracy in the mid-1980s. This history has shaped the conservation status and capacity in the region, setting the stage for today's situation (for more historical background, see Bartolomé 1982; UNESCO/ICOMOS 1995; Schiavoni 1998). The following timeline shows a few of the landmark events that shaped the current status of conservation in Misiones.

1934: First protected area is created in the Paraná forest, Iguazú National Park.

1948: National Forestry Administration is created under Law 13,273 on the Protection of Forest Wealth, the first to include the concept of a forest management plan.

1953: Federal Territory of Misiones becomes a province and becomes responsible for managing its own resources.

1964: Pine Law (Law 251) introduces the concession of forestlands to owners of industries that focus on the harvesting of other species.

1982: Islas Malvinas Provincial Park (100.37 km^2) is created, the first protected area under provincial jurisdiction.

1983: Democracy is restored in Argentina.

1984: Ministry of Ecology and Renewable Natural Resources is created.

1985: The province of Misiones proceeds with the construction of a hydroelectric dam on the Urugua-í River, a tributary of the Paraná River, endangering many sensitive species and habitats.

1990: Urugua-í Provincial Park (840 km^2) is created, the largest in Misiones.

1992: Law on Natural Protected Areas in the Province of Misiones is enacted (Law 2932); Earth Summit is held in Rio de Janeiro.

1993: Yabotí Biosphere Reserve is created in Misiones (2,400 km^2).

1995: Yabotí Biosphere Reserve is incorporated into the World Network of Biosphere Reserves of the United Nations Educational, Scientific and Cultural Organization (UNESCO).

1999: Green Corridor of the Province of Misiones Law (Law 3631) is approved, creating an integrated conservation and sustainable development area linking special conservation and sustainable use areas to maintain connectivity of the remaining forest of Misiones.

Today, in keeping with Argentina's constitution and its federal system of government, the province of Misiones has full control over its natural resources and jurisdiction over the protected areas located within in its boundaries. However, some land within the federal system of protected areas, including national parks and reserves, remains under federal jurisdiction. A variety of mechanisms exist to foster coordination between provincial and federal agencies that manage and control protected areas.

In this chapter, we briefly review the status of local government, nongovernment, and academic institutions, the status of protected areas, and other factors that provide the foundation for the conservation capacity of the Misiones province. We also review the regulatory and legislative cornerstones supporting conservation in the province and discuss trends and indicators in conservation capacity. We close with some recommendations that would strengthen conservation throughout Misiones.

Institutional Capacity

A variety of government and nongovernment institutions on various levels are currently charged with working on conservation issues in Misiones. National government organizations include the National Parks Administration (Administración de Parques Nacionales), which has jurisdiction over Iguazú National Park, Iguazú National Reserve, and San Antonio Strict Nature Reserve, and the National Institute of Agricultural Technology (Instituto Nacional de Tecnología Agropecuaria), which has a presence in Misiones and serves as an important liaison to rural communities.

Provincial government organizations include the Ministry of Ecology and Renewable Natural Resources, with undersecretariats for ecology, forestry, and bioregional planning. Although these offices have great potential to create and support conservation measures, their current budgets severely restrict their ability to carry out conservation-related activities. The province also has the Yabotí Biosphere Reserve Commission, which coordinates activities in the reserve and manages a cooperative project to establish a biological station in Esmeralda Provincial Park, at the core of the reserve.

Nongovernment organizations (NGOs) are also extremely active in the province of Misiones and include national and local organizations, academic institutions, and a trinational initiative. National NGOs include the Argentine Wildlife Foundation (Fundación Vida Silvestre Argentina or FVSA) representing the World Wildlife Fund (WWF) in Argentina, which runs the Paraná Forest Program in Misiones; the Argentine Bird Association/River Plate Ornithology Association (Asociación Aves Argentinas/Ornitológica del Plata), which represents BirdLife International and operates the Güira Oga Center for the Recovery of Endangered Birds in Port Iguazú; the Foundation for the Conservation of Species and the Environment (Fundación para la Conservación de Especies y el Medio Ambiente), which studies nontimber resources in Misiones and coordinates the National Biodiversity Group project of the Argentine Committee of the World Conservation Union (IUCN); and the Social Institute for Human Development and Advancement (Instituto de Desarrollo Social y Promoción Humana), which promotes rural development with an agroecological focus.

Local NGOs are networked and very active throughout Misiones. The Network of Ecology Associations (Red de Asociaciones Ecologistas) links several local organizations devoted to promoting local environmental education initiatives and carrying out public advocacy on environmental issues, including the Tamanduá Ecology Association (Asociación Ecologista Tamanduá) in Eldorado, the Cuña Pirú Ecology Group (Grupo Ecologista Cuña Pirú) in Aristóbulo del Valle, the Caá Porá Ecology Group (Grupo Ecologista Caá Porá) in Montecarlo, and the Our Environment Foundation (Fundación Nuestro Ambiente) in Posadas.

Another NGO, the Trinational Green Corridor Initiative, was established in 1999 to foster collaboration and coordination of the numerous sectors involved in use and management of the natural resources of the Paraná Forest region. The initiative is composed of elected national representatives of the administrations of protected areas, national and international conservation organizations from the private business and community sectors, rural development and grassroots organizations, and representative of scientific institutions and international organizations (WWF and FVSA 1999).

Finally, academic institutions also play an important role in developing conservation plans. The National University of Misiones is the most important academic institution in the province, with branches in Posadas, Oberá, and Eldorado and offering courses in the natural sciences, economics, humanities and social sciences, forestry and engineering, and the arts. The university owns a protected area, the Guaraní Experimental Reserve, which is classified as a multiple-use reserve and is located within the Yabotí Biosphere Reserve. The research conducted on this reserve consists mainly of studies of sustainable use of timber resources. Overall, however, considering the serious lack of information and understanding about the ecology of the Atlantic Forest region, the few programs and courses of study that exist related to conservation fall far short of educating and creating a core group of trained biologists and specialists in other disciplines, which is essential to study and promote applied conservation science in the region.

Status of Protected Areas

The Provincial System of Protected Natural Areas (created by Provincial Law 2932) includes private, municipal, provincial, and national conservation units that are classified into nine management categories, based on their conservation objectives and the degree of state intervention (Figure 20.1).

The province of Misiones has 57 conservation units with a total area of 4,498.52 km^2—15.1 percent of the total area of the province—that are legally registered, protected natural areas, and new areas are continuously being added. In all, 93.2 percent of the total protected area in the province is managed and administered by the national and provincial governments. As shown in Figure 20.1, the majority of these protected areas fall into a very conservative management category (IUCN Category II), the aim of which is to ensure the integrity of the conservation unit.

Privately and municipally owned natural areas, though numerous, account for little of the total protected area in Misiones (Figure 20.2). The municipal areas, in particular, generally are small, and their protected status is somewhat precarious. However, there are good prospects for the addition of some 200 km^2 more of private reserves.

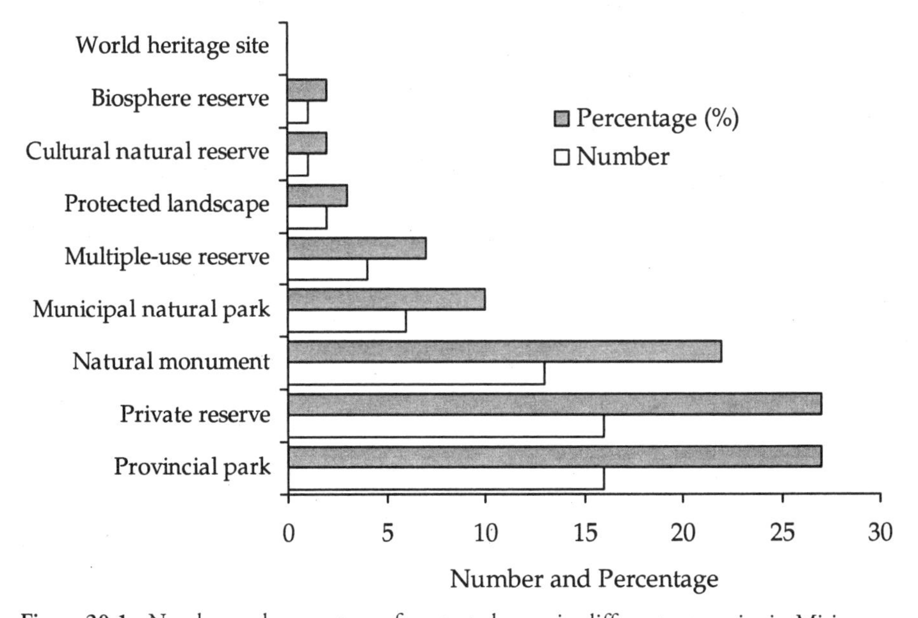

Figure 20.1. Number and percentage of protected areas in different categories in Misiones.

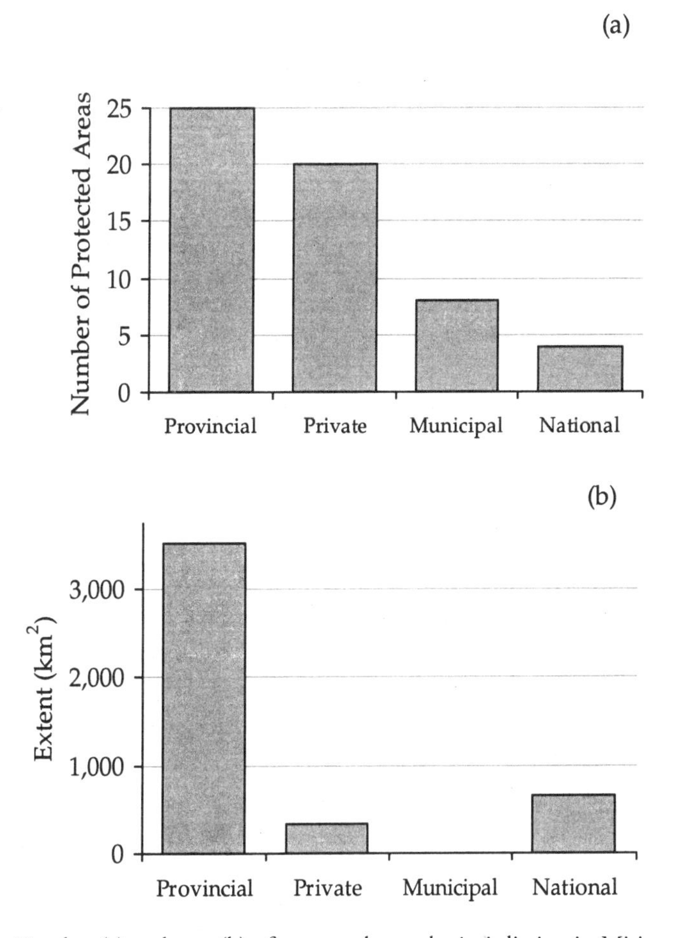

Figure 20.2. Number (a) and area (b) of protected areas by jurisdiction in Misiones.

Conservation Capacity Issues

Effective conservation in any setting requires that several factors be in place and functioning. In Misiones, the specific factors relevant to setting the stage for conservation include:

- development and use of management planning instruments,
- availability of financing and trained enforcement and management personnel,
- accessibility of research collections and information resources,
- availability of public education,
- infrastructure and equipment, and
- vulnerability status of protected areas.

All of these factors currently pose significant problems that affect the conservation units in the Paraná Forest (Chalukian 1999; Chapter 21, this volume).

Table 20.1. Status of management plans of the provincial protected areas in Misiones, Argentina.

Provincial Park	Department	Location	Status
De la Sierra	Apóstoles	South Zone	Completed
Moconá	San Pedro	Yaboti Biosphere Reserve	Completed
Cruce Caballero	San Pedro	Green Corridor	Completed
Urugua-í	Manuel Belgrano	Green Corridor	Completed
Guardaparque H. Foerster	Manuel Belgrano	Green Corridor	Completed
Salto Encantado del Valle del Cuñá-Pirú	Cainguás	Green Corridor	Completed
Araucaria	San Pedro	Green Corridor	Near completion
Piñalito	San Pedro	Green Corridor	In the process of elaboration

Planning Instruments

Although some management plans are in place throughout Misiones, they are not consistent, coherent, or well networked. At the national level, only the Iguazú National Park is developing a customized management plan, whereas all other parks are managed ad hoc with operational plans that are mostly budget-driven. However, since 1997 Misiones has created conservation management plans for seven protected areas under provincial jurisdiction, all developed and agreed upon with the communities directly associated with each area (Table 20.1). Among municipal protected areas, only one has a management plan and one has an annual plan of operations. About 60 percent of private reserves have some sort of planning instrument in place, perhaps because of tax incentives.

Funds and Financing

Like management plans, sources and mechanisms for funding conservation efforts throughout Misiones are also uneven and unstable. Currently, 80 percent of the province's conservation units suffer from significant financial shortfalls (Chalukian 1999). For example, entry fees to provincial protected areas are not uniformly required or collected, public budget lines are only sporadically allocated to fund salaries and other expenses, and few mechanisms are in place to ensure future budgets for conservation or planning. One exception is Iguazú National Park, which is the most visited protected area in the country. Iguazú National Park generates significant revenues, which are invested in the national system outside the province. Although some development funds have been established to collect resources to support the provincial system of protected natural areas, this mechanism is not yet fully operational; even when it is, resources may not be used for their original intended purposes but instead may be used to offset financial imbalances elsewhere.

Many potential sources exist to finance protected areas, from properly collected entry fees, to conservation concessions, to income from tourism, filmmaking, and international funding. However, few people who control these potential resources are aware of the stunning environmental settings in Argentina, and Misiones in particular. In addition to this lack of awareness, Argentina often is overlooked when cooperation initiatives for conservation are devised because technically Argentina cannot be classified as a developing country, or as a tropical country. However, as information about the forests of Misiones and the environment in Argentina is further disseminated around the world, more extensive funding for conservation initiatives may become available.

Enforcement and Management

National and provincial protected areas devote much of their limited resources to enforcement and control and both areas have a standing staff of park rangers. Although about 46 rangers are now employed in 15 stations or operations centers across protected areas, more than half the areas have many fewer rangers than needed. Properly trained technical personnel responsible for management planning and administration are also in short supply. The Ministry of Ecology lacks an administrative structure devoted exclusively to managing the protected natural areas in the province. The ministry also lacks an interdisciplinary team that can develop a technical approach to managing the Provincial System of Natural Areas, address the problems associated with increasing numbers of conservation units, implement management plans, or respond to the growing need for information from the field.

Research Collections and Information Resources

The absence of cultural, social, and biological information about the protected areas is one of the biggest impediments to management planning and decision making. In 71 percent of the conservation units, basic information is lacking, incomplete, or insufficiently organized (Chalukian 1999). This situation is exacerbated not only by the underdeveloped capacity for academic training for researchers mentioned earlier, but also by the unfortunate tradition of limited communication between the academic institutions and the agencies responsible for protected areas in the region.

Although the biodiversity of the Paraná Forest attracts numerous national and international researchers, the information generated as a result of their studies is not often made available in the region. Provincial legislation requires that reports on any studies be submitted, but this does not occur in the majority of cases because of the lack of oversight by the authorities charged with enforcing the law and neglect on the part of researchers.

In addition, although some conservation-related research is being done at national and provincial levels, little of this work is being published or distributed in ways that can be accessed and used by academics or conservation managers. The biological information currently available consists mainly of descriptive papers and inventories, and very few ecological studies have been done that include population, community, or landscape analysis or that examine ecosystem dynamics. Consequently, knowledge of the workings of the majority of natural systems in the region is limited, which in turn affects resource management for conservation purposes.

Similarly, the biological collections of Misiones are small and scattered throughout the province, in part because there seems to be no major natural history museum that can take biological collections and keep them in good condition. For this reason, much of the material collected in the course of the studies conducted in the province is placed in biological collections elsewhere in the country. Material from Misiones is found in the collections of museums and universities in the cities of Corrientes, La Plata, the Federal Capital, and Tucumán (Table 20.2).

However, the collections in these museums are not well organized or maintained, and the reference information of exhibited specimens is not complete or systematic, which greatly diminishes their scientific and biological value (Table 20.3).

Table 20.2. National biological collections housing specimens from the Atlantic Forest of Misiones, Argentina.

Taxonomic Group	Collection and Institution	City and Province
Flora	Herbario, UNaM (National University of Misiones [Universidad Nacional de Misiones], FCEQN (Faculty of Natural, Exact and Chemical Sciences [Facultad de Ciencias Exactas, Químicas y Naturales])	Posadas, Misiones
	Herbario, UNaM, FCF (Faculty of Forest Sciences [Facultad de Ciencias Forestales])	Eldorado, Misiones
	HerbNE (Northeast Botanical Institute [Instituto Botánico del Noreste])	Corrientes, Corrientes
Flora of Iguazú National Park	Herbario, CIES, IBONE (in duplicate)	Puerto Iguazú, Misiones
	Herbario, Instituto Botanico Darwinion	San Isidro, Buenos Aires
Vertebrates	MACN (Natural Sciences Museum of Argentina [Museo Argentino de Ciencias Naturales]) Bernardino Rivadavia	Buenos Aires, Buenos Aires
	Elio Massoia	Buenos Aires, Buenos Aires
	Museo de Ciencias Naturales	La Plata, Buenos Aires
	Universidad Nacional La Plata	

Table 20.2. Continued

Taxonomic Group	Collection and Institution	City and Province
Mammals	Facultad de Ciencias Exactas y Naturales e Instituto Miguel Lillo, Universidad Nacional de Tucumán	San Miguel de Tucumán, Tucumán
	CIES, Administracion de Parques Nacionales	Puerto Iguazú, Misiones
Bats	UNaM, FCEQN	Posadas, Misiones
	Facultad de Ciencias Exactas y Naturales e Instituto Miguel Lillo, Universidad Nacional de Tucumán	S.M. de Tucumán, Tucumán
Birds	Museo de Ciencias Naturales	La Plata, Buenos Aires
	Universidad Nacional La Plata	
	Facultad de Ciencias Exactas y Naturales e Instituto Miguel Lillo, Universidad Nacional de Tucumán	S.M. de Tucumán, Tucumán
	Museo de Ciencias Naturales	La Plata, Buenos Aires
	Universidad Nacional La Plata	
	CIES, Administracion de Parques Nacionales	Puerto Iguazú, Misiones
Amphibians and reptiles	Universidad Nacional del Nordeste	Corrientes, Corrientes
	Roberto Stetson, Facultad de Ciencias Exactas, Químicas y Naturales de la Universidad Nacional de Misiones	Posadas, Misiones
	Museo Argentino de Ciencias Naturales "Bernardino Rivadavia"	Buenos Aires, Buenos Aires
Fish	Instituto de Limnología "Dr. Raúl A. Ringuelet"	La Plata, Buenos Aires
Insects	Cátedra de Ecología, Facultad de Ciencias Exactas, Químicas y Naturales, Universidad Nacional de Misiones	Posadas, Misiones
Threatened fauna	Collection "Andrés Giai" of Chudy Mietes	Gobernador Lanusse, Misiones

Environmental Education

Little formal environmental education is offered through the schools because of the lack of training for educators and the scarcity of reference materials with local content. For the most part, environmental education in schools is provided on an ad hoc basis and is limited to activities carried out on special dates, such as Tree Day or Environment Day. There is no planned long-term program with specific objectives involving a particular geographic area.

In the system of protected areas, programs of environmental education envisaged for areas with management plans have yet to be formally implemented. The

Table 20.3. National museums with material collected in the Atlantic Forest of Misiones, Argentina.

Name	Responsible Institution	Location	Management Status
Natural Sciences Museum	Municipality of Oberá	Oberá	Threatened
Natural Sciences Museum Juan Foerster/Forest Museum Juan Foerster	Private	Previously, Dos de Mayo; today, Miramar, Buenos Aires	Uncertain
Iguazú National Park Visitor Center	National Park Administration	Puerto Iguazú	Will be donated to other institutions
Natural and Historical Sciences Museum	Institute Antonio Ruiz de Montoya	Posadas, Misiones	Regular
Regional Museum Aníbal Cambas	Misiones Province	Posadas, Misiones	Deficient

activities planned center mainly around environmental interpretation and community outreach, but to date very few such activities have been carried out.

Infrastructure and Equipment

The availability and condition of infrastructure and equipment in protected areas under provincial and national jurisdiction are markedly different. National protected areas have basic resources, including housing, transportation, and communication equipment for rangers, whereas provincial areas are significantly less developed and vary greatly from location to location. Even in the most visited provincial protected areas, indicative, regulatory, and interpretative signage remains deficient.

Vulnerability

The vulnerability of protected areas depends on numerous ecological, biological, socioeconomic, financial, and administrative variables or factors. Studies by Chalukian (1999) and Krauczuk (2000) conclude that areas under municipal jurisdiction are most vulnerable, mainly because of their small size and lack of trained managers, whereas areas under private and provincial jurisdiction are moderately vulnerable.

The greatest threats to protected areas of any kind are development projects (e.g., hydroelectric dams, road construction and improvement, agricultural development, and expansion of large forest-based enterprises) and illegal activities including hunting of animals and harvesting of plants, from palms to pine nuts.

Another important factor contributing to vulnerability is isolation. Sixty-seven percent of the protected areas are located in regions in which less than the 69 percent of the original forest cover remains. For 50 percent of the protected areas, land use in surrounding areas—for activities such as forest planting or replanting with exotic species, timber extraction, cattle ranching, and intensive or extensive but high-density agriculture—is a problem. Other areas are affected by the fact that they are located close to urban or industrial centers. Other critical aspects of the protected area system are discussed by Giraudo et al. (Chapter 21, this volume).

Regulatory Framework and Legislation Relating to Environmental Issues

Misiones Province has developed an extensive regulatory framework in the environmental area since the 1960s, when the large natural forest masses that covered almost the entire province began to be converted rapidly for agricultural use (Table 20.4).

One of the province's foremost environmental laws (1977 Provincial Law 854) establishes that it is in the public interest to ensure optimum use, protection, enrichment, and expansion of natural resources, optimum use of forestlands, and development of forest plantations and the timber industry. The law adopts the same forest classification as the National Law on the Protection of Forest Wealth (No. 13,273). The provincial law regulates the use of privately owned and provincially controlled forests and establishes forest reserves and Paraná pine seed reserve areas on public forest lands.

The purpose of creating forest reserves was to secure public lands with the greatest forest wealth and make the land available only for regulated forestry uses to ensure a continuous supply of wood. However, without enforcement, the law eventually lost its protective power. Some forests were divided up and turned over to new owners, other areas were ceded to indigenous communities, and others

Table 20.4. Conservation-related legislation for Misiones, Argentina.

Law 1040	Fishing	Regulates commercial and sport fishing; establishes regulations on methods and seasons.
Law 1279	Wildlife conservation	Declares as a public interest the protection, conservation, propagation, and rational use of wildlife. Absolutely prohibits commercial hunting.
Law 2980	Agrotoxic substances	Establishes control on the use of agrochemicals and their components in relation to human health, ecosystem protection, correct use, education, and information.
Law 3231	Soil conservation	This law, sanctioned in 1995, has not been approved yet.
Law 3337	Biodiversity conservation and sustainable development	Related to biodiversity preservation, sustainable use, and just retribution of genetic resources according to the Convention on Biological Diversity. Has not been enacted yet.
Law 3352	Medicinal plants	Focuses on preservation, conservation, and rational use of medicinal plant species.

were designated as protected natural areas, including the area that is now the Urugua-í Provincial Park.

Provincial Law 854 also created six seed reserves in the natural distribution area of the Paraná pine, which has been subject to intense exploitation since the 1950s. This law aimed at ensuring the availability of seeds of this endangered species in its natural distribution area and at increasing the genetic material of the region's forests. Sadly, these reserves were not properly protected, and the lands gradually lost their status and were taken over by farmers. Paradoxically, the only seed reserve currently protected was not originally covered under the law. Article 27 of Law 854 also establishes some controversial economic incentives by permitting the conversion of forests where exploitation of the native forest in perpetuity is not profitable, provided that the forest in question is not a protected or permanent forest.

In 1998, Law 25,080 was passed to regulate investment in cultivated forests. Under the law's legal framework, the Forestry Development Program of the national Secretariat of Agriculture, Cattle, Fisheries, and Food now provides economic support for the development of commercial forest plantations. In Misiones, developers can get the equivalent of approximately US$500 per hectare of developed plantation land.

Law 25,080 introduces several innovations that represent important advances in addressing the environmental impact of commercial forestry. One is the definition of a planted forest as a forest that, at the time the law was approved, was not covered by native trees masses or permanent or protective forests. This clause was included precisely to avoid the application of perverse incentives that would affect local biological diversity. However, this restriction is not always heeded by local forestry authorities, who often authorize the payment of incentives for plantations in areas that should be ineligible for receipt of this benefit, based on the legal definition of planted forests. This law also required that every forestry or forestry industrial undertaking include an environmental impact assessment and adopt measures that will ensure the maximum forest protection contemplated under the legislation.

Provincial Law 3426 reaffirms the prohibition on conversion of protective forests, expands the scope of protected forests, and mandates the maintenance of connectivity between protected forests (ecological buffers) in the design of commercial plantations. The function of this kind of reticulated design is promoted as a natural buffer that will prevent diseases and pests from damaging homogeneous forest crops.

Another important law, the 1991 Environmental Impact Law (Law 3079), was finally given enforcement authority in 1998, when regulations were established requiring that environmental impact assessments be carried out for all public or private projects, undertakings, or initiatives that might alter the physical, chemical, or biological properties of the environment (PRODIA 1999b).

Finally, the 1999 Green Corridor Law (Law 3631) created an area known as the Green Corridor Integrated Conservation and Sustainable Development Area of the Province of Misiones (see Figure 22.1, Chapter 22, this volume). This law establishes an area approximately 11,000 km^2, located in the upper river basin, where the largest section of the Paraná Forest is concentrated. The area encompasses 22 *municipios* (a political-administrative subdivision roughly equivalent to a county). As one of its principal components, the law creates three entities to support the forest. First is a Special Management Unit, which coordinates and designs strategies for ensuring the effective functioning of the Green Corridor. Second is an Advisory Commission, composed of individuals from various sectors who collaborate and advise on actions for the corridor's development. And third is a Special Joint Ecology Fund, which receives 1 percent of provincial tax revenues and may also receive funds from other potential sources.

The regulations recently approved under the law call for policies to be applied throughout the area to encourage restoration of the native forest cover and promote agricultural activities that will not entail replacement of native forests or break the connection between forest masses (Chapter 19, this volume).

Trends and Indicators in Conservation Capacity

Many factors necessary for effective conservation of biodiversity in Misiones are already in place. For example, the area of land with protected status has increased substantially over the past 10 years, and government initiatives could add another 100 km^2 by 2004 and another 200 km^2 by 2010 to official protection. However, budget cuts and policy changes since 1994 have reduced the number, size, and capacity of the entities responsible for conservation. A good example of this is the Ministry of Ecology, where funds allocated to conservation are grouped together with central administration expenses, with no provision for separating them within the ministry's budget.

The nominal budget (i.e., the budget approved under the General Budget Law) increased during the periods in which the protected areas underwent their greatest growth (1993–1995). Since then, however, this budget has gradually declined, with a particularly sharp reduction in the period 2000–2001 (Figure 20.3).

The operating budget, out of which the ministry's operating expenses are covered, has decreased even more. This budget is funded with moneys collected through fees, taxes, assessments, licenses, and fines levied by the Ministry of Ecology. It will be further reduced in coming years because it is derived only from extractive activities carried out in native forestlands, which are shrinking rapidly because of overexploitation and conversion to agriculture and plantations.

The lack of personnel with adequate technical skills in the protected areas is even more disheartening. At present, within the Ministry of Ecology, only three trained technicians work in the area of conservation, with two biologists focused primarily on surveying biodiversity and developing management plans for protected areas and one veterinary specialist devoted to the Center for Wildlife Breeding. Although the projected increase of 380 km^2 of protected land every year is a very positive trend, only five new park rangers for those areas will be added each year, certainly not enough to keep up with management and enforcement work (Figure 20.4). This woeful insufficiency of human resources for monitoring and control will only worsen as the area under protection is expanded. The shortage of park rangers and other workers limits the ability to implement

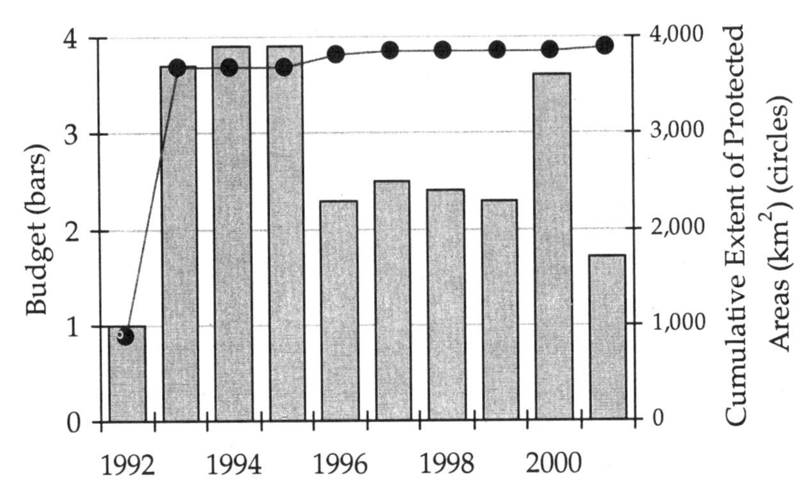

Figure 20.3. Trends in the area of protected areas (solid circles) and in their budget (bars) from 1992 to 2001 in Misiones.

both routine and urgent conservation measures. The situation is further exacerbated by the fact that the ministerial structure includes no technical entity for planning and managing protected areas (PRODIA 1999a). Moreover, budget constraints prevent park rangers from performing their monitoring and control functions, and the low salaries they receive hinder the execution of management plans and reduce the capacity to control illegal hunting and fishing.

Finally, to monitor the progress of conservation efforts and identify issues that need attention, periodic evaluations of management components must be performed using methodical, structured, and sequential procedures. Various potential indicators, structured as areas, variables, and subvariables, may be used to carry out these evaluations (Table 20.5). This particular structure can be adapted to the quality and availability of information (Cifuentes et al. 2000).

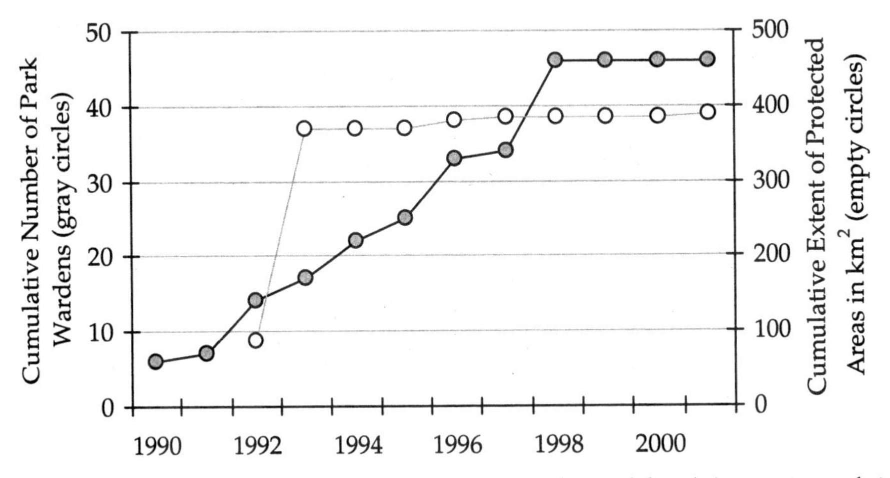

Figure 20.4. Despite an increase in the number of park wardens (solid circles) in Misiones, their numbers are inadequate to effectively manage protected areas (empty circles).

Table 20.5. Potential indicators to monitor institutional capacity for the Interior Atlantic Forest in Misiones, Argentina.

Areas	Variable	Subvariable	Information Availability
Administration	Personal	Technical, administrative, operative	Hiring rate, personnel roles
	Financing	Operative budget, special financing, fundraising capacity, capacity to hire additional personnel	Budgets of nonpermanent personnel, daily income, use of specific funds, fundraising projects
	Organization	Organizational structure, internal communication, files	Role in the structure, communication regulations, files
	Infrastructure and equipment	Infrastructure and basic equipment	Equipment department
		Mobility	Administrative office
		Communications	
		Investment	
Political	Support and participation	Programs with local organizations and municipalities	Agreements Strategic alliances
	Intrainstitutional support	Relations with other offices within the agency	Joint operations
	Interinstitutional support	Relations with other national ministries	Agreements
	External support	Relations with national and international agencies	Agreements Projects
Legal	Land ownership, general laws and regulations, laws for the creation of protected areas	Regularization of land tenure in protected areas, potential or present conflicts, coverage, enforcement	Property records, illegal occupation cases, law projects sent, generated, approved
Planning	Management plans, annual management plan, planning level, zoning, limits	Annual plans, availability and currency, coverage of areas, execution level, availability and currency	Annual planning of the Dirección General de Ecología
Knowledge	Socioeconomic information, bio physical information, legal information, monitoring	Information database	Environmental Information System Node Misiones
Special programs	Research, environmental education, environmental interpretation, protection, maintenance, community outreach	Design, execution, coordination, monitoring, assessment	Annual planning of the Dirección General de Ecología

Looking to the Future: Recommendations for Strengthening Capacity

This chapter underscores the marked imbalance between the biodiversity of the Paraná Forest and its conservation capacity. The reality is that Misiones possesses enormous biodiversity but, paradoxically, has limited capacity for basic and applied conservation research. This lack of capacity severely hinders conservation efforts, especially by government agencies faced with growth in private investment and development projects that pose a threat in the medium and long terms to conservation of these fragile and little-known natural systems.

A strong tradition of commercial forestry in the province of Misiones has led to the well-developed systems supporting the extraction of native timber and the conversion of native forests to planted species, without due regard for the complex ecological relationships that exist at various levels. The challenge now is to address these issues with the tools currently available and to devise plans that will strengthen local conservation capacity in the future.

Misiones, now recognized as a pioneer in building progressive conservation management, has clearly defined the conservation plans and goals it needs to accomplish now. In fact, Misiones's system of protected areas has a larger proportion of areas managed with progressive planning instruments than does the national system. The well-developed system of protected areas in Misiones is a tribute to the foresight and laudable priorities of the province's leaders, particularly bearing in mind that it has been in place for only 10 years and that the National Parks Administration is older, has a larger technical staff, and has the greatest control with regard to management of protected areas in the country.

Although a great deal of effort goes into preparing and documenting management planning tools, the resulting proposals and guidelines often are not properly implemented because of problems of a cyclical and structural nature, due to the perennial shortage of financial and human resources. In addition, the primary agent of conservation—the provincial government—is undergoing a budgetary adjustment that is further weakening its capacity to respond to the demands of a complex social and economic context. To accomplish its conservation goals, Misiones must establish alliances with international conservation organizations that will sustain the accomplishments that have been made to date and strengthen local and provincial capacities to carry on further conservation work.

Although international organizations will surely help, some actions by the government are clearly needed and should be prioritized. Specifically, to strengthen conservation capacity, the Ministry of Ecology and Renewable Natural Resources should take the following steps:

- Create a technical unit devoted to managing the provincial system of protected natural areas. This unit should be staffed by a variety of professionals, including biologists, anthropologists, geologists, and sociologists, who will work as a unit and with other institutions to monitor existing management plans and formulate new plans. Limited budgets seem to be the main obstacle to creating such a unit at present.

- Establish a provincial strategic program for conservation of the Paraná Forest that sets general management guidelines for the system of protected natural areas, as well as for strengthening institutional and financial capacities and coordinating legislative directives for municipal governments and other institutions involved in conservation and management of natural resources.
- Invest in training and hiring technical personnel and park rangers, equipment, and infrastructure for monitoring and control, mechanisms to ensure collection of fees and promotion of ecotourism, and funds and programs for public environmental education (Bertolini and Cinto 1999).

Because it may be unrealistic to expect an increase in public budget allocations for conservation activities and projects in a context of structural adjustment within the government, every effort should be made to seek additional resources from international organizations.

In addition to efforts that should be made by government institutions, other actions in scientific and academic realms should also be prioritized. Scientific and academic institutions should

- Create graduate programs geared specifically toward conservation and management of biodiversity and restoration of degraded environments.
- Focus existing research agendas on applied research on biodiversity conservation and compilation and management of biological information on the Paraná Forest.
- Create research centers focused on collecting, analyzing, and distributing information on conservation of this biome.

References

Bartolomé, L. 1982. *Colonos y colonizadores en Misiones.* Posadas (Misiones), Argentina: Instituto de Investigación, Impreso de la Facultad de Humanidades y Ciencias Sociales, Universidad Nacional de Misiones.

Bertolini, P. and Cinto, J. P. 1999. *Programa "Apoyo al plan estratégico para el Corredor Verde de la Provincia de Misiones."* Posadas, Misiones, Argentina: Ministerio de Ecología y Recursos Naturales Renovables de la Provincia de Misiones.

Chalukian, S. C. 1999. *Cuadro de situación de las unidades de conservación de la Selva Paranaense.* Buenos Aires: Fundación Vida Silvestre Argentina.

Cifuentes, M. A., Izurieta, A., and de Faria, H. H. 2000. *Measuring protected area management effectiveness.* Technical Series No. 2. WWF, GTZ, and IUCN.

Krauczuk, E. 2000. *Diagnóstico del Sistema Provincial de áreas Naturales Protegidas.* Informe técnico. Posadas (Misiones), Argentina: Ministerio de Ecología y Recursos Naturales Renovables.

PRODIA (Programa de Desarrollo Institucional Ambiental). 1999a. *Diagnóstico institucional y plan de acción para el Ministerio de Ecología y Recursos Naturales Renovables.* Secretaría de Recursos Naturales y Desarrollo Sustentable, Programa de Desarrollo Institucional Ambiental, Subprograma A: Final report. Iguazú: Ministerio de Ecología y Recursos Naturales Renovables. (Report prepared by Juan P. Cinto.)

PRODIA (Programa de Desarrollo Institucional Ambiental). 1999b. *Sistema de control ambiental, Núcleo Provincial Misiones, síntesis.* Iguazú: Ministerio de Ecología y Recursos Naturales Renovables.

Schiavoni, G. 1998. *Colonos y ocupantes: parentesco, reciprocidad y diferenciación social en la frontera agraria de Misiones.* Buenos Aires: Editorial Universitaria.

UNESCO (United Nations Educational, Scientific and Cultural Organization) and ICOMOS (International Council on Monuments and Sites). 1995. Las misiones Jesuíticas del Guayra. In: *La Herencia de la Humanidad.* Vol. II. Buenos Aires: Martínez Zago ediciones.

WWF (World Wildlife Fund) and FVSA (Fundación Vida Silvestre Argentina). 1999. *Memorias 3er. Taller Trinacional Conservación y Uso Sustentable de la Selva Paranaense (Argentina), Bosque Atlántico Interior (Paraguay), Mata Atlântica (Brasil).* Eldorado (Misiones), Argentina: WWF, FVSA.

Chapter 21

Critical Analysis of Protected Areas in the Atlantic Forest of Argentina

Alejandro R. Giraudo, Ernesto Krauczuk, Vanesa Arzamendia, and Hernán Povedano

Protected Areas: Size, Area, Percentage, and Representation

The province of Misiones has 60 protected areas with a total area of 4,597.66 km^2 (Table 21.1). These protected areas cover 15 percent of the province and encompass approximately 31 percent of the remaining forest (Rolón and Chebez 1998; Chalukian 1999; Krauczuk 2000). The largest area is under private jurisdiction (2,412.77 km^2, 52.7 percent of the total). As Figure 21.1 shows, provincial reserves are the next largest in terms of size (combined area of 1,493.15 km^2, 32.5 percent of the total), followed by national protected areas (680.20 km^2, 14.8 percent) and municipal protected areas (2.27 km^2, 0.05 percent).

Figure 21.2 shows the three large expanses of united or connected protected areas in the Atlantic Forest, which include:

- The Yabotí Biosphere Reserve (including Moconá and Esmeralda provincial parks, the Guaraní Experimental Area, and the Papel Misionero Natural Cultural Reserve) and the Premidia Private Reserve, with a total combined area of 2,418.13 km^2.
- The Urugua-í Provincial Park and Urugua-í Wildlife Reserve (872.43 km^2), which is connected to the system formed by Iguazú National Park and Reserve, the provincial parks Yacuy and Ingeniero Agrónomo Cametti (625.04 km^2), and the Caá Pora Private Reserve (0.41 km^2), which together cover 1,497.84 km^2.
- The Salto Encantado del Valle del Cuñapirú Provincial Park and Cultural Reserve, Cuñapirú Private Reserve, and Cuñapirú Municipal Ecological Reserve, with a total continuous surface area of 192.90 km^2.

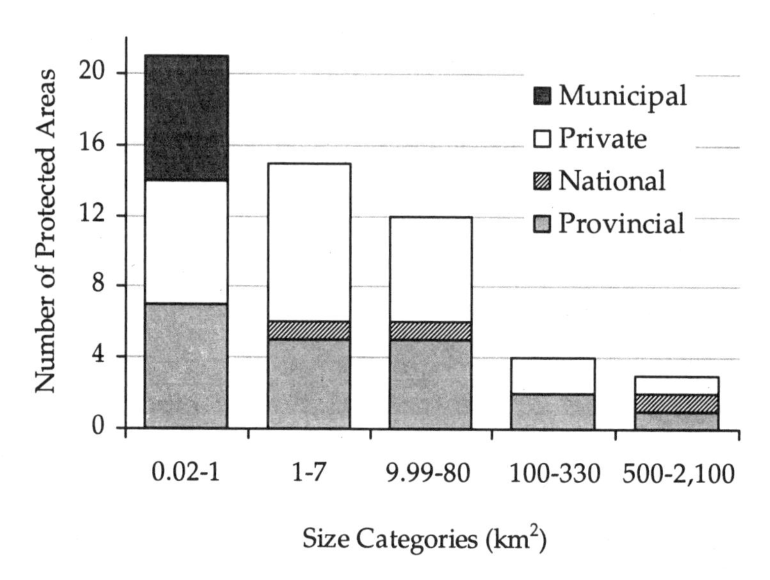

Figure 21.1. Protected areas of Misiones in different size categories by jurisdiction.

Table 21.1. List of protected areas in the Province of Misiones, Argentina.

	Protected Area	Jurisdiction	Legal Basis	Provincial Record	Extent (km^2)	Biogeographic Subregion
1	National Park Iguazú	National	Law N. No. 12103. Year 1934.	1	599.45	Rosewood and Assai Palm
2	National Reserve Iguazú	National	Law N. No. 18801. Year 1971.	1	76.75	Rosewood and Assai Palm
3	Forest Reserve Guaraní	Private	Provincial Law No. 628-854-2421. Years 1975, 1977, and 1987.	1	160.00	Montane
4	Experimental Area Guaraní	Private	Law Decree No. 26. Year 1975.	1, 2	53.43	Montane
5	Fish Reserve of Corpus, Río Paraná kms 1653–1667	Provincial	Provincial Law No. 1040. Year 1978.	1		Lotic aquatic (river)
6	Provincial Park Urugua-í	Provincial	Provincial Law No. 2794. Year 1990.	1	840.00	Rosewood and Assai Palm, Montane, and Montane Araucaria
7	Provincial Park Moconá	Provincial	Provincial Law No. 2854. Year 1991.	1, 2	9.99	Montane

Table 21.1. Continued

	Protected Area	Jurisdiction	Legal Basis	Provincial Record	Extent (km^2)	Biogeographic Subregion
8	Private Reserve Aguaray-mí	Private	Decree No. 1531. Year 1988.	1	30.50	Laurel and Guatambú
9	Private Reserve Itacuarahyg	Private	Decree No. 1647. Year 1989.	1	2.50	Laurel and Guatambú
10	Natural Municipal Park Paraje Los Indios	Municipal	Order No. 13/89 Municipality General Alvear.	1	0.11	Montane
11	Natural Municipal Park Amado Bompland	Municipal	Order No. 13/89 Municipality General Alvear.	1	0.02	Montane
12	Provincial Park Yacuy	Provincial	Provincial Law No. 2876. Year 1991.	1	3.47	Rosewood and Assai Palm
13	Provincial Park Esperanza	Provincial	Provincial Law No. 2876. Year 1991.	1	6.86	Laurel and Guatambú
14	Provincial Park Araucaria	Provincial	Provincial Law No. 2876. Year 1991.	1	0.92	Montane Araucaria
15	Provincial Park Cruce Caballero	Provincial	Provincial Law No. 2876. Year 1991. Disposition No. 2352. Year 1995.	1	5.22	Montane Araucaria
16	Provincial Park Teyú Cuaré	Provincial	Provincial Law No. 2876. Year 1991.	1	0.78	Campos
17	Provincial Park and Natural Cultural Reserve Salto Encantado del Valle del Cuñá-Pirú	Provincial	Provincial Law No. 2854. Year 1991. Provincial Law No. 3065. Year 1993.	1, 3	133.28	Montane
18	Wildlife Private Refuge Chachí	Private	With Fundación Vida Silvestre Argentina and Part. Year 1990.	1	0.18	Montane
19	Strict Natural Reserve San Antonio	National	National Decree No. 2149. Year 1990.	1	4.00	Montane Araucaria
20	Wildlife Private Refuge Caá-Porá	Private	With Fundación Vida Silvestre Argentina and Part. Year 1990.	1	0.41	Rosewood and Assai Palm
21	Provincial Park Isla Caraguatay	Provincial	Provincial Law No. 2876. Year 1991.	1	0.32	Laurel and Guatambú
22	Provincial Park Cañadón de Profundidad	Provincial	Provincial Law No. 2876. Year. 1991.	1	0.19	Campos

Continued

Table 21.1. Continued

	Protected Area	Jurisdiction	Legal Basis	Provincial Record	Extent (km^2)	Biogeographic Subregion
23	Private Reserve Tomo	Private	Decree No. 219. Year 1991.	1	14.41	Montane
24	Wildlife Private Refuge Timbó Gigante	Private	With Fundación Vida Silvestre Argentina and Part. Year 1991.	1	1.99	Laurel and Guatambú
25	Wildlife Private Refuge Chancay	Private	With Fundación Vida Silvestre Argentina and Part. Year 1991.	1	2.63	Montane
26	Wildlife Private Refuge Lapacho Cué	Private	With Fundación Vida Silvestre Argentina and Part. Year 1991.	1	1.60	Laurel and Guatambú
27	Natural Reserve Municipal Mbotaby	Municipal		1	0.14	Montane
28	Fish Reserve Caraguatay, Paraná River kms 1773–1786	Provincial	Provincial Law No. 1040. Year 1978.	1		Lotic aquatic (river)
29	Multiple Use Reserve EEA (Agricultural Experimental Station [Estación Experimental Agropecuaria]) Cerro Azul	Private	Resolution No. 7 INTA (Instituto Nacional de Tecnología Agropecuaria). Year 1992.	1	0.20	Montane
30	Multiple Use Reserve EEA Cuartel Victoria	Private	Resolution No. 7 INTA. Year 1992.	1	0.10	Montane
31	Private Reserve San Miguel de la Frontera-Premidia S.A. (Anonymous Society [Sociedad Anónima])	Private	Decree No. 92. Year 1993.	1	55.00	Montane
32	Biosphere Reserve Yabotí	Private and provincial	Provincial Law No. 3401. Year 1993.	1	2,363.13	Montane and Montane Araucaria
33	Provincial Park Sierra	Provincial	Decree No. 2402. Year 1993.	1	10.88	Campos
34	Natural National Monument Saltos del Moconá	National	Law N. No. 24288 Year 1993.	1	No data	Lentic aquatic (lake)
35	Natural Cultural Reserve Papel Misionero	Private	Provincial Law No. 3256. Year 1995.	1, 2	103.97	Montane

Table 21.1. Continued

	Protected Area	Jurisdiction	Legal Basis	Provincial Record	Extent (km²)	Biogeographic Subregion
36	Natural Park Municipal Lote C. Huerto Municipal	Municipal	Order No. 4. M. Puerto Esperanza Year 1995. Guatambú	1	0.84	Laurel and
37	Natural Municipal Park Luis Honorio Rolón	Municipal	Order No. 27–20 Municipality Puerto Iguazú. Years 1995 and 1996.	1	0.07	Rosewood and Assai Palm
38	Natural Municipal Park Salto Kuppers	Municipal	Order No. 23 Municipality Eldorado. Year 1995.	1	0.64	Laurel and Guatambú
39	Natural Municipal Park Yarará	Municipal	Order No. 24 Muncipality Puerto Esperanza. Year 1995.	1	No data	Laurel and Guatambú
40	Multiple Use Reserve Engineer (Ingeniero) Florencio Basaldúa	Provincial	Provincial Law No. 3376. Year 1996.	1	2.49	Rosewood and Assai Palm
41	Provincial Park Fachinal	Provincial	Provincial Law No. 3358. Year 1996.	1	0.51	Campos
42	Provincial Park Warden (Guardaparque) Horacio Foerster	Provincial	Provincial Law No. 3359. Year 1996.	1	43.09	Rosewood and Assai Palm
43	Protected Landscape Lake Urugua-í	Provincial	Provincial Law No. 3302. Year 1996.	1	79.80	Lentic aquatic
44	Natural Monument Provincial Isla Palacios	Provincial	Provincial Law No. 3302. Year 1996	1	0.20	Rosewood and Assai Palm
45	Private Reserve Yacutinga	Private		1	5.50	Rosewood and Assai Palm
46	Provincial Park Piñalito	Provincial	Provincial Law No. 3467 Year 1997.	1	37.96	Montane–Montane Araucaria
47	Multiple Use Reserve Alejandro Orloff (Saltito 0, 1, and 2)	Provincial	Provincial Law No. 3447. Year 1997.	1	No data	Montane
48	Provincial Park Esmeralda	Provincial	Provincial Law No. 3469. Year 1997.	1, 2	315.96	Montane
49	Protected Landscape Andrés Giai	Provincial		1	0.20	Rosewood and Assai Palm

Continued

Table 21.1. Continued

	Protected Area	Jurisdiction	Legal Basis	Provincial Record	Extent (km²)	Biogeographic Subregion
50	Wildlife Reserve Urugua-í	Private		1	32.43	Montane
51	Private Reserve Puerto San Juan	Private		1	2.50	Campos
52	Natural Park Municipal Cuñá Pirú	Municipal			0.45	Montane
53	Private Reserve Santa Rosa	Private		1	4.39	Campos
54	Private Reserve Santa María del Aguaraí-miní	Private		1	0.64	Laurel and Guatambú
55	Private Reserve Yaguaroundi	Private		1	4.00	Rosewood and Assai Palm
56	Private Reserve UNLP (Universidad Nacional de La Plata) Valle del Arroyo Cuña Pirú	Private		1	60.35	Montane
57	Provincial Park Engineer (Ingeniero). Agrónomo Roberto Cametti	Provincial		1	1.03	Rosewood and Assai Palm
58	Private Reserve El Paraíso	Private		1	4.40	Montane
59	Private Reserve Ingeniero Barney	Private		1	0.50	—
60	Arboretum Reserve Leandro Nicéforo Alem	Private	Agreement between Ministry of Ecology and Natural Resources of Misiones (MERNR) and National Institute of Agricultural Technology (Instituto Nacional de Tecnología Agropecuaria)	1	0.36	—
	Total extension				4,597.66	

[1]Included in the provincial records of Natural Protected Areas of Misiones.

[2]The extent of Yabotí Biosphere Reserve includes the Provincial Park Moconá, Esmeralda, Cultural Natural Reserve Papel Misionero, and the Guaraní Experimental Area. Therefore, these four areas are not included in the total sum.

[3]The Provincial Park Salto Encantado was created in 1991 and had an addition of the Cuña-Pirú Valley, through the project Natural and Cultural Reserve and Provincial Park Salto Encantado del Cuñá-Pirú, with the modification of Law 2.932 of 1994. It was included in the list of provincial protected areas (Cinto, pers. comm. 2002). Today, both areas function as a unit, and the project is expected to be approved soon.

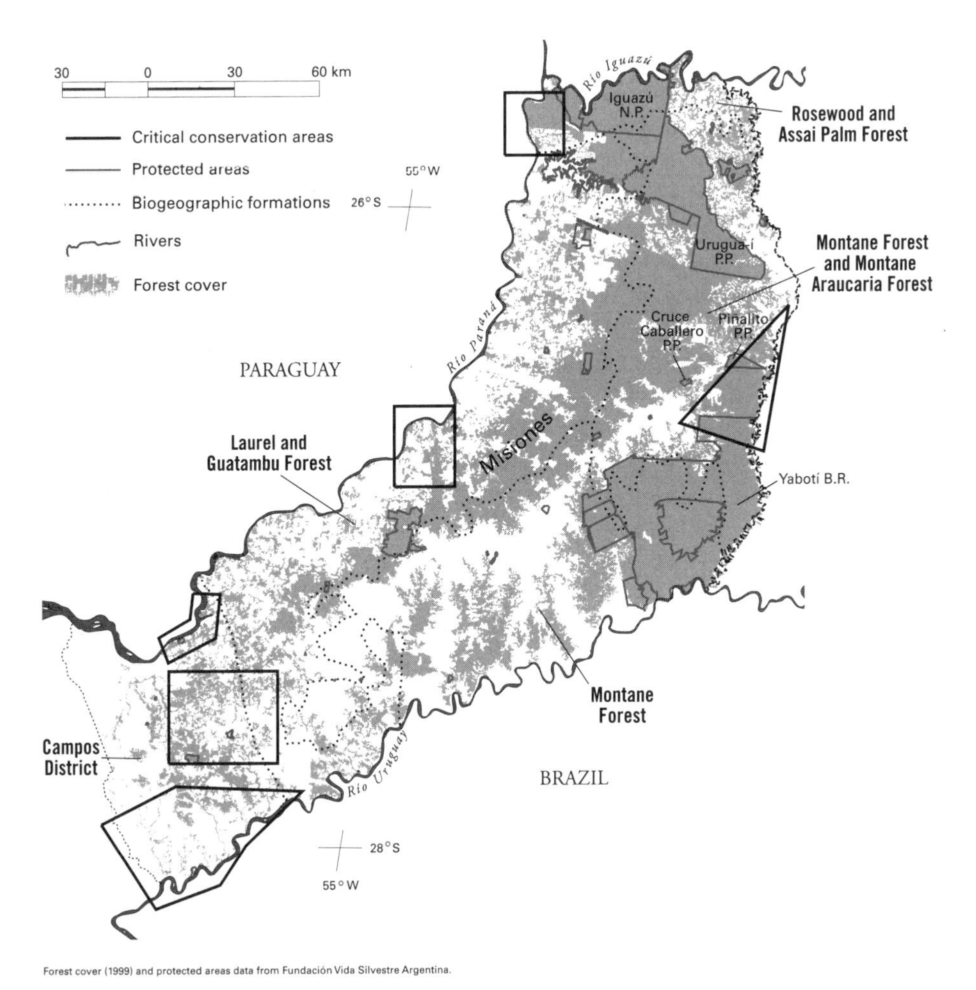

Forest cover (1999) and protected areas data from Fundación Vida Silvestre Argentina.

Figure 21.2. Critical regions for creating protected areas according to biodiversity patterns (endemism and species richness) and representativeness in relation to biogeographic subunits in Misiones.

However, as shown in Figure 21.3, various biogeographic subdivisions are quite unequally represented within the system of protected areas. The Montane Forest is the most extensively represented subdivision (3,285.1 km^2, 61.7 percent), followed by the Rosewood and Assai Palm Forests (1,815 km^2, 36.8 percent). The Laurel and Guatambú Forests (45.89 km^2, 0.9 percent), the Campos (grasslands) District (14.86 km^2, 0.4 percent), and the Montane Araucaria Forest (10.14 km^2, 0.2 percent) are poorly represented. Two reserves protect lotic aquatic communities along the Paraná River and one along the Uruguay River, and another protects Lake Urugua-í, an artificial lake (Table 21.1).

Of the protected areas, 63 percent (34 areas) are very small (between 0.2 and 7 km^2) (Figure 21.1), which, because of edge effects, fragmentation, altered demographic processes, and random variations or factors, limits their effectiveness for conservation purposes (Soulé 1980; Terborgh and Winter 1980; Quijano

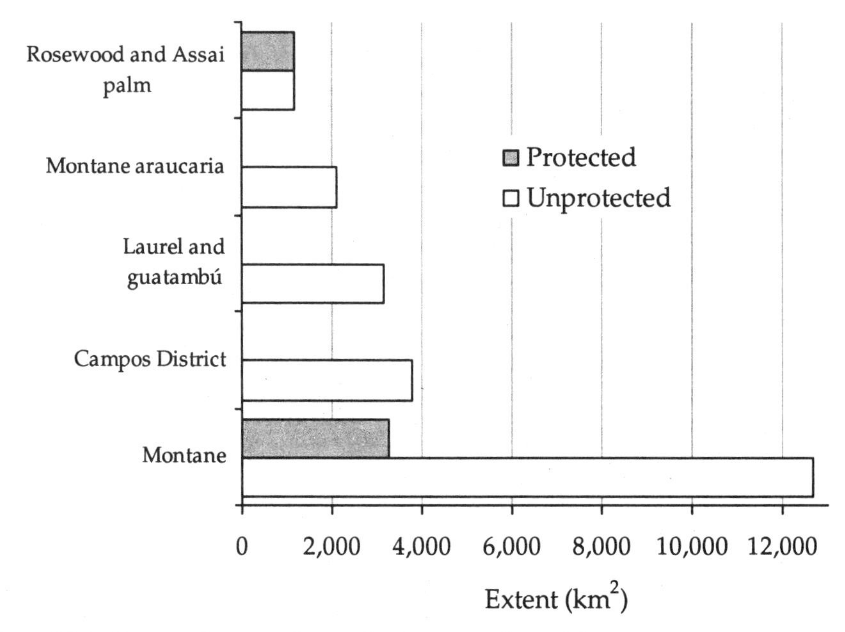

Figure 21.3. Extent of protected areas by biogeographic subdivision in Misiones. Montane Araucaria, Laurel and Guatambú, and Campos District each have less than 50 km^2 of protected lands.

1992; Chapter 31, this volume). Many species with large space needs are unable to maintain viable populations in such small areas (Gilpin and Soulé 1986). Furthermore, the disappearance of key species may subsequently affect other elements, given the complex biotic and abiotic relationships that exist in tropical forests (Terborgh 1992; Phillips 1997). Nevertheless, some researchers stress the importance of small areas, noting that fragments of at least 0.1 km^2 can maintain a significant proportion of the biodiversity within them for decades (Turner and Corlett 1996). The sustaining capacity of small forest fragments has been confirmed in intensive studies of bird communities in fragments located in southern and central Misiones, where protected areas are small but highly valuable for the populations they protect (Krauczuk and Giraudo, pers. obs. 2002).

Critical Analysis of the Protected Areas

The total protected area in Misiones has grown significantly thanks to the concerted effort of government, academic, conservation, and even private organizations (Chapter 14, this volume). Often, however, protected areas are not selected based on scientific criteria, which means they do not necessarily cover the areas with the greatest biodiversity or endemism or fully represent biogeographic subregions. Therefore, to succeed in consolidating existing protected areas and creating new ones in underrepresented locations, researchers need to conduct a critical analysis of the shortcomings of the current system, including financing. At present, 59 percent of the reserves that exist on paper have not yet been put into place, 30 percent of the protected areas are very vulnerable, and another 40 percent are somewhat vulnerable (Chalukian 2000). In large part, existing protected areas cannot fulfill their functions because necessary investments in infrastructure, personnel, and planning have not grown fast enough to keep up with needs.

We assessed the principal protected areas of Misiones based on a range of factors that influence reserve management, including administration, legal tools, planning, knowledge, management programs, present uses, biogeographic characteristics, and threats (de Faria 1994; Table 21.2). On average, we found that the protected areas of Misiones had unsatisfactory scores when they were rated on the basis of planning (34 percent), knowledge (31.5 percent), management programs (31 percent), and current uses (20 percent), although the average in the legal sphere was moderately satisfactory (70.3 percent). It is true that, when considered alone, the values for national reserves are satisfactory because all are protected by laws that provide for their creation and preservation. However, some reserves, especially private and municipal ones, are not currently covered under a legal framework, which makes their future uncertain.

In the area of administration, private and federal protected areas generally scored in the satisfactory range (>50 percent of optimal), but 90 percent of the provincial protected areas ranked as only marginally satisfactory (<50 percent of optimal), primarily because of lack of funding, administrative staff, and technical personnel. Moreover, many reserves lack planning mechanisms, particularly private reserves (45 percent).

Sixty percent of the reserves had poor scores for reserve management, with 95 percent of the private reserves ranked as unsatisfactory. In terms of threats, 72 percent of the reserves were unsatisfactory, 18 percent were marginally satisfactory, and only 10 percent scored in the moderately satisfactory range. The overall totals indicate that 49 percent of the protected areas were found to be unsatisfactory, 41 percent marginally satisfactory, and only 10 percent moderately satisfactory.

Finally, researchers and planners alike are hindered by the serious deficiencies in baseline knowledge regarding the protected areas: three-quarters of the reserves ranked very low in this area (<35 percent of optimal). Although some biophysical information is available on all the reserves, very little monitoring or data collection is being done on an ongoing basis (16 percent of optimal). From the biogeographic standpoint, 40 percent of the reserves we analyzed had unsatisfactory scores, and half of the protected areas we evaluated were too small to maintain a representative sample of their biotas, with 12 small reserves (30 percent of the total) also significantly isolated.

Table 21.2. Preliminary assessment of the main protected areas in Misiones. Percentages from the optimal situation: <35% indicates inadequate management, 36%–50% low adequate, 51%–75 % moderately adequate, 76%–89% adequate management, and >90% very adequate. (R. N. (Natural Reserve), P. N. (National Park), P. P. (Provincial Park), R. B. (Biosphere Reserve), A. E. (Experimental Area), R. P. (Private Reserve), R. N. C. (Natural and Cultural Reserve), M. N. N. (National Natural Monument).

Protected Area	Administration	Policy	Legal	Planning	Knowledge	Management Programs	Present Use	Biogeographic Characteristics	Threats	General Total
R. N. Iguazú	91.7%	85.0%	100.0%	75.0%	55.0%	65.0%	33.3%	83.3%	55.0%	66.5%
P. N. Iguazú	91.7	85.0	100.0	75.0	55.0	65.0	33.3	91.7	50.0	66.5
P. P. Moconá	45.8	70.0	100.0	83.3	50.0	70.0	27.8	58.3	35.0	52.8
R. N. Estricta San Antonio	87.5	70.0	83.3	75.0	55.0	50.0	16.7	33.3	35.0	52.3
P. P. Urugua-í	50.0	60.0	100.0	66.7	35.0	55.0	22.2	91.7	30.0	49.3
R. B. Yaboty	50.0	70.0	58.3	58.3	50.0	30.0	41.7	100.0	15.0	48.9
A. E. Guaraní	66.7	75.0	75.0	58.3	55.0	25.0	25.0	75.0	15.0	47.7
P. P. De la Araucaria	25.0	45.0	100.0	66.7	55.0	55.0	19.4	33.3	70.0	46.6
P. P. Cañadón de Profunidad	66.7	45.0	100.0	0.0	30.0	65.0	27.8	0.0	60.0	44.3
P. P. De la Sierra "Ing. Raúl Martínez Crovetto"	50.0	45.0	100.0	75.0	30.0	45.0	16.7	58.3	30.0	43.2
P. P. Salto Encantado del Valle del Cuñapirú	33.3	45.0	91.7	75.0	35.0	50.0	22.2	66.7	35.0	42.8
R. P. Puerto San Juan	66.7	45.0	50.0	75.0	30.0	25.0	30.6	41.7	40.0	42.6
R. P. Valle del Cuña Pirú	62.5	55.0	83.3	0.0	45.0	25.0	16.7	66.7	40.0	40.9
R. P. Yacutinga	83.3	45.0	16.7	50.0	20.0	50.0	27.8	41.7	30.0	40.9
P. P. Cruce Caballero	25.0	40.0	100.0	75.0	40.0	50.0	11.1	58.3	40.0	40.0
P. P. Horacio Foerster	41.7	45.0	100.0	75.0	30.0	45.0	8.3	33.3	30.0	38.6
R. P. Santa Rosa	75.0	35.0	50.0	75.0	20.0	15.0	25.0	33.3	40.0	38.6
R. P. Urugua-í	87.5	65.0	0.0	50.0	25.0	30.0	11.1	58.3	25.0	38.1

R. P. Tomo	70.8	45.0	50.0	66.7	20.0	15.0	13.9	66.7	20.0	36.4
R. P. Yaguarundi	70.8	55.0	50.0	8.3	20.0	25.0	19.4	58.3	30.0	36.4
P. P. Teyú-Cuaré	16.7	35.0	100.0	0.0	30.0	55.0	27.8	16.7	55.0	35.8
R. P. Aguarai-mí	66.7	40.0	50.0	25.0	30.0	20.0	30.6	41.7	20.0	35.8
P. P. Isla Caraguatay	16.7	40.0	100.0	16.7	55.0	30.0	22.2	25.0	40.0	35.2
P. P. Esmeralda	16.7	45.0	100.0	41.7	25.0	40.0	8.3	83.3	20.0	33.3
P. P. El Piñalito	16.7	45.0	100.0	50.0	35.0	45.0	8.3	41.7	25.0	33.3
R. P. Itacuarahyg	58.3	20.0	50.0	0.0	25.0	15.0	22.2	33.3	30.0	28.4
R. P. Caa Pora	66.7	40.0	25.0	0.0	20.0	20.0	19.4	33.3	15.0	27.8
P. P. Fachinal	16.7	25.0	100.0	0.0	30.0	20.0	16.7	25.0	40.0	27.3
R. P. Anna Park	50.0	20.0	50.0	0.0	20.0	15.0	22.2	8.3	35.0	25.6
R. P. Los Paraisos	41.7	30.0	25.0	8.3	20.0	15.0	13.9	50.0	30.0	25.0
R. P. Timbó Gigante	54.2	25.0	25.0	0.0	20.0	15.0	13.9	16.7	35.0	23.9
R. N. C. De Papel Misionero	12.5	15.0	41.7	0.0	35.0	20.0	22.2	66.7	15.0	23.3
R. P. Chancay	50.0	20.0	25.0	0.0	20.0	10.0	19.4	8.3	35.0	22.7
P. P. Ingeneiro Cametti	12.5	20.0	100.0	0.0	20.0	5.0	5.6	58.3	30.0	22.2
P. P. Yacuy	4.2	25.0	100.0	0.0	25.0	10.0	5.6	58.3	20.0	21.6
R. P. San Miguel de la Frontera	0.0	20.0	50.0	0.0	20.0	5.0	25.0	58.3	30.0	21.0
R. P. Lapacho Cué	54.2	20.0	25.0	0.0	20.0	10.0	13.9	8.3	20.0	20.4
P. P. Esperanza	0.0	10.0	100.0	0.0	25.0	5.0	5.6	16.7	35.0	17.2
M. N. N. Saltos del Moconá	0.0	0.0	66.7	0.0	0.0	0.0	36.1	33.3	16.7	15.3

Improving the System of Protected Areas

Representativeness, connectivity, financing, and engagement with local populations are all issues that must be addressed to improve the system of protected areas in Misiones.

Representativeness

First, more extensive and representative protected areas must be created in the Campos District, the Laurel and Guatambú Forests, and the Montane Araucaria Forest (Figure 21.3). The Campos District is the highest priority because it is one of the most species-rich units and has a substantial number of taxa exclusive to this biogeographic region (Giraudo 2001). In addition, the district has several endemic species that are not being adequately protected, perhaps because it is not a homogeneous area but rather a mosaic of forests, grasslands, and semixerophilous woodlands. However, these mixed features endow the Campos District with greater biodiversity than continuous forests (Laurance et al. 1997). The peripheral populations also merit special protection because they may be isolated for long periods and therefore are likely to be genetically different from central populations (Stotz et al. 1996).

The largest reserve in the Campos District of 10.88 km^2 (Table 21.1) is insufficient to conserve some fauna, such as large vertebrates, or the great diversity of habitats in the region. Other very small reserves include Teyú Cuaré Park (provincial) and Campo San Juan (private), on the Paraná River. It has been proposed that a provincial park be created on the latter site (Figure 21.2) to add about 50 km^2 that are well conserved. This initiative would also improve the representation of habitats and species distributed along the Paraná River such as the band-tailed manakin (*Pipra fasciicauda*), the yellow anaconda (*Eunectes notaeus*), and populations of threatened or endemic birds and plants. This area, which includes a great variety of habitats representative of the Campos District, will be partially affected by the Yacyretá Dam. In fact, the Ministry of Ecology of Misiones proposed the area as a compensatory reserve for the Yacyretá Dam (Bosso 1993; Chebez and Gil 1993). As shown in Figure 21.2, the Fachinal region also has areas that should be evaluated.

The second highest conservation priority in Misiones is the Rosewood and Assai Palm Forests in the northeast, which have vertebrate and plant species with high conservation value for Argentina (Giraudo 2001; Chapter 15, this volume). Approximately 1,356.3 km^2 (59 percent of the protected region) of these forests are adequately protected, including the Iguazú National Park and Reserve, which has a strong conservation infrastructure. Nevertheless, compared with the Iguazú National Park and Reserve, biological diversity is actually greater in the vicinity of the Paraná River because some species use the river as a dispersal corridor and others find attractive open habitats in this area (Giraudo 2001). The area around Puerto Península and the adjacent private lands (owned by Pérez Companc; Figure 21.2), are particularly important to preserve because they contain one of the last natural remnants along the Paraná River.

The third highest priority is the area in the southeastern tip of Misiones and a portion of northeastern Corrientes, which falls within the Campos District on the Uruguay River and contains species exclusive to this part of Argentina, including the mottle piculet woodpecker (*Picumnus nebulosus*) (Partridge 1962; Castelino, pers. obs. 2001). The only protected area in this region is the small (4.39 km^2) private reserve of Santa Rosa, although interesting sites remain between Concepción de la Sierra (Misiones) and the area around Garabí, Garruchos, and Santo Tomé (Corrientes). These areas have excellent forest islets that form galleries along the Uruguay River and extensive grasslands that should be assessed as part of an effort to create protected areas (Figure 21.2).

Some of these areas are being replaced by vast plantations of exotic species, as has occurred in Rincón de las Mercedes (Corrientes), which could alter the landscape quickly if conservation actions are not soon put in place. Moreover, the Uruguay River itself contains endemic species and serves as an effective corridor for the movement of Atlantic Forest flora and fauna to regions far south of Misiones (Menalled and Adámoli 1995; Giraudo 2001).

Only a small proportion of the Montane Araucaria Forests is protected (0.5 percent of their area), despite the fact that they contain some very important species from a biogeographic and genetic standpoint, including the Paraná or araucaria pine (*Araucaria angustifolia*) and pino del cerro (*Podocarpus lambertii*), the frog *Hyalinobatrachium uranoscopum,* the snakes *Bothrops cotiara* and *Pseudoboa haasi,* and the bird known as the araucaria tit-spinetail (*Leptasthenura setaria*).

An important protected area project is the triangle located north of and linked to the Yabotí Biosphere Reserve, which contains montane araucaria forests stretching from the region of the streams Toro, la Macaca, and Liso to Pepirí Guazú (Giraudo, pers. obs. 2001), with an approximate area of 500 km^2 (Figure 21.2). Paraná pines also still grow in islets in the region of Dos Hermanos, Tobuna, and Bernardo of Irigoyen, areas that also should be thoroughly assessed for potential protection.

Only 1.4 percent of the Laurel and Guatambú Forests, along with the forests along the Paraná River, is protected, mainly because these areas have the most fertile soils and the greatest agricultural development in the province. Very few natural areas remain in this biogeographic subdivision. Of those that remain, several are ideal candidates to become protected areas, such as the region of Alcázar, between Puerto Paraná and Puerto Rico (Figure 21.2), which has a forest islet that connects with the provincial park of Salto Encantado del Valle del Cuñapirú and areas around Puerto Península.

As mentioned earlier, the unevenness of the protected areas reflects an imbalance between the weight given to scientific criteria guiding the selection of areas and that given to other criteria, such as the aesthetic appeal of the landscape, the potential for tourism, the presence of land unsuitable for agriculture, and the proximity of public lands (Pressey and Tully 1994; Pressey 1995; Galindo-Leal et al. 2000). The application of scientific criteria for the selection of reserves would be more pervasive if there were greater interaction between government agencies, scientific and academic institutions, and nongovernment organizations devoted to conservation.

Connectivity

Another issue that has not been adequately addressed in most conservation activities is the need for corridors to be established between protected areas to prevent their isolation (Chebez and Gil 1993). When habitats are fragmented and no connectivity exists between reserves, ecosystems become isolated, with serious impacts on biological diversity (Saunders et al. 1991; Quijano 1992; May 1994; Chapter 31, this volume). The importance of connectivity is underscored by numerous research studies, such as Anjos and Boçon's (1999) finding that small forest areas connected to larger areas have more similar bird communities than small, isolated fragments (see also Chapter 31, this volume).

The Green Corridor Law recently approved in Misiones provides a legal framework for connectivity in biodiversity conservation efforts (Chapter 20, this volume). Until this legislation is fully implemented, however, the application and enforcement of existing legislation concerning forest remnants along rivers and streams can help alleviate the problem of isolation. Corridors should also be established in grassland areas, leaving belts of native grasses between cropland and forest plantations. This action is urgently needed to link the protected areas of differing size that exist in northern and central Misiones, thus enabling the survival of significant populations of large mammals such as the jaguar (*Panthera onca*), the tapir (*Tapirus terrestris*), the white-lipped peccary (*Tayassu pecari*), and large eagles such as the harpy eagle (*Harpya harpyja*), all species with broad space needs (Eisenberg 1980; Terborgh 1992; Chapter 16, this volume). The routes used by humans and the expansion of their activities are leading to isolation of the largest wildlife nuclei in Misiones.

In many areas no basic scientific studies have been conducted to ascertain which species and communities are being protected or to identify the functional characteristics and management problems of the area. Generally speaking, the agencies that administer protected areas do not have clear policies to promote and facilitate research. Lack of basic information undoubtedly will create real problems and challenges for achieving sustainable resource management in the largest protected area, the Yabotí Biosphere Reserve.

Financing

A third issue regarding protected areas is the lack of financial resources. To date, the growth of protected areas has not been accompanied by the financial investment needed to build a proper infrastructure (e.g., personnel, transportation resources, ranger stations) for protecting the ecosystems within area boundaries (Chapter 20, this volume). Although the Ministry of Ecology and Renewable Natural Resources of Misiones has begun to implement management plans in some areas, this effort has just begun, and progress has been hampered by the profound socioeconomic crisis in Argentina at both the provincial and national levels.

The financing for protected areas in Misiones and across Argentina is tremendously uneven. For instance, whereas some protected areas that have been created by law or decree have never really been implemented, largely because of lack of funding, some national parks, such as Iguazú National Park, have large budgets and staff and strong infrastructure and planning. Although the province of Misiones has made a great effort to implement large reserves, limited human and financial resources have made them difficult to manage. Many of the protected areas mentioned earlier suffer from a wide array of pressures, ranging from the intrusion of human settlements, commercial forestry, and cattle grazing to the numerous and unpredictable instances of fire and illegal hunting. Misiones has many private reserves, and although some are not particularly well run or have been abandoned by their owners (Chalukian 1999; Krauczuk 2000), many persist and are inspiring examples of private efforts to conserve nature.

Local Population

Another problem limiting the effectiveness of the protected areas in the region is that many conservation efforts have not engaged local populations. Many researchers have underscored that sociological and ecological factors must be taken into account for conservation to succeed (May 1994; Giraudo and Abramson 1998; Chapter 32, this volume). In the case of creating protected areas, local populations must understand the objectives of the reserve and the benefits it can bring to the region. The administering agency, in turn, must foster community involvement and find ways to enable the population to carry out productive activities that will not conflict with the objectives of the protected area, particularly in places with severe socioeconomic problems (Giraudo and Abramson 1998). This is especially urgent in the protected areas inhabited by the Mbyá indigenous people, such as Cuñapirú Provincial Park and Cultural Reserve, Cuñapirú Private Reserve, and Yabotí Biosphere Reserve. Many conflicts currently exist between local populations and the agencies responsible for protected area administration in Misiones (Giraudo and Abramson 1998). Many residents view the areas as an imposition that interferes with their productive or development interests. The long-term success of protected areas and conservation plans depends on how well the region's population is included in education and sustainable development programs.

Some protected areas established on a basis of substantial ecological research and with strong infrastructure and operating budgets have nonetheless failed to give adequate attention to the needs of the local population. This has led to a variety of problems, such as illegal extraction of firewood and palm hearts, poaching, and similar activities that thwart conservation efforts and cannot be prevented completely by control activities because the principal underlying cause is the existence of unfavorable socioeconomic conditions in Misiones. Although tourism is growing, it has not brought about any significant improvements for rural and low-income populations because the industry is concentrated in large and medium-sized companies that use foreign guides, although their knowledge of the forest is inferior to that of the people of the region (Giraudo and Abramson 1998). Activities such as touring that are led by those from outside the region should be regulated by the state to foster the training and involvement of a larger proportion of local residents.

Final Thoughts

Although protected areas are important elements in conservation policies, a strategy that relies entirely on designation of protected areas has obvious limitations. Such a strategy leaves the rest of the landscape open to rapid modification that may reduce or eliminate a large part of the biodiversity found outside the designated areas (Halffter 1994). It should not be forgotten that 69 percent of the remaining Interior Atlantic Forest of Argentina is outside currently protected natural areas.

The province of Misiones has a sizable system comprising 60 protected areas (Table 21.1) that cover 15 percent of the provincial area and encompass approximately 31 percent of the remaining forest. To achieve yet broader coverage, strategies aimed at expanding and enhancing the system of protected areas in Misiones should be carried out in a coordinated, multisectoral manner with the cooperation of government agencies, research institutions, private concerns, and the various segments of society.

Acknowledgments

The authors thank Carlos Galindo-Leal, Juan Pablo Cinto, Paula Bertolini, Alejandro Garello, and José Benitez for their valuable contributions and suggestions.

References

Anjos, L. D. and Boçon, R. 1999. Bird communities in natural forest patches in southern Brazil. *Wilson Bulletin* 11(3): 397–414.

Bosso, A. 1993. Yacyretá, el año en que vivimos en peligro. *Nuestras Aves* 11(28): 5–9.

Chalukian, S. C. 1999. *Cuadro de situación de las unidades de conservación de la Selva Paranaense.* Buenos Aires: Fundación Vida Silvestre Argentina.

Chalukian, S. C. 2000. La protección de la Selva Paranaense en Argentina. In: Bertonatti, C. and Corcuera, J. (eds.). *La situación ambiental Argentina 2000.* pp: 383–386. Buenos Aires: Fundación Vida Silvestre Argentina.

Chebez, J. C. and Gil, G. 1993. Misiones hoy: al rescate de la selva. *Nuestras Aves* 11(29): 5–9.

Eisenberg, J. F. 1980. The density and biomass of tropical mammals. In: Soulé M. E. and Wilcox, B. A. (eds.). *Conservation biology: an evolutionary-ecological perspective.* pp. 35–55. Sunderland, MA: Sinauer Associates.

de Faria, H. H. 1994. Evaluación de la efectividad de manejo de áreas protegidas. *Flora, Fauna y áreas Silvestres* 8(29): 15–19.

Galindo-Leal, C., Fay, J. P., Weiss, S., and Sandler, B. 2000. Conservation priorities in the Greater Calakmul Region, Mexico: correcting the consequences of a congenital illness. *Natural Areas Journal* 20(4): 376–380.

Gilpin, M. E. and Soulé, M. E. 1986. Minimum viable populations: processes of species extinction. In: Soulé, M. E. (ed.). *Conservation biology: the science of scarcity and diversity.* pp: 19–34. Sunderland, MA: Sinauer Associates.

Giraudo, A. R. 2001. *La Diversidad de serpientes de la Selva Paranaense y del Chaco Húmedo: taxonomía, biogeografía y conservación.* Buenos Aires: Editorial LOLA.

Giraudo, A. R. and Abramson, R. R. 1998. Usos de la fauna silvestre por los pobladores rurales en la Selva Paranaense de Misiones: tipos de uso, influencia de la fragmentación y posibilidades de manejo sustentable. *Boletín Técnico de la Fundación Vida Silvestre Argentina* 47: 1–41.

Halffter, G. 1994. Conservación de la biodiversidad: un reto del fin de siglo. *Bulletí Institució Catalana d'Història Natural* 62: 137–146.

Krauczuk, E. R. 2000. *Caracterización del sistema de conservación en la Provincia de Misiones.* Unpublished paper. Posadas, Argentina: Cátedra de Biogeografía Regional, Profesorado en Biología, Universidad Nacional de Misiones.

Laurance, W. F., Bierregard, Jr., R. O., Gascon, C., Didham, R. K., Smith, A. P., Lynam, A. J., Viana, V. M., Lovejoy, T. E., Sievin, K. E., Sites, Jr., J. W., Andersen, M., Tocher, M. D., Kramer, E. A., Restrepo, C., and Moritz, C. 1997. Tropical forest fragmentation: synthesis of a diverse and dynamic discipline. In: Laurance, W. F. and Bierregard, Jr. R. O. (eds.). *Tropical forest remnants: ecology, management, and conservation of fragmented communities.* pp. 502–514. Chicago: University of Chicago Press.

May, R. M. 1994. Ecological science and the management of protected areas. *Biodiversity and Conservation* 3: 437–448.

Menalled, F. D. and Adámoli, J. M. 1995. A quantitative phytogeographic analysis of species richness in forest communities of the Paraná River Delta, Argentina. *Vegetatio* 120: 81–90.

Partridge, W. H. 1962. Dos aves nuevas para la fauna argentina. *Neotrópica* 8(25): 37–38.

Phillips, O. L. 1997. The changing ecology of tropical forest. *Biodiversity and Conservation* 6: 291–311.

Pressey, R. L. 1995. Conservation reserves in NSW: crown jewels or leftovers? *Search* 26(2): 47–51.

Pressey, R. L. and Tully, S. L. 1994. The cost of ad hoc reservation: a case study in western New South Wales. *Australian Journal of Ecology* 19: 375–384.

Quijano, R. O. 1992. Modelos de extinción y fragmentación de hábitats. In: Halffter, G. (ed.). *La diversidad biológica de Iberoamérica.* pp. 25–38. México: Instituto de Ecología.

Rolón, L. H. and Chébez, J. C. 1998. *Reservas naturales Misioneras.* Posadas, Misiones (Argentina): Editorial Universitaria, Universidad Nacional de Misiones.

Saunders, D. H., Hobbs, R. J., and Margules, C. R. 1991. Biological consequences of ecosystem fragmentation: a review. *Conservation Biology* 5(1): 18–32.

Soulé M. E. 1980. Thresholds for survival: maintaining fitness and evolutionary potential. In: Soulé, M. E. and Wilcox, B. A. (eds.). *Conservation biology: an evolutionary-ecological perspective.* pp. 151–170. Sunderland, MA: Sinauer Associates.

Stotz, D. F., Fitzpatrick, J. W., Parker III, T. A., and Moskovits, D. F. 1996. *Neotropical birds: ecology and conservation.* Chicago: University of Chicago Press.

Terborgh, J. 1992. Maintenance of diversity in tropical forest. *Biotropica* 24(2b): 283–292.

Terborgh, J. and Winter, B. 1980. Some causes of extinction. In: Soulé M. E. and Wilcox, B. A. (eds.). *Conservation biology: an evolutionary-ecological perspective.* pp: 151–170. Sunderland, MA: Sinauer Associates.

Turner, M. I. and Corlett, R. T. 1996. The conservation value of small, isolated fragments of lowland tropical rain forest. *Trends in Ecology and Evolution* 11(8): 330–333.

Chapter 22

Last Opportunity for the Atlantic Forest

Luis Alberto Rey

Misiones is a province in northeastern Argentina, wedged like an arm up between Brazil and Paraguay. Similar to Belgium in size, with an undulating topography, it is framed by three major Latin American rivers: the Paraná on the northwest interior, the Uruguay to the east, and the Iguazú running along the short northern boundary. Settlement patterns here differed from those in neighboring regions, as a result of which almost half the original surface remains covered with forests (Chapter 19, this volume). Shockingly, these 15,000 km^2 today make up the largest continuous, unfragmented mass of the Interior Atlantic, or Paraná, Forest.

This natural marvel, unique and irreplaceable, is inhabited by more than 540 bird species (half the total diversity of Argentina), 124 mammal species (43 percent of the total), and some 180 species of amphibians and reptiles, which makes this one of the richest and most biodiverse regions on earth (Chapter 15, this volume).

In contrast, in eastern Paraguay and southern Brazil intense human activity nearly eliminated the forest and much of its contents in less than a century (Chapters 4 and 24, this volume). Today, only small fragmented areas remain, and even these are degraded and isolated from one another. The rest has been taken over by large-scale agriculture and stockbreeding.

Misiones, perhaps because it has several major tourist attractions—including the famous Iguazú Falls—developed a rooted environmental consciousness early on. Evidence of this is the creation of the first national park in the province in 1937, which has grown over time into an entire provincial system of natural protected areas was established. This process culminated in 1999 with the enactment of the Green Corridor Law, an ambitious legal instrument that provides for the conservation of this unparalleled natural heritage (Chapters 14 and Chapter 20, this volume; Figure 22.1).

The Green Corridor is to be followed by what is called the Trinational Initiative, which will entail making the decision to enlarge the area in Misiones as a core while also conserving and jointly managing important sites in Brazil and Paraguay. Iguazú National Park and Turvo State Park, adjacent to Moconá Falls in Brazil, and the Moisés Bertoni Reserve in Paraguay are examples.

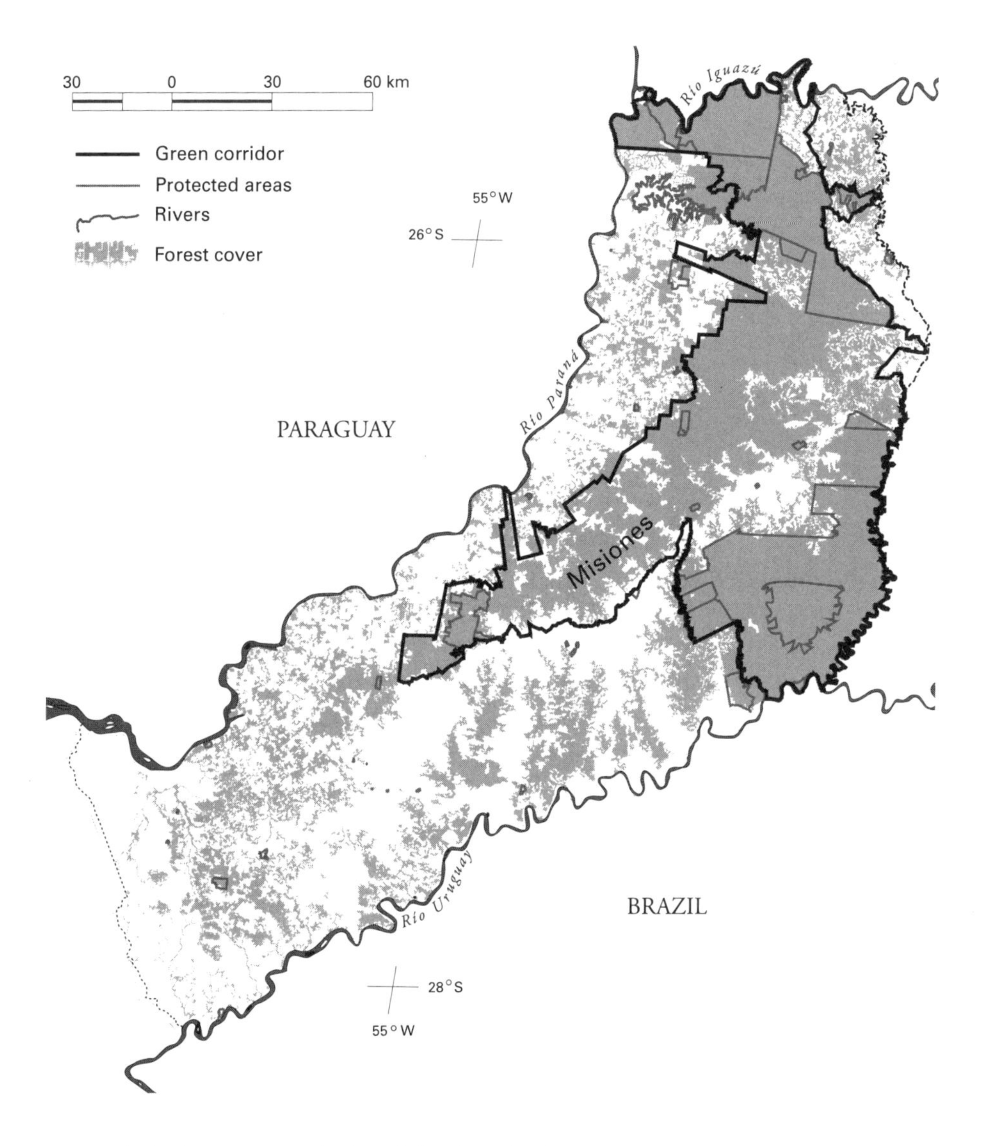

Forest cover (1999) and protected areas data from Fundación Vida Silvestre Argentina.

Figure 22.1. Boundaries of the Green Corridor created in Argentina specifically to conserve biodiversity in Misiones.

Each country has adapted its reality to its needs. Argentina, which has extensive flatlands with fertile soils (the Humid Pampas), did not need to develop agriculture on the red soils of Misiones. Brazil, on the other hand, and later Paraguay did have to develop farmland in forest areas.

The primary problem for the conservation of the Atlantic Forest in Misiones is the culture that is being absorbed from the neighboring countries, with which Misiones has sometimes had closer ties than with Buenos Aires. This culture views the whole forest as a waste of virgin soils that should be converted to farmland by burning down the forest and producing "crops from ashes." Implicit in this practice is abandonment of the land no more than 3 or 4 years after the forest has been destroyed, thus creating a farmer-nomad who moves from place to place, unable to escape absolute poverty. This culture is derived from the conviction that the forests are limitless. In fact, in less than a century their geographic limits have been reached.

The other major threat to the forest is the replacement of native trees with exotic species. In Misiones, there are three paper and pulp plants and numerous sawmills, which in recent years have attracted significant foreign investment as a result of great success in growing both pine and eucalyptus.

It is still possible to save the large mammals, such as the jaguar, the giant anteater, and the tapir (Chapter 16, this volume), as well as majestic birds such as the harpy eagle and the various species of parrots, macaws, toucans, and hummingbirds that live in the forest. But this can be achieved only by saving a continuous expanse of forest that will allow the survival of all and avoid genetic erosion. The Green Corridor and the Trinational Initiative offer the best hope of accomplishing this objective.

This enormous conservation responsibility cannot be entrusted to a single province, especially one that is also grappling with social problems and extreme poverty. The issues confronting the province include the survival of the Mbyá and Guaraní peoples (Chapter 18, this volume), the preservation of their natural and cultural environment, and an economic crisis created by production models that are not profitable in some cases and not environmentally sustainable in others.

As the problems multiply, the implementation of ecological laws is delayed by more urgent needs, some, in fact, more pressing than others. Greater effort is needed, but not only on the part of the inhabitants of Misiones. Outside help is also needed. Saving this sanctuary of biodiversity for humanity is a task for everyone, and this is our last chance. We await the presence of people of goodwill who fully understand the horror that the word *extinction* signifies. Time is running out, and decisions must be made quickly.

PART IV

Paraguay

Chapter 23

Dynamics of Biodiversity Loss in the Paraguayan Atlantic Forest: An Introduction

José Luis Cartes and Alberto Yanosky

Paraguay is one of the least-known South American countries. It is small in comparison to it neighbors, Brazil, Bolivia, and Argentina, and its history of war and dictatorships kept it from developing until the early 1980s. However, the human and natural resources of Paraguay hold great potential.

The destructive economic and political impacts of the 1865 War of the Triple Alliance in Paraguay can still be seen today (Chapter 24). After the war, large transnational companies took hold, setting cycles of underdevelopment into motion through their rampant use of public lands and extraction of natural resources. Autocratic governments were supported for many years by this model of large landowners who controlled localities, people, and resources. Not only did this political model stall the economy of Paraguay, but it also stymied many other important activities, including science and research. As a result, the status of biological diversity and threatened species in Paraguay remains very poorly known and understood. Despite a lack of baseline knowledge, however, it is evident that biodiversity is being lost at an accelerated pace, as discussed by authors in this section.

Paraguay's biodiversity is unique because of its strategic location, where the Atlantic Forest, Cerrado, Pampa (savannas), and the Great Chaco ecoregions merge. In Paraguay, the Atlantic Forest biome does not have as high biodiversity as areas of the forest in other countries, but it does house species whose future is at risk on a global level (27 birds, 8 mammals, 5 plants, and 1 fish) (Chapter 25). In addition, the Atlantic Forest lies over part of the largest underground freshwater reserve in the world, the Guaraní aquifer, which is the most economical and flexible source of water supply for human consumption in the region (Chapter 27). Unfortunately, this aquifer is highly threatened because the land above it is where Paraguay's rural extractive development is most vigorously played out.

Paraguay has little industrial development and relies on an economy that is rooted in extractive practices and raw goods such as wood, grain, cotton, oils, and meat. Colonization of land by peasants has collided with ownership of large private holdings, generating land conflicts, which the government handles poorly. Rural poverty, lack of public land availability, and lack of technological assistance make natural resources the only source of income for both peasants and indigenous people (Chapter 26). This situation has spurred many illegal activities, such as the invasion of private and public lands, traffic of illegally taken wood, and, more recently, growing illegal cannabis crops. These activities are associated with powerful groups, so authorities seem unable to take actions that can ensure the protection and conservation of biodiversity.

Conservation in Paraguay is also hampered by inadequate institutional capacity. In the last 10 years, as the capacity for conservation has increased in the private sector, it has decreased in government agencies. The Ministry of Environment is responsible of enforcing environmental laws, and although it has recently reorganized the government structure related to natural resources, the country still lacks a national environmental policy and continues to be particularly weak at state and local government levels.

In addition, in Paraguay as elsewhere, poverty is closely linked to development and in turn to loss of biodiversity and rapid environmental degradation. Investigations into the specific dynamics of that relationship in the country are important to develop partnerships, plans, and processes that can work together to relieve pressure on the natural resources. However, although international donors direct their support to such applied initiatives, research and socioeconomic development take time. To be truly effective, the government must play a central role in building and enforcing the dynamic. However, for the Atlantic Forest in Paraguay, there is no time to wait for the results of long-term research or for adequate regulatory frameworks or institutional collaborations to be put in place. With the private sector increasingly taking a leading role in conservation, international funding organizations have begun to recognize the ability of these organizations, alone or in partnership, to implement effective conservation projects (Chapter 28). In fact, some believe that it is the determined actions of the private sector that are saving the last forest remnants, at least in Paraguay.

Whether with private or public funding, all those distressed by the astonishing rate of biodiversity loss in the Atlantic Forest must work together to devise creative ways to implement development that is sustainable in real contexts and to continue to develop the overall capacities of government institutions. If we meet these goals—research, institutional capacity, education, and development—we will be able to maintain the important areas of Paraguay and throughout the Interior Atlantic Forest.

Chapter 24

Brief History of Conservation in the Interior Atlantic Forest

José Luis Cartes

Historical Overview, Colonization, and Land Use

Paraguay is a landlocked country with an area of 406,752 km^2, located in south-central South America. Geographically, the country is divided by the Paraguay River into two distinct regions: the western region (62 percent of the total area), which forms part of the Gran Chaco region (a semidesert, sparsely vegetated, and largely uninhabited plain) and the eastern region (38 percent of the area), which comprises parts of the Atlantic Forest, the Cerrado (a dry biome of forests, woodlands, and open savannas), and extensive wetlands. This chapter focuses on the Atlantic Forest.

Before the sixteenth century the area known today as Paraguay was inhabited almost entirely by Guaraní natives, mainly members of the Mbyá Guaraní and Ava Guaraní groups, but also Aché and Pai Tavytera peoples (who belong to the same linguistic family). The earliest migration to the region occurred more than 10,000 years ago in the part of Paraguay inhabited by the Láguido and Arawak groups and, more recently (500 B.C.), the Guaraní (Chase-Sardi 1992). A review of the main historical events that have influenced the status of the environment in Paraguay is presented in Table 24.1.

In more than 100 years of history the great problem of land tenure was never resolved, and to this day the distribution of land remains extremely uneven, with an estimated 77 percent in the hands of only 1 percent of the population. Moreover, the country has undergone a series of severe financial and economic crises since 1980, when it entered a period of recession (gross domestic product [GDP] growth shrank to 2.5 percent a year between 1982 and 1989 after growing at an annual rate of 10 percent between 1976 and 1981). In the 1980s, GDP growth was insufficient to keep up with the population growth rate, which was 3 percent annually (Nikiphoroff 1994). In this context, the area originally covered by the Atlantic Forest was devastated as a result of squatting, clearing of new farmland out of fear of expropriation, and illegal logging (because prices for wood, before it became scarce, rose to very high levels).

Table 24.1. History of principal contributors to forest deterioration and conservation in Paraguay.

Year	Event
1500s	The Guaraní are known to have grown corn and cassava as staple crops. When the soil became depleted, new plots of land were cleared, which generally meant migrating to another area (Súsnik and Chase-Sardi 1995). When the Spanish colonists arrived, large portions of the region were already inhabited. Researchers estimate that in 1596 the Indian population living in the earliest Jesuit *reducciones,* or mission settlements, numbered close to 50,000, distributed among eight villages (Carbonell de Masy 1986). The first human impacts on the forest probably derived from the agricultural practices of the Guaraní. The Spaniards also introduced livestock (cattle and horses), which undoubtedly had the greatest impact on natural environments, especially those associated with the savannas.
1700s	Based on Carbonell de Masy (1986), the largest productive centers in colonial Paraguay were located in the area of Asunción. The Jesuit *reducciones* colonial systems had great impacts on the forest because their sphere of influence was so large (700,000 km^2). The native forest served as a source of *yerba mate* (*Ilex paraguariensis*), which was harvested directly from the forest and sold to miners in Chile and Peru, as it was popular for its allegedly energizing properties. Initially, a large portion of the Paraguayan native forest was subject to *mate* harvesting, although later *mate* crops were grown in the *reducciones.* It is estimated that an average of 162.4 tons of *mate* were sold and 340 tons were consumed on the *reducciones* annually during the period 1729–1733. These amounts could be produced on little farmland. Livestock production was also an important activity in the Jesuit settlements and was perhaps the activity that had the greatest impact on the forest because cattle were not restricted to savannas. One estimate of livestock production in 1716 puts the number of cattle at 100,000 head, distributed among 30 communities. Another agricultural activity was cotton farming, mainly to make fabrics for use by the colonists. By the end of the European colonial era and the beginning of the independence movements (early nineteenth century), the main economic activities were linked to cattle ranching, at the expense of the land.
1840–1860	The progressive government of Carlos A. López (1840–1860) took advantage of the enormous accumulated wealth and government control over extensive lands (public lands) that resulted from the dictatorship of José Gaspar Rodríguez de Francia (1820–1840). Government policies promoted settlement of areas that became known as *estancias de la patria* ("ranches for the fatherland"). Railway lines and an iron foundry were also established in this period (Rubiani 2001).
1865–1870	The "War of the Triple Alliance," between Paraguay and Argentina, Brazil, and Uruguay (the Triple Alliance) devastated the Paraguayan economy and natural resources. The population of Paraguay was decimated (58 percent died in the war), and most of the remaining inhabitants were women, children, and older adults (Rubiani 2001). It is considered one of the bloodiest genocides in the history of the Americas. To pay off the debts from the war, the government was forced to sell large amounts of land. The largest *latifundios* (estates) in the world were thus established. These properties were generally held by foreigners (from England, Argentina, and the United States), who devoted themselves almost exclusively to extracting forest products—including first-quality wood and *yerba mate*—and stockbreeding.

Table 24.1. Continued

Year	Event
1883	The sale of land was systematized with legal reforms (such as the Law of 1883), which made vast regions of forestland, *mate* groves, and pasturelands available to foreign investors. Immense *latifundios* were established in Paraguay between 1885 and 1897. For example, a total of 124,821.43 km^2 was sold in the Chaco region alone. Only 27 land owners held 96,268.80 km^2 (one, Casado, owned 31,500 km^2, an area larger than Belgium). The multinational company Industrial Paraguayan held 26,477.27 km^2 in the eastern region, of which 8,850 km^2 was natural forest composed chiefly of *yerba mate* trees. From the end of the War of the Triple Alliance to 1940, the Paraguayan economy remained extremely fragile, with only two land-use profiles: commercial forestry and traditional cattle ranching. During this period, settlement was encouraged, especially among Europeans (Germans, Russians, Poles, Ukrainians, and members of the Mennonite faith), who used traditional agricultural models that were not very dynamic (Carter and Galeano 1995).
1932–1935	Paraguay experienced another great war (Chaco War), during which its economy suffered, though to a lesser extent than in the war of 1870. This marked the start of a series of efforts to create a rural settlement system, which ultimately served only to reinforce the *latifundista* system of land use. As a result, there were frequent land disputes, stemming from of the lack of public lands available for settlement (Carter and Galeano 1995).
1940–1975	Some land was expropriated, some was distributed, *latifundios* were restored, unplanned settlements sprang up, and the agricultural frontier expanded.
1975–1983	Rapid expansion of the agricultural frontier and capitalist modernization. The construction of the Itaipú hydroelectric dam (which inundated 1,460 km^2) and the enormous capital investment that accompanied it had a massive impact on the national economy (Nikiphoroff 1994), leading to reduction in demand for land among poor farmers who found good work in public works, construction, or industry. The agricultural sector benefited from an upsurge in prices, with tremendous growth in cotton and soy exports. The government strengthened the export agriculture production model, based on large capital investment and private settlement (generally by Brazilians), with a restoration of the *latifundios* in exchange for political favors. The impact on Paraguay's natural resources, especially in the area of the Atlantic Forest, where the best farmlands were located, was devastating. Although there was little growth in productivity, the agricultural frontier expanded at an unprecedented rate during that decade, creating environmental problems that are now irreversible (Nikiphoroff 1994).
1983 to present	Fierce competition for land, especially since the fall of the Stroessner regime in 1989 and the ensuing political ups and downs. This period has been marked by a new land crisis, falling international prices, and the final phase in the construction of several dams (Carter and Galeano 1995). Peasant farmers are reasserting their demands for land in an adverse context, in which the expansion of the agricultural frontier has been halted. Speculation on land prices has also intensified, which resulted in a 66 percent increase between 1975 and 1986. The overthrow of Stroessner prompted mass invasion of the *latifundios* established by associates of those in power. In response, the large land owners cleared forestlands and converted them to farmland to avoid their expropriation.

Ecosystems of the Interior Atlantic Forest

The Interior Atlantic Forest is the innermost portion of the Atlantic Forest region, which is known for its rich biological diversity and high degree of endemism. It is considered a region of diversity comparable to the Amazon Forest, the moist forest of the Choco-Darién region, and the forests of the Peninsular Malaysia–North Borneo system, which are also critically endangered (Olson and Dinerstein 1998). The evolutionary history of the Interior Atlantic Forest has been unique because of its Pleistocene refugia (Haffer 1974), and as a result it has been given many different names (Table 24.2).

Originally, the Atlantic Forest in Paraguay stretched over a large part of the eastern region, encompassing all of the departments of Alto Paraná, Canindeyú, Itapúa, and Caaguazú; a large share of the departments of Amambay, eastern San Pedro, and Caazapá; and parts of Concepción and Paraguarí. The forest in Paraguay currently covers an area of about 20,800 km^2 (Barboza et al. 1997) and consists mostly of fragments of nondegraded forest (11,618.20 km^2) and many fragments of degraded forest (9,267.61 km^2) (Figure 24.1).

The flora of the Atlantic Forest in Paraguay includes dense wooded areas of subtropical semideciduous vegetation. However, the eastern region of Paraguay is a convergence point for diverse floras distributed across the country and neighboring countries, including the following: strictly Paraná flora, or Lauriflora, composed of the moistest and densest forests, which are mostly distributed in the Alto Paraná ecoregion and are the areas that have suffered most from deforestation; and seasonal semideciduous flora from the Pleistocene, which consists of elements with fewer ecological needs, occupying most of the eastern region and stretching westward into the moist Chaco region. Floras from other regions—such as the Cerrado, the savannas, and to a lesser extent the Chaco—also converge in eastern Paraguay. The overlap of these floras with plant life more typical of the Interior Atlantic Forest greatly increases floristic diversity and therefore biological diversity.

The Interior Atlantic Forest in Paraguay harbors some 70 species of threatened plants and about 50 species that are considered vulnerable. Many species are

Table 24.2. Classifications of the Interior Atlantic Forest of Paraguay.

	Classification	Reference
Flora	Moist temperate forest and hygrophilous forest	Holdrige 1947
	Forest district of the phytogeographic province of Paraná, Amazon domain	Cabrera and Willink 1973
	Southern subtropical forest	Hueck 1978
	Central forest and Alto Paraná ecoregions	CDC 1990
	Interior Atlantic Forest characterized as a tropical moist broadleaf forest	Dinerstein et al. 1995
Fauna	Central Paraguay and Alto Paraná	Hayes 1995

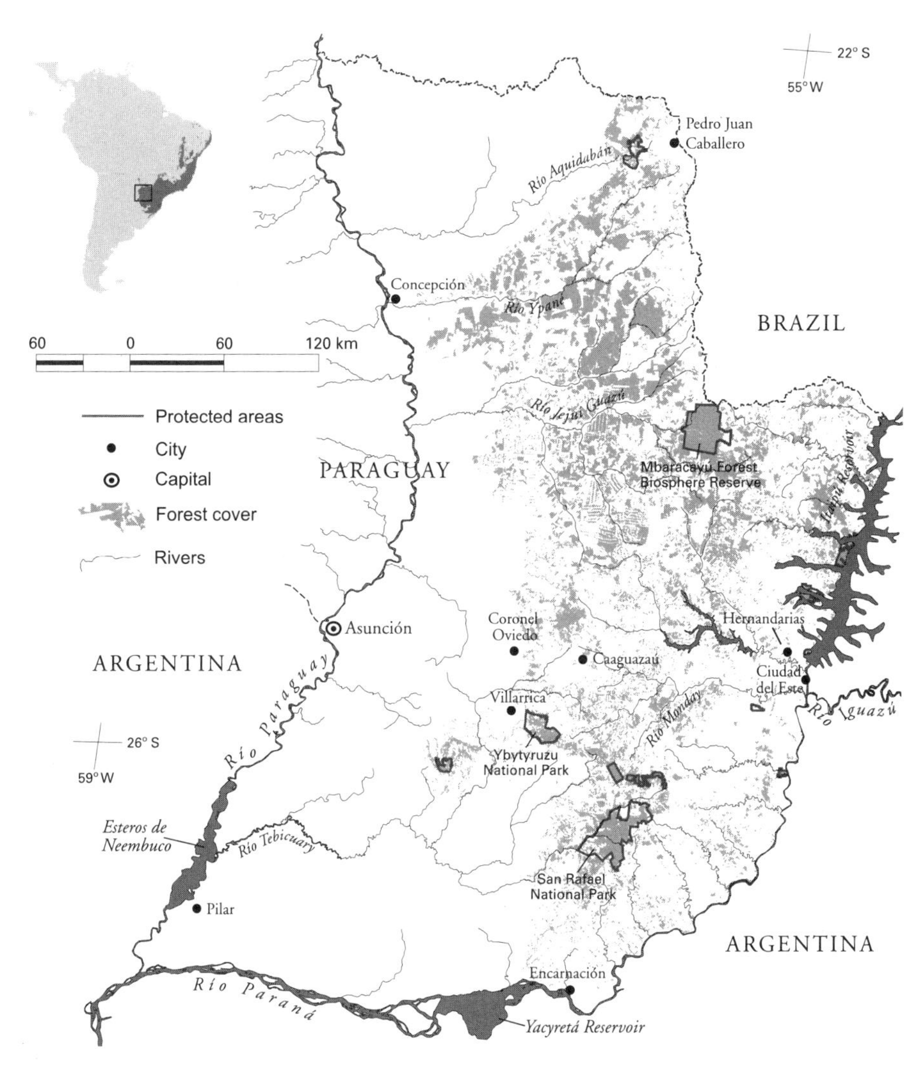

Figure 24.1. Distribution of the Interior Atlantic Forest in eastern Paraguay.

Atlantic Forest endemics of great commercial value (FMB 1994) (Table 24.3). The predominant vegetation in the Interior Atlantic Forest landscape is tall forest (Jiménez and Espinoza 2000), which includes wooded masses of varying size and density depending on the amount of timber extraction that has occurred and the extent of regeneration that has taken place. The forest comprises three well-differentiated strata, the tallest one reaching or exceeding 25 m (Table 24.4).

Interspersed in the tall forest, often associated with slopes, streams, and particular soil conditions, are stands or groves composed almost exclusively of arborescent ferns of the species *Alsophila cuspidata* and *Cyathea atrovirens,* locally known as *chachi.* These plants may grow to heights of up to 6 to 7 m, interspersed with a few other species, such as *yvyra pytã* or copperpod (*Peltophorum dubium*),

Table 24.3. Endemic species of great commercial value from the Interior Atlantic Forest of Paraguay.

Common Name	Scientific Name
Rosewood	*Aspidosperma polyneuron*
Urunday (also known as *Gonçalo Alves,* zebrawood, and tigerwood)	*Astronium fraxinifolium*
Aratiku guasu	*Annona amambayensis*
Black *lapacho*	*Tabebuia heptaphylla*
Incienso	*Myrocarpus frondosus*
Cedar	*Cedrela fissilis*
Cancharana	*Cabralea canjerana*
Guatambu	*Balfourodendron riedelianum*
Species typical of the coastal Atlantic Forest whose distribution area extends into Paraguay	
Ka'I kaigua	*Cariniana estrellensis*
Yvyra katu	*Xylopia brasiliensis*
Kuri'y or Paraná pine	*Araucaria angustifolia*
Assai palm	*Euterpes edulis*
Queen palm	*Syagrus romanzoffiana*

amba'y (*Cecropia pachystachya*), and *ka'a oveti* (*Luehea divaricata*). These stands have open undergrowth. Many epiphyte species (ferns, cactus, and orchids) find in the *chachi* the perfect support for their development, and they endow these environments with a very special beauty.

The Interior Atlantic Forest in Paraguay never formed a continuous mass but rather constantly alternated with savannas (low savannas, high savannas, and "dirty" savannas) and other types of forest formations (medium forest, gallery forest). Alternating blocks of savanna of different sizes are very common and generally follow a south-north and east-west gradient. In the north flora from the Cerrado encroaches into the forest, whereas in the west and south flora from the Chaco and from the grasslands of the Pampas, respectively, are commonly recorded. There is also an east-west gradient of soils, ranging from clay to pure sand.

The Atlantic Forest zoogeographic region, which includes Interior Atlantic Forest, is a center of biological endemism for plants, birds, mammals, reptiles, and butterflies. It contains approximately 200 species of endemic birds and some 60 species classified as threatened at the global level (Wege and Long 1995). In Paraguay, approximately 82 species endemic to the region have been recorded, but the true number of endemic species is probably higher. Birdlife International has identified six areas of endemic birds in the region, two of which are in Paraguay (Wege and Long 1995). The Interior Atlantic Forest has more endemic birds than any other zoogeographic subregion in the neotropics, with 101 endemic species (followed by the central Andes, with 70 endemic species) (Stotz et al. 1996). There are 403 bird species in the Atlantic Forest in Paraguay, of which

Table 24.4. Common species of the different strata of the Interior Atlantic Forest of Paraguay.

Forest Strata	Common Name	Scientific Name
Canopy (35–40 m in height)	Black *lapacho*	*Tabebuia heptaphylla*
Most prized from the commercial standpoint	*Guatambu*	*Balfourodendron riedelianum*
	Viraro	*Pterogyne nitens*
	Copperpod	*Peltophorum dubium*
	Yvyra pere	*Apuleia leiocarpa*
	Guajayvi	*Patagonula americana*
	Peterevy	*Cordia trichotoma*
	Cedar	*Cedrela fissilis*
	Vilca (also known as *cebil, huilca, yopo,* and *cohoba*)	*Anadenanthera colubrina*
	Queen palm	*Syagrus romanzoffiana*
Intermediate stratum (10–20 m in height)	*Yvyra pepe*	*Holocalyx balansae*
Many of these species produce edible fruits and are important resources for the forest fauna.	River laurel	*Nectandra angustifolia*
	Yvaporoity	*Plinia rivularis*
	American muskwood	*Guarea guidonia*
	Aguai	*Chrysophyllum gonocarpum*
	Guavira	*Campomanesia xanthocarpa*
	Yvyra piu	*Diatenopteryx sorbifolia*
	Ambay	*Cecropia pachystachya*
Understory tree species	*Ñandyta* or *ñandypa mi*	*Sorocea bonplandii*
	Jaborandi	*Pilocarpus pennatifolius*
	Jacaratia	*Jacaratia corumbensis*
	Ñandypa rá	*Hennecartia omphalandra*
	Catuaba	*Trichilia catigua*
	Katigua	*Trichillia elegans*
	Inga'i (ice cream bean tree)	*Inga affinis*
Understory shrub species	Pepper family (Piperaceae)	*Piper medium*
	Pepper family (Piperaceae)	*Piper hispidum*
Bamboo and related plants	*Takuarembo*	*Chusquea ramosissima*
	Takuapi	*Merostachys claussenii*
	Takuare'I	*Olyra micrantha*
Rubiaceae (bright white or blue flowers)		*Hamelia patens*
	Mborevi rembi'u	*Faramea porophylla*
		Psychotria paracatuensis
Terrestrial ferns	*Amambái*	*Didymochlaena truncatula*
	Amambái	*Blechnum brasiliense*
	Amambái	*Pteris denticulate*
	Amambái	*Adianthopsis radiata*

Continued

Table 24.4. Common species of the different strata of the Interior Atlantic Forest of Paraguay.

Forest Strata	Common Name	Scientific Name
Epiphyte species Orchids		*Miltonia flavescens* *Oncidium jonessianum* *Pleurothallis* spp. *Catasetum fimbriatum*
Bromeliads	Queen's tears (friendship plant) Air plants	*Billbergia nutans* *Tillandsia* spp.
Araceae	Tree philodendron Prickly pear or cactus	*Philodendron bipinnatifidum*
Bignoniaceae	Lianas Climbing ferns	*Microgramma vaccinifolia* *Pleopeltis lattipes* *Polypodium hirsutissimun*

many are endangered, as well as mammals, reptiles, and amphibians (Chapter 25, this volume).

Current Conservation Status and Trends

The Interior Atlantic Forest is the region most affected by deforestation processes in Paraguay. Various factors, including the suitability of the red clay soils for agriculture and the high quality of the timber found in the forest, have conspired to make this region the target of deforestation. The issue of deforestation in the Interior Atlantic Forest is extremely complex. Here, we look at only the most important causal factors. At present, 11,618.20 km^2 of native forest remains out of the 88,050 km^2 that originally existed in Paraguay's eastern region. Most of these remnants are part of the Interior Atlantic Forest (Barboza et al. 1997; Figure 24.1).

As was mentioned in Table 24.1, until 1940 deforestation processes were associated mainly with selective logging and *yerba mate* harvesting activities, which had minor impact on the forest cover. After 1940, with the establishment of settlements and expansion of the agricultural frontier, the practice of clear-cutting accelerated in the forest, especially on soils considered good for farming. Few of the trees cut down were used for timber because large portions of the forest had already been selectively logged, and the primary objective was to clear the land for agricultural purposes. This course of action reflected the prevailing view, still widely held today, that the forests are "unproductive lands" that yield only vermin and diseases.

Following the processes of agricultural expansion and, to a certain extent, rural settlement, deforestation intensified, averaging more than 2,000 km^2/year—among the highest rates in the world. This occurred in response to land tenure and distribution conflicts that arose in the nineteenth century and the lack of resources (caused by unemployment and low agricultural prices) that reinforced a

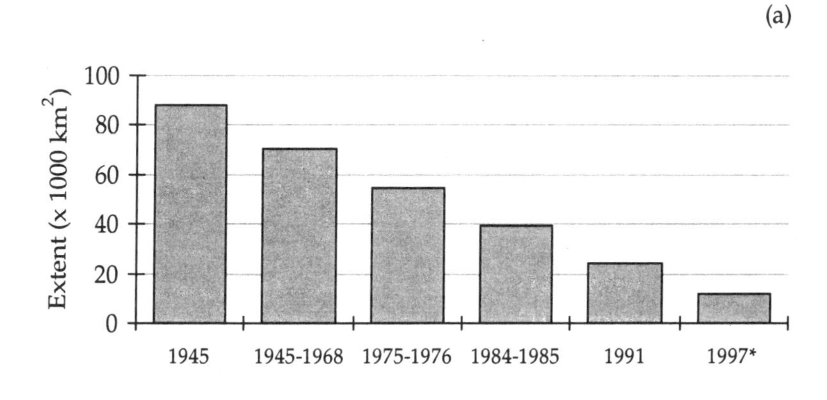

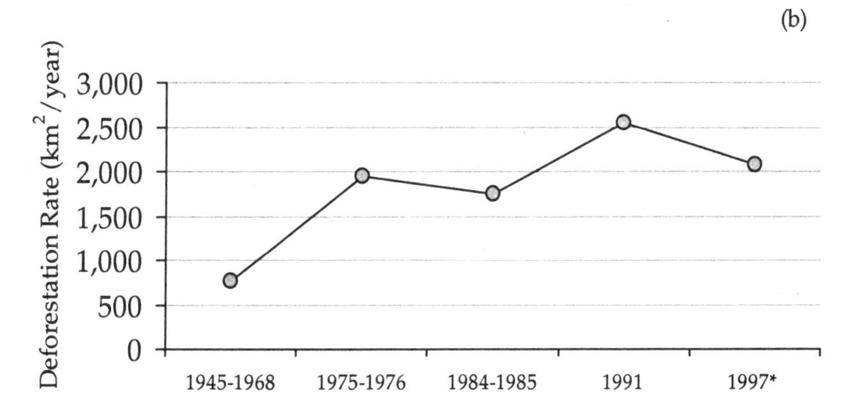

Figure 24.2. Changes in forest extent (a) and deforestation rates (b) in the eastern region of Paraguay. *For 1997, only the extent of nondegraded forest is considered (Bozzano and Weik 1992; Barboza et al. 1997).

whole industry of illegal trafficking in timber destined for Brazil. In 1989, after the fall of the Stroessner regime, deforestation reached a record level of over 2,000 km^2/year (Figure 24.2). Moreover, as a result of these problems, the processes of settlement and clearing of agricultural land took place in a very disorganized and inappropriate manner. For example, numerous rural settlements were created as a result of squatting on private lands that were not suitable for agriculture. Similarly, on the *latifundios* large pasturelands were created out of fear of squatter invasions and subsequent expropriation. In both cases, the destruction of the forest represented a true affront to conservation in that it was aimed at converting forestlands for agricultural use when they were clearly not suited to that purpose.

The 1950s marked the start of a more intensive settlement process that was generally not planned and arose as a result of illegal squatting on private lands (Carter and Galeano 1995). Of all the new rural settlements that originated between 1884 and 1974, 50 percent were established between 1950 and 1974 under state-sponsored settlement programs. This trend continued and, in the 1990s, increased substantially. Between 1989 and 1993, the Rural Welfare

Institute (Instituto de Bienestar Rural), the agency responsible for promoting rural settlement, supported settlement of a total of 97.69 km^2 (816 land titles in 3 settlements). During the period 1994–1996, this figure increased to 1,237.44 km^2 (9,779 land titles in 50 settlements), which amounts to an average growth rate of 1,200 percent (IBR 1997). Many of the settlements began when squatters invaded lands, some of which were located within the boundaries of officially protected wild areas.

The creation of pastureland for cattle grazing has also contributed to deforestation. Artificial pastureland is generally cultivated in deforested areas, after the removal and burning of trees. Several grass varieties have been introduced on these pasturelands. The most common ones are the African varieties *Brachiaria decumbens, Panicum maximum,* and *Cynodon plectostrachyus.* The latter grass has had a great impact on the forest because of its ability to spread into the forest, in combination with periodic burning. The rate of pasture creation kept pace with the deforestation rate, averaging approximately 1,350 km^2 per year (Figures 24.2 and 24.3). This process is very interesting because it can be related to the aforementioned fear of squatting and expropriation on the part of large land owners.

The deforestation process in the Interior Atlantic Forest of Paraguay eliminated large masses of continuous forest that existed in 1945. Of the nine departments that the forest originally encompassed, the principal remaining fragments are located in Canindeyú, San Pedro, Guairá, and Caazapá. In the area there are five national parks (Cerro Corá, Yvytyruzú, Ybycuí, Caaguazú, and San Rafael) and three natural monuments (Kuri'y, Ñacunday, and Puerto Bertoni), with a total area of 1,392 km^2, of which only 690 km^2 are nondegraded forests (Barboza et al. 1997). However, these public protected areas suffer from serious implementation problems, mainly because of disputes with land owners over land and land use. In fact, most are parks on paper only (Cartes 2000) because, among other reasons, only 302 km^2 (22 percent) of the total 1,392 km^2 is owned by the government.

The situation of protected areas is better in the private sector. The privately held areas include the Mbaracayú Forest Natural Reserve and the reserves of the Itaipú binational hydroelectric power plant (Tati Yupi, Itabó, and Limoy), which encompass a total area of 794 km^2 of nondegraded forest that is being effectively conserved. However, the two systems together account for only 2.4 percent of the original area of the Interior Atlantic Forest, which is too small to ensure conservation of all the biodiversity of this ecosystem. Other private land owners have protected areas within the bounds of their properties. These include six private reserves (Tapyta, Ypeti, Itabó, Kaaguy Rory, Morombi, and Ka'i Rague), with a total area of 550 km^2, and two ecological easements (San Isidro and San Pedromi), with a combined area of 23 km^2. Clearly, given the prevailing land tenure situation in Paraguay, private conservation systems must be considered key elements in conservation efforts. The incorporation of these private lands not only has substantially increased the protected portions of the Interior Atlantic Forest but also has enhanced interconnection of the remaining fragments, facilitating biological exchange between them (Cartes 2000). Unfortunately, resource constraints and

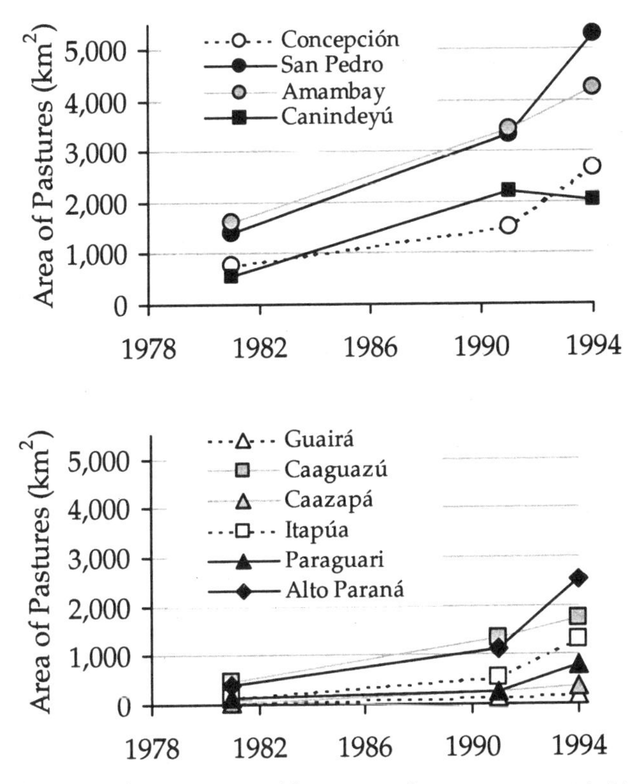

Figure 24.3. Changes in the area occupied by pastures from 1981 to 1994; (a) northern provinces and (b) southern provinces of Paraguay (Molas et al. 1995).

political and economic instability have limited further development of private conservation initiatives.

The Alto Paraná ecoregion has the greatest biological diversity in the country (CDC 1990), but rapid loss of the forest cover has put numerous animal species in this ecoregion at risk. The National System of Protected Natural Areas (DPNVS/FMB 1993) estimates that in the country as a whole, 100 to 300 vertebrate species and 200 flora species are threatened to some degree.

The threatened species originally found in the Interior Atlantic Forest are declining or are now limited to small areas within forest remnants (CDC database, Fundación Moisés Bertoni and Guyra Paraguay, unpubl.). Some bird species that were formerly very common are the glaucous macaw (*Anodorhynchus glaucus*), which is now extinct in the wild, and the Brazilian merganser (*Mergus octosetaceus*), very few of which have been sighted for some time (Chapter 16, this volume). It is estimated that some 9 species are threatened and 13 species are near-threatened in the Interior Atlantic Forest in Paraguay (Chapter 25, this volume), all of which are dependent on the forest. Many are species typical to savannas because patches of this ecosystem exist in the Atlantic Forest of Paraguay. Therefore, in reality there are even more threatened species in the forest, given the presence of fauna from other ecoregions.

The mammal species most characteristic of the Interior Atlantic Forest are the dwarf brocket (*Mazama nana*), a deer species currently limited to certain forest fragments but not classified as threatened; the porcupine *Coendou paraguayensis* (= *Sphiggurus spinosus*), very few of which have been recorded; and another 11 mammal species. A total of 22 species are considered threatened or near-threatened. Some mammal species that were widely distributed in the past are now considered critically endangered in the Paraguayan portion of the Atlantic Forest, including the giant armadillo (*Priodontes maximus*) and the giant otter (*Pteronura brasiliensis*).

History of Use and Abuse

Some of the factors causing rapid biodiversity loss have historical roots. Among them, land distribution and mistaken development schemes have had constant negative impacts on the forest ecosystem.

Inequalities in Land Tenure

The various land distribution schemes implemented in Paraguay since the end of the nineteenth century exacerbated the tremendous disproportion in land tenure and resulted in the elimination of many public lands. The productive systems associated with these forms of land tenure, which were based on selective logging and *mate* harvesting, were the first source of extensive damage to the Interior Atlantic Forest, although selective logging did leave large continuous forest masses that remained essentially unchanged until the 1950s.

The great inequality in land tenure and the lack of public lands to be offered under settlement plans also ultimately led to the current land disputes. Illegal "invasion" of private lands (squatting) caused widespread deforestation out of fear of expropriation. Moreover, in many instances squatting and deforestation were strongly encouraged by groups linked to illicit trafficking in wood. Therefore, two distinct deforestation processes occurred: one at the hands of land owners who, fearing that their lands would be expropriated, cleared extensive areas for agriculture, and the other by landless farmers who took over forest areas on private lands.

Threats arise constantly, and in some cases whole settlements can be established through squatting. Rural groups are now well organized, and they represent large numbers of people. The squatters usually occupy forestlands located within large properties (more than 150 km^2), and they tend to select lands that are the object of some sort of tenure dispute (e.g., mortgaged or untitled lands, lands on which property taxes are overdue). This strategy makes it difficult to evict them and facilitates expropriation. Such lands generally are first settled by groups who extract all the timber, followed by groups of farmers who move in and settle on plots of 10 ha each. The availability of technical assistance and credit for these settlements usually is quite limited, and the only source of livelihood is cotton farming, which is subject to unfavorable price fluctuations caused by a centralized bulking and marketing system. These colonies typically spring up in "comb" form: perpendicular access roads leading to a strip of deforested land. Still, there is always a small residual forest, which could serve as a corridor if it were not subject to excessive hunting.

Many of the new settlements and lands are then sold to foreign investors. Many land owners, discouraged by current socioeconomic problems, are selling their lands to investors, mostly Brazilians. This promotes the establishment of productive systems designed to yield quick profits, which include intensive agricultural practices or overgrazing of pasturelands, with no provision for ensuring sustainable land use.

Development Problems: The Perception of the Forest as Unproductive

Adding to the conflicts described in this chapter is the critical misconception that the forest is unproductive. It is ironic that a country with such rich forest resources has always considered the forest unproductive, viewing it as an obstacle to agricultural production and a land of wild animals and vermin. The concept of a productive forest and of environmental services and profitable nonforest products emerged only in the 1980s, when studies documented the degree of deforestation. Only recently has Paraguayan legislation incorporated the idea of protective forests (i.e., forests that protect river basins), and not until 1986 was a law enacted that made it mandatory to conserve at least 50 m of forestland on both sides of rivers (Decree 18.831). Another factor that contributed to the perception of the forest as unproductive was low prices for wood on the domestic market, which began to rise only after the elimination of all the timber sources in Brazil, the main consumer of wood from the Atlantic Forest.

This notion of lack of productivity also led to underuse of the native forest, most of which was simply burned down. In the 1980s and part of the 1990s, deforestation processes were aimed primarily at opening up new agricultural lands. Often this was a strategic measure taken to avoid illegal occupation and possible expropriation, and quick (emergency) methods therefore were used, namely felling the trees with tractors and chains and then burning them.

As an example of the extent to which the forest was underused, the annual rate of deforestation between 1945 and 1985 was equivalent to around 100 million m^3 (8 million m^3 of timber and 92 million m^3 of firewood). However, only 7 million m^3 were consumed annually (1.8 million m^3 of timber and 5.5 million m^3 of firewood), which means that the equivalent of 92,650,000 m^3 of wood was wasted, resulting in losses equivalent to US$173,514,000 for timber and US$360,208,000 for fossil fuels consumed in place of the unused firewood (Bozzano and Weik 1992). For many years, the total amount being lost was equivalent to one-third of Paraguay's foreign debt.

Currently, because wood prices are high, contracts are awarded through a competitive process for the complete extraction of all timber before any lands are cleared for agriculture. However, most of the wood is trafficked illegally outside the country, and most of the capital (financing and profits) therefore comes from and returns to foreigners who are motivated by a desire for quick profits.

Inadequate Legal and Institutional Framework for Management and Oversight of Natural Resources

Numerous allusions have been made to the illegal activities that led to deforestation of the Interior Atlantic Forest. Although Paraguay has many environmental laws, most of which originated in the 1990s, they are weakened by some serious legal and institutional problems (Chapter 28, this volume). The legal problems arise from contradictions and ambiguities in the interpretation of the laws. The institutional problems stem from an overlapping of environmental agencies, each of which carries out similar activities but with different objectives.

A clear example of this is the recent creation (2000) of the Secretariat of the Environment, which is responsible for enforcing most of the environmental laws but has no operating budget. Moreover, the secretariat does not have authority over the forestry sector, which still falls within the purview of the Ministry of Agriculture, to which the National Forest Service is attached. The Rural Welfare Institute continues to disregard the reserve areas and promote the creation of rural settlements in protected areas (such as the San Rafael, Cerro Corá, and Ybycui National Parks). The Office of the Attorney General of Paraguay has recently taken a harder line on environmental issues, modifying the Directorate of Environmental Affairs and, in 1999, giving it the authority to prosecute environmental offenses (Cañiza, pers. comm. 2001). In 2000, the first environmental crime unit was created, with four environmental prosecutors, which will foster institutional reform and facilitate prosecution in situ of those who engage in illegal activities. Another weakness in the legal framework is total ignorance of environmental laws and procedures on the part of the national police force, especially in the rural areas where illicit activities are occurring.

The country's political and socioeconomic instability has exacerbated the institutional problems. Between 1998 and 2001, 11 different individuals were responsible for environmental and conservation-related matters (vice-ministers and, now, ministers of the environment), each of whom served for an average of less than 1 year (the majority of the changes occurred in the last 4 years, when each vice-minister or minister averaged fewer than 5 months in office). This has led to a great deal of insecurity among subordinate officials and an exodus of qualified personnel who have left in search of more promising opportunities.

Lack of Capacity to Implement Viable Incentives for Conservation

Neither the government nor illicit activities are entirely responsible for the degradation of the Interior Atlantic Forest. The private sector has also played a part. In addition, nongovernment organizations (NGOs) have had limited success in putting in place sustainable development alternatives that could be effectively replicated. Although there have been a number of sustainable development initiatives that have sought to conserve the forest, and many of them have yielded positive results, they have not been designed in ways that are feasible to replicate. Most of those involved in deforestation processes therefore continue to opt for illegal economic activities.

Protection and Conservation

The history of wild area protection initiatives in Paraguay began late in comparison with other Latin American countries. The first agency charged with implementing and administering public protected areas was the Ministry of Defense in 1963.

The development of protected areas began in 1945 with Decree 9,535, which established "reserve areas" along national highways. However, it took a number of years for the first reserve areas to be created, and even then the process was disorganized and technically weak, as with the case of Guayakí Park (1962). Not until 1966 was a protected area established that was large enough to fulfill its conservation objectives: Tinfunque National Park, which encompassed 2,000 km^2.

In 1973, the Ministry of Agriculture was given responsibility for administering protected areas through its National Forest Service and the Department of National Parks and Forest and Wildlife Management. After that year several other conservation entities were created, including the Jacuy and Ñacunday Protective Forests and seven national parks (as of 1990) (CDC 1990).

Several technical missions intervened in this process, attempting to organize and structure a system for managing protected areas. All these missions pointed out the need to identify the sites for future parks and recreation areas before more agricultural land was cleared (CDC 1990). An effort was also made to organize and establish a national system for classifying protected areas based on sound technical criteria. In fact, the lack of uniform criteria has been one of the main obstacles to developing and strengthening protected natural areas.

Several changes occurred in the 1980s that were propitious for the conservation of biodiversity and the establishment of protected areas. In 1980, the National Biological Inventory (now the National Museum of Natural History of Paraguay) began collecting biological information under an agreement between the Paraguayan National Forest Service and Institute of Basic Sciences and the U.S. Fish and Wildlife Service and the Peace Corps (Celeste Acevedo, pers. comm. 2001; also described in CDC 1990). In that decade, responsibility for managing protected areas was also transferred to the Directorate of National Parks and Wildlife, which was initially an independent authority and then, in 1987, became an agency of the Undersecretariat of State for Natural Resources and the Environment of the Ministry of Agriculture. This change was important because it created an entire agency for environmental issues. In addition, the Nature Conservancy and the Peace Corps provided support for the development of the Conservation Data Center (Centro de Datos para la Conservación), whose function is to organize the country's biological information (Acevedo, pers. comm. 2001).

At the same time, two new kinds of natural protected areas were created. The first was associated with the Itaipú binational hydroelectric plant, which, as an independent entity, administers and regulates three protected areas—Itabo, Limoy Refugio Mbaracayu, and Tati Yupi—which are intended to offset the damage caused by the dam. The other new kind of protected area was the private reserve, which has come to be recognized as an important alternative in a country with little public land.

In 1990 a major step forward was taken with the effort to assess and determine the need for conservation of protected areas. At that time, only 3 percent of the country's total area was protected. That effort resulted in publication of a document that identified the priority areas for conservation (CDC 1990), which provided the framework for conservation of 10 percent of the major wild areas.

This strategic process culminated in 1993 with the creation of the National System of Protected Wild Areas (Sistema Nacional de Áreas Silvestres Protegidas [SINASIP], Law 352, 1994). The system comprised three subsystems: the public subsystem, administered by the Directorate of Parks (now the Secretariat of the Environment); the subsystem administered by the Itaipú binational entity; and the subsystem under private control. SINASIP identified 16 potential areas that should be protected, 3 of which are in the Interior Atlantic Forest and would add some 26,000 km^2 to the system.

Thereafter, the institutional and administrative problems intensified because of tremendous sociopolitical and economic instability. Between 1995 and 2000 there were two major bank crises and one major political crisis that resulted in replacement of the country's president, in addition to several corruption scandals involving politicians and judges. One consequence of this instability was frequent changes in public authorities. For example, over a 14-year period (1987–2001) nine different individuals served as vice-ministers of the environment before the creation of the Secretariat of Environment, after which there were two ministers of the environment and 11 directors of national parks and wildlife.

The immediate result of the problems associated with political instability was inaction by enforcement authorities, which ultimately weakened institutional capacity and led to the loss of previously trained personnel. During this period, several events occurred that had a negative impact on the management and conservation of protected areas, such as the revocation of the official designation of one area as a national park (Cerro Sarambi, one of the priority areas for conservation of the Interior Atlantic Forest) and lack of attention to the implementation of private reserves. With regard to the latter, although the law creating private reserves was enacted in 1994 (Law 352/94), only in 2000 did the Directorate of Parks issue a resolution setting out the procedures for registering and obtaining official recognition of such areas, and to date none of the private protected areas has been officially designated as such (although three have been approved). Therefore, none of the owners of private reserves has yet begun to accrue benefits as a result of conservation.

Past, Present, and Future Trends

Current trends in the conservation of the Interior Atlantic Forest and the evolution of the associated social problems are not very encouraging. It is unlikely that the land dispute problems will be resolved any time soon; on the contrary, rural groups continue to take over the lands of large owners. Moreover, it is not "socially correct" to base conservation of the Interior Atlantic Forest on maintenance of a system of private *latifundios*. Large portions of the Interior Atlantic Forest remain in Paraguay, and from a biodiversity standpoint there are still some species that are very important because of their rarity or degree of endangerment. But the range and numbers of some species continue to diminish, and the problems that threaten this ecosystem are extremely complex and difficult to solve.

Socioeconomic and political prospects are uncertain, given the lack of state guarantees for domestic and foreign investment, Paraguay's membership as a minor partner in a regional trading block (Mercosur) that has no systems for compensation among its members, and lack of land reform. All these factors lead to poverty, which is almost always bad for conservation.

The main sources of monetary wealth for the country are its hydroelectric dams, which have an enormous potential for sales of electric power, especially to Brazil. Unfortunately, in some cases very high construction costs largely eliminated economic returns (Chapter 35, this volume). During 2001, the construction of a third generating plant on the middle Paraná River was approved, although its exact location is still being debated. This new dam could seriously damage forest areas on both sides of the river.

State-led conservation actions are insufficient, and even with the recent creation of the Secretariat of the Environment, the government lacks the capacity and resources needed to address priority issues (Chapter 28, this volume). It is not expected that the problems associated with the secretariat's lack of resources will be resolved before the next presidential elections in 2003. Social pressures are very strong, and conservation issues therefore are often relegated to a position of secondary importance in decision makers' discussions.

On the other hand, NGOs have done some good and have played a decisive role in achieving several important successes for conservation. For example, two of the most notable achievements are the social mobilization activities carried out by the NGO Sobrevivencia (Survival) to organize people against dam construction, including protests, petitions, education, and information, and the purchase, implementation, and maintenance of the Mbaracayú Forest Natural Reserve by the Mbaracayú and Moisés Bertoni foundations. Nevertheless, the development of conservation areas patterned after the Mbaracayú model is almost impossible in the Interior Atlantic Forest. For that reason, the current trend is to seek to involve the residents of the areas in a participatory approach that includes the creation of management committees, as is the case in the area of the San Rafael National Park.

NGOs also engage in a number of extension and social assistance activities aimed at alleviating rural poverty and encouraging ecologically sustainable productive activities. One example is the work being carried out by organizations like

Alter Vida, Moisés Bertoni Foundation, and Servicios Ecoforestales para Productores (Ecoforestry Services for Producers). Unfortunately, although these efforts have yielded positive results, the process of replicating the models is moving very slowly.

In conclusion, present trends indicate that if the remnants of the Interior Atlantic Forest in Paraguay are to be conserved, programs must be put in place to fully implement and safeguard the protected areas that already exist. It will also be necessary to pay to obtain the last remaining important areas of the forest. The process of paying for conservation should involve all social strata and assistance providers (donors, government, NGOs, land owners, producers, indigenous groups), which corresponds perfectly to the objectives of the Management Committees for the development of protected areas. Moreover, the protected areas cannot survive in isolation, which makes it crucial to develop systems of compensation and incentives to encourage conservation of areas that serve as corridors or links between protected areas. It is here that rural extension activities and private conservation mechanisms become particularly important. In a scenario in which all these conditions are met, Paraguay still has a chance to conserve a large part of its biological diversity.

References

Barboza, F., Pinazzo, J., and Fracchia, F. 1997. *Bosque Atlántico Interior 1997.* Mapa. Proyecto Sistema Ambiental de la Región Oriental (SARO). Asunción: Ministerio de Agricultura y Ganadería and World Wildlife Fund.

Bozzano, B. and Weik, J. H. 1992. *El avance de la deforestación y el impacto económico.* Asunción: Ministerio de Agricultura y Ganadería/Deutsche Gesellschaft für Technische Zusammenarbeit.

Carbonell de Masy, R. S. J. 1986. Organización administrativa y contabilidad en las reducciones Guaranies. Instituto Anchietano de Pesquisas, Rio Grande do Sul, Brasil, Pesquisas História n. 25, 185 pp.

Cabrera, A. L. and Willink, A. 1980. *Biogeografía de América Latina.* Washington, DC: Secretaría General de la Organización de los Estados Americanos, Programa Regional de Desarrollo Científico y Tecnológico. (Serie de Biología, Monografía N. 13).

Carter, M. and Galeano, L. A. 1995. *Campesinos, tierra y mercado.* Asunción: Centro Paraguayo de Estudios Sociológicos/University of Wisconsin Land Tenure Center.

Cartes, J. L. 2000. Strategic private efforts for the conservation of Paraguayan biodiversity. *Ecological Tropical Forest Research Network News* 31: 25–30.

CDC (Centro de Datos para la Conservación). 1990. *Areas prioritarias para la conservación en la región oriental del Paraguay.* Asunción: CDC/DPNVS (Sistema Nacional de Áreas Silvestres Protegidas)/MAG (Ministerio de Agricultura y Ganadería).

Chase-Sardi, M. 1992. Ligeras notas sobre la prehistoria de los Arawak, Karibe y Tupí-Guaraní. *Suplemento Antropológico* XXVII(2): 7–30.

Dinerstein, E., Olson, D. M., Graham, D. J., Webster, A. L., Primm, S. A., Bookbinder, M. P., and Ledec, G. 1995. *Una evaluación del estado de conservación de las eco-regiones terrestres de América Latina y el Caribe.* Washington, DC: WWF/World Bank.

DPNVS/FMB (Sistema Nacional de áreas Silvestres Protegidas/Fundación Moisés Bertoni). 1993. *SINASIP: plan estratégico del Sistema Nacional de Areas Silvestres Prote-*

gidas. Asunción: Ministerio de Agricultura y Ganadería, Dirección de Parques Nacionales y Vida Silvestre/Fundación Moisés Bertoni.

FMB (Fundación Moisés Bertoni). 1994. Datos no publicados de botánica (unpublished botanical data). Asunción: FMB.

Haffer, J. 1974. Avian speciation in tropical South America, with a systematic study of the toucans (Rhamphastidae) and jacamars (Galbulidae). *Bulletin of the Nuttall Ornithological Club* 14: 3.

Hayes, F. E. 1995. *Status, distribution and biogeography of the birds of Paraguay.* New York: American Birding Association. (Monographs in Field Ornithology No. 1).

Holdrige, L. R. 1947. Determination of world plant formations from simple climatic data. *Science* 105(2727): 367–368.

Hueck, K. 1978. *Los bosques de Sudamérica: ecología, composición e importancia económica.* Eschborn: Deutsche Gesellschaft für Technische Zusammenarbeit.

IBR (Instituto de Bienestar Rural). 1997. *Campesinos, economía cultura, modernidad.* Asunción: Instituto de Bienestar Rural/Proyecto ALA/90/24 Unión Europea.

Jiménez, B. and Espinoza, C. 2000. *Manual de plantas útiles de Ypeti.* Asunción: Fundación Moisés Bertoni.

Molas, O., Heyn, R., and Arias, R. 1996. Documento base sobre el sector pecuario y su impacto ambiental. Paraguay: ENAPRENA, Asunción

Nikiphoroff, B. 1994. *El subdesarrollo rural Paraguayo: la problemática algodonera.* Asunción: Fundación Moisés Bertoni/Intercontinental Editora.

Olson, D. M. and Dinerstein, E. 1998. The *Global 200: a representation approach to conserving Earth's distinctive ecoregions.* Draft manuscript. Washington, DC: Conservation Science Program, WWF.

Stotz, D. F., Fitzpatrick, J. W., Parker, T. A., and Moskovits, D. K. 1996. *Neotropical birds: ecology and conservation.* Chicago: University of Chicago Press.

Súsnik, B., Chase-Sardi, M. 1995. *Los indios del Paraguay.* Editorial Mapfre. Madrid: Colección Indios de América.

Wege, D. C. and Long, A. J. 1995. *Key areas for threatened birds in the neotropics.* Cambridge, UK: BirdLife International.

Chapter 25

Biodiversity Status of the Interior Atlantic Forest of Paraguay

Frank Fragano and Robert Clay

The history of the rainforests of eastern Paraguay is the history of the country itself, for there the European colonizers had their first contact with the original inhabitants of the jungle and founded the capital of Asunción. In this corner of the continent, where there was no promise of Eldorado, the colonial economy was forced to rely on products extracted from the forest (Chapter 24, this volume).

Over the years, Paraguay has been the subject of little in-depth study by scientists and explorers. Perhaps the isolation imposed by centuries of dictatorial regimes made it less inviting for scientific research and exploration. Much of what is known is owed to the work of early explorers and priests who took an interest in nature (de Azara 1998) and later such scientists as the Swiss émigré Moisés Santiago Bertoni and his son Arnaldo de Winkelried Bertoni.

These early works, though highly laudable, tended to be broad and superficial. Only recently, especially after Paraguay's transition to democracy in 1989, did scientists turn attention to biodiversity in the Paraguayan Atlantic Forest. The National Biological Inventory Project was launched in 1980 under the joint efforts of the Ministry of Agriculture and Livestock, the U.S. Peace Corps, the World Wildlife Fund, and the U.S. Fish and Wildlife Service. This project, in which young Paraguayan biologists worked with specialized Peace Corps volunteers, led to establishment of the National Museum of Natural History, which is almost the only institutional repository for scientific research in the country. Itaipú Binational also owns a scientific collection developed during construction of the Itaipú hydroelectric power facility.

Later, with support from the Nature Conservancy, the Conservation Data Center (CDC) was established and developed within the Bureau of National Parks and Wildlife under the Ministry of Agriculture and Livestock (now under the Secretariat for the Environment). The CDC has promoted the systematic collection of information on biodiversity and has made technical proposals for creating a National System of Protected Forest Areas (Chapter 28, this volume).

Unfortunately, the advent of democracy in Paraguay has not significantly improved the economic situation. Research is still neglected by the government and poses a major challenge for Paraguayan academic institutions. However, thanks to the work of a few local and foreign academic institutions, it has been possible to gradually collect data and increase the knowledge about biodiversity within the country. Nongovernment organizations in Paraguay—most notably the Moisés Bertoni Foundation and Guyra Paraguay—have also provided substantial support for biological inventories and the conservation of critical areas.

Biogeographic Subdivisions and Criteria

The Atlantic Forest in southeastern Brazil, northeastern Argentina, and eastern Paraguay is a unique region composed mainly of tropical rainforests separated from the Amazon forests by the more arid Caatinga and Cerrado regions (Chapter 5, this volume). It comprises various subdivisions, or forest ecoregions, based on geomorphologic and ecologic characteristics. The most extensive of these ecoregions encompasses the Paraná and Paraíba River basins, where the convergence of the forests of southeastern Brazil, northeastern Argentina, and eastern Paraguay is commonly called the Interior Atlantic Forest (Dinerstein et al. 1995) or the Paraná-Paraíba Interior Forests (Olson and Dinerstein 1998).

The Interior Atlantic Forest is a semideciduous forest characterized by predominance of a single species and distinguished by a group of species typical of warm temperate climates that are limited to this region (Mori et al. 1981; López et al. 1987; Keel et al. 1993). Together with the more diverse tropical evergreen forests in the coastal and montane regions of Brazil, it constitutes the region known broadly as the Atlantic Forest (Dinerstein et al. 1995; Davis et al. 1997).

The Atlantic Forest region of Paraguay is located in the eastern region, an area of 159,800 km^2, or approximately 40 percent of the national territory. This region is bordered on the east and south by the Paraná River and the Mbaracayú and Amambay Mountains, on the north by the Apa River, and on the west by the Paraguay River. The entire region is humid (average annual rainfall between 1,300 and 1,800 mm) (Fariña Sánchez 1973), mildly hilly, and irrigated by numerous rivers and streams that empty into the Paraguay and Paraná. A discontinuous mountain range divides the waters between the two principal rivers. The forests in the eastern region of Paraguay have been regarded as part of the Paraná Province of the Amazon Domain (Cabrera and Willink 1973). This province includes various districts, and the forest comes under the tropical forest district.

In terms of aquatic ecosystems, the Paraguayan Atlantic Forest belongs to the ecoregion known as the Paraguay-Paraná Complex (Olson et al. 1998). The fauna of this aquatic ecoregion shares many elements with Amazonia, including some of the same species. Flooding of the Sete Quedas falls when the Itaipú Dam was completed permitted the exchange of fish species between the Paraguay-Paraná Complex and the Upper Paraná ecoregions (Olson et al. 1998).

Geologic and edaphic conditions influence the composition and characteristics of the predominant vegetation. Rich basaltic soils, which predominate in the east, especially in the Paraná River basin, have sustained the tallest forests and the most

Table 25.1. Classifications of the subregions of the Paraguayan Atlantic Forest.

Spichiger et al. (1995)	Veloso and Góes-Filho (1982)
Flora from the Brazilian south and southeast	Paraná Province (according to Cabrera and Willink 1973)
Seasonal deciduous flora of the Pleistocene	Seasonal deciduous forest
Flora from Paraná: lauriflora	Mixed ombrophilous forest
Flora from the southern plateau: pine flora	Mixed montane ombrophilous forest
Atlantic rain flora or Atlantic forest flora	Dense ombrophilous forest or Atlantic Province (according to Cabrera and Willink 1973)
Humid neotropical flora with southern limits in Paraguay	Dense alluvial ombrophilous forest

diverse in terms of both flora and fauna. In the central and western parts of the region in the Paraguay River basin the soils are sandy and the forests are less dense, with a canopy of 25–35 m. Although there are fewer tree species, several are important for their wood. This plain also has areas with humid soil and pasturelands.

In short, the Paraguayan Atlantic Forest has been divided into subregions according to several classifications (Tortorelli 1967; Ferreiro 1981; CDC 1990; Spichiger et al. 1995; Sanjuro and Gauto 1996). One study (Spichiger et al. 1995) has identified five floristic trends (Table 25.1), but the tendency has been to use simpler classifications that divide the Paraguayan Atlantic Forest into three biogeographic regions on the basis of broad characteristics (Table 25.2).

Table 25.2. Distribution and characteristics of the three ecoregions of the Interior Atlantic Forest of Paraguay.

Ecoregion	Distribution	Characteristics
Amambay/ Montane Forests	Bounded by the Estrella Arroyo on the north, the Amambay and Mbaracayú ranges on the east (up to the Brazilian border), and the Jejuí River and the northern limit of basaltic soil formations on the south. The western boundary is difficult to delineate because there is no natural barrier and the forests are, or were, continuous and transition gradually.	It is the most important in the country, both for its species and for its large productive volume. This is the tallest and the densest forest formation in the country, reaching 35–40 m in height. Approximately 60 tree species make up the forest canopy, plus 20 additional species of emerging trees. In the tallest forests, located in the Amambay and Mbaracayú mountains in the northeast, there is a predominance of *peroba rosa* (*Aspidosperma polyneuron*), which accounts for as much as 50 percent of the canopy in some places. The wood of this tree is highly sought after, and the species is overexploited. Other commercially important species found in this ecoregion include *yvyra pyta,* or yellow poinciana, "yellow flame tree," or "copperpod" (*Peltophorum dubium*) and *guatambu* (*Balfourodendron riedelianum*) (IIED/USAID 1985). In addition, some of the predominant plants limited to this ecoregion include the custard apple *Annona amambayensis; kai kay'gua* or *jequitiba branco* (*Cariniana estrellensis*); and *kuati'y* (*Vochysia tucanorum*). The understory is very dense and characterized by an abundance of tree ferns.

Table 25.2. Continued

Ecoregion	Distribution	Characteristics
Upper Paraná/ Paraná Forests	Bounded on the north by the Mbaracayú range, on the south and east by the Paraná River, and on the west by the Joaquín highlands and the Ybytyruzú and San Rafael ranges, or basically the dividing line between the basins of the Paraná and Paraguay rivers.	A forest covers most of the ecoregion, which is less homogeneous in height and density than that of the Amambay ecoregion. The top stratum consists largely of trees over 30 m in height, reaching even 35 or 40 m in well-drained areas. Each stratum has a large number of species, the most common ones being cedar (*Cedrela tubiflora*); *aju'y,* or laurel (*Nectandra* spp.); *lapacho,* or trumpet tree (*Tabebuia* spp.); *yvyrá peré,* or garapa (*Apuleia leiocarpa*); *guatambu* (*Balfourodendron riedelianum*); *incienso,* or *cabreuva* (*Myrocarpus frondosus*), *yvyrá pytá,* or yellow poinciana (*Peltophorum dubium*), the laurel *guaicá* (*Ocotea* spp.), and *timbó,* or earpod tree (*Enterolobium contortisiliquum*). The forests in this ecoregion are also characterized by a large number of lianas, epiphytes, tree ferns, and palms, including the *palmito,* or assai palm (*Euterpe edulis*).
Central Jungle/ Central Forests	This ecoregion shares borders with the Aquidabán ecoregion in the north (consisting mainly of a continuation of the Cerrado) and the Amambay and Upper Paraná ecoregions in the east. In the west and south it is coextensive with the jungle until it transitions into the pasturelands of the Central Coast and Ñeembucú ecoregions.	Most of the ecoregion consists of tall forest interspersed with pasturelands. The forest has fewer species and is more open than those of the Amambay and Upper Paraná ecoregions. There are more species of deciduous trees and more species with needles. The canopy reaches up to 35 m in height, and the predominant species are *lapacho,* or pink trumpet tree (*Tabebuia ipe*) in the east; *trébol,* or *cumbaru* (*Amburana cearensis*) in the north; and *manduvi'ra,* or rain tree (*Samanea saman*) in the east; as well as cedar (*Cedrela* spp.); *ybyrá pytá,* or yellow poinciana (*Peltophorum dubium*); *yvyrá ró,* or *guiaro* (*Pterogyne nitens*); *incienso,* or *cabreuva* (*Myrocarpus frondosus*); *guatambu* (*Balfourodendron riedelianum*); the silk tree *yvyrá yú* (*Albizia hassleri*); and *cancharana* (*Cabralea* spp.) in certain parts.

Trends in Biodiversity

There is little information on biodiversity trends in the Atlantic Forest of Paraguay; however, in the following section we describe some of the available information on terrestrial and aquatic ecosystems, as well as on some of the endemic and threatened species.

Terrestrial Ecosystems

The Atlantic Forest is recognized as one of the ecosystems undergoing the most severe deforestation in the world. It is also a center of endemism for both flora and fauna (Müller 1973; Mori et al. 1981; Brown 1982; Mittermeier et al. 1982;

Table 25.3. Deforestation by departments in Paraguay, from 1984 to 1991 (Huespe Fatecha et al. 1994).

Department	Forest Cover in 1991 (km²)	Deforestation from 1984 to 1991		Annual Deforestation Rate	
		km²	%	km²/year	%
Alto Paraná	3,902.26	4,637.17	54	662.45	7.7
Itapúa	2,657.76	3,088.12	54	441.16	7.7
Caaguazú	2,344.61	1,804.77	44	257.82	6.3
Central	54.31	40.56	43	5.79	6.1
San Pedro	4,603.17	3,330.80	42	475.83	6
Caazapá	2,114.76	1,290.76	38	184.39	5.4
Canindeyú	6,699.64	3,015.38	31	430.77	4.4
Paraguarí	334.02	137.76	29	19.68	4.1
Guairá	538.68	224.61	29	32.09	4.1
Cordillera	259.99	105.37	29	15.05	4.1
Amambay	4,423.90	1,454.00	25	207.71	3.6
Misiones	168.31	51.92	24	7.42	3.4
Concepción	5,245.07	1,006.73	16	143.82	2.3
Ñeembucú	76.80	10.63	12	1.52	1.7

Cracraft 1985; Haffer 1985; Oliver and Santos 1991; Bibby et al. 1992; Davis et al. 1997; Stattersfield et al. 1998), so the conservation of its biodiversity has high priority at the world level (Myers 1988; Dinerstein et al. 1995; Brooks and Balmford 1996; Olson and Dinerstein 1998; Myers et al. 2000).

The Paraguayan Atlantic Forest was once one of the largest subtropical rainforest formations in South America (Davis et al. 1997), covering an estimated 94,000 km², or 58.8 percent of the land area of the eastern region (Sanjurjo and Gauto 1996); in 1945 it was calculated at 88,000 km² (Bozzano and Weik 1992). Not long ago the percentage of unexplored virgin forest was estimated at more than 37 percent (Gorham 1973), but today Paraguay has almost nothing left of its forest frontier (Table 25.3; Bryant et al. 1997).

Estimates of the rate of deforestation vary greatly. Although two studies (Bozzano and Weik 1992; Huespe Fatecha et al. 1994) have made an especially thorough review of changes in the forest surface of the eastern region (Chapter 26, this volume; Table 25.3), effective conservation has been hampered by disagreement over appropriate analytical methods and by long gaps between studies.

The Paraguayan Atlantic Forest is being fragmented rapidly. Only 29 remaining fragments are larger than 100 km², and only 12 are more than 200 km² (Table 25.4). The largest of these—the 590-km² Mbaracayú Natural Forest Reserve—is the only one under effective protection. It is unlikely that any of these fragments is large enough for the long-term survival of large mammals such as the lowland tapir (*Tapirus terrestris*), jaguar (*Panthera onca*), or birds of prey such as the harpy eagle (*Harpia harpyja*) (Chapter 31, this volume).

Table 25.4. Fragments of the Atlantic Forest of Paraguay larger than 120 km^2.

Fragment	Status	Department	Forest Extent (km^2)	Protected Forest Area
Mbaracayú	Natural forest reserve	Canindeyú	1,146.87	590.56
Ypané Forest	Without protection	San Pedro	801.10	0.00
San Rafael	Without protection	Itapúa/Caazapá	786.43	0.00
Yaguarete Forest	Without protection	San Pedro	611.71	0.00
San Pedro Occidental	Without protection	San Pedro	466.08	0.00
Amambay Forest	Without protection	Amambay	441.83	0.00
Itabó	Private protection	Canindeyú	403.44	30.00
San Joaquín	Without protection	Caaguazú	259.41	0.00
North Canindeyú	Without protection	Canindeyú	246.25	0.00
Cerro Guasu	Indigenous reserve	Amambay	210.70	110.00
Cerro Corá-Arroyo Blanco	National park–private reserve	Amambay	207.07	110.35
Caaguazú	National park	Caazapá	202.22	127.35
Total			5,783.11	968.26

Aquatic Ecosystems

The aquatic ecosystems of the Paraguayan Atlantic Forest are also highly threatened, especially by construction of hydroelectric dams (Chapter 35, this volume), loss and degradation of riverine forests, and contamination by agriculture, industry, and urban wastewater (Olson and Dinerstein 1998).

For example, the damming of the Paraná for the Itaipú hydroelectric project was followed by a massive die-off of fish in the Iguaçu River below the dam. At the same time, peak levels of phenol were detected, either from the breakdown of lignin in flooded plants or from the degradation of herbicides on inundated cropland (ENAPRENA 1995b). High levels of pesticides and heavy metals have also been documented in tributaries of both the Paraná and the Paraguay rivers (ENAPRENA 1995b; Carlini et al. 2001).

Erosion from deforestation and other activities is degrading the river basins, and soil loss in many cultivated areas exceeds allowed limits (ENAPRENA 1995b). In the area affected by the Itaipú project, the water in some of the tributaries (e.g., the Piratiy and Charapa Rivers) is extremely turbid (less than 0.5 m transparency). By contrast, in forested areas of the Limoy and Itabó Rivers, transparency levels were 2 m or more (Hermosa 1999). Eroded sediment is retained in the Itaipú reservoir and could have long-term effects on both the biodiversity of the basin and the potential effectiveness of the hydroelectric project itself. Serious erosion is also occurring along the Itaipú reservoir itself, where more than 13 m^3 of soil per meter of shoreline is washed away each year (Acha Navarro 1999).

Species Diversity

Complete species lists are lacking both for Paraguay and, with the exception of birds, for its Atlantic Forest region. As of 1994 there was no up-to-date information on the diversity of the country's amphibians and reptiles (Iriarte Walton 1994). Recent estimates indicate considerable biodiversity: about 13,000 species of vascular plants, 100,000 species of invertebrates (including 765 butterfly species [Barrios and Kochalka 1993]) 46 amphibians, 100 reptiles, and 167 mammals (ENAPRENA 1995a). Bats and rodents have received special attention thanks to studies of vectors of disease agents (Williams et al. 1997). Estimates of fish species are more variable, ranging from 182 to 390 (ENAPRENA 1995a; Mandelburger et al. 1996; DPNVS/FMB 1998; Frost 2000; Guyra Paraguay, unpubl. data). Guyra Paraguay, an organization devoted to bird conservation, has been compiling information from recent studies, earlier expeditions, and museum collections throughout the world. To date, 688 avian species have been documented, with another 113 species sighted but not confirmed.

For the aquatic ecosystems there is very little information available. In a pilot analysis of freshwater ecosystems of the world, Revenga et al. (2000) identified the Upper Paraná River basin as both a freshwater ecoregion of global significance and an important area for biodiversity of freshwater mollusks. Olson and Dinerstein (1998) have given priority to the Paraguay-Paraná Complex ecoregion because of its more than 300 species of fish, large invertebrates such as river crab, and mollusks.

Endemic Species

The Paraguayan Atlantic Forest harbors a variety of subtropical endemic genera and also several species characteristic of tropical forest and of the Cerrado in its southernmost area (Davis et al. 1997).

A total of 7,851 vascular plant species are known to exist in Paraguay, including 350 fern species. However, knowledge about the number of endemic species is hampered by the absence of a detailed registration system. At least 73 species have ranges known to be limited to Paraguay (Pérez de Molas, pers. comm. 2002). A study conducted in the Cerro Corá National Park, located in the Amambay ecoregion, documented 10 endemic species from different families (Basualdo and Soria 1994). In addition, several plant species are known to be endemic in the northeastern part of the eastern region, especially in the Amambay mountains: *Schinus weinmanniifolia* var. *hassleri* (the mastic tree), six species of *Annona* (custard apple), and *Callistene hassleri, Peltastes stemmadeniiflorus,* and *Rhodocalyx rotundifolius.*

Paraguay has an estimated 100,000 species of invertebrates, possibly 5,000 of which are unknown (ENAPRENA 1995a). Most invertebrates have been collected only once or from a single place in the country, and distribution data are lacking. Indeed, some species have been collected from only one site in the world,

Table 25.5. Endemic amphibian species of the Atlantic Forest recorded in Paraguay.

Guaraní or Spanish Common Name	English Common Name	Scientific Name
Kururu or *sapo franjeado*	Striped toad	*Bufo crucifer*
Kururu or *sapo amarillo*	Yellow Cururu toad	*Bufo ictericus*
Sapo pigmeo	Rio Parahyba toad	*Bufo pygmaeus*
Rana trepadora punteada	Spotted treefrog	*Hyla albopunctata*
Herrero or *rana herrero*	Blacksmith treefrog	*Hyla faber*
Rana trepadora	Cope's eastern Paraguay treefrog	*Hyla polytaenia*
Escuercito	Gunther's smooth horned frog	*Proceratophrys appendiculata*

such as the spiders *Alpaida alto* and *A. itapua,* both apparently endemic to the Paraguayan Atlantic Forest (Levi 1988). Of Paraguay's 61 known species of social wasps (Garcete-Barrett 1999), at least five—*Polistes obscurus, P. bequaertianus, Miscocyttarus parallelogrammus, M. araujoi,* and *M. villarricanus*—may be endemic to the Atlantic Forest, and *M. villarricanus* appears to be endemic to Paraguay (Garcete-Barrett, pers. comm. 2001).

Of the 355 known fish species in the Paraná River basin, 32 have been introduced; the Paraguay River basin has 254 documented species, 85 of which are endemic (World Resources Institute 1998).

Seventy-five amphibian species have been confirmed in Paraguay (Aquino et al. 1996; Frost 2000), of which two appear to be limited to the country, and seven probably are endemic to the Atlantic Forest (Table 25.5).

Among the approximately 100 known reptile species in Paraguay, 4 appear to be endemic—the lizards *Tropidurus guarani* and *Colobosaura kraepelin* and the snakes *Simophis rohdei* and *Phalotris nigrilatus*—and at least the last of these is thought to be endemic to the Atlantic Forest region. In addition, at least 11 species are believed to be limited to the Atlantic Forest region in general, including two amphisbaenas known as *yvy'ja* (*Amphisbaena mertensii* and *A. prunicolor*), the thirst snake (*Dipsas albifrons*), the swampsnakes (*Liophis frenatus* and *L. viridis*), the coral snake (*Micrurus tricolor*), the slender blindsnake (*Liotyphlops wilderi*), the pit vipers (*Bothrops jararaca* and *B. mojen*), and the toad-headed turtles (*Phrynops tuberculatus* and *P. vanderhaegei*).

Paraguay has no endemic avian species of its own; however, 80 species endemic to the Atlantic Forest have been documented in the country (Table 25.6), and the Paraguayan Atlantic Forest harbors populations of several species endemic to the ecoregion that are important from the conservation standpoint, such as the *parakáu keréu,* or vinaceous Amazon parrot (*Amazona vinacea*), the *ypeku akâ mirâ,* or helmeted woodpecker (*Dryocopus galeatus*), the *karichu de oreja negra,* or São Paulo tyrannulet (*Phylloscartes paulistus*), and the *jurupe,* or russet-winged spadebill (*Platyrinchus leucoryphus*).

Although Paraguay has only one or two mammal species found nowhere else, it has 11 species—mainly marsupials, rodents, and bats—that are endemic to the Atlantic Forest region and at least 4 rodent species that are now limited to the Atlantic Forest (Table 25.7).

Table 25.6. Endemic bird species of the Atlantic Forest recorded in Paraguay.

Guaraní or Spanish Common Name	English Common Name	Scientific Name
Ynambu kagua or *macuco*	Solitary tinamou	*Tinamus solitarius*
Taguato'i or *azor grande*	Gray-bellied hawk	*Leucopternis polionota*
Jaku po'i or *yacupoí*	Rusty-margined guan	*Pipile jacutinga*
Jakutinga or *yacutinga*	Black-fronted piping-guan	*Odontophorus capueira*
Sarakura, ypaka'a ka'aguy, or *saracura*	Slaty-breasted wood-rail	*Aramides saracura*
Pararu or *palomita morada*	Purple winged ground-dove	*Claravis godefrida*
Chiripepe, arivaya, or *chiripepé cabeza verde*	Reddish-bellied parakeet	*Pyrrhura frontalis*
Tu'i guembe or *lorito cabeza roja*	Red-capped parrot	*Pionopsitta pileata*
Chorao or *charao*	Red-spectacled parrot	*Amazona pretrei*
Parakáu keréu or *loro vinoso*	Vinaceous-breasted parrot	*Amazona vinacea*
Kavure or *lechucita grande*	Black-capped screech-owl	*Otus atricapillus*
Urukure'a mini or *urucureá chico*	Tawny-browed owl	*Pulsatrix koeniswaldiana*
Suinda ka'aguy or *lechuza listada*	Rusty-barred owl	*Strix hylophila*
Mainumby ruguaitî or *picaflor ermitaño grande*	Scale-throated hermit	*Phaethornis eurynome*
Mainumby or *picaflor negro*	Black jacobin	*Florisuga fusca*
Mainumby apiratî or *picaflor copetón*	Black-breasted plovercrest	*Stephanoxis lalandi*
Mainumby or *picaflor verde frente azul*	Violet-capped woodnymph	*Thalurania glaucopis*
Mainumby pyti'a morotî or *picaflor garganta blanca*	White throated hummingbird	*Leucochloris albicollis*
Suruku'a or *surucuá*	Surucua trogon	*Trogon surrucura*
Guyra paja or *burgo*	Blue-crowned motmot	*Momotus momota*
Arasari pôka or *arasarí chico*	Spot-billed toucanet	*Selenidera maculirostris*
Arasari pakova or *arasarí banana*	Saffron toucanet	*Baillonius bailloni*
Tukâ-í or *tucán pico verde*	Red-breasted toucan	*Ramphastos dicolorus*
Ypeku ne'i or *carpinterito cuello canela*	Ochre-collared piculet	*Picumnus temminckii*
Kurutu'i or *carpintero arco iris*	Yellow-fronted woodpecker	*Melanerpes flavifrons*
Ypeku para or *carpinterito barrado*	White-spotted woodpecker	*Veniliornis spilogaster*
Ypeku hovy or *carpintero verde*	White-browed woodpecker	*Piculus aurulentus*
Ypeku akâ mirâ or *carpintero cara canela*	Helmeted woodpecker	*Dryocopus galeatus*
Ypeku guasu ka'aguy or *carpintero grande*	Robust woodpecker	*Campephilus robustus*
Arapasu hovy or *trepador pardo*	Plain-winged woodcreeper	*Dendrocincla turdina*
Arapasu pini or *chinchero escamado*	Scaled woodcreeper	*Lepidocolaptes falcinellus*
Arapasu'i or *chinchero enano*	Lesser woodcreeper	*Lepidocolaptes fuscus*
Ypeku juru karapâ or *picapalo oscuro*	Black-billed scythebill	*Campylorhamphus falcularius*
Tacuarero	Canebrake groundcreeper	*Clibanornis dendrocolaptoides*
Turu kue or *pijuí corona rojiza*	Rufous-capped spinetail	*Synallaxis ruficapilla*
Kurutie hovy or *curutié oliváceo*	Olive spinetail	*Cranioleuca obsoleta*
Titiri or *titiri ceja blanca*	White-browed foliage-gleaner	*Anabacerthia amaurotis*
Ka'a'i guyra or *ticotico cabeza negra*	Black-capped foliage-gleaner	*Philydor atricapillus*
Ka'a'i guyra, titiri, or *ticotico ocráceo chico*	Ochre-breasted foliage-gleaner	*Philydor lichtensteini*

Table 25.6. Continued

Guaraní or Spanish Common Name	English Common Name	Scientific Name
Tiatui or *ticotico ojo blanco leucophthalmus*	White-eyed foliage-gleaner	*Automolus*
Arapasu'I rá or *picolezna estriado*	Sharp-billed treehunter	*Heliobletus contaminatus*
Ogaraiti or *raspahojas*	Rufous-breasted leaftosser	*Sclerurus scansor*
Mbatara or *batará goteado*	Spot-backed antshrike	*Hypoedaleus guttatus*
Chororo or *batará punteado*	Large-tailed antshrike	*Mackenziaena leachii*
Akâ botô or *batará copetón*	Tufted antshrike	*Mackenziaena severa*
Takuari pytâ or *batará coludo canela*	Bertoni's antbird	*Drymophila rubricollis*
Takuari or *batará coludo estriado*	Dusky-tailed antbird	*Drymophila malura*
Mbatara'i, tiluchi para, or *batará enano*	Streak-capped antwren	*Terenura maculata*
Mbatara chioro or *batará negro*	White-shouldered fire-eye	*Pyriglena leucoptera*
Chululu'i or *chululú chico*	Speckle-breasted antpitta	*Hylopezus nattereri*
Tokotoko or *chupadientes*	Rufous gnateater	*Conopophaga lineata*
Mosqueta corona oliva	Greenish tyrannulet	*Phyllomyias virescens*
Tachuri or *ladrillito*	Gray-hooded flycatcher	*Mionectes rufiventris*
Karichu or *mosqueta media luna*	Southern bristle-tyrant	*Phylloscartes eximius*
Karichu or *mosqueta de oreja negra*	São Paulo tyrannulet	*Phylloscartes paulistus*
Karichu or *mosqueta ojo colorado*	Bay-ringed tyrannulet	*Phylloscartes sylviolus*
Tachuri or *mosqueta enana*	Eared pygmy-tyrant	*Myiornis auricularis*
Mosqueta de anteojos	Drab-breasted pygmy-tyrant	*Hemitriccus diops*
Jurupe or *picochato grande*	Russet-winged spadebill	*Platyrinchus leucoryphus*
Viudita coluda	Shear-tailed gray-tyrant	*Muscipipra vetula*
Tiotoi, bailarín oliváceo, or *flautín*	Greenish manakin	*Schiffornis virescens*
Saraki hovy, bailarín, or *saltarín azul*	Blue manakin	*Chiroxiphia caudata*
Jaku toro, guyra toro, or *yacutoro*	Red-ruffed fruitcrow	*Pyroderus scutatus*
Guyra pong, guyra campana, or *pájaro campana*	Bare-throated bellbird	*Procnias nudicollis*
Aka'ê hovy or *urraca azul*	Azure jay	*Cyanocorax caeruleus*
Tacuarita blanca	Creamy-bellied gnatcatcher	*Polioptila lactea*
Korochire chiâ, havía ñakyrâ, or *zorzal herrero*	Slaty thrush	*Turdus nigriceps*
Chivi akâ pytâ or *chiví coronado*	Rufous-crowned greenlet	*Hylophilus poicilotis*
Arañero silbón	White-browed warbler	*Basileuterus leucoblepharus*
Sai hovy or *tangará arcoiris*	Green-headed tanager	*Tangara seledon*
Tietê or *tangará cuello castaño*	Red-necked tanager	*Tangara cyanocephala*
Tietê or *tangará alcalde*	Chesnut-bellied euphonia	*Euphonia pectoralis*
Teî teî or *tangará picudo*	Green-chinned euphonia	*Euphonia chalybea*
Tie hu, jurundi, or *frutero coronado*	Ruby-crowned tanager	*Tachyphonus coronatus*
Pioro or *frutero cabeza castaña*	Chesnut-headed tanager	*Pyrrhocoma ruficeps*
Guaranichinga or *pepitero negro*	Black-throated grosbeak	*Saltator fuliginosus*
Reinamora enana or *arrocero azul*	Blackish-blue seedeater	*Amaurospiza moesta*
Pichochô, katatáu, or *corbatita oliváceo*	Buffy-fronted seedeater	*Sporophila frontalis*
Katatáu, pichochô, or *corbatita picudo*	Temminck's seedeater	*Sporophila falcirostris*
Pichochô, cigarra, or *afrechero plomizo*	Uniform finch	*Haplospiza unicolor*

Table 25.7. Endemic and restricted mammal species of the Atlantic Forest recorded in Paraguay.

Guaraní or Spanish Common Name	English Common Name	*Scientific Name*
Endemic		
Myikurê hû or *comadreja orejuda*	Opossum	*Didelphis aurita*
Anguja myikurê or *colicorto rojizo*	Hensel's short-tailed opossum	*Monodelphis sorex*
Mbopi or *murciélago frutero grande oscuro*	Bat	*Artibeus fimbriatus*
Mbopi or *murciélago de ojos grandes*	Bat	*Chiroderma doriae*
Mbopi or *murciélago acanelado de Azara*	Bat	*Myotis rubber*
Guasu pororo or *mbororo*	Dwarf red brocket	*Mazama nana*
Anguja or *rata arrocera*	Rice rat	*Oryzomys intermedius*
Anguja Kaaguy, anguja pytâ, or *rata arrocera*	Rice rat	*Oryzomys ratticeps*
Anguja or *ratón subterráneo*	Mouse	*Akodon nigrita*
Rata tacuarera	Rat	*Kannabateomys amblyonyx*
Restricted		
Myikurê or *comadrejita cola larga*	Rat	*Thylamys macrura*
Anguja Kaaguy, anguja pytâ, or *rata pigmea*	Rice rat	*Oligorizomys nigripes*
Anguyá pepé or *ratón de monte*	Mouse	*Akodon cursor*
Rata guira or *ratón espinoso*	Rat	*Euryzygomatomys spinosus*

Population Trends

Population studies are a luxury in a country that has been explored as little as Paraguay. Most scientific expeditions have only recorded the presence of species and collected specimens for museums. One study in the Mbaracayú Natural Forest Reserve has enlisted members of the local Aché group to collect information on the populations sizes and exploitation of mammals and large birds (Hill et al. 1997; Hill and Padwe 2000). Early results indicate that exploitation of three mammal species—the *tatu hû,* or nine-banded armadillo (*Dasypus novemcinctus*); the *akutipak,* or paca (*Agouti paca*); and the *ka'i paraguay,* or brown capuchin monkey (*Cebus apella*)—would be sustainable even though their population sizes are vastly different in areas with and without hunting (Hill and Padwe 2000).

Another study focused on status of the *yakutinga,* or black-fronted piping-guan (*Pipile jacutinga*), in the Paraguayan Atlantic forest. This species, classified as vulnerable (BirdLife International 2000; IUCN 2000), was once abundant and wide ranging in the eastern region, where it now numbers fewer than 2,000 individuals and is limited to a few forest fragments (Clay et al. 1999; Clay 2001). The vinaceous Amazon parrot (*Amazona vinacea*) is another once-common bird that is now in danger of disappearing from the Paraguayan Atlantic Forest. Bertoni (1927) noted that in the 1890s large flocks of migrating vinaceous Amazon parrots (*Amazona vinacea*) (Chapter 15, this volume) darkened the sky, and as recently as 1978 local populations of thousands of individuals were reported (Collar et al. 1992). Today, however, only a few isolated populations remain, apparently none with more than 80 individuals (Guyra Paraguay, unpubl. data). The Brazilian merganser (*Mergus octosetaceus*) is now regarded as locally extinct.

Species with Global Conservation Problems

Five plant species endemic to the Atlantic Forest are on the world list of threatened species (IUCN 2000); three are globally threatened, one is near-threatened, and one is in the category of "insufficient data" (Table 25.8). In addition, two laurel species (*Ocotea* spp.) that may range into Paraguay are considered globally threatened.

The mollusks *Aylacostoma chloroticum, A. guaraniticum,* and *A. stigmaticum,* which had been limited to the Paraná River in the area of now-flooded Yacyretá Island, are now extinct in the wild (IUCN 2000; Chapter 35, this volume).

The short-tailed river stingray (*Potamotrygon brachyura*) is the only fish on the world list, under the heading of "insufficient data" (IUCN 2000). In Paraguay its distribution is limited to the Paraná River basin.

Of the 80 known avian species endemic to the Atlantic Forest in Paraguay, 22 are either in danger of extinction or near-threatened at the world level (BirdLife International 2000; Table 25.9). Five species that are not endemic to the Atlantic Forest but in Paraguay are limited to that region are threatened or near-threatened (Table 25.9).

Table 25.8. Globally threatened plant species in the Atlantic Forest of Paraguay.

Guaraní or Spanish Common Name		Scientific Name	Category
Endemic species of the Atlantic Forest	*Guatambu*	*Balfourodendron riedelianum*	Endangered
	Kuri'y or Paraná pine	*Araucaria angustifolia*	Vulnerable
		Myrciaria cuspidata	Vulnerable
	Katigua moroti	*Trichilia pallens*	Near threatened
	Yvyra paje or *incienso*	*Myrocarpus frondosus*	Insufficient data
Potentially occurring in Paraguay		*Ocotea porosa*	Vulnerable
		Ocotea pretiosa	Vulnerable
Nonendemic species but limited to this region in Paraguay	*Ygary* or cedar	*Cedrela fissilis*	Endangered
	Yvyra ro'mi or *peroba*	*Aspidosperma polyneuron*	Endangered

Table 25.9. Globally threatened bird species in the Atlantic Forest of Paraguay.

Atlantic Forest Endemic Birds				
	Common Name	**Order**	**Family**	**Number of Species**
Endangered	Doves and pigeons	Columbiformes (5)	Columbidae	5
	Parrots and parakeets	Psittaciformes (21)	Psittacidae	21
Vulnerable	Curassows	Galliformes (2)	Cracidae	1
	Quails		Odontophoridae	1
	Crakes and rails	Gruiformes (26)	Rallidae	22
	Finfoots		Heliornithidae	1
	Limpkins		Aramidae	1
	Seriemas		Cariamidae	2
	Plovers and lapwings	Charadriiformes (36)	Charadriidae	5
	Stilts		Recurvirostridae	1
	Jacanas		Jacanidae	1
	Painted snipe		Rostratulidae	1
	Sandpipers, snipes, and allies		Scolopacidae	22
	Gulls and terns		Laridae	5
	Skimmers		Rynchopidae	1
	Pigeons and doves	Columbiformes (16)	Columbidae	16
	Parrots and parakeets	Psittaciformes (21)	Psittacidae	21
	Cuckoos	Cuculiformes (12)	Cuculidae	12
	Barn owls	Strigiformes (17)	Tytonidae	1
	Owls		Strigidae	16
	Nightjars and nighthawks	Caprimulgiformes (18)	Caprimulgidae	15
	Potoos		Nyctibiidae	3
	Swifts	Apodiformes (6)	Apodidae	6
	Hummingbirds	Trochiliformes (17)	Trochilidae	17
	Trogons	Trogoniformes (3)	Trogonidae	3
	Motmots		Momotidae	2
	Kingfishers	Coraciiformes (7)	Alcedinidae	5
	Puffbirds	Piciformes (30)	Bucconidae	4
	Toucans		Ramphastidae	5
	Woodpeckers		Picidae	21
	Woodcreepers	Passeriformes (353)	Dendrocolaptidae	12
	Ovenbirds and allies		Furnariidae	43
	Antbirds		Thamnophilidae	22
			Formicariidae	3
	Gnateaters		Conopophagidae	1
	Tapaculos		Rhinocryptidae	3
	Tyrants		Tyrannidae	103
	Manakins		Pipridae	6
	Cotingas		Cotingidae	3
	Sharpbills		Oxyruncidae	1
	Plantcutters		Phytotomidae	1
	Crows and jays		Corvidae	4
	Swallows		Hirundinidae	13
	Wrens		Troglodytidae	6
	Gnatcatchers		Sylviidae	2
	Thrushes		Muscicapidae	7

Table 25.9. Continued

Atlantic Forest Endemic Birds				
	Common Name	Order	Family	Number of Species
Vulnerable	Mockingbirds	Passeriformes	Mimidae	2
	Pipits		Motacillidae	6
	Vireos		Vireonidae	3
	Wood warblers		Parulidae	7
	Tanagers, euphonias, cardinals		Emberizidae	84
Near-threatened	Tinamous			
	Grebes	Podicipediformes (5)	Podicipedidae	5
	Cormorants	Pelecaniformes (2)	Phalacrocoracidae	1
	Darters		Anhingidae	1
	Bitterns, herons, egrets	Ardeiformes (23)	Ardeidae	14
	Ibises and spoonbills		Threskiornithidae	6
	Storks		Ciconiidae	3
	Flamingos	Phoenicop Teriformes (1)	Phoenicotperidae	1
	Screamers	Anseriformes (21)	Anhimidae	2
	Ducks, swans, and geese		Anatidae	19
	Vultures	Falconiformes (53)	Cathartidae	5
	Kites, harriers, hawks, eagles		Accipitridae	35
	Caracaras and falcons		Falconidae	12
	Guans	Galliformes (7)	Cracidae	6
	Quails		Odontophoridae	1
	Crakes and rails	Gruiformes (26)	Rallidae	22
	Finfoots		Heliornithidae	1
	Limpkins		Aramidae	1
	Seriemas		Cariamidae	2
	Plovers and lapwings	Charadriiformes (36)	Charadriidae	5
	Stilts		Recurvirostridae	1
	Jacanas		Jacanidae	1
	Painted snipe		Rostratulidae	1
	Sandpipers, snipes, and allies		Scolopacidae	22
	Gulls and terns		Laridae	5
	Skimmers		Rynchopidae	1
	Pigeons and doves	Columbiformes (16)	Columbidae	16
	Parrots and parakeets	Psittaciformes (21)	Psittacidae	21
	Cuckoos	Cuculiformes (12)	Cuculidae	12
	Barn owls	Strigiformes (17)	Tytonidae	1
	Owls		Strigidae	16
	Nightjars and nighthawks	Caprimulgiformes (18)	Caprimulgidae	15
	Potoos		Nyctibiidae	3
	Swifts	Apodiformes (6)	Apodidae	6
	Hummingbirds	Trochiliformes (17)	Trochilidae	17
	Trogons	Trogoniformes (3)	Trogonidae	3
	Motmots		Momotidae	2
	Kingfishers	Coraciiformes (7)	Alcedinidae	5
	Puffbirds	Piciformes (30)	Bucconidae	4
	Toucans		Ramphastidae	5

Table 25.9. Continued

Atlantic Forest Endemic Birds				
	Common Name	**Order**	**Family**	**Number of Species**
Near-threatened	Woodpeckers		Picidae	21
	Woodcreepers	Passeriformes (353)	Dendrocolaptidae	12
	Ovenbirds and allies		Furnariidae	43
	Antbirds		Thamnophilidae	22
	Antbirds		Formicariidae	3
	Gnateaters		Conopophagidae	1
	Tapaculos		Rhinocryptidae	3
	Tyrants		Tyrannidae	103
	Manakins		Pipridae	6
	Cotingas		Cotingidae	3
	Sharpbills		Oxyruncidae	1
	Plantcutters		Phytotomidae	1
	Crows and jays		Corvidae	4
	Swallows		Hirundinidae	13
	Wrens		Troglodytidae	6
	Gnatcatchers		Sylviidae	2
	Thrushes		Muscicapidae	7
	Mockingbirds		Mimidae	2
	Pipits		Motacillidae	6
	Vireos		Vireonidae	3
	Wood warblers		Parulidae	7
	Tanagers, euphonias, cardinals		Emberizidae	84

Table 25.10. Globally threatened mammal species of the Atlantic Forest of Paraguay.

Atlantic Forest Endemics			
Vulnerable	*Anguja myikurê* or *colicorto rojizo*	Hensel's short-tailed opossum	*Monodelphis sorex*
	Mbopi or *murciélago acanelado de Azara*	Bat	*Myotis ruber*
Near-threatened	*Myikurê* or *comadrejita cola larga*	Rat	*Thylamys macrura*
	Mbopi or *murciélago frutero grande oscuro*	Fruit bat	*Artibeus fimbriatus*
Insufficient data	*Guasu pororo* or *mbororo*	Dwarf red brocket	*Mazama nana*
Nonendemic Species but Limited to the Region			
Vulnerable	*Akuti sayju* or *agutí*	Agouti	*Dasyprocta azarae*
	Jagua yvyguy, jagua turuñe'e, or *zorro vinagre*	Bush dog	*Speothos venaticus*
Near-threatened	*Tiríka, mbaracaja'i,* or *chivi'i*	Little spotted cat	*Leopardus tigrinus*

The Paraguayan Atlantic Forest supports populations of four mammal species that are endangered globally (Table 25.10): two that are endemic to the region and two that are limited in Paraguay to the Atlantic Forest. Also, there are populations of three near-threatened species (two of which are endemic to the region). And finally, another species endemic to the Atlantic Forest, the dwarf brocket (*Mazama nana*), is in the category of "insufficient data," but in light of its rarity and the disappearance of its habitat, it probably should be regarded as at least near-threatened at the global level.

Species with National Conservation Problems

Evaluation of the conservation status of Paraguayan flora reveals that 15 species are endangered, 125 are vulnerable, 115 are rare, and 24 are of indeterminate status (Bertoni et al. 1994). Of the 15 imperiled species, 6 are endemic to the Atlantic Forest: the *kuri'y*, or candelabra tree (*Araucaria angustifolia*); the *jejy'y*, or *palmito* (Assai palm) (*Euterpe edulis*); the snakeroot known as *kino* (*Rauvolfia sellowi*); the orchid known as *chachi poty* (*Zygopetalum maxillare*); and the bromeliads known as *ka'avo tyre'y* (*Tillandsia arhiza* and *T. esseriana*), both of which are endemic to Paraguay.

A similar evaluation of Paraguayan fauna shows that 50 invertebrate species, 7 reptiles, 86 birds, and 38 mammals are threatened to some degree at the national level (DPNVS/FMB 1998). Of these species, at least 4 invertebrates, 30 birds, and 5 mammals are endemic to the Atlantic Forest. None of the 89 fish species or 7 amphibian species are considered threatened or in danger of extinction at the national level. There are 765 known butterfly species in Paraguay, and 470 of these are so rare that they were found only once or not at all during an intensive year of collection and in 10 previous years of sporadic collection (Barrios and Kochalka 1993). Many butterflies are endemic to the Atlantic Forest, and it may be assumed that many are threatened to some degree at the national level (Barrios and Kochalka 1993).

Loss of Unique Populations

Paraguay is a confluence point for some of the most diverse and also the most threatened ecosystems of the neotropics. Indeed, it may be regarded as a country of ecotones, or transitional areas. The eastern region comprises four ecoregions: the Interior Atlantic Forest in the eastern part, the Cerrado in the north, natural grasslands and humid soils in the south, and the flooded savanna of the Lower Chaco, along the Paraguay River.

Recent studies (e.g., Smith et al. 1997; Schneider et al. 1999) have demonstrated that ecotones are potentially important for differentiation and speciation. It has been argued that because of their capacity to produce biodiversity, special emphasis should be given to the conservation of ecotones (Schneider et al. 1999). One unique site in the ecotone between the Atlantic Forest and the Chaco is the Acahay Hills, located in the department of Paraguari, with its rocky terrain and endemic species of cacti and bromeliads (Pérez de Molas 1998). This area is under intense pressure from agriculture, extraction of firewood and medicinal plants, mining, and expanding urbanization.

Cultural Biodiversity

Humans have also contributed to shape biological diversity by selecting traits in plants and animals and producing domesticated varieties.

Loss of Domesticated Species and Varieties

Little information exists regarding the status of wild crop relatives in Paraguay. However, the CDC, with funding from the U.S. Department of Agriculture, is mapping the distributions of wild species that form part of cultivated crop gene pools. Even without the results of this project, it is possible to identify Paraguay, and its eastern region in particular, as important for wild relatives of crop plants, including custard apple, guava, cassava, papaya, peanut, peppers, guayaba, pineapple, potatoes, rice, and tomatoes (Mereles 2001). Sixteen species—wild relatives of the guayaba, the custard apple, the papaya, and cassava—are threatened with national or global extinction (Bertoni et al. 1994).

Another genus of economic importance is *Prunus* (plums and their relatives). Fruits of this genus provide the basis for a large food and liquor industry, and several species bear edible almonds. Some species also yield a fine oil used in cooking and the manufacture of cosmetics, and the beautiful flowers and foliage are used for ornamental purposes. Paraguay has 11 tree species in the genus *Prunus,* known by the general name *yvaro.* Five of these are endemic, and four are in danger of extinction at the national level (Bertoni et al. 1994).

A near-endemic to Paraguay, *Stevia rebaudiana,* with a distribution centered near the Monday River, is potentially a major source of high-potency sweetener for the growing natural food market (Soejarto et al. 1982, 1983). But it has proved difficult to cultivate, and natural populations are now considered nationally threatened (Bertoni et al. 1994).

The snakeroot known as *kino* (*Rauvolfia sellowi*) is endemic to the Atlantic Forest and ranks in the imperiled category at the national level. In Paraguay its range is limited to the Paraná River basin and the area around the headwaters of the Jejuí River. The *kino* bark is used in traditional medicine to treat malaria; scientific studies are needed to determine its pharmaceutical potential (Bertoni et al. 1994).

Two threatened native forest species of economic and cultural importance are *jejy'y,* or palmito (Assai palm) (*Euterpe edulis*), and *ka'a,* or *yerba mate* (*Ilex paraguayensis*). Both are endemic or near-endemic to the Atlantic Forest, and both are considered to be threatened with extinction at the national level. In addition, *I. paraguayensis* is considered near-threatened at the global level (Hilton-Taylor 2000). The leaves of *ka'a* are the basis for a very important industry, and drinking tea brewed from *ka'a* leaves is a custom deeply entrained in Paraguayan society.

The *palmito* (*Euterpe edulis*) produces a food for national and international markets, but because of overexploitation and destruction of the species' habitat in the Paraná River basin, the palmito industry is declining, and the species is in serious danger of extinction in Paraguay (Chapter 34, this volume).

Another species in danger of disappearing from Paraguay is the orchid known as *chachî poty* (*Zygopetalum maxillare*), an Atlantic Forest endemic recognized as imperiled at the national level (Bertoni et al. 1994). This epiphyte prefers the trunk of the tree fern known as *chachî* (*Cyathea atrovirens*), which is also endemic to the Atlantic Forest and threatened at the national level. The orchid *chachî poty* is of mainly ornamental value, but it is especially useful because of the ease with which it hybridizes.

Availability and Quality of Information

As is typical for areas throughout the neotropics, gaping holes exist in the knowledge of Paraguay's flora and fauna. The only taxonomic group studied in any depth is birds, and new species are registered every year even as their habitat is being steadily impoverished. The discovery curve for invertebrates is still on the rise, but there are no local invertebrate zoologists, and even most visiting foreign scientists concentrate in botany, herpetology, ornithology, and mammalogy.

The Amambay area, in the department of Concepción near the Brazilian border, has been especially neglected, mainly because field studies have been hampered by safety concerns arising from political instability, drug trafficking, and other illegal activities.

Transfer of information from abroad is another problem, as most relevant lists and academic journals reach only a few professionals. The *Revista de la Sociedad Científica del Paraguay* is one of the few Paraguayan journals that has persisted, but publication is irregular.

Conclusions

Although Paraguay was opened up for democracy more than 10 years ago, it continues to be terra incognita for national and foreign scientists. The expansion of mechanized agriculture, the lack of economic and social progress, and conflicts between land owners and peasants are propelling the biodiversity of the Paraguayan Interior Atlantic Forest toward the brink of extinction.

Thanks largely to efforts by the private sector, particularly nonprofit conservation organizations, it has been possible to save some important examples of what was once the great wealth of the Gran Provincia del Paraguay. So far, the Paraguayan State has failed to take any decisive, far-reaching actions to halt deforestation and create areas for biodiversity conservation or models for sustainable resource use.

Creation of the Secretariat of the Environment and the mobilization of sizable resources from groups such as the Global Environment Facility and the Inter-American Development Bank were initiatives that foundered because of political instability in the country and inadequate management. Specifically, the structural problems have included successive changes in the administration of environmental agencies, lack of decentralization, shortage and poor quality of human resources, and lack of funds. Overall, both the state and the public as a whole seem apathetic toward Paraguay's biodiversity crisis. The inaction stems from failure to recognize the situation as a problem that affects the national economy, the health of the people, and the conservation of species at the world level. The global conservation community needs to invest more in learning about a country that is the ecotone for several regions of great biological diversity.

References

Acha Navarro, J. A. 1999. Erosión costera en el embalse de Itaipú (margen derecha). *Biota* 10: 1–28.

Aquino, A. L., Scott, N. J., and Motte, M. 1996. Lista de anfibios y reptiles del Museo Nacional de Historia Natural del Paraguay. In: Romero Martínez, O. (ed.). *Colecciones de flora y fauna del Museo Nacional de Historia Natural del Paraguay.* pp. 331–400. Asunción: MNHNP.

Barrios, B. B. and Kochalka, J. A. 1993. *Mariposas del Paraguay: informe final.* Unpublished report. Asunción, Paraguay: Sección Invertebrados del Museo Nacional de Historia Natural del Paraguay.

Basualdo, I. and Soria, N. 1994. Algunas especies endémicas de Paraguay en el Parque Nacional Cerro Corá, Amambay, Paraguay. *Cuarta Jornada de Biología* (Asunción).

Bertoni, A. de W. 1927. Nueva forma de psitácidos del Paraguay. *Revista de la Sociedad Científica del Paraguay* 2(3): 149–150.

Bertoni, S., Duré, R., Florentín, T., Pin, A., Pinazzo, J., Quintana, M., Ríos, T., and Rivarola, N. 1994. *Flora amenazada del Paraguay.* Asunción: Dirección de Parques Nacionales y Vida Silvestre.

Bibby, C. J., Collar, N. J., Crosby, M. J., Heath, M. F., Imboden, Ch., Johnson, T. H., Long, A. J., Stattersfield, A. J., and Thirgood, S. J. 1992. *Putting biodiversity on the map: priority areas for global conservation.* Cambridge, U.K.: ICBP.

BirdLife International. 2000. *Threatened birds of the world.* Barcelona and Cambridge, UK: Lynx Editions and BirdLife International.

Bozzano, B. E. and Weik, J. H. 1992. *El avance de la deforestación y el impacto economico.* Proyecto de Planificación del Manejo de los Recursos Naturales, Serie no. 12. Asunción: MAG/GT-GTZ.

Brooks, T. M., and Blamford, A. 1996. Atlantic Forest extinctions. *Nature* 380: 115.

Brown, K. S. Jr. 1982. Paleoecology and regional patterns of evolution in Neotropcial butterflies. In: Prance, G. T. (ed.). *Biological diversification in the tropics.* pp. 255–308. New York: Columbia University Press.

Bryant, D., Nielsen, D., and Tangley, L. 1997. *Last frontier forests: ecosystems and economies on the edge.* Washington, DC: World Resources Institute.

Cabrera, A. L., and Willink, A. 1973. *Biogeografía de América Latina.* Monografía No. 13. Washington, DC: Organización de Estados Americanos.

Cabrera, A. L., and Willink, A. 1980. Biogeografía de América Latina. Washington, DC: Secretaría General de la Organización de los Estados Americanos, Programa Regional de Desarrollo Científico y Tecnológico. Serie de Biología, Monografía N. 13.

Carlini, A., Coronel, L. M., Farías, F., Hillner, U., Hoffmann, R., König, W., Kruck, W.,

Mereles, F., Pasig, R., and Rojas, C. 2001. *Inventario y evaluación de los recursos naturales de la región oriental del Paraguay.* San Lorenzo, Paraguay: SEAM/BGR.

CDC (Centro de Datos para la Conservación). 1990. *Áreas prioritarias para la conservación en la región oriental del Paraguay.*

Clay, R. P. 2001. The status and conservation of cracids in Paraguay. In: Brooks, D. M. and Gonzalez, G. F. (eds.). *Ecology and conservation of cracids in the new millennium.* pp. 124–138. Misc. Pub. no. 2. Houston, TX: Houston Museum of Natural Science.

Clay, R. P., Madroño, N. A., and Lowen, J. C. 1999. A review of the status and ecology of the black-fronted piping-guan *Pipile jacutinga* in Paraguay. In: Brooks, D. M., Begazo, A. J., and Olmos, F. (eds.). *Biology and conservation of the piping-guans.* pp. 14–25. Special Pub. CSG no. 1. Houston: CSG.

Collar, N. J., Gonzaga, L. P., Krabbe, N., Madroño-Nieto, A., Narango, L. G., Parker, T. A. III, and Wege, D. C. 1992. *Threatened birds of the Americas: the ICBP/IUCN red data book.* Cambridge, UK: International Council for Bird Preservation.

Cracraft, J. 1985. Historical biogeography and patterns of differentiation within the South American avifauna: areas of endemism. *Ornithological Monograph* B: 49–84.

Davis, S. D., Heywood, V. H., Herrera-MacBryde, O., Villa-Lobos, J., and Hamilton, A. C. 1997. *Centres of plant diversity: a guide and strategy for their conservation.* Vol. 3: *The Americas.* Cambridge, UK: WWF/IUCN.

de Azara, F. 1998. *Viajes por la América Meridional.* Vols. 1 and 2. Buenos Aires: Editorial El Elefante Blanco.

Dinerstein, E., Olson, D. M., Graham, D. J., Webster, A. L., Primm, S. A., Bookbinder, M. P., and Ledec, G. 1995. *A conservation assessment of the terrestrial ecoregions of Latin America and the Caribbean.* Washington, DC: International Bank for Reconstruction and Development/World Bank.

DPNVS/FMB (Dirección de Parques Nacionales y Vida Silvestre/Fundación Moisés Bertoni). 1998. *Fauna amenazada del Paraguay.* Asunción: Dirección de Parques Nacionales y Vida Silvestre.

ENAPRENA (Estrategia Nacional para la Protección de los Recursos Naturales). 1995a. *Documento base sobre biodiversidad.* Asunción: SSERNMA/MAG-GTZ.

ENAPRENA (Estrategia Nacional para la Protección de los Recursos Naturales). 1995b. *Documento base sobre el sector agrícola y su impacto ambiental.* Asunción: SSERNMA/MAG-GTZ.

Fariña Sánchez, T. 1973. The climate of Paraguay. In: Gorham, J. R. (ed.). *Paraguay: ecological essays.* pp. 33–38. Miami: Academy of the Arts and Sciences of the Americas.

Ferreiro, O. 1981. *Aproximación hacia una clasificación de las formaciones forestales del Paraguay.* Turrialba, Costa Rica: CATIE.

Frost, D. R. 2000. *Amphibian species of the world: an online reference.* Online: http://research.amnh.org/herpetology/amphibia (ver. 2.20, 1 September 2000).

Garcete-Barrett, B. R. 1999. *Guía ilustrada de las avispas sociales del Paraguay (Hymenoptera: Vespidae: Polistinae).* London: The Natural History Museum.

Gorham, J. R. 1973. *Paraguay: ecological essays.* Miami, FL: Academy of the Arts and Sciences of the Americas.

Hermosa, A. J. L. 1999. Calidad de agua del embalse de Itaipú. *Biota* 9: 7–36.

Haffer, J. 1985. Avian zoogeography of the Neotropical lowlands. In: Buckley, P. A., Foster, M. S., Morton, E. S., Ridgely, R. S., and Buckely, F. C. (eds.). *Neotropical ornithology.* pp. 113–146. Washington, DC: American Ornithologist's Union.

Hill, K. and Padwe, J. 2000. Sustainability of Aché hunting in the Mbaracayú Reserve,

Paraguay. In: Robinson, J. G. and Bennett, E. L. (eds.). *Hunting for sustainability in tropical forests.* pp. 79–105. New York: Columbia University Press.

Hill, K., Padwe, J., Bejyvagi, C., Bepurangi, A., Jakugi, F., Tykuarangi, R., and Tykuarangi, T. 1997. Impact of hunting on large vertebrates in the Mbaracayu Reserve, Paraguay. *Conservation Biology* 11: 1339–1353.

Huespe Fatecha, H., Spinzi Mendonca, L., Curiel, M. V., Burgos, S., and Rodas Insfrán, O. 1994. *Uso de la tierra y deforestación en la región oriental del Paraguay, Periodo 1984–1991.* Vol. 1. San Lorenzo, Paraguay: UNA/FIA/CIF.

IIED/USAID. 1985. *Environmental profile of Paraguay.* Washington, DC: USAID (United States Agency for International Development).

Iriarte Walton, A. 1994. *Estado de conservación de la fauna silvestre del cono sur Sudamericano.* RLAC Documento Técnico no. 13. Santiago, Chile: Oficina Regional de la FAO para América Latina y el Caribe.

IUCN. 2000. The 2000 IUCN red list of threatened species. http://www.redlist.org.

Keel, S., Gentry, A. H., and Spinzi, L. 1993. Using vegetation analysis to facilitate the selection of conservation sites in eastern Paraguay. *Conservation Biology* 7: 66–75.

Levi, H. W. 1988. The Neotropical orb-weving spiders of the genus Alpaida (Araneae: Araneidae). *Bulletin of the Museum of Comparative Zoology* 151: 365–487.

López, J. A., Little, E. L. Jr., Ritz, F. G., Rombold, J. S., and Hahn, W. J. 1987. *Arboles comunes del Paraguay: Ñyvyra Mata Kuera.* Washington, DC: Peace Corps.

Mandelburger, D., Medina, M., and Romero Martínez, O. 1996. Los peces del Inventario Biológico Nacional. In: Romero Martínez, O. (ed.). *Colecciones de flora y fauna del Museo Nacional de Historia Natural del Paraguay.* pp. 285–330. Asunción: MNHNP.

Mereles, H. M. F. 2001. Recursos fitogenéticos: plantas útiles de las cuencas del Tebicuary mí y Capiíbary, Paraguay oriental. *Rojasiana* 6: 4–142.

Mittermeier, R. A., Coimra-Filho, A. F., Constable, I. D., Rylands, A. B., and Valle, C. 1982. Conservation of primates in the Atlantic Forest of Eastern Brazil. *International Zoo Yearbook* 22: 17–37.

Mori, S. A., Boom, B. M., and Prance, G. T. 1981. Distribution patterns and conservation of eastern Brazilian coastal forest tree species. *Brittonia* 33: 233–245.

Müller, P. 1973. *The dispersal centres of terrestrial vertebrates in the Neotropical realm.* The Hague: W. Junk.

Myers, N. 1988. Threatened biotas: "hotspots" in tropical forests. *Environmentalist* 8: 1–20.

Myers, N., Mittermeier, R. A., Mittermeier, C. G., da Fonseca, G. A. B., and Kent, J. 2000. Biodiversity hotspots for conservation priorities. *Nature* 403: 853–858.

Oliver, W. L. R., and Santos, I. B. 1991. *Threatened endemic mammals of the Atlantic forest region of southeast Brazil.* Special Scientific Report 4. Les Augres Manor, Trinity, Jersey (Channel Islands): Jersey Wildlife Preservation Trust.

Olson, D. M. and Dinerstein, E. 1998. *The global 200: a representation approach to conserving the earth's distinctive ecoregions.* Unpublished draft. Washington, DC: World Wildlife Fund.

Olson, D. M., Dinerstein, E., Canevari, P., Davidson, I., Castro, G., Moriste, V., Abell, R., and Toledo, E. (eds.). 1998. *Freshwater biodiversity of Latin America and the Caribbean: a conservation assessment.* Washington, DC: World Wildlife Fund Biodiversity Support Program.

Pérez de Molas, L. F. 1998. *Proyecto de plan de manejo del Monumento Natural Macizo Acahay, Departamento de Paraguari, Paraguay.* Asunción: Convenio Paraguay EDAN.

Revenga, C., Brunner, J., Henninger, N., Kassem, K., and Payne, R. 2000. *Pilot analysis of global ecosystems: freshwater systems.* Washington, DC: World Resources Institute.

Sanjurjo, M. and Gauto, R. 1996. Paraguay. In: Harcourt, C. S. and Sayer, A. J. (eds.). *The conservation atlas of tropical forests: the Americas.* pp 286–293. Singapore: Simon & Schuster.

Schneider, C. J., Smith, T. B., Larison, B., and Moritz, C. 1999. A test of alternative models of diversification in tropical rainforests: ecological gradients vs. rainforest refugia. *PNAS* 96: 13869–13873.

Smith, T. B., Wayne, R. K., Girman, D. J, and Bruford, M. W. 1997. A role for ecotones in generating rainforest biodiversity. *Science* 276: 1855–1857.

Soejarto, D. D., Compadre, C. M., Medon, P. J., Kamath, S. K., and Kinghorn, A. D. 1983. Potential sweetening agents of plant origin. II. Field search for sweet-tasting *Stevia* species. *Economic Botany* 37: 71–79.

Soejarto, D. D., Kinghorn, A. D., and Farnsworth, N. R. 1982. Potential sweetening agents of plant origin. 3. Organoleptic evaluation of *Stevia* leaf herbarium samples for sweetness. *Journal of Natural Products* 45: 590–599.

Spichiger, R., Palese, R., Chautems, A., and Ramella, L. 1995. Origin, affinities and diversity hot spots of the Paraguayan dendrofloras. *Candollea* 50(2): 517–537.

Stattersfield, A. J., Crosby, M. J., Long, A. J., and Wege, D. C. 1998. *Endemic bird areas of the world: priorities for biodiversity conservation.* BirdLife Conservation Series No. 7. Cambridge, U.K.: BirdLife International.

Tortorelli, L. 1967. Formaciones forestales y maderas del Paraguay. *Boletín del Instituto Forestal Latinoamericano* (Mérida) 18: 3–34.

Veloso, H. P., and Góes-Filho, L. 1982. Fitogeografia Brasileira-Classificação Fisionômico-Ecológica da Vegetação Neo-tropical. *Projeto RADAM Brasil, Brasília, Boletim Técnico, Série Vegetação* 1: 3–79.

Williams, R. J., Bryan, R. T., Mills, J. N., Palma, R. E., Vera, I., and de Velasquez, F. 1997. An outbreak of hantavirus pulmonary syndrome in western Paraguay. *The American Journal of Tropical Medicine and Hygiene* 57:274–282.

World Resources Institute 1998. *Water resources and freshwater ecosystems: Paraná watershed.* Online: http://www.igc.org/wri/watersheds/ww-same.html.

Chapter 26

Socioeconomic Drivers in the Interior Atlantic Forest

Ana María Macedo and José Luis Cartes

New developments in a variety of spheres are catapulting the Republic of Paraguay toward major changes as the twenty-first century begins. On the political front, the country has embraced democracy and adopted a new national constitution. In the economic sphere, two major hydroelectric dams (Itaipú and Yacyretá) have been constructed. Changes in the social realm include organization of rural movements and frequent illegal occupation of private properties. Finally, in the cultural domain, a new system of primary and secondary education (BCP 2000b) is being put in place. These developments and the often related demographic changes over the last 40 years, such as population growth and migration, are intertwined with the way natural resources are used, and the result has been a significant negative impact on the environment (FMB 1994; Chapter 25, this volume). All these factors have dampened the potential for economic development, increasing poverty and creating substandard living conditions for the people of Paraguay.

The development model that prevailed in the 1960s and 1970s in Paraguay incorporated the products of the green revolution in agriculture. With the introduction of high-yield crop varieties, mechanized farming equipment, and chemical pesticides and fertilizers, large agribusinesses were established, and medium-sized farming operations (50–200 ha) proliferated. Most farms were owned by people of Brazilian and Germano-Brazilian origin, and most of their mechanized production was intended for sale in foreign markets, especially Brazil and Argentina (STP 2000). Production generally was concentrated on a single crop, principally soybeans or cotton. The use of tractors and combine harvesters made it possible to take advantage of economies of scale, but only by clearing large areas of land. The result was rapid deterioration of soil and a loss of biodiversity; many agribusinesses switched to planting aggressive grasses of African origin (e.g., *Brachiaria brizanta*) for grazing cattle, especially as crop yields declined (STP 2000; Chapter 33, this volume).

This revolution in agricultural practices was accompanied by large-scale land purchases, leading to an exodus of the rural population as impoverished small farmers moved to other small rural settlements or to urban centers.

Socioeconomic Roots of Biodiversity Loss

Development efforts in the country have been oriented toward the pursuit of so-called modernization (Masulli et al. 1996). With its industrial base still nascent and its economy geared largely toward subsistence, Paraguay's entry into the market economy has depended mainly on its forest resources and the fertility of its soils. However, no development planning system has taken this into account. The government has emphasized the development of export products, and public efforts and financial resources have been channeled into profitable monoculture crops such as soybeans and cotton, leading to the expansion of the agricultural frontier (ENAPRENA et al. 1995). Even today, both the public and private sectors are not fully aware of the social and environmental benefits of the forest in terms of erosion control, biodiversity protection, and preservation of ecosystem integrity and even landscape value.

Evolution of the Forestry Sector and Deforestation

Commercial forestry companies emerged after the government of General Bernardino Caballero ordered the sale of public lands in the late 1800s. They focused on selective extraction of a few species, with little thought to the need for effective management of forest resources. The consequences were a reduction in the supply of logs, higher prices, and waste of small trees that could have commanded a high price on the market had they been allowed to reach maturity.

The forestry industry began to grow in 1967 with the arrival of new timber companies and expansion of existing sawmills. Installed capacity mushroomed, and by 1979 the industry included more than 600 sawmills and employed more than 10,000 people (ENAPRENA et al. 1995). As growth outpaced consumption, however, this boom was followed by more than a decade of stagnation and then reversal as manufacturing plants and sawmills closed. The forestry sector had failed to plan its development.

In the 1950s and 1960s, commercial cultivation of exotic species, such as eucalyptus (*Eucalyptus camandulensis*) and pine (*Pinus taeda* and *P. elliotti*), began. Some attempts were made to grow native species such as *Tabebuia* spp., *Cordia trichotoma, Peltophorum dubium,* and *Parapiptadenia rigida* on plantations, but they were insignificant.

Logging and wood processing have been concentrated mainly in the forests of the eastern region, where the most valuable timber species are found (ENAPRENA et al. 1995). As usual, road construction and other infrastructure works soon brought with them new settlements, accelerating deforestation. Completion of the road between Asunción and Ciudad del Este in 1965 contributed to gradual destruction of the forest in the surrounding areas (Bozzano and Weik

Table 26.1. Forest cover and land use (area in km²) in Paraguay in 1991 (ENAPRENA/GTZ/SSERNMA 1995; FAO 2000).

Department	Conserved Tall Forest	Degraded Tall Forest	Clumped Woodland	Flooded Grassland	Seasonally Flooded Grassland	High Grassland	Agriculture and Livestock	Deforestation	Water	Total
Canendiyú	6,351.60	348.04	0.00	0.00	245.37	0.00	4,706.61	3,015.38	0.00	14,667.00
San Pedro	3,550.49	1,052.68	0.00	751.41	3,565.39	3,494.09	4,257.14	3,330.80	0.00	20,002.00
Amambay	3,132.26	926.41	365.23	0.00	330.83	2,820.66	3,903.61	1,454.00	0.00	12,933.00
Alto Paraná	2,951.07	951.19	0.00	0.00	512.87	0.00	5,676.98	4,637.17	165.72	14,895.00
Itapuá	1,852.70	805.06	0.00	206.77	1,649.64	1,500.34	7,422.37	3,088.12	0.00	16,525.00
Caazapá	1,539.42	575.34	0.00	176.01	1,940.02	1,280.57	2,693.88	1,290.76	0.00	9,496.00
Caaguazú	1,183.90	1,160.71	0.00	13.00	876.93	930.44	5,179.33	1,804.77	324.92	11,474.00
Concepción	1,161.33	421.49	3,662.25	60.62	629.58	8,011.27	3,097.73	1,006.73	0.00	18,051.00
Guairá	247.66	291.02	0.00	0.00	265.73	189.77	2,627.21	224.61	0.00	3,846.00
Total	21,970.43	6,531.94	4,027.48	1,207.81	10,016.36	18,227.14	39,564.86	19,852.34	490.64	121,889.00

Figure 26.1. Remnant forest in different departments of Paraguay.

1992). From 1945 to 1985, approximately 49,000 km^2 of forestland, an average of about 1,230 km^2/year, was eliminated in the eastern region. Since the late 1960s, the rate of deforestation has increased steadily, surpassing 4,000 km^2 per year in 1989 (ENAPRENA et al. 1995). Almost 90 percent of the native forest has been cut down to make way for cattle ranching, migratory farming, and urbanization in the most densely populated region of the country (MAG et al. 1999; Table 26.1 and Figure 26.1).

Demographic Trends, Population Distribution, and Migrations

Although Paraguay's population is small compared with that of other countries, in the last 40 years (1959–2000) it has quadrupled to roughly 5.5 million (Figure 26.2). With a surface area of 406,752 km^2, the population density now stands at 12.04 people per square kilometer. The growth rate is expected to slow in the first half of the twenty-first century, but the total population will continue to increase, reaching about 13 million in 2050 (Figure 26.2).

One of Paraguay's most noteworthy demographic phenomena of the past 30 years is rapid, unplanned urbanization (Figure 26.3). In that period, the country's urban population increased from 32 percent to 50.5 percent (the total urban population was approximately 2 million in 1992; STP 2000), fueled in part by the building of the Itaipú Dam. By 2018 city dwellers are expected to constitute 62 percent of the total population (SPT 2000).

Settlement since the colonial period has been most intensive in the eastern region, where the Interior Atlantic Forest is located. This region accounts for 40 percent of the country's total surface area but houses more than 95 percent of its total population. The principal cities and population centers are located here, as are the main communication arteries and systems of basic services (Figure 26.1). The departments of Asunción and the Central and Alto Paraná are the most urbanized areas in the country, together accounting for 47 percent of the total population (more than 2.6 million inhabitants). The urban populations in Itapúa and Caaguazú follow with a combined population of more than 950,000. These four departments and the national capital occupy 11 percent of the country's surface area but account for 64.3 percent of its total population.

International migration has had a great impact on settlement, but no accurate information exists on the influx and outflow of people. Paraguay generally has

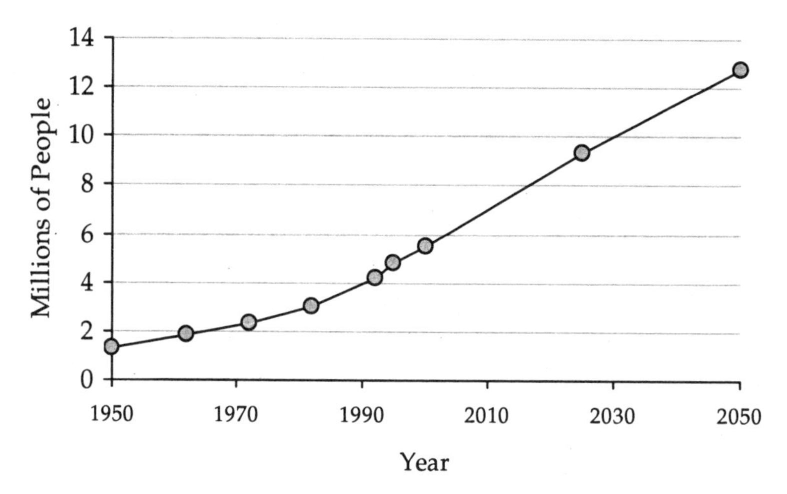

Figure 26.2. Population growth in Paraguay from 1950 to 2000 and projections to 2050 (STP 2000).

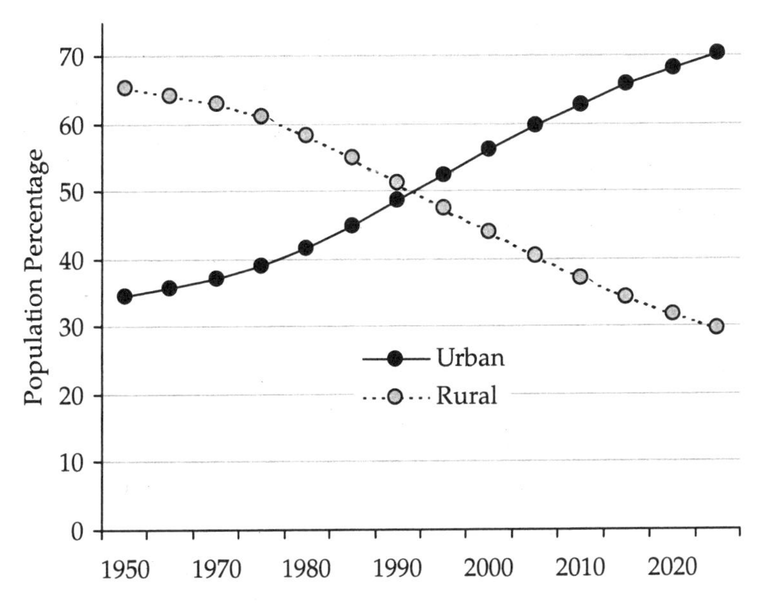

Figure 26.3. Trends in rural and urban populations in Paraguay from 1950 to 2025 (STP 2000; CEPAL-CELADE 1999).

been a net "exporter" of people to neighboring countries that enjoy higher standards of living. Currently, however, emigration is being offset by the contingents of immigrants coming into the country, especially in the border areas. According to the 1992 census, most immigration is from Brazil and Argentina, with small percentages of immigrants from Asia, especially Korea, Japan, and China. The majority of foreign immigrants settle in Alto Paraná, Asunción, Canindeyú, Central, and Itapuá.

Without a doubt, the largest volume of immigration recently has been from Brazil to the Paraná River region, a movement spurred by construction of the Asunción–Ciudad del Este highway, the international bridge between Paraguay and Brazil, and the Itaipú hydroelectric plant and by the settlement process known as the "March to the East" and the expansion of the agricultural frontier.

An increase in internal migration (rural and urban) in the last decade has also affected the urbanization rate (STP 2000). From 1987 to 1992, some 314,000 people changed residence, whereas in a single year in the late 1990s (1997–1998), internal migrants numbered slightly more than 546,000 (DGEEC 1996; DGEEC/FNUAP 1999). Migration and population distribution in Paraguay are closely interlinked with socioeconomic status. Asunción and the surrounding metropolitan area, Ciudad del Este, Pedro Juan Caballero, and Encarnación are the biggest magnets for the rural poor, who flock to cities in search of better economic conditions (STP 2000).

In the departments of the Interior Atlantic Forest, agricultural activities, especially growing soybeans and cotton and cattle ranching, are predominant. In the departments of Caazapá and San Pedro, for example, approximately 80 percent of the population makes a living by farming, ranching, hunting, and fishing (Figure

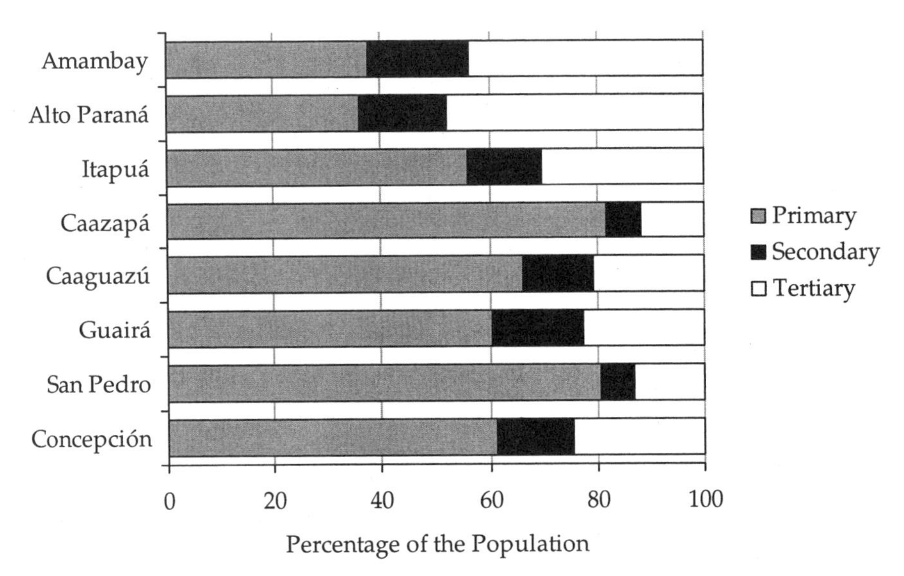

Figure 26.4. Percentage of the population engaged in primary (e.g., agricultural), secondary (e.g., manufacturing), and tertiary (e.g., service) sectors in different departments of eastern Paraguay.

26.4). The secondary sector (mining, manufacturing, and construction) employs an average of 10–15 percent of the population, with higher rates in the departments of Amambay, Alto Paraná, and Guairá (GTZ 1994).

Sociocultural Factors

Among the main ultimate causes of biodiversity loss are socioeconomic factors such as rural and urban poverty, health care and gender inequities, and an inadequate economic model.

Poverty

The model of economic development applied in the past has led to increased poverty and growing social and economic inequity (PNUD 2001). Poverty has increased almost 4 percent in Paraguay since 1995 and now affects one-third of the total population, or 1.9 million people. Almost half of these people live in extreme poverty and are unable to meet their basic needs.

Whereas economic growth has benefited large rural businesses, 30 percent of the population remains landless. The rural poor, 42 percent of the rural population, are beset with inadequate education, poor nutrition, rudimentary production techniques, and a reliance on too few crops.

Urban poverty is associated with the influx of large numbers of immigrants moving to the cities from belts of poverty in the areas around Asunción and the cities of Central department. More than a quarter of the urban population lives in poverty as well.

Inequities in Public Health

Major advances in public health in Paraguay have reduced mortality by 56 percent over a 30-year period (GTZ 1994). Despite this progress, marked inequities in coverage and access to public health services for poor and marginalized groups remain.

The leading causes of death are heart disease, cerebrovascular disease, and cancer, followed by accidents and respiratory infections. For children under 1 year of age, the leading causes of death are infections, followed by birth injuries, congenital anomalies, premature birth, and diarrheal diseases. Higher infant mortality is associated with higher rates of poverty and lower levels of education. One-third of the population lacks access to health services, and more than 80 percent cannot afford health insurance.

There is little education with regard to reproductive health and, consequently, little family planning and high maternal mortality from abortion, hemorrhage, sepsis, and other pregnancy-related complications. In fact, education, both formal and informal, is one of the greatest challenges facing Paraguayan society (Chapter 28, this volume; PNUD 2001).

Gender Issues

Especially in rural areas and indigenous populations, women play an active role in managing natural resources, such as water, coal or firewood, and medicinal plants, which means that the environment and development crisis has a direct impact on them (Masulli et al. 1996). In recent years, Paraguayan women have gained greater access to educational and employment opportunities, which has helped create the conditions necessary for broadening women's role in society. Nevertheless, Paraguay's low ranking on a United Nations index that measures gender equality indicates that the country is far from achieving equality for women in political and economic spheres (Prieto 1999).

Economic Considerations

The main development problems for Paraguay are related to the failings of the economic model. The foundation of the Paraguayan economy has traditionally been its natural resources: good farm and ranch land and forests that offered great timber wealth. But excessive demand for forest resources destined for export and, to a lesser extent, for local consumption, combined with an economic strategy that promoted agricultural and agroindustrial exports in the 1970s, have led to systematic deforestation of large expanses of forestland (Chapter 25, this volume). Moreover, much of the forestry activity was carried out illegally, so it yielded very low revenues for the state. For example, in 1997 production of services accounted for a large proportion of the gross domestic product (GDP), followed by mining, construction, manufacturing, and agriculture (Figure 26.5). The GDP share of the forestry sector was minimal (BCP 2000a).

The primary products of the forest sector include logs, poles, railroad ties,

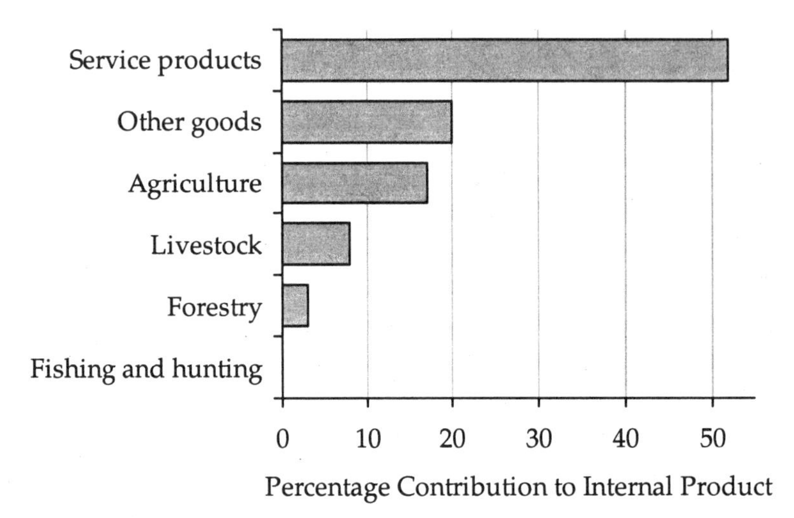

Figure 26.5. Participation in economic sectors in Paraguay, 1997 (BCP 2000a).

firewood, and palms. The volume of wood used for fuel is much higher than that of wood in log or pole form (Figure 26.6), which indicates that the resource capacity of the forest system is being underused (Bozzano and Weik 1992). The total GDP share of the forestry sector remained fairly stable at around 2.8 percent through the period 1993–1999.

The country imports a large volume of paper and paperboard because it lacks the industrial capacity to produce the raw materials needed to manufacture them domestically. It exports large amounts of sawnwood and wood boards to other markets. In 1997, the principal destinations for forest exports such as sawnwood were Brazil (224,068 tons, or 69.8 percent) and Argentina (68,821 tons, or 21.55 percent). Manufactured wood products were exported primarily to Argentina

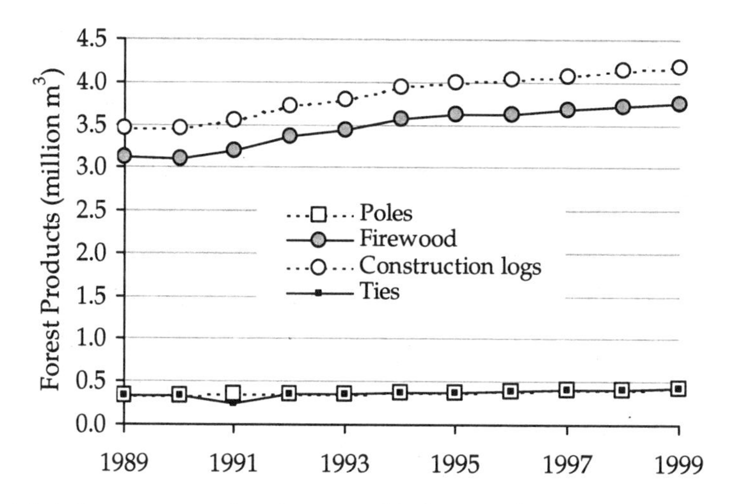

Figure 26.6. Primary products from forest wood in Paraguay from 1989 to 1999.

(26.4 percent), followed by the United States and Brazil (STP 2000). The most important nontimber forest products include tanning agents, seeds, honey, and a variety of other plant products for domestic consumption and export.

The Growth Model and Its Economic and Socioenvironmental Implications

Paraguay is an eminently agricultural country. Almost half its population resides in rural areas, and 37 percent of the employed population is engaged in farming and ranching activities. More than one-quarter of the country's output originates in the agricultural sector, and a significant portion of its foreign currency income from exports is derived from two agricultural crops: soybeans and cotton (Fogel 1998). Because the Paraguayan identity is closely linked to rural culture, the future of agriculture is a key element in improving quality of life for more than 2 million people who earn their livelihood from agriculture (Fogel 1998).

Until the mid-1960s the nation relied on subsistence agriculture, an economy that had stagnated, had very low productivity, and produced almost no surplus for the internal or the external market. International financial institutions, especially the World Bank and the Inter-American Development Bank, were financing large integrated rural development projects. These projects sought to improve basic infrastructure to increase production in the rural sectors and facilitate access to basic services, in particular health care, education, and technical assistance for the rural population (Fogel 1998).

A hallmark of the prevailing economic policy was strong intervention by the state, which controlled exchange rates, interest rates, and wages, either directly or indirectly. However, monetary, fiscal, exchange, and credit policies were implemented without applying suitable technical criteria, and the management of public income and spending was not guided by clear economic objectives. Only in the 1990s, after the fall of the dictatorship, were some very moderate economic reforms introduced (Fogel 1998). Still, it is important to point out that Paraguay has never suffered serious economic instability. Inflation consistently has been far lower than in neighboring countries, and the guarani-dollar exchange rate has remained constant for several years.

Politically, on the other hand, Paraguayan history has been characterized by instability as a result of frequent coups d'état and several wars and armed revolutions. The political system has functioned on the basis of favors offered to possible partners, voters, or social sectors, coupled with *caudillismo* and political clientelism, which intensified under the government of President Alfredo Stroessner (1954–1989). The economic reforms of the early 1990s were accompanied by several political reforms, among them the enactment of the new national constitution in 1992. A new supreme court was also created to strengthen the judicial branch.

In the political and administrative domain, new legislation provided for electing departmental officials, replacing the presidentially appointed delegates. This legislative change helped to strengthen local governments because previously the department and municipal government authorities had been chosen by the president.

Recent Economic and Social Trends

The effect of agriculture on Paraguay's basic structural features has profoundly influenced the country's society and economy and distinguished it from other countries in the region (Fogel 1998). The replacement of traditional small production of a variety of crops for the internal market by large agribusinesses concentrating on only a few for export has had profound effects on the social and economic trends of the country. One of the most visible of these is an ever more pronounced social and economic stratification, which has consequences for both society and the environment.

Economically, Paraguay experienced a brief period of growth associated with large influxes of foreign currency while the Itaipú Dam was being built. Following that period, the country barely succeeded in maintaining its GDP, which in recent years has registered negative growth. Sustained economic expansion remains an elusive goal. Paraguayan industry has experienced considerable growth, especially in production of textiles, fruit juices, and oils, but even so its development has been limited because of a lack of incentives. Attempts to attract *maquila* industries (foreign-owned export assembly plants) have not yielded fruit to date. Moreover, the dollar has fluctuated tremendously since 1985 (DGEEC/FNUAP 1999).

Because agribusinesses often operate without regard for the environmental damage caused by their activities (Fogel 1998), their effect has been to weaken or eliminate local rural communities. Indiscriminate deforestation is the result as agribusinesses overuse resources and small farmers and indigenous groups transform the forest to sell forest products as a source of fast cash (Chapter 2, this volume). The greatest impact is on households that rely on natural resources for their subsistence. This deterioration of natural resources is closely linked to poverty, migration, and other social issues discussed above.

These social and economic trends—expansion of the agricultural frontier, rapid population growth, internal and international migration, and illegal occupation of private lands—are the leading threats to the conservation of the Interior Atlantic Forest. The challenge for conservation is made more difficult by the lack of effective land management and enforcement on the part of the authorities.

Future Infrastructure Projects

Three infrastructure projects may exacerbate the problems of the Interior Atlantic Forest of Paraguay. The extension of Route 10 to facilitate access to forest remnants crosses a large forested area in the departments of Caaguazú and Canindeyú. Mitigation and compensation measures—especially the purchase of lands for conservation purposes—will be essential.

Construction of the Corpus Dam, the third large dam on the middle Paraná River, between Itaipú and Yacyretá, would lead to the loss of the small forest remnants in the area from flooding, installation of high-voltage lines, and other dam-related activities (Chapter 35, this volume).

The proposed *hidrovía* (waterway) project would create and maintain a shipping channel on the Paraguay and Paraná rivers for transporting export goods, especially soybeans, rice, and meat. The waterway would stretch for 3,400 km, from Cáceres, Brazil, to Nueva Palmira, Uruguay, and would be deep and wide enough to allow continuous navigation throughout most of the year by vessels with a draft of 3.3 m. It would affect an area of about 720,000 km^2 and a population of 40 million people. The project's negative effects will include alteration of the river system, loss of wetlands, loss of the regulatory effect of the Pantanal, increase in flow velocity, changes in water quality, and loss of fish diversity and productivity and of biodiversity in general. It would affect both rural populations and those living in cities, especially the marginal urban populations close to the river.

International Agreements

Several international agreements exist between Paraguay and Brazil, such as for the regulation of fishing and logging traffic. However, in the absence of practical instruments, regulations, and the participation of local and regional agencies, the possibilities for their full implementation are limited. Cross-border environmental management is still incipient or nonexistent, except for isolated activities to address specific issues and emergencies. Feasible, comprehensive programs and projects on natural resources must be developed involving municipal and department governments and cross-border entities and providing the necessary legal, institutional, operational, and financial mechanisms and instruments. Economic effects and environmental repercussions of Mercosur are analyzed elsewhere in this book (Chapters 28 and 37, this volume).

New Conservation Initiatives

Recently, the private sector has become increasingly involved in natural resource conservation initiatives. The Law on Protected Natural Areas (Law 352/94) establishes a subsystem of private protected areas under which private land owners may set aside part of their property for conservation and sustainable use. In exchange, they receive certain tax benefits and are protected from expropriation of their lands (Chapter 28, this volume).

The regulations for applying this law have never been established, but several private protected areas have been designated by resolution of the Secretariat of the Environment. Two reserves were created by executive decree on October 11, 2001: Morombi Reserve (250 km^2), in the Department of Caaguazú, and Arroyo Blanco (57.15 km^2), in the Department of Amambay. This marked a milestone in conservation on private lands in Paraguay, and it is hoped that more such areas will be created in the near future.

Another private conservation instrument that may aid in the preservation of forest remnants is the conservation easement (Chapter 28, this volume). Numerous other private conservation instruments commonly used in other areas, such as usufruct, land grants, and trusts, are also available. In the case of both the private natural reserves and the conservation easements, economic incentives must be created to encourage conservation.

Indicators for Monitoring Socioeconomic Factors

In addition to the socioeconomic factors and infrastructure projects discussed in this chapter, mining, petroleum extraction, and especially forestry activities pose additional risks to Paraguay's natural resources. An additional impact to consider is that of Mercosur, in light of the proposal for a common external tariff and the difficulties in harmonizing macroeconomic policies (Chapter 37, this volume). These problems encompass many variables, and indicators must be identified that can be evaluated periodically to determine the need for and effect of new activities. The best indicators for that purpose are presented in Table 26.2.

Table 26.2. Indicators to monitor socioeconomic drivers in the Interior Atlantic Forest of Paraguay.

Indicator	Rationale
Types and amounts of agricultural products marketed through rural cooperatives	This indicator would provide information on the economic viability of rural activities. Alternatives to cotton and soybean agriculture are needed, and some, such as sesame, are already being promoted.
Level of technology used in production	The quality of the technological tools and procedures used on small farms would indicate the need for greater or lesser use of arable lands.
Organizational capacity	Rural cooperatives and associations generally do not function effectively. The constancy and economic growth of these institutions would indicate improvements in the profitability of farming and quality of life.
Trend of agricultural prices	Price trends are closely associated with incentives for the development of new product categories or diversification of production.
Infrastructure developed	Settlements are in need of access by gravel or paved roads and development of schools, local government institutions, courts, and banks.
Diversification of products for personal consumption	Currently, items for personal consumption are limited to a handful of farm products and domestic animals. Greater diversification would strengthen rural development.
Marketing of legal forest and nonforest products	There are still no clearly identified categories or marketing chains for nonforest products. Rising prices of forest products could accelerate deforestation.
Expansion of the agricultural frontier	The amount of farmland currently existing appears to be sufficient. Law 816/01 prohibits the conversion of land in the departments of Concepción and Amambay, creating a protective band along the land border with Brazil.
Growth of poverty belts around cities	Increases in this demographic index would indicate underemployment and a need for urban planning.
School enrollment and dropout rates	This is an indicator of quality of life. As family income rises, children are able to complete higher levels of education. These rates also indicate the need for labor because during harvest seasons children who are needed to work in the fields do not attend school.

Conclusions

Analysis of the country's socioeconomic indicators reveals that the majority of the population suffers from limited access to education, health care, and clean drinking water as well as from lack of employment opportunities, and the effects of environmental deterioration. Economic development in Paraguay has overlooked the rights of vulnerable groups and progress has been slow toward achieving a model of sustainable development that harmonizes protection of natural resources and control of environmental pollution with economic and demographic growth.

Conservation of the Interior Atlantic Forest faces huge obstacles. Every day forest areas are shrinking. A small minority continues to hold large expanses of land, while the majority of the Paraguayan people struggle to carve out a minimal living from a small plot of land.

Indigenous populations of Paraguay are important to protect as well, for, although small in absolute numbers, they have been an important group in Paraguayan history, with cultural and material heritage (Chapter 32, this volume). Government and nongovernment programs and projects to protect and preserve indigenous peoples and culture are needed, and these groups need to help design and implement such efforts in keeping with the international agreements ratified by Paraguay.

Achieving sustainable use and long-term conservation of the forest ecosystems of Paraguay, especially the Interior Atlantic Forest, will not be an easy task. Corrective measures must be taken immediately to forestall further deterioration of human, social, economic, and environmental capital, and only a combination of actions strategically coordinated at the national and regional levels can achieve this objective.

Acknowledgments

The authors are grateful to Dr. Carlos Galindo-Leal, Lucía Bartrina, Heidi Goerzen, and Ramón Sienra Zavala for their assistance in writing this chapter.

References

BCP (Banco Central de Paraguay). 2000a. *Boletín cuentas nacionales,* no. 34. Asunción: Banco Central de Paraguay.

BCP (Banco Central de Paraguay). 2000b. *Boletín cuentas nacionales,* no. 36. Asunción: Banco Central de Paraguay.

Bozzano, B. and Weik, J. H. 1992. *El avance de la deforestación y el impacto económico,* Serie no. 12. Asunción: Ministerio de Agricultura y Ganadería, Deutsche Gesellschaft für Technische Zusammenarbeit.

DGEEC (Dirección General de Estadísticas, Encuestas y Censos). 1996. *Programa MECOVI: migración interna en Paraguay. Análisis económico de la encuesta de hogares.* Asunción: Dirección General de Estadísticas, Encuestas y Censos.

DGEEC/FNUAP (Dirección General de Estadísticas, Encuestas y Censos/Fondo de las Naciones Unidas para Actividades en Materia de Población). 1999. *Población en el Paraguay.* Asunción: Dirección General de Estadísticas, Encuestas y Censos.

ENAPRENA/GTZ/SSERNMA (Proyecto Estrategia Nacional para la Protección de los

Recursos Naturales/Deutsche Gesellschaft für Technische Zusammenarbeit/Subsecretaría de Estado de Recursos Naturales y Medio Ambiente). 1995. *Diagnóstico del sector forestal Paraguayo.* Asunción: Makrografics.

FMB (Fundación Moisés Bertoni). 1994. Diagnóstico de los recursos socio-ambientales. Proyecto Trinacional de Manejo del Bosque Atlántico Interior. Capítulo Paraguay. Vol. I. Asunción, Paraguay: Fundación Moises Bertoni.

FAO. 2000. *Informes nacionales de los países.* Bogotá: Latin American and Caribbean Forestry Commission (LACFC).

Fogel, R. 1998. *La investigación acción participativa: lecciones aprendidas en Paraguay.* Asunción: CERI (Centro de Estudios Rurales Interdisciplinarios) and CEPADES (Centro Paraguayo de Estudios de Desarrollo Económico y Social).

GTZ (Deutsche Gesellschaft für Technische Zusammenarbeit). 1994. *Perfil del país: informaciones y comentarios relacionados al desarrollo económico y social.* Asunción: GTZ.

MAG/SSERNMA/GTZ/ENAPRENA (Ministerio de Agricultura y Ganadería/Subsecretaría de Estado de Recursos Naturales y Medio Ambiente/Deutsche Gesellschaft für Technische Zusammenarbeit/Proyecto Estrategia Nacional para la Protección de los Recursos Naturales). 1999. *Diagnóstico del sector forestal: actualización.* Asunción: Makrografics.

Masulli, B., Mereles, F., Aquino, A. L., de Fox, I. G., de Medina, F. A., Rossato, V., Sottoli, S., and Monge, V. V. 1996. *El rol de la mujer en la utilización de los recursos naturales en el Paraguay: un enfoque multidisciplinario.* Asunción: Editora Litocolor S.R.L.

PNUD (Programa de las Naciones Unidas para el Desarrollo). 2001. *Visión conjunta de la situación de Paraguay.* Asunción: PNUD.

Prieto, E. 1999. *Mujer y desarrollo en síntesis: 1988–1998.* Secretaría de la Mujer, la Coordinadora de Mujeres de las Naciones Unidas en el Paraguay. Asunción: Runner Gráfica.

STP (Secretaría Técnica de Planificación). 2000. *Diagnóstico sociodemográfico del Paraguay.* Asunción: Makrografics.

STP/DGEEC (Secretaría Técnica de Planificación/Dirección General de Estadísticas, Encuestas y Censos). 1999. *Paraguay: indicadores socioeconómicos y demográficos: atlas temático departamental del Paraguay.* Asunción: Dirección General de Estadísticas, Encuestas y Censos.

Chapter 27

The Guaraní Aquifer: A Regional Environmental Service

Juan Francisco Facetti

The Guaraní aquifer is the largest underground freshwater reserve in the world; coincidentally, it runs under a sizable portion of the Interior Atlantic Forest. The aquifer extends from the Paraná River basin, shared by Paraguay, Brazil, and Uruguay, to the Paraguay-Paraná basin between Paraguay and Argentina, stretching from 14°S to 35°S and from 47°W to 60°W (Figure 27.1).

Some 15 million people live in the region associated with the aquifer, which includes the cities of Campo Grande, Londrinas, and Foz de Iguazú in Brazil; Ciudad del Este, Salto del Guairá, Coronel Oviedo, and Encarnación in Paraguay; Posadas and Yguazú in Argentina; and Saltos in Uruguay.

The aquifer is made up of strata deposited some 140 to 210 million years ago in the sedimentary basin of the Paraná River. The aquifer region includes several different geological formations: the Pirambóia in Brazil, the Buena Vista in Uruguay, the Misiones in Paraguay, and the Tacuarembó, shared by Uruguay and Argentina. The Pirambóia and Rosário do Sul formations in Brazil and the Buena Vista formation in Uruguay date from the Triassic period (220 million years ago), and the wind-generated sand deposits of the Jurassic period (170 million years ago) created the Botucatu formation in Brazil, the Misiones in Paraguay, and the Tacuarembó in Uruguay and Argentina (Rocha 1996).

The Guaraní aquifer extends over an area of approximately 1.1 million km^2: 770,000 km^2 in Brazil, 220,000 km^2 in Argentina, 80,000 km^2 in Paraguay, and 60,000 km^2 in Uruguay. Its depth ranges from just a few meters to nearly 1,000 m. The volume of its permanent water reserves is estimated to be on the order of 45,000 km^3 (Fili et al. 1998).

The volume of naturally recharged reserves (i.e., the potential renewable water circulating in the aquifer) is estimated to be 160 km^3 per year, or 5,000 m^3 per second. Confined water pockets are believed to exist that are not subject to replenishment through infiltration or vertical filtration (Facetti and Stichler 1995). Therefore, forest conservation plays a key role in maintaining the quality and quantity of this water resource.

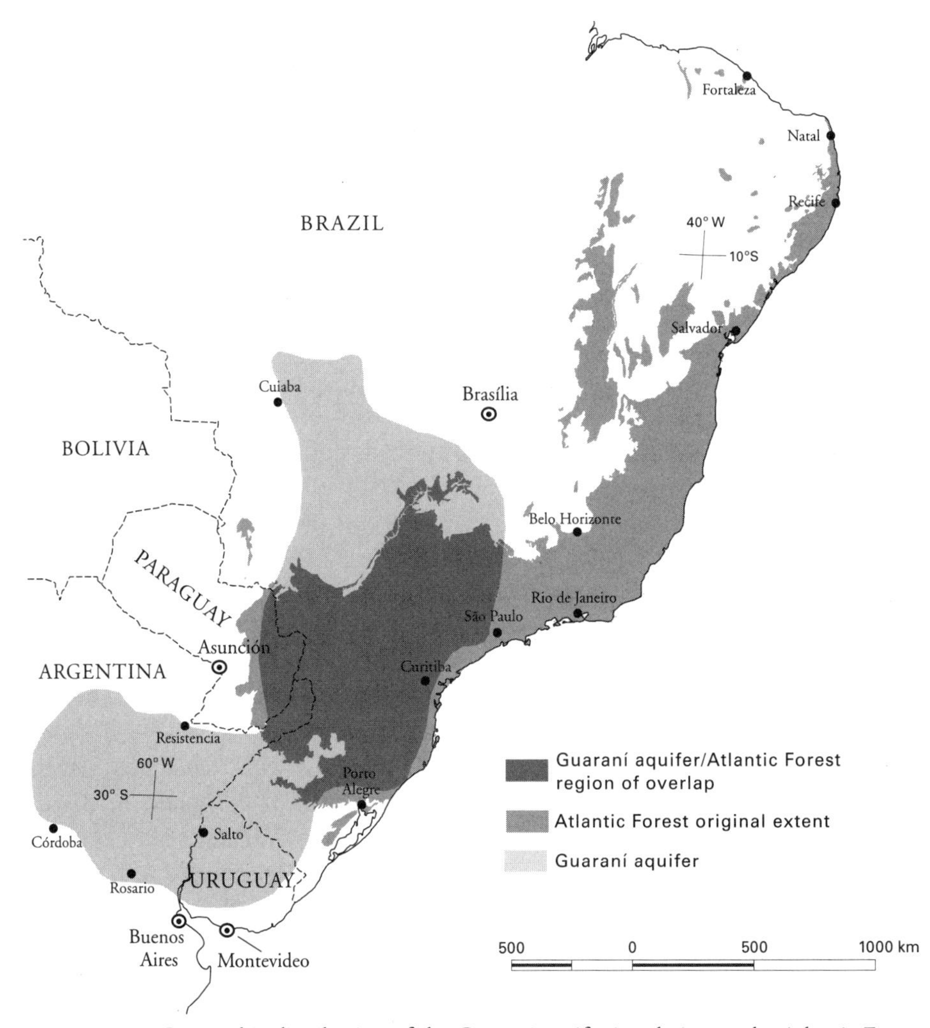

Figure 27.1. Geographic distribution of the Guaraní aquifer in relation to the Atlantic Forest hotspot.

Because the Guaraní aquifer has been well protected and generally unexploited, many regard it as the most economical and flexible source of water supply for human consumption in the region. However, exploitation is bound to increase greatly to meet rising demand. The rapid growth of the region's primary production and industrial sectors and the cities' needs for potable water are expected to spur the installation of deep well pumping systems.

Particularly in Paraguay the aquifer has had a major socioeconomic impact because it is a source of potable water that does not need chemical purification. Because of its geological and morphological characteristics, which up to now have ensured good water quality, the World Bank designated the aquifer as the primary water source over rivers and streams in its International Bank for Reconstruction and Development (IBRD-IV) project on potable water and rural sanitation (World Bank 1997).

If the aquifer is to be protected, it must be used sustainably, and the conditions that ensure good water quality must be maintained. The latter requirement is closely linked to preservation of the Interior Atlantic Forest because the forest overlies much of the aquifer. Recently the government's aquifer monitoring program (Project PAR/8/005) found chemical and bacteriological contamination of the aquifer in densely populated urban areas (PMAP 2000). Failure to conserve the last remaining areas of the Interior Atlantic Forest could trigger a series of physical and chemical phenomena that would alter the infiltration capacity of the soils, the time needed for the water to replenish itself, and, ultimately, the quality of the water in the aquifer. Specific potential risks include microbiological contamination by fecal coliform bacteria, chemical pollution by nitrates from chemical fertilizers, and organic microcontaminants from agrotoxins. Moreover, these risks would be exacerbated by overexploitation of the aquifer for irrigation, industrial use, or human water supply systems.

Currently the four countries that share the aquifer are working to protect it through a project with the Global Environment Facility (http://www.worldbank.org/guarani). This project, which also has the support of the World Bank and the Organization of American States, will provide tools for decision makers in the integrated environmental and hydrogeological management of the Guaraní aquifer.

The water of the Guaraní aquifer should be regarded as a flexible fund of environmental, social, and economic capital that can be drawn on for human use. Each of the four nations that use the aquifer has its own institutional system for regulating water resources. These systems vary in complexity and in the extent to which they are decentralized. Ideally, existing initiatives for conserving and protecting the Interior Atlantic Forest and initiatives to protect the Guaraní aquifer should be coordinated through appropriate institutional mechanisms.

References

Facetti, J. and W. Stichler. 1995. *Analysis of concentration of environmental isotopes in rainwater and groundwater from Paraguay.* International Seminar of Isotopic Hydrology. Vienna: IAEA.

Fili, M., Rosa Filho, E. F., Auge, M. and Xavier, J. M. 1998. Acuífero Guarani: un recurso compartido por Argentina, Brasil, Paraguay y Uruguay (América del Sur). *Boletín del Instituto Tecnológico Geominero de España* (Madrid) 109(4): 73–78.

PMAP (Programa de Monitoreo de Acuíferos del Paraguay). 2000. Asunción: Corporación de Obras Sanitarias/International Atomic Energy Agency.

Rocha, G. A. 1996. *Mega reservatório de água subterrânea do Cone Sul: bases para uma política de desenvolvimento e gestão.* Seminário e Workshop "Internacional do Aquífero Gigante do Mercosul," Curitiba. Universidade Federal do Paraná (Brazil), Universidad de la República Oriental del Uruguay (Uruguay), Universidad Nacional de la Plata (Argentina), SENASA (Servicio Nacional de Sanidad y Calidad Agroalimentaria), and International Development Research Centre (Canada).

World Bank. 1997. Project appraisal document report no. 16770 PA. *Finance, private sector and infrastructure.* Country Management Unit 6, Latin America and the Caribbean Region. Washington, DC: The World Bank.

Chapter 28

Conservation Capacity in the Interior Atlantic Forest of Paraguay

Alberto Yanosky and Elizabeth Cabrera

Paraguay currently lacks a comprehensive environmental policy, sufficient environmental legal norms, and the resources needed to implement conservation actions effectively. Environmental management in Paraguay is also hampered by the contradictory approaches taken by official institutions. However, the current democratic process in Paraguay, initiated in 1989, has made the conditions favorable for adding environmental issues to the government's agenda. The new democratic process has also brought more technical expertise to the public sector, which, in turn, has helped bring added recognition to the effects of environmental deterioration in Paraguay.

The Interior Atlantic Forest became nationally recognized as an important ecosystem in 1993–1994 when several sectors started to see the similarities between the Brazilian, Argentine, and Paraguayan components of this shared ecosystem. Interest in conserving this region on the part of national and international as well as local institutions started gaining momentum in 1994 (Table 28.1).

Status and Capacity of Organizations

During the past years, a few government, nongovernment, and academic institutions have been involved in the environmental issues of eastern Paraguay. In the following section we describe the evolution and main capacity difficulties of these institutions.

Federal Government Organizations

As of July 2000, Paraguay did not have a centralized institution responsible for environmental planning and management, nor did it have a national environmental action plan, making it the only Mercosur country without a separate ministry for the environment. Recently, however, a secretariat was created to oversee the execution of environmental policy. Three institutions have been

Table 28.1. Recent relevant events related to conservation of the Interior Atlantic Forest of Paraguay.

Year	Event	Objective	Organization
1994, Paraguay	Socioeconomic, cultural, and productive assessment of natural resources in the Interior Atlantic Forest	Basis for the sustainable management plan for the Interior Atlantic Forest	Fundación Moisés Bertoni
1995, Asunción, Paraguay	"National Action Plan for the Conservation of the Interior Atlantic Forest" workshop	Analysis of the present status of the Interior Atlantic Forest and areas of influence in Paraguay, National Action Plan definition for conservation.	State Undersecretary of Natural Resources and Environment, Fundación Moisés Bertoni, World Wildlife Fund (WWF)
1995, Misiones, Argentina	National workshop for the conservation of the Paraná Forest	Review and update of the Action Plan	Financed by Fundación Vida Silvestre Argentina (FVSA) and WWF
1995, Hernandarías, Paraguay	National Workshop for the Conservation of the Interior Atlantic Forest	Analysis of biological, institutional, economic, and political aspects related to biodiversity. Recommendation of seven regional units of high conservation priority.	Fundación Moisés Bertoni, Ministry of Natural Resources and Environment
1996	"Implementation of a Trinational Plan for the Conservation of the IAF in Argentina, Brazil and Paraguay" project		WWF
1996, San Bernardino, Paraguay	"Regional Consultation for the Southern Cone" within the South American Strategy for Forests Conservation program	Promote conservation in the Interior Atlantic Forest	International Union for the Conservation of Nature (IUCN)

(Continued)

Table 28.1. Continued.

Year	Event	Objective	Organization
1996, Hernandarías, Paraguay	First Trinational Workshop for Planning in the Interior Atlantic Forest	Definition of proposals on protected areas, sustainable use, and policy and strategies	Financed by WWF
1996, Misiones, Argentina	First Technical Workshop on Protected Area Management and Conservation of the Interior Atlantic Forest, Argentina, Brazil, and Paraguay	Discuss common problems of control, management, research, and interpretation. Search for solutions.	Technical Regional Office of Northeast Argentina, National Parks Administration.
1997, Curitiba, Brazil	Second Trinational Workshop, "Use and Conservation of the Interior Atlantic Forest"	Support preliminary project on the Green Corridor law (Chapters 14 and 19, this volume) in Misiones, Argentina	Financed by Fundación Boticario, IUCN, and WWF
1999, El Dorado, Misiones, Argentina	Third Trinational Workshop on Conservation and Sustainable Use of the Forest	Update trinational region diagnosis. Assess themes and interests. Define future steps.	Financed by WWF-US through FVSA, IUCN, INTA (National Institute of Agricultural Technology) [Instituto Nacional de Tecnología Agropecuaria]), Facultad de Ciencias Forestales

1999	Production of the Interior Atlantic Forest Map 1977	Elaboration of the status of the Interior Atlantic Forest in Paraguay	Project Environmental System of the Eastern Region [(Sistema Eastern Region (SARO [Environmental System of the Eastern Region (Sistema Ambiental de la Región Oriental)]) (MAG [Ministry of Agriculture and Livestock (Ministerio de Agricultura y Ganadería)]), SERNMA [Undersecretary of the State of Natural Resources and Environment (Subsecretaría de Estado de Recursos Naturales y Medio Ambiente)], SFN [National Forest Service (Servicio Forestal Nacional)], DNPVS [Office of National Parks and Wildlife (Dirección de Parques Nacionales y Vida Silvestre)], DOA [Office of Environmental Planning (Dirección de Ordenamiento Ambiental)], DOT [Office of Territorial Planning (Dirección de Ordenamiento Territorial)], UNA National University of Asunción (Universidad Nacional de Asunción): Faculty of Agricultural Sciences, and Fundación Moisés Bertoni. Financed by WWF.
2000	Meeting of the Trinational Commission of the Green Corridor Initiative	Proposal draft of the mission. Definition of functions of the Trinational Commission and areas of interest.	WWF-US
2000, Foz do Iguaçu, Brazil	Trinational Workshop for the Biological Vision for the Interior Atlantic Forest	Gathering of scientific, biological, social, and economic information from the three countries to identify conservation priority areas, threats, and opportunities.	Coordinated by WWF and FVSA

given authority on environmental matters, a situation that has resulted in legal contradictions and a tendency among these institutions to dodge responsibility.

The Technical Secretariat for Planning (STP) reports directly to the Presidency of the Republic. It is responsible for making sure that environmental issues are incorporated into development policies. However, the low priority given to the environment in the policy-making process has been an obstacle even for the secretariat.

The Under Secretariat for Natural Resources and the Environment (SSERNMA), under the Ministry of Agriculture and Livestock, used to be the authority responsible for environmental protection. It had various subdivisions, or bureaus, such as the Bureau of National Parks and Wildlife, the Bureau of Environmental Policy, the National Forestry Service (SFN), and the Bureau of Fishing. However, SSERNMA has been replaced by the Secretariat for the Environment (SEAM).

The National Commission for the Defense of Natural Resources was created in 1990 under the National Congress. This commission has been tasked with developing plans for protecting ecosystems and for coordinating public agencies in this area. The commission will conduct audits and studies and issue opinions regarding ecosystem defense, although it is not an executive agency.

Several efforts at environmental oversight have been successful, as in the case of pollution of Lake Ypacaraí (Central Department) and, more recently, the discovery of toxic waste in the national territory (*Diario Noticias* 2001). Law 716 on Ecological Offenses also is being implemented and, where the political will exists, enforced.

The best that can be said now is that Paraguay's institutional framework for environmental management is in a state of transition. Among the Mercosur nations, Paraguay's current system is certainly the most outdated; indeed, it is one of the most outdated in all of Latin America. So far, several agencies have shared environmental responsibilities, with no single institution leading the state and executing environmental policy. However, the task of providing environmental leadership is now in the hands of the new Secretariat for the Environment.

Creation of the Secretariat for the Environment

SEAM was created under Law No. 1561 (dated July 21, 2000) to regulate national environmental policy. SEAM replaces the Under Secretariat for Natural Resources and the Environment under the Ministry of Agriculture and Livestock and is responsible for administering all national laws and international agreements relevant to the environment that are currently in effect.

Previously, and even today as SEAM works to get established, several different institutions have addressed environmental problems, but with little coordination. With a view to improving this situation, a National Council on Sustainable Development is being formed, which will serve as a forum for reaching consensus on development issues and a nexus for decisions on economic, social, and environmental policy. The Inter-American Development Bank, the German Technical Cooperation Agency, and the Pan American Health Organization are providing technical and financial support for this process.

Other government organizations involved in Interior Atlantic Forest conservation include the recently created National Council for the Environment, through SEAM; the Ministry of Agriculture and Livestock, through its several agencies, including the Agricultural Extension Bureau; the Ministry of Public Works and Communications, especially its Department for the Environment; and the Ministry of Public Health and Social Welfare, particularly its National Environmental Health Service. Furthermore, the Ministry of Foreign Affairs has recently asked organized civil society to assist the central government in formulating a presentation and accompanying the delegation representing Paraguay at the Conference of the Parties to the United Nations Framework Convention on Climate Change. The Ministry of Education and Culture has also changed the teaching curriculum to include the development of values that promote the recovery, conservation, and protection of the environment and culture.

The SFN, originally a part of SSERNMA, remains under the Ministry of Agriculture and Livestock and has not been incorporated into SEAM. It is responsible for overseeing compliance with the law on the promotion of forestation and reforestation (Law 536/94); the law that prohibits the exportation of or traffic in wooden logs, chumps, or beams (Law 515/94); and the forestry law (Law 422/73). The SFN also has the authority to grant permits for use of the forests for certain periods of time under stated conditions. However, the continuing deforestation that has critically endangered Paraguay's natural resources is evidence that the SFN has failed to fulfill its responsibilities.

Departmental and Municipal Organizations

Outside the central government agencies, environmental issues are either poorly addressed or ignored completely. Decentralization and state reform are the topics of the moment in Paraguay, but no concrete action has yet been taken (Freaza 1996). However, a few autonomous agencies carry concrete positive or negative actions in the Interior Atlantic Forest: the Rural Welfare Institute (IBR), the National Indian Institute (INDI), and the Itaipú Binational Hydroelectric Authority.

The IBR, responsible for agrarian reform, is in the unique position of having promoted forest destruction. It has developed its settlements on forestland without any planning and with no regard for the ecological characteristics of the affected areas, and it has not provided support for the relocated families. Moreover, on a much larger scale, it has promoted the sale of millions of hectares of government land, much of it forested, for the development of agribusinesses and large-scale livestock production (DPNVS 2000). The INDI purports to deal with the country's indigenous issues, but in reality all it has done is secure land for reservations, without providing for their harmonious development. Finally, Itaipú Binational, for its part, reshaped the landscape and modified the climate of the transfrontier area shared with Brazil to build the largest hydroelectric dam in the world. However, they have attempted to mitigate their impact by creating their own protected areas.

Academic Institutions

The only academic institution formally committed to conserving the Interior Atlantic Forest is Asuncion National University (UNA), through such programs as forestry engineering, exact and natural sciences, and chemistry. The UNA has sponsored research projects, internships, agreements for managing natural areas, and training courses at various levels on sustainable management. In particular, its Multidisciplinary Center for Technological Research has spurred conservation research and developed a study area in the Interior Atlantic Forest.

Other national academic institutions have done little to address environmental issues. Those that have broached the subject have not made research contributions. For example, the Technical University of Marketing and Development offers environmental career programs but lacks the technical or scientific personnel to carry out research. The Universidad Americana encourages the inclusion of environmental topics in its undergraduate and postgraduate curricula, but so far the university has not carried out any research on the Interior Atlantic Forest in particular.

National Nongovernmental Organizations

There are over 200 nongovernmental organizations (NGOs) in Paraguay, dozens of them working in areas related to sustainable development. Although NGOs existed under the dictatorship, under the democratic government their number has increased to take up work in areas that the state has neglected. NGOs make an important contribution to national processes and are especially effective on the consultative committees in the National Congress. The soon-to-be-created National Council on Sustainable Development envisages the active participation of NGOs in formulating and implementing sustainable development policies.

Three NGOs in particular have played key roles in assisting the state with its responsibilities and strengthening civil society's role in conserving the Interior Atlantic Forest, namely Alter Vida (since 1985), Sobrevivencia (since 1986), and the Moisés Bertoni Foundation (FMB, since 1988).

International Nongovernmental Organizations

The World Conservation Union (IUCN) holds regular meetings in Paraguay. Their recent workshop on the impact of World Bank intervention and the new forest policy (IUCN 2000) included as priority topics for 2000–2005 deforestation, genetic erosion, degradation of humid soils and water sources, and desertification.

The Latin American Forest Network brings together the Latin American and Caribbean NGOs that work on preserving the region's forests. Its mission is to contribute to forest conservation for the benefit of present and future generations. Currently about 60 NGOs in the region belong to this umbrella organization.

In the past, international NGOs had no presence in Paraguay, but this has been changing. In 1986 The Nature Conservancy supported creation of the Conservation

Data Center and more recently, the Mbaracayú Program and the FMB. Subsequently, BirdLife International formed a partnership with FMB (which has since been moved to Guyra Paraguay). The World Wildlife Fund (WWF), which established an office in Paraguay in 2000, has also begun to work with FMB. Conservation International is showing interest and is establishing more formal ties with Guyra Paraguay and FMB, as are other national and international organizations.

The Trinational Commission on the Interior Atlantic Forest and the Green Corridor Initiative were recently established to follow up on action plans developed at various workshops held over the last 2 years (Table 28.1). The Trinational Commission is committed to mutual cooperation and the participation of all three members through horizontal communication and operations in the form of a flexible and dynamic network. Its mission is to martial forces to ensure the conservation of biodiversity and sustainable development in the Interior Atlantic Forest. It comprises one representative each from the national commissions of Paraguay, Brazil, and Argentina, plus a representative of the WWF and one from the IUCN.

Organized civil society has carried out many concrete actions in the Interior Atlantic Forest (Table 28.2), to a large extent making up for the state's failure to

Table 28.2. Brief synthesis of interests and activities of nongovernment organizations (NGOs) working in the area of the Interior Atlantic Forest of Paraguay.

Organization	Activities
Guyra-Paraguay	Conservation of birds and their habitats through research, interpretation, and actions. Has identified conservation priority areas and supports efforts to consolidate San Rafael National Park. Supports strengthening of ProCosara, a local NGO in Itapúa. Supports scientific research on status and distribution of birds, education, and communication and provides advice on several environmental issues.
Moisés Bertoni Foundation (Fundación Moisés Bertoni)	Works for socioeconomic sustainable development through nature conservation and organized population action. Has developed a sustainable development model, the Mbaracayú Program, providing most current information on the Interior Atlantic Forest.
Instituto de Derecho y Economía Ambiental (IDEA)	Created to promote sustainable development by improving public policies, laws, and regulations.
Alter Vida	Center of studies and capacity for ecodevelopment. Works for sustainable development through appropriate ecosystem management, with strong social participation.
Asociación para la Protección del Department. Medio Ambiente del Amambay (AMPA)	Promotes education and environmental education. Conducts activities to protect historical, cultural, and natural heritage of Amambay.
Acción Axial Naturaleza y Cultura	Restores, protects, and conserves natural resources.

Continued

Table 28.2. Continued

Organization	Activities
Center for Education, Capacity Building and Rural small Technology (Centro de indigenous Educación, Capacitación y Tecnología Campesina) (CECTEC)	Research on practical and theoretical knowledge on rural productive processes, and increased awareness of rights and obligations of young peasants. Promotes integration and active and efficient participation in improving family and community life quality. Provides capacity to agriculture businesses in production, management, commercialization, information, technology, and traditional practices of peasants and people.
Paraguayan Center for Sociological Studies (Centro Paraguayo de Estudios Sociológicos) (CPES)	Conducts sociological research in several departments of Paraguay, working with peasants. Publishes the *Sociological Journal of Paraguay.*
Fundación Yvyturuzu	Promotes sustainable development, conservation, and environmental education.
SOBREVIVENCIA (SURVIVAL)	Environmental research and action of problems related to native Paraguayan communities: development of settlements, known as "Ecourbanistic"; development of an ecological corridor network; environmental service law proposal; and program to stabilize agricultural frontiers.
Procordillera San Rafael (PROCOSARA)	Activities limited to conservation actions in the natural environment of the San Rafael mountains.
Natural Land Trust	Works for nature conservation by involving organizations, communities, and land owners. Establishes ecological easements and promotes forestry certification. It is the first conservation endowment in south America and is establishing the first conservation easement in Paraguay in the area reserved for San Rafael National Park.
Servicios Ecoforestales para Agricultures (SEPA)	Promotes women's participation in programs of sustainable agriculture, short-term planning, and capacity building of professionals as trainers.
International Center for Capacity Building of Environmental and Development Organizations (Centro International de Capacitación para Organizaciones Ambientalistas y de Desarrollo) (CICOAM)	Managerial training for leaders in civil society organizations.

halt the destruction of this ecosystem in Paraguay. The inefficiency of some of the public agencies and the corruption to which Paraguayan people are subjected with impunity has led to the development of organized units such as PROCOSARA (Procordillera San Rafael) in the reserve area of San Rafael National Park, which, together with the Mbaracayú, are the last two large remaining forest areas (500 to 650 km^2).

Status of the Protected Areas

Paraguay's Protected Forest Areas (ASPs) constitute an essential strategy for maintaining biodiversity (Yanosky and Escalante 2000). In 1993 Paraguay sought to increase its protected forest area from 4 to 9.8 percent, as recommended by international organizations (SINASIP 1993). However, in the year 2000 the total protected area was only 17,111.85 km^2, or 4.2 percent of the national territory, and most of it was located in the eastern region. This demonstrates the state's inability to fulfill its own objectives. Moreover, the protected areas themselves are exposed to major threats. For example, in 2000 the integrity of Chaco National Park was seriously threatened when government officials attempted to take possession of a parcel more than 300 km^2, thus placing their personal interests above those of Paraguayan people. Faced with this threat, several sectors of the community (NGOs, environmentalists, and youth) came together to form the citizens' council Let's Defend the Defenders (Defendamos los Defensores).

Creating or consolidating protected forest areas in the eastern region is more difficult because of the land ownership situation and the great pressure being exerted on natural resources. More than 95 percent of the country's total population lives in the eastern region (FMB and USAID 2000).

The fundamental objective of the strategic plan for the National System of Protected Forest Areas of Paraguay (SINASIP), developed in 1993 by the Bureau of Natural Parks and Wildlife with the support of the FMB, is to provide maximum protection of the country's biogeographic diversity. In addition, it is supposed to provide guidelines for policy on the management of ASPs over the next 10 years (SINASIP 1993). SINASIP is divided into areas under public management (SEAM), areas under Itaipú Binational, and areas administered by private parties (FMB and land owners). In 1993 there were 16 ASPs within the Interior Atlantic Forest area of influence, covering a total of 2,728.03 km^2 (FMB 1994), or 3.09 percent of the area originally covered by the Atlantic Forest in Paraguay.

The recently created Pikyry Protected Area, Tapyta Private Natural Reserve, and Arroyo Blanco Private Natural Reserve are being incorporated into the Atlantic Forest area of influence and will bring the number of protected areas to 19 (Table 28.3). The Arroyo Blanco Private Natural Reserve will be the first private reserve in Paraguay. As of June 2001, according to information from SEAM, the declaration of its establishment was awaiting signature by Paraguay's president.

Currently, less than 20,000 km^2 of the Interior Atlantic Forest is in a

Table 28.3. List of protected areas in the Atlantic Forest of Paraguay.

Name	Category	IUCN Category	Extent (km^2)	Land Ownership (P = private; S = state)	Part of the National System of Protected Areas (SINASIP)	Level of Management (Bragayrac et al. 1998)	Decrees	Reference
Cerro Corá	National park	II	120.38	60% S,40 % P	Yes (state)	High	N 20.698/1976; N 6.090/1990	Bragayrac et al. 1998
Ybycuí	National park	II	50.00	70% S, 30% P	Yes (state)	High	N 32.772/1973	Bragayrac et al. 1998
Serranía de San Rafael	National park	II	780.00	100% P	Yes (state)	Low	N 13.680/1992	Bragayrac et al. 1998
Caaguazú	National park	II	160.00	80% S, 20% P	Yes (state)	High	N 20.933/1976; N 5.137/1990	Bragayrac et al. 1998
Ñacunday	National park	II	20.00	100% P	Yes (state)	Low	N 17.071/1975; N 16.146/1993	Bragayrac et al. 1998
Ybytyruzu	Resource management reserve	II	240.00	100% P	Yes (state)	Low	c14945/2001	Bragayrac et al. 1998
Kuri´y	National reserve	III	20.00	100% P	Yes (state)	Low	N 30.956/1973	Bragayrac et al. 1998
Bella Vista	National park	II	73.11	100% S	Yes (state)	Moderate	N 20.713/1998	Bragayrac et al. 1998
Moisés Bertoni	Scientific monument	III	2.00	100% S	Yes (state)	High	N 11.270/1995	Bragayrac et al. 1998
Itabó	Biological reserve	IV	178.80	100% P	Yes (Itaipú)	High	RDE N 052/84/1984	Itaipú Binacional. 1999b

Limoy	Biological reserve	IV	133.96	100% P	Yes (Itaipú)	High	RDE N 052/84/1984	Itaipú Binacional. 1999b
Tatí Yupí	Biological refuge	IV	19.16	100% P	Yes (Itaipú)	High	RDE N 052/84/1984	Itaipú Binacional. 1999b
Mbaracayú	Biological refuge	IV	14.37	100% P	Yes (Itaipú)	High	RDE N 052/84/1984	Itaipú Binacional. 1999b
Carapá	Biological refuge	IV	29.16	100% P	Yes (Itaipú)	High	1993	Itaipú Binacional. 1999b
Pikyry	Biological refuge	IV	11.10	100% P	Yes (Itaipú)	High		Itaipú Binacional. 1999b
Yui-Rupá	Biological refuge	IV	7.50	100% P	Yes (Itaipú)	High		http://www.itaipu.gov.br
Bosque del Mbaracayú	Biosphere reserve	*	644.05	100% P	Yes (private)	High	N 112/1991	FMB 2000
Itabó	Natural private reserve	IV	30.00	100% P	No (proposal)	Low		FMB 2000
Ypetî	Natural private reserve	IV	100.00	100% P	No (proposal)	Moderate		FMB 2000
Morombi	Natural reserve	IV	250.00	100% P	Yes (private)	Low	N 14940/2001	FMB 2000
Tapytá	Natural private reserve	IV	40.85	100% P	No	Moderate		FMB 2000

Continued

Table 28.3. Continued

Name	Category	IUCN Category	Extent (km^2)	Land Ownership (P = private; S = state)	Part of the National System of Protected Areas (SINASIP)	Level of Management (Bragayrac et al. 1998)[1]	Decrees	Reference
Arroyo Blanco	Natural reserve	IV	57.14	100% P	Yes (private)	Low	N 14944/2001	Resolución DPNVS No. 96
Ka'i Rague	Natural private reserve	IV	14.95	100% P	No	Low		FMB 2000
San Antonio	Natural private reserve	IV	5.00	100% P	No	Low		FMB 2000
Pindo'y	Natural private reserve	IV	30.00	100% P	No	Low		FMB 2000
C. Pfannl	Natural private reserve	IV	5.00	100% P	No	Low		FMB 2000
Sta. Rita de Cassia	Natural private reserve	IV	Undefined	100% P	No	None		FMB 2000
Pindoty	Natural private reserve	IV	Undefined	100% P	No	None		FMB 2000
San Francisco	Natural private reserve	IV	Undefined	100% P	No	None		FMB 2000
Natividad	Natural private reserve	IV	Undefined	100% P	No	None		FMB 2000
Ka'aguy Rory	Natural private reserve	IV	Undefined	100% P	No	None		FMB 2000

San Isidro	Ecological easement	IV	Undefined	100% P	No (private contract)	Low	FMB 2000
San Pedro Mí	Ecological easement	IV	10.32	100% P	No (private contract)	Low	FMB 2000
Caaguy Rory	Ecological easement	IV	7.89	100% P	No (private contract)	Low	
San Clemente	Ecological easement	IV	7.90	100% P	No (private contract)	Low	
San Pablo	Ecological easement	IV	6.30	100% P	No (private contract)	Low	
Mamorei	Ecological easement	IV	0.80	100% P	No (private contract)	Low	
Acaray-Mí	Ecological reserve	IV	250.00	100% P	No (proposal)	None	Bragayrac et al. 1998
Cerro Sarambi	National park	II	300.00	100% P	No (proposal)	None	Bragayrac et al. 1998
Laguna Blanca	Ecological reserve	IV	300.00	100% P	No (proposal)	None	Bragayrac et al. 1998
Estero Milagro	National park	II	250.00	100% P	No (proposal)	None	Bragayrac et al. 1998
Total extent			4,169.74				

Number of Areas: 19 protected areas are in the National System; 22 are not recognized.

*It complies with international conditions for category II, but that category is for exclusive government use. Furthermore, Bosque del Mbaracayú is classified as a Protected Area under private domain and does not fit into that category.

conserved state (MAG et al. 2000). The public system of Atlantic Forest protected areas covers 1,392.38 km^2. Another 644.05 km^2, corresponding to the Mbaracayú Forest Natural Reserve, is under private administration, 483.14 km^2 is natural reserves in private hands, and 386.54 km^2 comes under the Itaipú Binational Hydroelectric Authority, for a total of 2,906.11 km^2 (Table 28.3). In other words, all the ASPs containing elements of the Atlantic Forest amount to 2,906.11 km^2, or only 3.3 percent of the original cover of this forest type.

The Mbaracayú Forest Natural Reserve

The best example of a protected area in Paraguay, and a model for Latin America, is the Mbaracayú Forest Natural Reserve (RNBM), created with international support and commitments from the government and local communities. This initiative is example of the private sector, the government, and international organizations working together to protect the Interior Atlantic Forest.

The RNBM represents an important milestone in Paraguay's conservation history. In June 1991 the government of Paraguay, the United Nations, the Nature Conservancy, and the FMB signed an agreement to establish and conserve in perpetuity the RNBM and the upper Jejuí watershed. This agreement was ratified by National Law 112/91, approved by the legislature on December 19, 1991, and enacted on January 3, 1992 (FMB and USAID 2000).

The RNBM is a protected forest area under private domain that, according to international criteria and because of its conservation objectives, still qualifies as a national park. From the start, the reserve's model of integrated management has served as the basis for the development and consolidation of support from private conservation initiatives in Paraguay. The RNBM is the only sizable (640 km^2) representative sample of the Interior Atlantic Forest that is effectively protected (Keel et al. 1993; FMB 1994; Madroño and Esquivel 1995; Lowen et al. 1996). Another large area in Paraguay with similar characteristics, San Rafael National Park, is being secured, and the prospects for its long-term protection look good.

Conservation of the RNBM is especially important because, in addition to the Atlantic Forest, the eastern part of its territory contains samples of *cerrado* vegetation that contribute significantly regional biodiversity. The RNBM thus protects another environment of very high conservation priority in the region (BSP et al. 1995). The RNBM also helps to reduce levels of carbon dioxide in the atmosphere, thus playing an important role in stabilizing the world's climate. The FMB, at the request of the donor corporation Applied Energy Systems, Inc., estimated that the vegetal biomass of the RNBM annually captures 1.5 tons of carbon (FMB 2001).

Private Natural Reserves

The land ownership problem and the prospect of unproductive *latifundia* has resulted in the concept of private natural reserves in Paraguay (Chapter 26, this volume). Private natural reserves, along with other private conservation concepts such as ecological easements (*servidumbres ecológicas*), are gaining impetus at the national level. Ecological easements are agreements between two or more land owners in which at least one voluntarily agrees to plan the future use of his or her property in a manner that conserves existing resources and turn over supervision of the lands so designated to a third party—in effect, a private conservation land trust. Once this agreement has been recorded in the General Property Register, future owners of the land are required to abide by it. Some international institutions are prepared to finance the implementation of ecological easements in the Atlantic Forest.

Private natural reserves have played a role in mitigating the grave status of some of Paraguay's public areas. However, the political will to give them official recognition has so far been lacking (FMB and USAID 2000).

The Sociedad Agrícola Golondrina S.A. of Grupo Espíriti Santo established their Ypetî Private Natural Reserve (formerly Golondrina III) with a strong commitment to conservation. This reserve is an example of a joint effort to achieve biological conservation in harmony with the sustainable use of natural resources. Its management plan includes a program of ongoing ecotourism (FMB and USAID 2000).

Public Protected Areas

The protected public forest areas that lie within the Paraguayan Interior Atlantic Forest have several problems. First, they have insufficient ecological coverage and fail to adequately represent the country's biodiversity and its ecoregions. In addition, management classifications are inadequate, and those that exist are applied without technical criteria. Some areas fail to fulfill their ecological function because of this lack of technical criteria, most do not have legally defined boundaries, and, with two exceptions, administration of the current protected forest areas is significantly hampered by the presence of private property within the areas' jurisdictions and by a scarcity of reliable information on land tenure.

Because of the state's weak role in maintaining protected forest areas, many exist only on paper. Meanwhile, the private sector is working to fill the void by promoting the use of voluntary mechanisms for the protection, regeneration, and sustainable management of private lands, especially in the deteriorated ecosystem of the Interior Atlantic Forest.

Scientific Capacity

The development of technical and scientific knowledge should be a priority for the national government. Recently, the National Council for Science and Technology (CONACYT) was created to strengthen institutions and secure greater resource allocations. Its mission is to promote research in scientific and technological areas, along with its partner agency, the National Fund for Science and Technology. However, two years after its establishment, CONACYT has still not

Table 28.4. Analysis of Paraguay's science and technology under the National Council for Science and Technology (CONACYT) framework (2001, unpublished).

Strengths	Opportunities	Weaknesses	Threats
Low industrial development	Development of ecotourism and job creation	Lack of natural resource policies	Politicized technical positions
Abundance of water resources	New system of worker capacity with the private sector	Inapplicability of environmental laws	Corruption and impunity
Low population	Development of nontimber forest products	Imported and inadequate laws	Lack of transparency in technical regulations of Mercosur
Capable personnel	Production and export of certificated organic products	Little communication of sustainable development concept	Emigration of trained personnel
Good legal framework, broad and modern	Biodiversity assessment for sustainable use	Lack of laboratory facilities to control environmental parameters	Lack of detailed technical regulations
Abundance of natural resources and low pollution	Cooperatives for productive activities	Lack of integration of universities with society	
New Secretariat for the Environment with specific functions		Degradation of natural resources with little economic value	
		Scarce land-use planning activities	
		Lack of scientific background in the media	
		Insufficient protection to private property	
		Enforcement problems	
		Lack of human resources in biotechnology	
		Lack of sustainable research, little political and financial support	

formulated its national policy on science and technology, much less put it into effect in the country's high levels of government.

Scientific research is limited to a few public and private institutions and has not been systematized as a structural concept. Research is needed in a number of fields, but typically the response to this need has been to appropriate foreign tech-

Table 28.5. Priority actions that resulted from an analysis of strengths, opportunities, weaknesses, and threats.

Strengths	Promote an environmental legal framework in the Secretariat of the Environment (SEAM), including transparency in commercial negotiations.
	Promote civil society participation to demand that public offices include technical expertise and periodic assessment to decrease corruption and impunity.
	Generate jobs for scientists and technicians to recruit capable personnel and bring back Paraguayan scientists from abroad.
Opportunities	Develop ecological and rural tourism focusing on natural resources, especially water resources.
	Coordinate working capacity-building institutions in the new approach to education.
	Promote development of certified organic products using nontimber forest products.
	Support the new SEAM to prioritize national biodiversity assessments as the basis of sustainable development.
	Promote human resource capacity building using production cooperatives as models of sustainable use of natural resources.
Weaknesses	Promote the incorporation of national policies in regional and international treaties.
	Establish the concept of sustainable development by certifying organic natural resources to increase their value.
	Generate jobs of laboratory professionals to improve biotechnology research and production.
	Support communication of environmental issues through the media and the training of environmental communication professionals through alliances with research centers.
	Increase fundraising for natural resource research and management based on regional planning.
Threats	Establish professional training programs for public servants in environmental issues.
	Create national incentives for public servants.
	Improve capacity for public agencies to develop natural resource conservation activities.

nologies for use in the national setting. No attempt is made to standardize research results so that they can be shared with the target groups. A recent analysis points out the strengths, weaknesses, opportunities, and threats associated with the development of science and technology in Paraguay (Table 28.4) and the needed priority actions (Table 28.5).

One of Paraguay's national priorities is the dissemination and transfer of technologies for cleaning the environment (air, water, and soil). However, the knowledge produced by academic institutions is generally unavailable to public and private institutions. Moreover, no incentives are in place to encourage communication of scientific findings at the national or international level.

In addition to its shortcomings with regard to technology transfer, Paraguay also has very few institutions with biological collections. The three major collections include: the herbarium of the Botany Department in the School of Chemical Sciences; the Museum of Natural History, also known as the National Biological Inventory (Museo de Historia Natural del Paraguay 1996); and the museum of Itaipú Binational (Itaipú Binacional 1999b). Nonetheless, museum quality specimens are not completely lacking as many international institutions have specimens from Paraguay in their collections, including the Museum of Natural History in Geneva, the National Museum of Natural History in Paris, the Missouri Botanical Garden, the Botanical Institute of the Northeast in Corrientes (Argentina), the Herbarium of Kansas State University, the University of Kansas, and the Cooperative Wildlife Collection at Texas A&M University.

A database containing more than 13,000 records on Paraguayan botanical specimens at the herbarium of the Natural History Museum, London, dating back to the mid-nineteenth century, has recently been compiled. This database is part of the country's extensive historical legacy and constitutes a valuable repatriation of historical data (Peña-Chocarro et al. 1998).

The lack of a clearinghouse for information on the Interior Atlantic Forest makes access more difficult. To promote public education and interest, information should be readily accessible. The National Commission for Conservation of the Interior Atlantic Forest should organize an information center on conservation and sustainable use.

Environmental Education and Public Participation

The 1992 educational reform in Paraguay incorporated environmental education as one of the basic components of the curriculum and integrated it into other areas as well. Under this new approach, the environment is no longer a circumscribed subject within the natural sciences, nor is it separate and independent. The textbooks now recommended by the Ministry of Education and Culture introduce the subject of the Interior Atlantic Forest through problems that the student is asked to solve. This new approach is a theoretical and methodological challenge, particularly for the teacher, for it is necessary to cultivate new values and attitudes about the environment not just in the curriculum but in all aspects of student life (MEC 1997).

Environmental education in Paraguay is going through a period of change. At the same time, government institutions are updating their actions to better reflect and respond to the needs of society. In 1995 the FMB conducted the first national survey on environmental issues, a milestone in the country's history of environmental education (FMB 1998).

Several environmental education initiatives have taken place in the context of the Interior Atlantic Forest, including work on sustainable management of natural resources with communities in the upper Jejuí watershed by the Rural Extension Department of the FMB; agroforestry education for relocated landless *campesinos* (Project ALA 90-24), funded by the European Union (FMB 2000); informal education activities by the U.S. Peace Corps; and environmental education workshops by the organization Tierra Nuestra. Itaipú Binational has undertaken other promising programs and initiatives in the Granja Mamorei (Itaipú Binacional 1999a).

Almost all NGOs in Paraguay offer some form of both general education and education for improving the quality of life. Also, many other public agencies have environmental units that provide education on environmental topics (e.g., the Ministry of Public Works and Transportation, with its Bureau of Environmental Affairs).

Until recently Paraguay had no environmental graduate studies programs, but this situation has changed. Now Asunción National University's School of Graduate Academic Studies offers master's degree programs in political sciences, environmental sciences and public policy, preparation and evaluation of investment projects, natural sciences (biology), and environmental management. It also offers a program on environmental impact assessment. Private universities (Technical University of Marketing and Development, Universidad Americana) have also begun to offer both bachelor's and master's programs related to environmental sciences. No doctoral programs are offered in Paraguay; all Paraguayans who hold doctoral degrees have earned them abroad.

Conservation Funding

Compared with the Interior Atlantic Forest in Brazil (Mata Atlântica) and Argentina (Selva Paranaense), the Paraguayan Atlantic Forest has received little attention. Brazil and Argentina mobilized international resources and have even managed to get the subject of the threatened ecosystem incorporated into national policies. In contrast, Paraguay has only come to accept the Interior Atlantic Forest concept in the last four years, although already in 1994 the FMB was making references to the Atlantic Forest ecosystem in the country's eastern region. Much of the funding mobilized by the FMB has been for the Atlantic Forest: more than US$15 million brought into the country, not including intangibles or the human resources that have arrived to provide development assistance.

Among international institutions, the WWF probably has provided the most funding in recent years for a number of activities including wildlife research, environmental impact assessments, priority workshops, national action plan development, remnant forest map updating, international information campaigns, development of large-scale funding proposals, creation of a biological vision for the future, evaluation of degrees of implementation in protected areas of Brazil and Argentina, and support for promoting the certification of timber products. Recently the WWF opened an office in Paraguay, having previously worked through the FMB or with its support.

The Institute for Leadership in Sustainable Development promotes individual leadership for sustainable development with a view to supporting grassroots management of development projects. The institute recently opened an office in Paraguay, representing the AVINA Foundation (a Swiss NGO that promotes leadership development), and one of its priority areas is the Interior Atlantic Forest. Currently it supports and finances projects undertaken by individual leaders in such organizations as Guyra Paraguay, Natural Land Trust, the Farmers' Ecoforest Service, Tierra Nuestra, and the Institute of Environmental Law and Economics.

Other international institutions providing funding for the Interior Atlantic Forest are the Nature Conservancy, the United States Agency for International Development, the Canadian International Development Agency, German Technical Cooperation, the European Union, and Fonds Français pour l'Environnement Mondial. The Global Environment Facility and United Nations Development Programme has allocated US$135,000 for formulation of the National Biodiversity Strategy and Action Plan. In addition, the German government has committed US$3.8 million to help develop a National Strategy for the Protection of Natural Resources and to strengthen the Paraguayan government's capacity for environmental management. The Japan International Cooperation Agency is investing US$7 million in reforestation projects in the eastern region of Paraguay and specifically in the Interior Atlantic Forest. The World Bank is investing US$55 million to improve the country's agricultural sector, with US$2–3 million earmarked for conservation activities. The government of France has committed US$1.2 million to expand infrastructure for ecotourism in the Mbaracayú Forest Natural Reserve (Morales, pers. comm. 2001).

Environmental Legislation

Conservation efforts in Paraguay must confront the widespread corruption that continues to grip the country. Although Paraguay has a broad and up-to-date legal framework covering the environment, there has been little oversight or monitoring of compliance, and political will for change is still lacking. For example, 7 years after the Law 352 (1994) on Protected Forest Areas was created, it still has no enabling regulations (Table 28.6), nor are there any regulations for Law 96 on Paraguayan Wildlife, decreed in 1992.

It is hoped that the recent creation of the Secretariat for the Environment (2000) will lead to coordination between conflicting and overlapping environmental authorities. The new SEAM is responsible for enforcing the laws governing conservation of the Interior Atlantic Forest (Table 28.6). More laws are not needed until those that already exist are enforced.

Paraguay's organized civil society has been involved in national actions that have relevance for citizens. It participated in the Framework Accord on the Environment for Mercosur, also known as Florianópolis Accord, in an effort to facilitate negotiations under the draft protocol of amendment to the Treaty of Asunción on the subject of the environment. Unfortunately, the accord, though based on good intentions, does not appear to be binding, is difficult to implement, and has done little for some countries in the region to strengthen weak environmental

Table 28.6. Laws related to conservation of the Interior Atlantic Forest in Paraguay.

Law	Subject
Law No. 369/72	Creation of National Service on Environmental Health
Law No. 583/76	Approval of Convention on International Trade of Endangered Species
Law No. 42/90	Use, disposal, and transportation of toxic industrial waste
Law No. 112/91	Approving establishment of the Natural Reserve Mbaracayú forest and Jejuí river watershed (Nature Conservancy and Fundación Moisés Bertoni)
Law No. 61/92	Ozone layer protection
Law No. 96/92	Wildlife
Law No. 232/93	Water quality in Paraguay and Brazil
Law No. 251/93	Approval of Climatic Change Convention
Law No. 253/93	Approval of Biodiversity Convention
Law No. 294/93	Environmental impact assessment
Law No. 350/94	Approval of Convention in Internationally Important Wetlands
Law No. 352/94	Protected areas
Law No. 799/96	Fishing
Law No. 970/96	Approval of the Convention on Desertification
Law No. 1314/98	Approval of the Convention on Conservation of Migratory Wildlife

The Secretariat of the Environment also has the authority to enforce the following laws:

Law	Subject
Law No. 369/72	Creation of the Environmental Sanitation Service and modification No. 908/96
Law No. 422/73	Forestry rules and regulations
Law No. 904/81	Indigenous communities statutes and modifications No. 919/96
Laws No. 60/90 and No. 117/91	Capital investment and related regulations
Law No. 123/91	Phytosanitation protection
Law No. 198/93	Border health issues between Paraguay and Argentina
Law No. 231/93	Approving Convention 169 on the independence of indigenous peoples adopted during the 76th International Conference in Geneva (June 7, 1989)
Law No. 751/95	Approving cooperation agreement against illegal timber traffic
Law No. 1344/98	Consumer and user protection rules and regulations

Table 28.7. Institutional capacity indicators for the Interior Atlantic Forest in Paraguay.

Changes in the Execution Organization	
Attainment of mission and objectives	Degree of attainment of mission, effective accomplishment within deadlines
Efficiency in the use of resources	Costs by client, management efficiency in financial resource use, donor information, financial audits, financial and management analysis
Participation of society and key actors	Social satisfaction; changes and trends among clients, competitors, donors; assessment of trends in programs and services related to the client needs
Evolution of financial viability	Number of new donors, increase in financial capacity, financial self-sufficiency to execute programs and projects
Changes in Execution Capacity	
Strategic orientation	Definition, existence, and use of strategies to guide the organization, identification of organization's niche, key roles and responsibilities
Structure and restructuring	Well-defined and clear structure, responsibility identification, planning and monitoring of processes that allow programs to function effectively
Financial system management	Increase in financial capacity, transparency, and accounting of invested resources, both capital and noncapital
Human resource management	Training programs, clear decision-making processes, and appropriate links and structures
Infrastructure access	Basic infrastructure available, plans for future needs
Management of organizational processes	Standardized working processes, appropriately motivated personnel, clear indicators of team and individual motivation
Organizational program and work management	Clear alliances and coalitions
Links with key agents	Level of participation of society in coalitions
Information management and use	Updated information and efficient information use
Changes in Motivation That Influence Execution	
Improvements in organizational culture	Institutional vision and mission present and accepted, definition of organizational principles and, clear definition of organization's clients
Improvements in internal incentive systems	Functional systems to support institutional values, incentive systems in accordance with institutional priorities
Ability to move the organization forward	Organization flexibility and history supporting changes and innovation
Mission improvement and vision to improve capacity	Definition and updating of principles and values to guide the mission's fulfillment

Table 28.7. Continued

Understanding of the Organizational and Work Sphere	
Administration and legal aspects	Appropriate national and regional legal framework
Political systems	Political resources to support action
Economic, social, and cultural systems	Civil society values in accordance with conservation objectives
Incentive systems	Market incentives provided for the workers, relationship between market orientation and institutional priorities
Market and technology systems	Understanding of international, national, and regional market trends and development infrastructure

structures. Furthermore, it fails to cover transfrontier environmental impacts, the adoption of informed decisions, and consultation processes for large infrastructure projects affecting the region (Vera and Yanosky 1998).

The Institute of Environmental Law and Economics has expressed concern about the Framework Accord's lack of clarity and potential efficacy and has therefore proposed that revised measures be developed to ensure that this legal instrument will respond to the environmental needs of the region (IDEA 1999).

Indicators and Capacity

The prevailing trend toward deforestation in Paraguay reflects a lack of conservation capacity and ineffective national public and transfrontier policies. Although the Paraguayan government appears to have assigned high priority to the preservation and sustainable use of the nation's rich national biodiversity and had created a legal framework for protected areas, deforestation continues even in the priority conservation areas. For example, the future San Rafael National Park has already been deforested from an original area of 780 km^2 to 580 km^2, which includes 340 km^2 of deteriorated forest (MAG et al. 2000).

Previous discussions here all underscore how slow Paraguay has been to develop and implement a national policy for conservation of the Interior Atlantic Forest. Notably, all funding for this purpose has come from foreign sources.

Institutions are made up of people, and it is these people who bring about change and who, through their priorities and actions, are responsible for the current situation. Therefore, indicators must be developed that make it possible to evaluate capacity at the individual level and that can be corroborated with other indicators to comprehensively evaluate societal trends with regard to conservation issues (Table 28.7).

Four critical aspects of organizations that should be periodically evaluated are changes in organizational execution, changes in execution capacity, changes in organizational motivation that affect execution, and level of understanding and interaction with the environment in which the organization works (Adrien

Table 28.8. Log frame analysis for conservation capacity in the Interior Atlantic Forest of Paraguay.

Capacity Objective	Baseline Indicators	Process Indicators	Outcome Indicators	Concrete Results
System indicators	Financial resources that support biodiversity research	Agreements and fundraising efforts to promote changes in environmental tax procedures	Resource use policies and regulations	National support for the conservation of threatened species
Organization indicators	Available equipment to obtain updated information	Organization leadership provides incentives to accomplish established goals	Organization with updated information for decision making	Increased capacity to update and provide information on national biodiversity
Individual indicators	Professional training	Training in latest technologies motivates learning and efficient problem solving	Amount of training and individual performance on the training topics	Results obtained through improved capacity such as environmental conflict resolution

and Morgan 2000). For each of these four parameters there are indicators that can be improved at the national level. The three levels at which capacity must be measured—the individual, the institution, and society—can be assessed by applying log frame analysis to conservation of the Atlantic Forest (Table 28.8).

Recommendations

The long-standing record of incapacity with regard to conservation of the Atlantic Forest means that more investment must be made in capacity and experience. It is of the utmost importance to create opportunities for continuing education and capacity building.

When the goal is to improve a government agency's ability help protect a given area, priority should be given to enlisting broad participation and developing new ways of thinking about problems and solutions.

The numerous programs and projects directed at the Paraguayan Interior Atlantic Forest have had little impact. None of these programs has seriously proposed long-term multi-institutional collaboration as its basic approach to designing, managing, monitoring, and evaluating conservation efforts. Since deforestation is proceeding rapidly and the impacts of policies take years to show effects, all involved need to work with an awareness of the urgency of the situation in the Atlantic Forest.

The indicators that provide a regional overview of the individual, institutions, and systems must be defined and put into effect during monitoring. Some organizations, such as Alter Vida and the FMB, have begun to adopt the indicators mentioned in this chapter to evaluate the actions and impact of organizations. However, this practice is only beginning, and the political will to introduce it to government agencies is still lacking.

Acknowledgments

We would like to thank Claudia Mercolli, Ana María Macedo, and Lucía Bartrina for their assistance in preparing this chapter.

References

Adrien, H. and Morgan, P. 2000. *Integrating capacity development into project design and evaluation: approach and frameworks. Monitoring and evaluation.* Working Paper 5. Washington, DC: Global Environment Facility.

Bragayrac, E., Sosa, W., and Rivarola, N. 1998. Informa Nacional. Sistema Nacional de Areas Silvestres Protegidas del Paraguay. SINASIP. Dirección de Parques Nacionales y Vida Silvestre, Subsecretaría de Estado de Recursos Naturales y Medio Ambiente, Ministerio de Agricultura, Asunción, 41 pp.

BSP (Biodiversity Support Program), CI (Conservation International), TNC (The Nature Conservancy), WCS (Wildlife Conservation Society), WRI (World Resources Institute), and WWF (World Wildlife Fund). 1995. *A Regional analysis of geographic priorities for biodiversity conservation in Latin America and the Caribbean.* Washington, DC: Biodiversity Support Program.

Diario Noticias. 2001. Descubren un cargamento de tóxico en Alto Paraná. February 5, 2001: 49.

DPNVS (Departamento de Parques Nacionales y Vida Silvestre). 2000. *Taller nacional sobre las causas subyacentes de deforestación y la degradación de los bosques en Paraguay.* Asunción: DPNVS, Ministerio de Agricultura y Ganadería, and Sobrevivencia, Amigos de la Tierra Paraguay, with support from UNDP.

FMB (Fundación Moisés Bertoni). 1994. *Proyecto trinacional del Bosque Atlántico Interior. 1ra etapa: diagnóstico de los recursos socio-ambientales, Paraguay.* Vols. 1 and 2. Asunción: FMB.

FMB (Fundación Moisés Bertoni). 1998. *Encuesta de información y opinión sobre el medio ambiente.* Asunción: FMB.

FMB (Fundación Moisés Bertoni). 2000. *Propuesta reserva de la biosfera: Bosque Mbaracayú.* Asunción: FMB.

FMB (Fundación Moisés Bertoni). 2001. *Estudio estratégico para la determinación de biomasa y carbono almacenado en la Reserva Natural del Bosque Mbaracayú y diseñado de un sistema de monitoreo.* Asunción: FMB.

FMB (Fundación Moisés Bertoni) and USAID (United States Agency for International Development). 2000. *Programa de apoyo a iniciativas privadas de conservación: una revisión de 10 años de experiencias.* Asunción: FMB and USAID.

Freaza, M. A. 1996. *Política de descentralización económica y administrativa.* Asunción: Ediciones y Arte.

IDEA (Institute of Environmental Law and Economics). 1999. *Guía de derecho ambiental del Paraguay.* Asunción: GG Servicios Gráficos.

Itaipú Binacional. 1999a. *Memoria anual, medio ambiente.* Hernandarias, Paraguay: Itaipú Binacional, Dirección de Coordinación/Superintendencia de Medio Ambiente.

Itaipú Binacional. 1999b. *Plan de manejo conceptual y operativo de las áreas silvestres protegidas 1999–2002.* Hernandarias, Paraguay: Itaipú Binacional, Dirección de Coordinación Ejecutiva/Superintendencia de Medio Ambiente.

Keel, S., Gentry, A. H., and Spinzi, L. 1993. Using vegetation analysis to facilitate the selection of conservation sites in eastern Paraguay. *Conservation Biology* 7: 66–75.

Lowen, J. C., Bartrina, L., Clay, R. P., and Tobias, J. A. 1996. *Biological surveys and conservation priorities in eastern Paraguay.* Cambridge, UK: CSB Conservation Publications.

Madroño, N. A. and Esquivel, E. Z. 1995. Reserva Natural de Mbaracayú: su importancia en la conservación de aves amenazadas, cuasi-amenazadas y endémicas del Bosque Atlántico Interior. *Cotinga* 4: 52–57.

MAG (Ministerio de Agricultura y Ganadería), UNA (Universidad Nacional de Asunción), FMB (Fundación Moisés Bertoni), and WWF (World Wildlife Fund). 2000. *Mapa de remanentes boscosos del BAI en Paraguay.* Asunción: MAG, UNA, FMB, and WWF.

Museo de Historia Natural del Paraguay. 1996. *Colecciones de flora y fauna del Museo Nacional de Historia Natural del Paraguay.* Asunción: Museo de Historia Natural del Paraguay.

Peña-Chocarro, M., Knapp, S., Jimenez, B., and Marín, G. 1998. Database: *Especimenes botánicos de Paraguay en el Museo de Historia Natural de Londres.* Asunción: Fundación Moisés Bertoni; London: Darwin Initiative, Natural History Museum.

SINASIP (Sistema Nacional de áreas Silvestres Protegidas). 1993. *Plan estratégico del sistema nacional de áreas silvestres protegidas.* Asunción: Dirección de Parques Nacionales y Vida Silvestre, Subsecretaría de Estado de Recursos Naturales y Medio Ambiente, Ministerio de Agricultura y Ganadería, Fundación Moisés Bertoni.

UICN (Unión Internacional para la Conservación de la Naturaleza). 2000. *Memorias VIII Reunión del Comité Regional Sudamericano de la Unión Mundial para la Naturaleza.* San Bernardino, Paraguay: Unión Internacional para la Conservación de la Naturaleza.

Vera, R. and Yanosky, A. A. 1998. Mercosur y medio ambiente en Paraguay. In: Blanco, H. and Borregarrd, N. (eds.). *Mercosur y medio ambiente.* pp. 142–149. Santiago: Publicaciones CIPMA.

Yanosky, A. A. and Escalante, A. 2000. Las reservas naturales privadas del Paraguay: asistencia al mantenimiento de la diversidad biológica Paraguaya. In: Cabrera, E., Mercolli, C., and Resquin, R. (eds.). *Manejo de fauna silvestre en Amazonía y Latinoamérica.* pp. 139–151. Asunción: CITES-PY, Fundación Moisés Bertoni and University of Florida.

PART V

Trinational Issues

Chapter 29

Dynamics of Biodiversity Loss: An Introduction to Trinational Issues

Thomas R. Jacobsen

The diverse ecosystems that make up the Atlantic Forest extend across the political boundaries of Brazil, Paraguay, and Argentina. Each of these modern nations has evolved from its own unique and rich history. Consequently, although these nations are becoming a part of today's modern, global society, they each still maintain distinct social, political, economic, and cultural realities. Effective conservation in the Atlantic Forest therefore entails exploring the links between these countries so that solutions to the problems they share overcome the differences that divide them. The chapters in this section identify components that are key to establishing these links.

One component that is critical to developing effective conservation measures in the Atlantic Forest is the basic task of assessing biodiversity loss. Alarmingly, habitat loss in this region is leading to mass species extinction. In Chapter 30, Brooks and Rylands examine the 28 species that are near extinction in the Atlantic Forests. Thankfully, the extinction of these species is not a foregone conclusion: it can be prevented if existing protected areas are managed effectively and new protected areas are selected carefully and created quickly.

The biodiversity of the Atlantic Forest once occupied a large, continuous ecosystem. Today, the region's flora and fauna are dispersed throughout hundreds of small, isolated forest fragments. Whereas some species need only a few hectares for survival, many others need several thousand square kilometers to achieve viable populations. In Chapter 31, Galindo-Leal explores the ecological dynamics of fragmentation and outlines the conservation measures necessary for maintaining the ecological processes that support biodiversity in these complex ecosystems.

In addition to its unique biodiversity, the Atlantic Forest is characterized by its unique cultural diversity. Throughout time, human societies have depended on these forests for their livelihoods and have developed an intimate knowledge of them. However, just like the region's biodiversity, the native cultures of the Atlantic Forest are highly threatened. In Chapter 32, Jacobsen provides an overview of the indigenous populations living in the region and underscores the importance of including these peoples and their knowledge of resource use in plans for the region's conservation.

Humans have affected the Atlantic Forest in many ways. Today, people continue to introduce plant and animal species that have far-reaching impacts on the biotic structure of the region. In Chapter 33, Reaser and colleagues discuss how the modern context of international trade, travel, and transport has promoted the spread of invasive alien species. Drawing on select examples, they review the biological, economic, and health impacts of invasive alien species in the Atlantic Forest and recommend five priorities for ameliorating their effect.

The daily need for income and subsistence among people living in the Atlantic Forest has placed unsustainable demands on biological resources. Effective conservation therefore should encourage sustainable use of natural resources over the long term. Chediack and Franco confront the issue of sustainability in Chapter 34, where they examine harvesting regimes of the *palmito,* or heart palm, and highlight the ecological effects and sustainability of the harvest. They argue that a well-planned forest management scheme that includes *palmito* extraction will contribute to conservation by helping to alleviate rural poverty and improving the link between people and their natural environment.

Humans in the Atlantic Forest also exert a great demand for energy. As a result, hundreds of dams have been established over the last 50 years, at a high social and environmental cost. In Chapter 35, Fahey and Langhammer review how dams have caused human populations to be displaced and resettled and to become ridden with disease. They have also caused the Atlantic Forest's biological communities to suffer from such factors as habitat loss, impeded migration, and invasive species. Despite these enormous negative impacts, dams are still being constructed in Brazil, Paraguay, and Argentina to meet seemingly insatiable demands for energy. Therefore, it is vital that alternative sources of energy be adopted.

As in all other parts of the world, human populations are growing in Brazil, Paraguay, and Argentina. Brazil, for instance, is the fifth most populous country in the world, and it contains 2 of the world's 15 largest cities. In Chapter 36, Jacobsen demonstrates that growth is particularly high in the Atlantic Forest of Paraguay and in the Argentinian province of Misiones. Because these areas also contain significant amounts of intact, continuous forest, they have high conservation priority. However, Jacobsen notes that although population density is a threat to biodiversity in the Atlantic Forest, it is not as great a threat as large-scale timber, mining, and farming interests.

The causes of biodiversity loss are not always local. International corporations, market demands, and trade often are responsible for habitat loss and species endangerment. International trade agreements, for instance, give low priority to environmental issues and consequently can be extremely harmful to biodiversity. To address this problem Leichner (Chapter 37) underscores the need to implement dynamic environmental policy instruments based on strong institutional structure.

Effective measures aimed at establishing protected areas are needed to safeguard the biodiversity of the Atlantic Forest. In Chapter 38, Lairana surveys the state of protection of the Atlantic Forest hotspot, noting that approximately 676 protected areas exist in the region, with highly variable effectiveness. Although this may seem like a high number of protected areas, about 30 percent are less than 1 km^2, and only 3 percent are larger than 1,000 km^2. Furthermore, fewer than 20 percent of the Atlantic Forest's protected areas are strictly protected. Therefore, effective conservation mechanisms must be implemented.

Chapter 30

Species on the Brink: Critically Endangered Terrestrial Vertebrates

Thomas Brooks and Anthony B. Rylands

Only 7.5 percent of the historical extent of the Atlantic Forests of Brazil, Argentina, and Paraguay remains (Myers et al. 2000; Chapters 5, 15, and 25, this volume). As a result many researchers expect a large number of species inhabiting these forests to have become extinct or to be threatened with imminent extinction (Brooks and Balmford 1996). The World Conservation Union (IUCN) Red Lists (Hilton-Taylor 2000; Table 30.1) bear out this prediction: the Atlantic Forest hotspot holds no fewer than 28 species of the terrestrial vertebrates that are listed as critically endangered, the most serious category of threat, indicating a 50 percent probability of extinction within a decade (Table 30.2). Half of these are birds (BirdLife International 2000), eight are primates (Rylands et al. 1996, 1997), and another four are frogs (Frost 1985). Only the Caribbean hotspot has more critically endangered vertebrates (Hilton-Taylor 2000).

That so many Atlantic Forest species are teetering on the brink of extinction sends two messages. The first is a gloomy one: the destruction of this hotspot's habitat is leading to a mass extinction. The second, however, is positive: these species may be critically endangered, but they are not yet extinct. With a concerted effort, it is still possible to achieve a goal of zero extinction in the Atlantic Forest.

The extreme northeast probably is the most threatened portion of the Atlantic Forest, with less than 5 percent of the original extent of the forests of Pernambuco, Alagoas, and Sergipe surviving (CI do Brasil et al. 2000) (Figure 30.1).

Tiny forest fragments in these three states are home to at least eight critically endangered terrestrial vertebrates. Tragically, one of these has now become extinct in the wild: the Alagoas curassow (*Mitu mitu*) was last seen in the mid-1980s and now survives only in a captive population in Rio de Janeiro (Teixeira 1986). A second, the recently discovered monkey Coimbra-Filho's titi (*Callicebus coimbrai*), persists only in six tiny forest fragments in Sergipe (Kobayashi and Langguth 1999) (Figure 30.1). Remarkably, the region's other six critically endangered species, all birds, occur in a single 15-km^2 forest—Murici, in Alagoas—which is

Table 30.1. Species of terrestrial vertebrates on the Red List for the Atlantic Forest hotspot (Hilton-Taylor 2000).

Category	Class	Common Name	Scientific Name	Category	Class	Common Name	Scientific Name
Extinct in the wild	Birds	Alagoas curassow	*Mitu mitu*	Vulnerable	Amphibians	Sazima	*Hylodes sazimai*
					Reptiles	Brazilian snake-necked turtle	*Hydromedusa maximiliani*
Critically endangered	Amphibians	Itatiaia highland frog	*Holoaden bradei*			Lizard	*Liolaemus occipitalis*
		Lutz's rapids frog	*Paratelmatobius lutzii*			Piraja's lancehead	*Bothrops pirajai*
		River frog	*Thoropa lutzi*			Snake	*Liophis atraventer*
		River frog	*Thoropa petropolitana*			Tree lizard	*Anisolepis undulates*
	Reptiles	Golden lancehead	*Bothrops insularis*			Tropical sand lizard	*Liolaemus lutzae*
	Birds	Alagoas antwren	*Myrmotherula snowi*		Birds	Black-backed tanager	*Tangara peruviana*
		Alagoas foliage-gleaner	*Philydor novaesi*			Black-capped manakin	*Piprites pileatus*
		Alagoas tyrannulet	*Phylloscartes ceciliae*			Black-fronted piping-guan	*Pipile jacutinga*
		Araripe manakin	*Antilophia bokermanni*			Black-headed berryeater	*Carpornis melanocephalus*
		Bahia tapaculo	*Scytalopus psychopompus*			Black-legged dacnis	*Dacnis nigripes*
		Brazilian merganser	*Mergus octosetaceus*			Blue-bellied parrot	*Triclaria malachitacea*
		Cherry-throated tanager	*Nemosia rourei*			Blue-chested parakeet	*Pyrrhura cruentata*
		Forbes's blackbird	*Curaeus forbesi*			Blue-winged macaw	*Propyrrhura maracana*
		Fringe-backed fire-eye	*Pyriglena atra*			Buff-breasted tody-tyrant	*Hemitriccus mirandae*
		Kinglet calyptura	*Calyptura cristata*			Buffy-fronted seedeater	*Sporophila falcirostris*

(Continued)

Table 30.1. Continued

Category	Class	Common Name	Scientific Name
Critically endangered	Birds	Plain spinetail	*Synallaxis infuscate*
		Rio de Janeiro antwren	*Myrmotherula fluminensis*
		Stresemann's bristlefront	*Merulaxis stresemanni*
		White-collared Kite	*Leptodon forbesi*
	Mammals	Black-faced lion tamarin	*Leontopithecus caissara*
		Brazilian arboreal mouse	*Rhagomys rufescens*
		Coimbra's titi	*Callicebus coimbrai*
		Golden lion tamarin	*Leontopithecus rosalia*
		Golden-rumped lion tamarin	*Leontopithecus chrysopygus*
		Muriquí	*Brachyteles arachnoides*
		Northern Bahian blond titi	*Callicebus barbarabrownae*
		Northern muriquí	*Brachyteles hypoxanthus*
		Yellow-breasted capuchin	*Cebus xanthosternos*
Endangered	Amphibian	Maldonada redbelly toad	*Melanophryniscus moreirae*
	Reptiles	Hoge's sideneck turtle	*Phrynops hogei*

Category	Class	Common Name	Scientific Name
Vulnerable	Birds	Canebrake groundcreeper	*Clibanornis dendrocolaptoides*
		Cinnamon-vented piha	*Lipaugus lanioides*
		Grey-winged cotinga	*Tijuca condita*
		Helmeted woodpecker	*Dryocopus galeatus*
		Narrow-billed antwren	*Formicivora iheringi*
		Pink-legged graveteiro	*Acrobatornis fonsecai*
		Plumbeous antvireo	*Dysithamnus plumbeus*
		Red-spectacled Amazon	*Amazona pretrei*
		Restinga tyrannulet	*Phylloscartes kronei*
		Salvadori's antwren	*Myrmotherula minor*
		Sao Paulo tyrannulet	*Phylloscartes paulistus*
		Shrike-like cotinga	*Laniisoma elegans*
		Temminck's seedeater	*Sporophila frontalis*
		Unicolored antwren	*Myrmotherula unicolor*
		White-bearded antshrike	*Biatas nigropectus*
		White-necked hawk	*Leucopternis lacernulata*

Group	Common name	Scientific name
Birds	Atlantic royal flycatcher	*Onychorhynchus swainsoni*
	Bahia spinetail	*Synallaxis whitneyi*
	Bahia tyrannulet	*Phylloscartes beckeri*
	Banded cotinga	*Cotinga maculate*
	Band-tailed antwren	*Myrmotherula urosticta*
	Black-hooded antwren	*Formicivora erythronotos*
	Brown-backed parrotlet	*Touit melanonota*
	Buff-throated purpletuft	*Iodopleura pipra*
	Fork-tailed pygmy-tyrant	*Hemitriccus furcatus*
	Golden-tailed parrotlet	*Touit surda*
	Hook-billed hermit	*Glaucis dohrnii*
	Kaempfer's tody-tyrant	*Hemitriccus kaempferi*
	Marsh antwren	*Stymphalornis acutirostris*
	Orange-bellied antwren	*Terenura sicki*
	Purple-winged ground-dove	*Claravis godefrida*
	Red-billed curassow	*Crax blumenbachii*
	Red-browned Amazon	*Amazona rhodocorytha*
	Red-tailed Amazon	*Amazona brasiliensis*
	Restinga antwren	*Formicivora littoralis*
	Scalloped antbird	*Myrmeciza ruficauda*

Group	Common name	Scientific name
Mammals	Atlantic giant tree rat	*Echimys thomasi*
	Bishop's fossorial spiny rat	*Clyomys bishopi*
	Black-fronted titi	*Callicebus nigrifrons*
	Bokermann's nectar bat	*Lonchophylla bokermanni*
	Brazilian big-eyed bat	*Chiroderma doriae*
	Brazilian three-banded armadillo	*Tolypeutes tricinctus*
	Brown howling monkey	*Alouatta guariba*
	Chestnut-striped short-tailed opossum	*Monodelphis rubida*
	Geoffroy's tufted-ear marmoset	*Callithrix geoffroyi*
	Hairy-tailed bat	*Lasiurus ebenus*
	Long-nosed short-tailed opossum	*Monodelphis scalops*
	Masked titi	*Callicebus personatus*
	Red myotis	*Myotis ruber*
	Shrewish short-tailed opossum	*Monodelphis sorex*
	Southern Bahian masked titi	*Callicebus melanochir*
	Southern three-stripped opossum	*Monodelphis unistriata*
	Strange big-eared brown bat	*Histiotus alienus*
	Theresa's short-tailed opossum	*Monodelphis theresa*
	Thin-spined porcupine	*Chaetomys subspinosus*
	White-lined bat species	*Platyrrhinus recifinus*
	Ruschi's Rat	*Abrawayaomys ruschii*
	Maned Sloth	*Bradypus torquatus*
	Buffy-tufted-ear Marmoset	*Callithrix aurita*
	Buffy-headed Marmoset	*Callithrix flaviceps*
	Golden-headed Lion Tamarin	*Leontopithecus chrysomelas*
	Rio de Janeiro Arboreal Rat	*Phaenomys ferrugineus*

Table 30.2. Critically endangered species in the Atlantic Forest hotspot. The Alagoas curassow is extinct in the wild.

Class	Common Name	Species	State
Amphibians	Lutz's river frog	*Thoropa lutzi*	Rio de Janeiro
	Lutz's rapids frog	*Paratelmatobius lutzii*	Rio de Janeiro
	Itatiaia highland frog	*Holoaden bradei*	Rio de Janeiro
	Petropolis river frog	*Thoropa petropolitana*	Rio de Janeiro
Reptiles	Golden lancehead	*Bothrops insularis*	São Paulo
Birds	Alagoas curassow	*Mitu mitu*	Pernambuco, Alagoas, Sergipe
	Foliage-gleaner	*Philydor noveasi*	Alagoas
	Alagoas antwren	*Myrmotherula snowi*	Alagoas
	White-collared kite	*Leptodon forbesi*	Alagoas
	Plain spinetail	*Synallaxis infuscata*	Alagoas
	Alagoas tyrannulet	*Phylloscartes cecilae*	Alagoas
	Forbes's blackbird	*Curaeus forbesi*	Alagoas, Minas Gerais
	Stresemann's bristlefront	*Merulaxis stresemanni*	Bahia
	Bahia tapaculo	*Scytalopus psychopompus*	Bahia
	Fringe-backed fire-eye	*Pyriglena atra*	Bahia, Sergipe
	Cherry-throated tanager	*Nemosia rourei*	Minas Gerais, Espírito Santo
	Forbes's blackbird	*Curaeus forbesi*	Minas Gerais
	Rio de Janeiro antwren	*Myrmotherula fluminensis*	Rio de Janeiro
	Kinglet cotinga	*Calyptura cristata*	Rio de Janeiro
	Brazilian merganser	*Mergus octosetaceus*	Rio de Janeiro, São Paulo, Santa Catarina
Mammals	Coimbra-Filho's titi	*Callicebus coimbrai*	Sergipe
	Northern Bahian blond titi	*Callicebus barbarabrownae*	Bahia, Sergipe
	Yellow-breasted capuchin	*Cebus xanthosternos*	Bahia
	Northern muriquí	*Brachyteles hypoxanthus*	Bahia, Minas Gerais, Espírito Santo
	Black-faced lion tamarin	*Leontopithecus caissara*	Paraná, São Paulo
	Black lion tamarin	*Leontopithecus chrysopygus*	São Paulo
	Golden lion tamarin	*Leontopithecus rosalia*	Rio de Janeiro
	Southern muriqui	*Brachyteles arachnoides*	São Paulo
	Brazilian arboreal mouse	*Rhagomys rufescens*	Rio de Janeiro

probably therefore the single most important forest in the entire world for bird conservation and was federally decreed by the president as an Ecological Station on May 29, 2001 (Figure 30.1). Two of these species, the Alagoas foliage-gleaner (*Philydor noveasi*) (Teixeira and Gonzaga 1983) and Alagoas antwren (*Myrmotherula snowi*) (Teixeira and Gonzaga 1985), are endemic to Murici, although they may yet be found in forests in the nearby Usina Serra Grande. The white-collared kite (*Leptodon forbesi*) (Teixeira et al. 1987), plain spinetail (*Synallaxis infuscata*) (Pinto 1950), and Alagoas tyrannulet (*Phylloscartes cecilae*) (Teixeira 1987) historically occurred a little more widely, but recent records are largely confined to Alagoas. Finally, the little-known and enigmatic Forbes's blackbird (*Curaeus forbesi*) survives in three populations totaling approximately 150 birds in

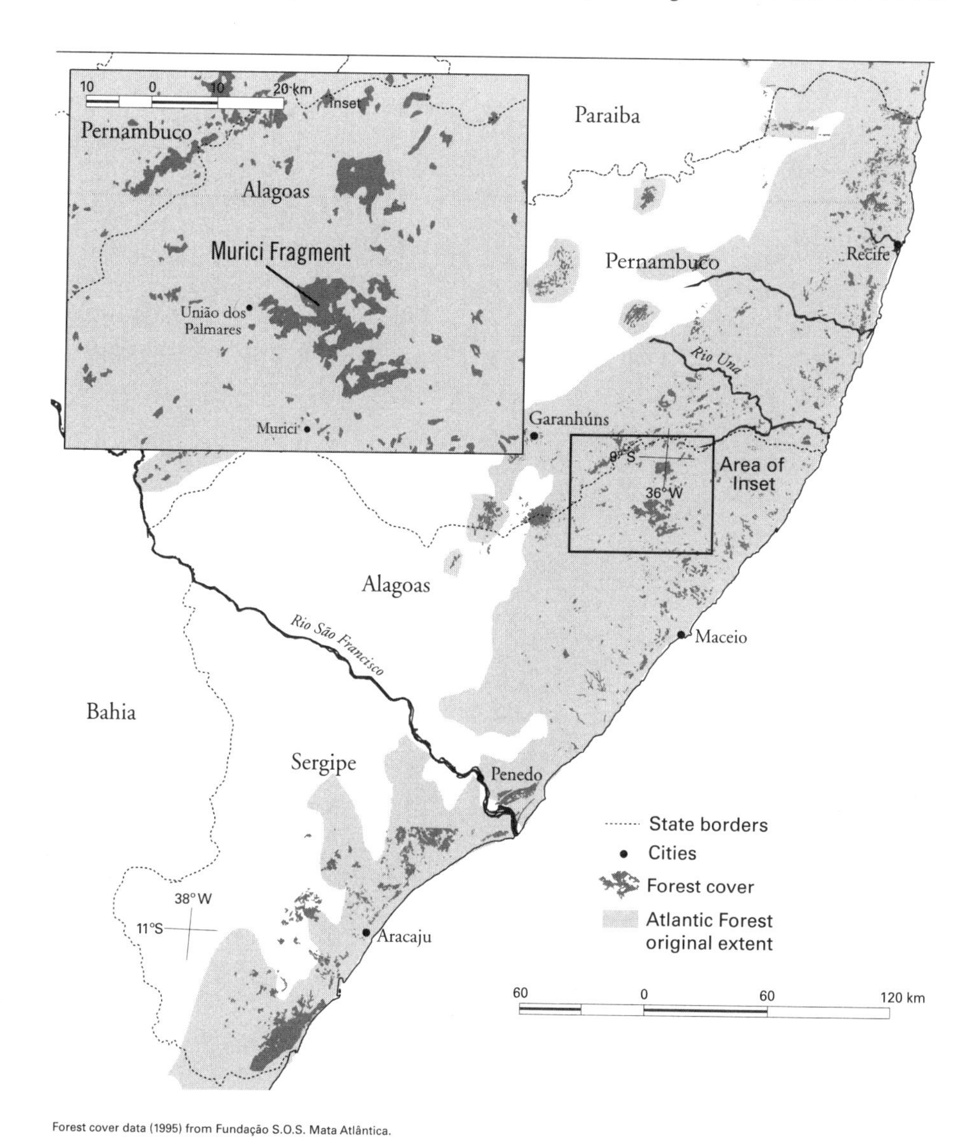

Forest cover data (1995) from Fundação S.O.S. Mata Atlântica.

Figure 30.1. Tiny forest fragments in the Brazilian states of Pernambuco, Alagoas, and Sergipe are home to at least eight critically endangered terrestrial vertebrates.

Alagoas, separated from the two sites at which it occurs in Minas Gerais by 1,400 km; historically it also occurred in Pernambuco (Short and Parkes 1979).

The Atlantic slope of Bahia, south into Espírito Santo and inland into Minas Gerais, retains just over 5 percent of its original forest cover (CI do Brasil et al. 2000), and as a consequence also holds numerous critically endangered birds and primates (Figure 30.2).

Two bird species are now almost completely unknown, each recorded historically from just two sites between Salvador and Ilhéus in Bahia: Stresemann's bristlefront (*Merulaxis stresemanni*) (Sick 1960) and Bahia tapaculo (*Scytalopus psychopompus*) (Teixeira and Carnevalli 1989). The scarce fringe-backed fire-eye

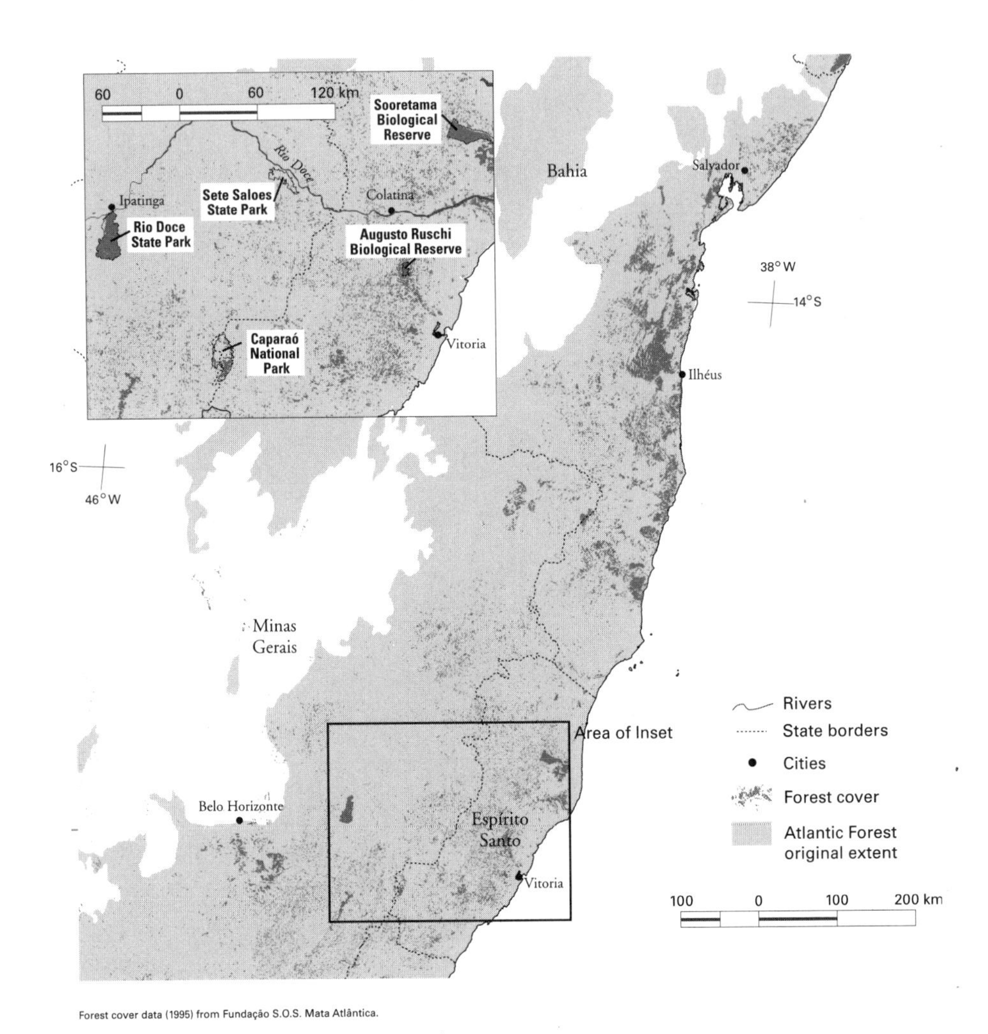

Figure 30.2. Forest fragments in the Atlantic slope of Bahia, Espírito Santo, and Minas Gerais, Brazil, hold numerous critically endangered species.

(*Pyriglena atra*) occurs in the same region of lowland forest near Salvador, Bahia (Willis and Oniki 1982) and has also recently been discovered in southern Sergipe (Pacheco and Whitney 1995), and the northern Bahian blond titi (*Callicebus barbarabrownae*) is found in the same area of Sergipe (Marinho-Filho and Veríssimo 1997). Further south, the yellow-breasted capuchin (*Cebus xanthosternos*) is endemic to the southern half of Bahia, where its populations are severely fragmented and hunted (Coimbra-Filho et al. 1992). The cherry-throated tanager (*Nemosia rourei*) is known from only five sites in Minas Gerais and Espírito Santo, with definite recent observations from one site only, Fazenda Pindobas IV (Bauer et al. 2000), and two populations of Forbes's blackbird (*Curaeus forbesi*) survive in Minas Gerais (Studer and Vielliard 1988). Finally, the northern muriqui (*Brachyteles hypoxanthus*), recently separated from its sister taxa in the Serra do Mar (Lemos de Sá et al. 1990), occurs slightly more widely in Bahia, at least historically (Santos et al. 1987), and in Minas Gerais (Stallings and Robinson 1991) and Espírito Santo (Mendes and Chiarello 1993). Most of the populations surviving today are limited to relict forests in remote montane regions, often in small private reserves (Strier and Fonseca 1996–1997). Even in larger reserves such as Caparaó National Park, Augusto Ruschi Biological Reserve, and Rio Doce State Park, northern muriqui are extremely scarce (Strier and Fonseca 1996–1997).

The largest extent of remaining Atlantic Forest survives in the Serra do Mar of São Paulo, extending west into Paraná and east into Rio de Janeiro (CI do Brasil et al. 2000). Nevertheless, at least 12 critically endangered terrestrial vertebrate species are endemic to the region (Figure 30.3). The best known of these are undoubtedly the endemic lion tamarins (*Leontopithecus*), divided into four species (Rosenberger and Coimbra-Filho 1994). The golden-headed lion tamarin (*L. chrysomelas*) occurs in southern Bahia (Pinto and Rylands 1997) and is classified as endangered. The remaining three tamarins are categorized as critically endangered and live in only three minute ranges, limited to very few forest fragments in southeast Brazil. The black-faced lion tamarin (*L. caissara*) occurs in coastal lowland and sandy soil restinga forests in northeast Paraná and the extreme southeast of São Paulo (Persson and Lorini 1993). The black lion tamarin (*L. chrysopygus*) inhabits remnant forests of the interior plateau of São Paulo (Valladares-Padua and Cullen 1994), and the golden lion tamarin (*L. rosalia*) is found only in coastal lowland forest fragments in Rio de Janeiro (Coimbra-Filho 1969). Another of the region's conservation flagships is the southern muriqui (*Brachyteles arachnoides*), which occurs in a number of protected forests in São Paulo state (Martuscelli et al. 1994) but is under heavy hunting pressure throughout (Mittermeier et al. 1987). Rio de Janeiro has one other endemic, critically endangered mammal: the Brazilian arboreal mouse (*Rhagomys rufescens*), known from only two century-old specimens (Fonseca et al. 1994).

Besides these mammals, the state of Rio de Janeiro also holds two critically endangered birds and four frogs. Astonishingly, both bird species were rediscovered in 1996, allaying fears that they had already become extinct, although they have not been recorded since. The first, the Rio de Janeiro antwren (*Myrmotherula fluminensis*), was only known from a single 1982 specimen from Santo Aleixo, Majé, which has since been deforested, until it was reported in the Serra do Mar (now Guapi-Açu) Ecological Reserve (Knapp 1997). Even more dramatic, the kinglet cotinga (*Calyptura cristata*) was fairly widespread in the nineteenth century in the foothills of Rio de Janeiro but was not recorded for more than a century before its rediscovery in the Serra dos Órgãos National Park (Gonzaga 1997). Of the amphibians, two frog species are endemic to the high elevations of Alto Itatiaia: Itatiaia highland frog (*Holoaden bradei*) and Lutz's rapids frog (*Paratelmatobius lutzii*) (Heyer 1976). Meanwhile the Lutz's river frog (*Thoropa lutzi*) is a lowland species, from Recreio dos Bandeirantes, just southwest of Rio de Janeiro itself, and the Petropolis river frog (*Thoropa petropolitana*) has a broader distribution, stretching from Rio de Janeiro north into Espírito Santo (Heyer 1979). Finally, to

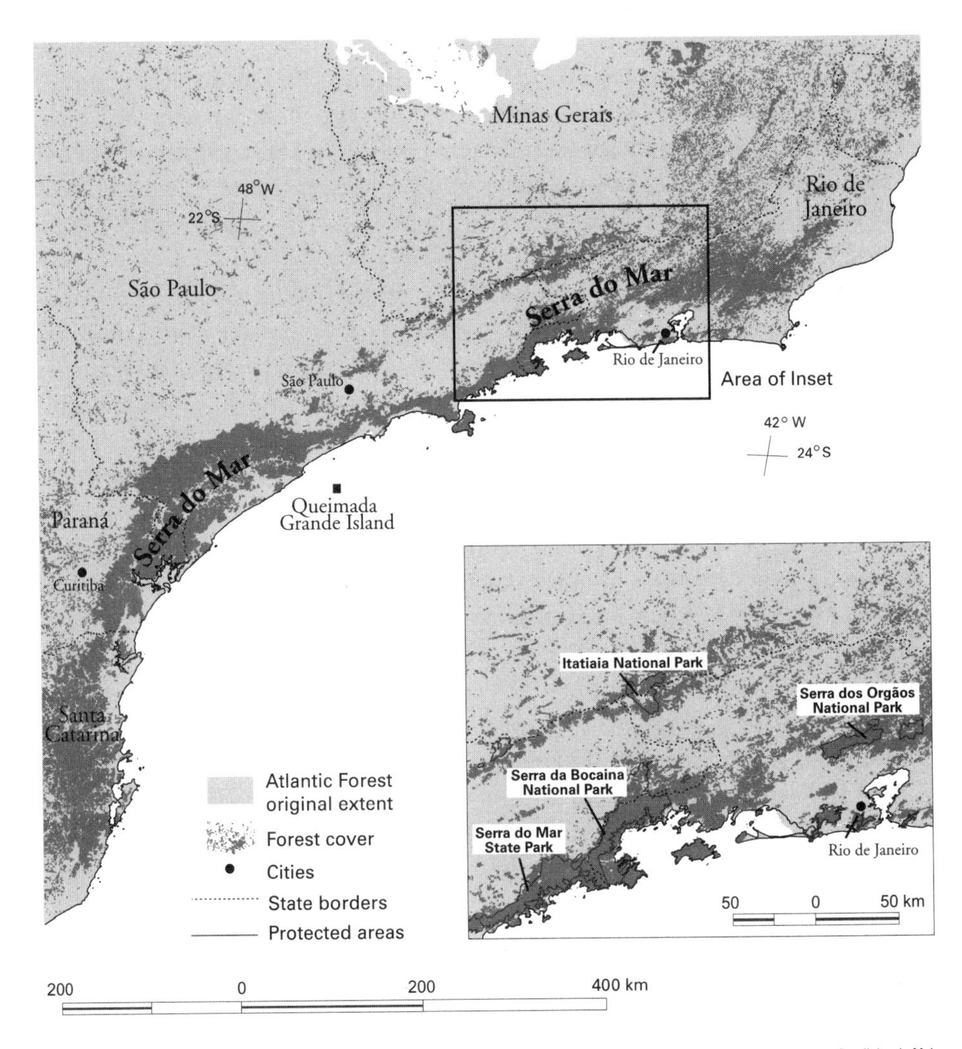

Figure 30.3. The remaining Atlantic forest in the Serra do Mar of São Paulo, Paraná, and Rio de Janeiro, Brazil, is home to at least 12 critically endangered species.

the south, São Paulo state holds a critically endangered snake, the golden lancehead (*Bothrops insularis*), endemic to Queimada Island (Duarte et al. 1995).

The final Atlantic Forest critically endangered terrestrial vertebrate is the Brazilian merganser (*Mergus octosetaceus*). It is not endemic to the hotspot because it also occurs to the north, at four localities in the Cerrado hotspot, which is fortunate because it is has been driven to extinction by disturbance, pollution, and damming of rivers across most of its former Atlantic Forest range in Rio de Janeiro, São Paulo, and Santa Catarina (Collar et al. 1992) and in eastern Paraguay (Brooks et al. 1993). However, tiny Atlantic Forest populations do survive on the Rio Tibagi in Paraná state, Brazil (Anjos et al. 1997) and the Río Urugua-í in Misiones province, Argentina (Benstead 1994; Chapter 16, this volume).

Without conservation intervention, 14 of these critically endangered species are likely to become extinct within the next decade and all 28 of them within 50 years. However, the Atlantic Forest is a hotspot where such interventions are eminently possible, through existing conservation capacity, funding potential, and political and social stability. Undoubtedly the most important single action to conserve the Atlantic Forest's critically endangered species is to establish protected areas, many of which will need to be privately owned reserves through the Private Reserves of Natural Heritage (RPPN) mechanism (Chapter 9, this volume). This will need to be supplemented with improved enforcement of existing, mainly government, parks, along with species-specific interventions such as reintroduction of the Alagoas curassow and control of hunting for the larger primates. If the Brazilian and international conservation communities can meet these challenges, then we still have the opportunity to stave off mass extinction in the Atlantic Forest.

References

Anjos, L., Schuchmann, K. L., and Berndt, R. 1997. Avifaunal composition, species richness, and status in the Tibagi River basin, Paraná state, southern Brazil. *Ornitología Neotropical* 8: 145–173.

Bauer, C., Pacheco, J. F., Venturini, A. C., and Whitney, B. M. 2000. Rediscovery of the cherry-throated tanager *Nemosia rourei* in southern Espírito Santo, Brazil. *Bird Conservation International* 10: 97–108.

Benstead, P. 1994. Brazilian merganser in Argentina: going, going. *Cotinga* 1: 8.

BirdLife International. 2000. *Threatened birds of the world.* Barcelona and Cambridge: Lynx Editions and BirdLife International.

Brooks, T. and Balmford, A. 1996. Atlantic forest extinctions. *Nature* 380: 115.

Brooks, T. M., Barnes, R., Bartrina, L., Butchart, S. H. M., Clay, R. P., Esquivel, E. Z., Etcheverry, N. I., Lowen, J. C., and Vincent, J. 1993. *Bird surveys and conservation in the Paraguayan Atlantic Forest. Project CANOPY '92: final report.* BirdLife International Study Report No. 57. Cambridge, UK: BirdLife International.

CI do Brasil, Fundação SOS Mata Atlântica, Fundação Biodiversitas, Instituto de Pesquisas Ecológicas, Secretaria do Meio Ambiente do Estado de São Paulo, and SEMAD/Instituto Estadual de Florestas-MG 2000. *Avaliação e ações prioritárias para a conservação da biodiversidade da Mata Atlântica e Campos Sulinos.* Brasília, Brazil: Ministério do Meio Ambiente/Secretaria de Biodiversidade e Florestas.

Coimbra-Filho, A. F. 1969. Mico-leão, *Leontideus rosalia* (Linneaus, 1766), situação atual da espécie no Brasil (Callithricidae: Primates). *Anais da Academia Brasileira de Ciências* 41(Suppl.): 29–52.

Coimbra-Filho, A. F., Rylands, A. B., Pissinattii, A., and Santos, I. B. 1992. The distribution and status of the buff-headed capuchin monkey, *Cebus xanthosternos,* in the Atlantic forest region in eastern Brazil. *Primate Conservation* 12–13: 24–30.

Collar, N. J., Gonzaga, L. P., Krabbe, N., Madroño Nieto, A., Naranjo, L. G., Parker, T. A. III, and Wege, D. C. 1992. *Threatened birds of the Americas. The ICBP/IUCN red data book* 3rd ed., part 2. Cambridge, UK: International Council for Bird Preservation.

Duarte, M. R., Puorto, G., and Franco, F. L. 1995. A biological survey of the pit viper *Bothrops insularis* Amaral (Serpentes, Viperidae): an endemic and threatened offshore island snake of southeastern Brazil. *Studies on Neotropical Fauna and Environment* 30: 1–13.

da Fonseca, G. A. B., Rylands, A. B., Costa, C. M. R., Machado, R. B., and Leite, Y. L. R. 1994. *Livro vermelho dos mamíferos brasileiros ameaçados de extinção.* Belo Horizonte, Brazil: Fundação Biodiversitas.

Frost, D. R. 1985. *Amphibian species of the world: a taxonomic and geographical reference.* Lawrence, KS: Association of Systematics Collections. Online: http://research.amnh.org/herpetology/amphibia/index.html. Accessed July 8, 2001.

Gonzaga, L. P. 1997. Kinglet calyptura survives in south-east Brazil! *Cotinga* 7: 9.

Heyer, W. R. 1976. The presumed tadpole of *Paratelmatobius lutzi* (Amphibia, Leptodactylidae). *Papéis Avulsos de Zoologia, Museu de Zoologia, Universidade de São Paulo* 30: 133–135.

Heyer, W. R. 1979. Natural history notes on *Craspedoglossa stejnegeri* and *Thoropa petropolitana* (Amphibia: Salientia, Leptodactylidae). *Journal of the Washington Academy of Sciences* 69:17–20.

Hilton-Taylor, C. 2000. *2000 IUCN red list of threatened species.* Gland, Switzerland: IUCN. Online: http://www.redlist.org. Accessed July 8, 2001.

Knapp, S. 1997. Rio de Janeiro antwren rediscovered. *Cotinga* 7: 9–10.

Kobayashi, S. and Langguth, A. 1999. A new species of titi monkey, *Callicebus* Thomas, from north-eastern Brazil (Primates, Cebidae). *Revista Brasileira de Zoologia* 16: 531–551.

Lemos de Sá, R. M., Pope, T. R., Glander, K. E., Struhsaker, T. T., and da Fonseca, G. A. B. 1990. A pilot study of genetic and morphological variation in the muriqui (*Brachyteles arachnoides*). *Primate Conservation* 11: 26–30.

Marinho-Filho, J. and Veríssimo, E. W. 1997. The rediscovery of *Callicebus personatus barbarabrownae* in northeastern Brazil with a new western limit for its distribution. *Primates* 38: 429–433.

Martuscelli, P., Petroni, L. M., and Olmos, F. 1994. Fourteen new localities for the muriqui (*Brachyteles arachnoides*). *Neotropical Primates* 2(2): 12–15.

Mendes, S. L. and Chiarello, A. G. 1993. A proposal for the conservation of the muriqui in the state of Espírito Santo, southeastern Brazil. *Neotropical Primates* 1(2): 2–4.

Myers, N., Mittermeier, R. A., Mittermeier, C. G., da Fonseca, G. A. B., and Kent, J. 2000. Biodiversity hotspots for conservation priorities. *Nature* 403: 853–858.

Pacheco, J. F. and Whitney, B. M. 1995. Range extensions for some birds in northeastern Brazil. *Bulletin of the British Ornithological Club* 115: 157–163.

Persson, V. G. and Lorini, M. L. (1993) Notas sobre o mico-leão-da-cara-preta, *Leontopithecus caissara* Lorini & Persson, 1990, no sul do Brasil (Primates, Callithrichidae). In Yamamoto, M. E. and Sousa, M. B. C. (eds.). *A primatologia no Brasil, 4.* pp. 169–181. Natal, Brazil: Editora Universitária, Universidade Federal do Rio Grande do Norte.

Pinto, L. P. de S., Costa, C. M. R., Strier, K. B., and da Fonseca, G. A. B. 1993. Habitats, density, and group size of primates in the Reserva Biológica Augusto Ruschi (Nova Lombardia), Santa Teresa, Brazil. *Folia Primatologica* 61: 135–143.

Pinto, L. P. de S. and Rylands, A. B. 1997. Geographic distribution of the golden-headed lion tamarin, *Leontopithecus chrysomelas:* implications for its management and conservation. *Folia Primatologica* 68: 161–180.

Pinto, O. 1950. Miscelânea ornitológica, V. Descrição de uma nova subespéie nordestina em *Synallaxis ruficapilla* Vieillot (fam. Furnariidae). *Papéis Avulsos de Zoologia, Museu de Zoologia, Universidade de São Paulo* 9: 361–364.

Rosenberger, A. L. and Coimbra-Filho, A. F. 1994. Morphology, taxonomic status and affinities of the lion tamarins, *Leontopithecus* (Callitrichinae, Cebidae). *Folia Primatologica* 42: 149–179.

Rylands, A. B., da Fonseca, G. A. B., Leite, Y. L. R. and Mittermeier, R. A. 1996. Primates of the Atlantic forest: origin, endemism, distributions and communities. In: Norconk, M. A., Rosenberger, A. L., and Garber, P. A. (eds.). *Adaptive radiations of the neotropical primates.* pp. 21–51. New York: Plenum Press.

Rylands, A. B., Mittermeier, R. A., and Rodríguez-Luna, E. 1997. Conservation of Neotropical primates: threatened species and an analysis of primate diversity by country and region. *Folia Primatologica* 68: 134–160.

Santos, I. B., Mittermeier, R. A., Rylands, A. B., and Valle, C. M. C. 1987. The distribution and conservation status of primates in southern Bahia, Brazil. *Primate Conservation* 8: 126–142.

Short, L. L. and Parkes, K. C. 1979. The status of *Agelaius forbesi* Sclater. *Auk* 96: 179–183.

Sick, H. 1960. Zur Systematik und Biologie der Bürzelstelzer (Rhinocryptidae), speziell Brasiliens. *Journal für Ornithologie* 101: 141–174.

Stallings, J. R. and Robinson, J. G. 1991. Distribution, forest heterogeneity, and primate communities in a Brazilian Atlantic forest park. In: Rylands, A. B. and Bernardes, A. T. (eds.). *A primatologia no Brasil-3.* pp. 357–368. Belo Horizonte, Brazil: Fundação Biodiversitas and Sociedade Brasileira de Primatologia.

Strier, K. B. and da Fonseca, G. A. B. 1996–1997. The endangered muriqui in Brazil's Atlantic Forest. *Primate Conservation* 17: 131–137.

Studer, A. and Vielliard, J. 1988. Premières données étho-écologiques sur l'Ictéridé brésilien *Curaeus forbesi* (Sclater, 1886) (Aves, Passeriformes). *Revue Suisse de Zoologie* 95: 1063–1077.

Teixeira, D. M. 1986. The avifauna of the north-eastern Brazilian Atlantic forests: a case of mass extinction? *Ibis* 128: 167–168.

Teixeira, D. M. 1987. A new tyrannulet (*Phylloscartes*) from northeastern Brazil. *Bulletin of the British Ornithological Club* 107: 37–41.

Teixeira, D. M. and Carnevalli, N. 1989. Nove espécie de *Scytalopus* Gould, 1837, do nordeste do Brasil (Passeriformes, Rhinocryptidae). *Boletim do Museu Nacional do Rio de Janeiro* 331: 1–11.

Teixeira, D. M. and Gonzaga, L. P. 1983. Um novo Furnariidae do nordeste do Brasil: *Philydor novaesi* sp. nov. (Aves, Passeriformes). *Boletim do Museu Emílio Goeldi, Série Zoologia* 124: 1–22.

Teixeira, D. M. and Gonzaga, L. P. 1985. Uma nova subespécie de *Myrmotherula unicolor* (Ménétries, 1835) do nordeste do Brazil. *Boletim do Museu Nacional do Rio de Janeiro* 310: 1–15.

Teixeira, D. M., Nacinovic, J. B., and Pontual, F. B. 1987. Notes on some birds of northeastern Brazil (2). *Bulletin of the British Ornithological Club* 107: 151–157.

Valladares-Padua, C. and Cullen, L., Jr. 1994. Distribution, abundance and minimum viable population of the black lion tamarin *Leontopithecus chrysopygus. Dodo, Journal of the Wildlife Preservation Trust* 30: 80–88.

Willis, E. O. and Oniki, Y. 1982. Behaviour of fringe-backed fire-eyes (*Pyriglena atra,* Formicariidae): a test case for taxonomy versus conservation. *Revista Brasileira de Biologia* 42: 213–223.

Chapter 31

Putting the Pieces Back Together: Fragmentation and Landscape Conservation

Carlos Galindo-Leal

The once continuous Atlantic Forest is now a collection of hundreds of small, isolated fragments in a complex matrix of land uses. Every fragment, regardless of size, may be important for the conservation of certain species (Turner and Corlett 1996). For example, viable populations of some insects and plants may require only a few hectares. Not every species will be conserved in small fragments, however. Raptors, large carnivores, and other species may need thousands of square kilometers to maintain healthy populations. To address the conservation needs of this region effectively, therefore, we must first assess what we know about fragmentation and its consequences in the Atlantic Forest. In this chapter, I review the lessons learned from research on fragmentation in this still threatened hotspot (Figure 31.1) and discuss recent landscape conservation initiatives.

The Anatomy and Physiology of Fragments

Fragments differ in many ways from the continuous forests of which they were once a part. First, of course, they are smaller and relatively isolated from other patches of forest. Second, fragments tend to possess distinctive characteristics that have allowed them to remain forested in the midst of deforestation. For instance, fragments may be located in areas of low productivity or on steep slopes, or they may be inaccessible. Third, changes in microclimate (e.g., temperature, humidity, wind velocity) at the edges of fragments influence their composition and structure to different degrees. Finally, compared to continuous forests, fragments have proportionally more edge and less interior habitat. For example, in the southeastern region of Pernambuco in northeastern Brazil, 1,839 forest fragments were analyzed within a matrix dominated by sugarcane fields (Ranta et al. 1998). The fragments were located on the tops of low hills surrounded by sugarcane fields in slopes and valleys. Their sizes ranged from less than 0.01 km^2 to 15.39 km^2, and the average size was 0.34 km^2. Almost half were smaller than 0.1 km^2, and only 7

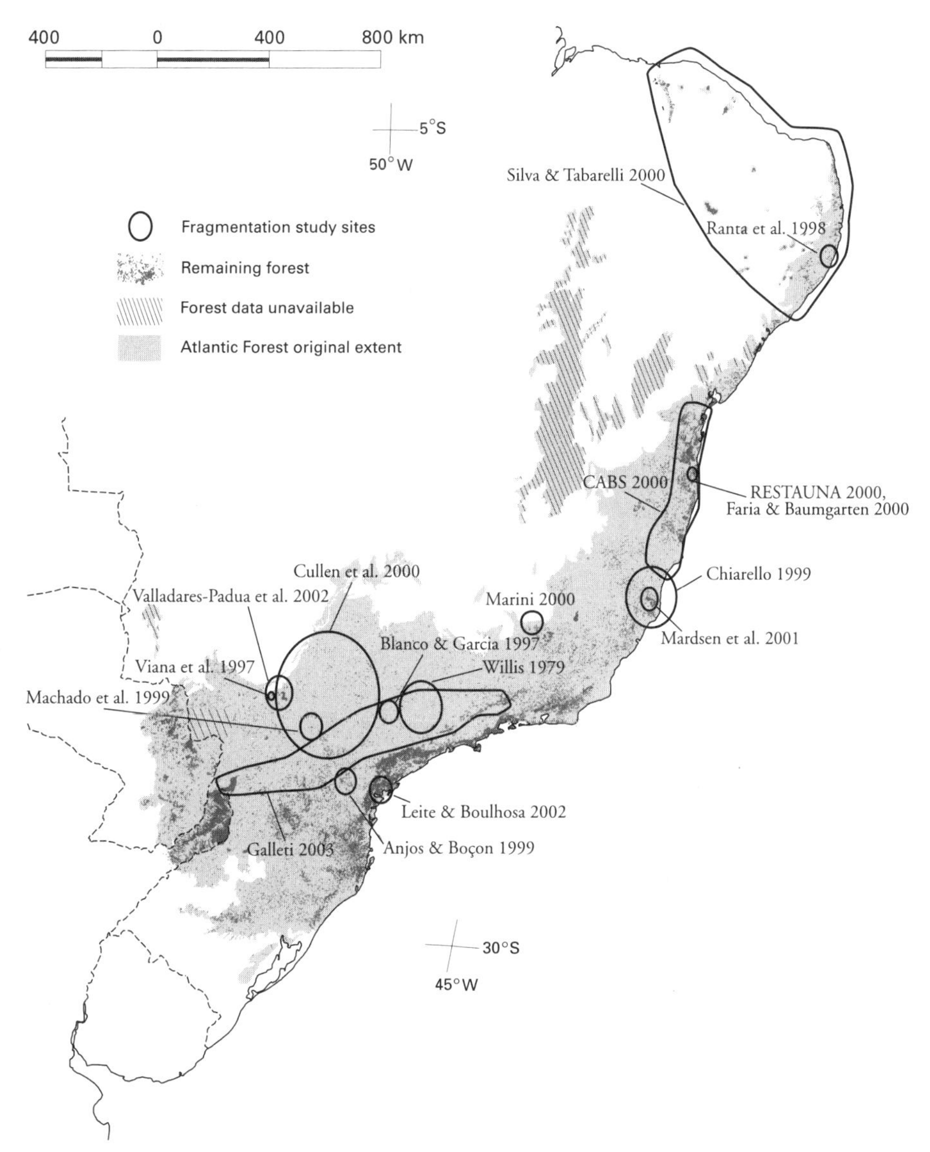

Figure 31.1. Areas of research on forest fragments reviewed in the text.

percent were larger than 1 km^2. The amount of edge habitat in these forest patches was comparatively large, ranging from more than 50 percent to as high as 94 percent of the total habitat for the patch. Similarly, in the municipal district of Botucatu (a study area of 157.74 km^2) in southeastern Brazil (São Paulo State), most patches were small (< 0.20 km^2) and few were large (> 0.60 km^2). Their average size was 0.11 km^2 (Blanco and Garcia 1997).

A fragment's size, shape, location, and proximity to other fragments affect its biological composition, structure, and function. Some species may benefit from the new geometry of a fragment; others may not. Because they include smaller numbers of individuals, populations in fragments are more susceptible to extinction due to genetic, demographic, or environmental factors. In addition, because fragment populations are isolated, opportunities for colonization, and the genetic variability that results, are usually unavailable. Of the species that fail to thrive in fragments, those with large area requirements and low densities are the first to disappear (Aleixo and Vielliard 1995; Christiansen and Pitter 1997; Marini 2000). Often forests are degraded near the edges of fragments. Trees at the edges may begin to have higher mortality rates, and invasive vegetation—such as lianas and vines—may start to proliferate (Viana et al. 1997; Chiarello 1999; Gascon et al. 2000). The structure of edges also has negative influences on some animals. Researchers have observed, for example, that bat richness significantly decreases at fragment edges, suggesting that bats are adversely affected by conditions there (Faria and Baumgarten 2000). Not all animals suffer from edge effects, however. Researchers have noted that small mammals and bird communities are richer on the edges than in interior transects of fragments (Faria et al. 2000). Finally, edge effects disturb a larger proportion of the remaining habitat of small fragments (Ranta et al. 1998).

Fragments are not static; their composition and structure continue to change long after they are formed. For example, in the Santa Rita forest fragment (0.95 km^2) in the state of São Paulo on the western slopes of the Atlantic Forest, after the forest was fragmented, populations of 16 tree species with less than 15 individuals continued to decline due to short life cycles, anthropogenic pressures, and higher vulnerability to windthrows (Viana et al. 1997). Forest edges, in particular, underwent a process of degradation that led to a lower density of trees.

Decreased Diversity in Fragments

Several studies in different regions of the Atlantic Forest have documented some of the biological consequences of habitat fragmentation (Willis 1979; Viana et al. 1977; Ranta et al. 1998; Chiarello 1999; Cullen et al. 2000). In fact, some of the earliest studies anywhere of the impacts of fragmentation were carried out in the Atlantic Forest hotspot. In the state of São Paulo in southern Brazil, for instance, Willis (1979) documented that smaller fragments had fewer bird species. He found that fragments of 14 km^2, 2.5 km^2, and 0.21 km^2 had 202, 146, and 93 bird species, respectively.

In Lagoa Santa, Minas Gerais, patterns of bird species loss and decline were studied by examining survey records from 1870. Thirteen forest species recorded in 1870 were absent from the same area in 1987. Some of these species had been common. This species loss was related to fragment size. In fragments of 0.24 km^2, 0.63 km^2, and 1.98 km^2, 20, 29, and 34 forest species were found, respectively. As a result of fragmentation, the diverse bird community of the original forest had been replaced by a few dominant species (Marini 2000). Similarly, in Londrina municipality, north of Paraná, the diversity of frog species was lower in altered areas (14 species in Estação de Piscicultura da Universidade Estadual de

Londrina) than in well-preserved areas (24 species in Parque Estadual Mata dos Godoy) (Machado et al. 1999).

In the Campos Gerais region, Paraná, bird species diversity was also found to be lower in smaller patches. However, smaller forest patches linked to a large patch (8.4 km^2) were observed to be more similar in species composition to this larger patch than to other isolated patches. The number of edge species increased with decreasing patch area, while the number of forest species declined (Anjos and Boçon 1999).

How Much Is Enough? Area Requirements for Demanding Species

A comparison of six fragments of different sizes (ranging from 2.1 km^2 to 242.5 km^2) surrounded by coffee and sugarcane fields in northern Espírito Santo revealed the local disappearance of some mammal species in small fragments (Chiarello 1999). Jaguars, pumas, tapirs, peccaries, giant armadillos, agouties, and the giant anteater were completely absent from smaller fragments (2.1 km^2 and 2.6 km^2). Jaguars and giant armadillos were not recorded even in medium-sized patches (15.04 km^2 and 24 km^2). According to this study, only reserves with approximately 200 km^2 or more are able to harbor intact wildlife communities (Chiarello 1999).

In the Mata de Planalto region of São Paulo, two heavily hunted fragments of approximately 20 km^2 have lost several species of such game mammals as tapirs, brocket deer, and white-lipped peccaries. Two fragments of similar size with low hunting pressure, however, retained most of their original fauna despite 50 years of isolation (Cullen et al. 2000). The largest fragment, Morro de Diabo State Park (350 km^2), had a faunal composition similar to that of the small, lightly hunted patches. However, several species—such as brocket deer, nine-banded armadillos, coatis, Guianan squirrels, brown capuchin monkeys, and brown howler monkeys—had inexplicably lower densities inside the park than in the smaller patches, regardless of hunting pressure.

It is surprising that in this last study, white-lipped peccaries were recorded in the small fragments (20 km^2), because this species is believed to require large areas. It is likely that the peccary populations in these fragments are not viable in the long term. Among large mammals, hunting exacerbates the effects of fragmentation and is probably the most important factor responsible for the eradication of populations (Cullen et al. 2000). Population estimates in the Pontal do Paranapanema region, an area consisting of approximately 350 km^2 of forest in Morro do Diabo State Park and some surrounding forests, suggest that there are no more than 20 jaguars, 30 pumas, and 250 tapirs there (IPE 2000).

Large predators such as jaguars and pumas are usually hunted to eliminate cattle predation. Those that do not fall victim to hunting still feel its effects indirectly, since their diet broadly overlaps that of the hunters, who have overhunted the prey of large predators. These animals are severely affected even within protected areas such as Superagüi National Park (224 km^2), Guaraqueçaba Environmental Protection Area (3,134 km^2), and the Marumbi Special Interest Area for Tourism

(667.32 km^2) in Paraná (Leite et al. 2002). Indeed, Superagüi National Park already is lacking large prey species and jaguars (Leite and Boulhosa 2002).

According to recent estimates of jaguar densities in the Atlantic Forest area, ranging from 1 to 3.7 jaguars per 100 km^2 (Crawshaw 1995; Leite and Boulhosa 2002), areas of more than 10,000 km^2 would be required to maintain populations of 500 to 1000 individuals. In the Atlantic Forest there are only two regions with areas larger than 10,000 km^2: Serra do Mar and the region of Misiones in Argentina and Iguaçu National Park, Brazil.

Cascading Consequences

Changes in faunal assemblages in forest fragments affect ecological processes. As various species diminish or disappear, such ecological interactions as pollination, dispersal, predation, and competition may be altered. In the Atlantic Forest, for example, animals disperse a large proportion of the seeds of trees. In northeastern Brazil, birds and mammals disperse the seeds of 71.4 percent of the 427 documented tree species. Of this proportion, 31.6 percent depend on wide-gape birds such as guans, chachalacas, toucans, aracaris, and cotingas for dispersal (Silva and Tabarelli 2000). Many of these birds, however, are threatened by a combination of habitat loss and overhunting, and their eventual disappearance will likely mean that forest patches will come to be dominated by wind-dispersed tree species. Moreover, without the appropriate dispersers, some tree species with large fruits may become locally extinct.

Species Loss in the Hotspot

As the Atlantic Forest was colonized, much of the accessible forest habitat was converted into agricultural land (see Figure 33.1), reducing the extent of habitat to the point where only 7.5 percent of the former native vegetation remains today (Myers et al. 2000). According to theory, a decrease in native habitat of this magnitude (over 90 percent) should have resulted in the extinction of as many as half the endemic species (Chapter 30, this volume). Although almost no extinction among bird species has been documented, the extent of deforestation nevertheless accurately predicts the documented numbers of threatened birds (Brooks et al. 1999) and mammals (Grelle et al. 1999). These results support the notion that a time lag exists between deforestation and extinction, giving us time to act.

Landscape Connectivity and Management: Putting the Pieces Back Together

Throughout the former Atlantic Forest range a landscape dominated by pastures, sugarcane, soy, coffee, cocoa, pine, and eucalyptus plantations surrounds forest fragments (Marsden et al. 2001). The research cited above shows that preserving the biological composition and structure of fragments is necessary to maintain biodiversity. However, fragments are temporally dynamic, heavily influenced by the surrounding matrix, and too small to maintain many ecological processes and viable populations of some species.

The ecosystem deficits caused by fragmentation suggest that a bioregional approach is required to accomplish long-term conservation (Silva and Tabarelli 2000). A regional approach focuses not only on conserving fragments but also on restoring the connectivity between them, using both natural and anthropogenic patches, such as shaded coffee and cocoa plantations). Shaded plantations can help restore connectivity between fragments because they resemble the vertical structure of tropical forests and are therefore used by diverse bird and bat species. Bat abundance and richness, for instance, were observed to be significantly higher in shaded cocoa plantations in the Una region of northeastern Brazil (Faria and Baumgarten 2000). In particular, *Phyllostominae* bats, species usually associated with pristine forests, were more frequent in shaded cocoa plantations than in either fragmented forest (<2 km^2) or continuous forest (>10 km^2) (Faria and Baumgarten 2000).

Although some plantations (e.g., cocoa) may provide vertical structures resembling those of original forest ecosystems, these plantations result from historical and social conditions that, in many cases, are no longer present. New economic and policy incentives, therefore, must be promoted to maintain such structures (Alger, pers. comm. 2002).

In contrast, sugarcane, eucalyptus, and pine plantations, as well as pastures, have little vertical or horizontal heterogeneity and thus support a very low number of species. For example, although a total of 111 bird species were documented in a forest reserve in east-central Espírito Santo, only 8 were documented in adjacent eucalyptus plantations (Marsden et al. 2001).

Patterns of extinction following habitat loss are now well documented in the Atlantic Forest (Chapter 30, this volume). The extent of the effects of fragmentation on ecological processes is only recently being discovered, however. Although biodiversity is continuing to disappear from fragments, the time lag between habitat loss and extinction documented in some studies grants us a precious opportunity to piece together fragmented landscapes before it is too late.

Several landscape-level research and conservation projects are currently under way in the Atlantic Forest, including:

- Projeto RestaUna (Projeto Remanescentes de Floresta da Região de Una) by Universidade Estadual de Santa Cruz (UESC) is designed to investigate the impact of forest fragmentation on biological communities in the Una region of northeastern Brazil. The project began in 1996 in southeastern Bahia, and the results will be used to develop conservation strategies (Projecto RestaUna 2000).
- Since 1996, researchers at the Instituto de Pesquisas Ecologicas (IPE) have been working to increase the viability of forest fragments in the Morro do Diabo State Park area in western São Paulo (Valladares-Padua et al. 2002). They have surveyed and assessed remaining forest fragments (> 400 ha in size) and promote their conservation through a participatory approach with stakeholders. They maintain a forest

fragment database (http://www.bdt.org.br/ipe) and also encourage the creation of a buffer zone to decrease edge effects in the state park. The buffer zone is a multiple-use zone where agroforestry projects will produce fruits, wood, firewood, honey, medicinal herbs, and critical ecosystem services such as wind barriers, live fences, and soil conservation (IPE 2000).

- The ecology department of the Universidade Estadual Paulista recently began a project in 12 areas, from Iguaçu National Park in southwestern Paraná to Itatiáia National Park in western Rio de Janeiro, to investigate game density in the conservation units (protected areas) of the Atlantic Forest (Galleti, pers. comm. 2002).
- The Instituto do Estudos Socio-ambientais do Sul da Bahia (IESB) and the Center for Applied Biodiversity Science (CABS) at Conservation International have been involved in planning for the Discovery or Bahia Corridor, a conservation corridor (a regional planning unit comprising a mosaic of land uses) on the coast of southern Bahia and northern Espírito Santo (CABS 2000; Chapter 11, this volume). This landscape mosaic is composed of cacao and rubber plantations, pastures, eucalyptus plantations, and forest fragments. A modeling tool (TAMARIN) has been used to assess the value of various fragments and the economic opportunities for maintaining and restoring connectivity.

Researchers are reaching consensus that to make fragments more effective reservoirs of biodiversity and to avoid impending extinctions, landscapes should incorporate three elements. First, large fragments or clusters of natural vegetation should be included in a protected area system. Second, a network of biological corridors between fragments should be maintained or reestablished to increase connectivity and allow biological processes to continue. Third, incentives should be sought to make current agricultural practices compatible with conservation (Valladares-Padua et al. 2002). Conservation and restoration planning should integrate both forest fragments and the land-use matrix in designing conservation landscapes that can sustain the ecological processes necessary to maintain viable populations and ecosystems.

Acknowledgments

I thank Penny Langhammer, Thomas Brooks, Paulo Inácio Prado, and Keith Alger for their suggestions for this chapter, as well as Suzana Padua and Laury Cullen, Jr., for helping with references.

References

Aleixo, A. and Vielliard, J. M. E. 1995. Composição e dinámica da avifauna da mata de Santa Genebra, Campinas, São Paolo, Brasil. *Revista Brasileira de Zoologia* 12: 493–511.

Anjos, L. and Boçon, R. 1999. Bird communities in natural forest patches in southern Brazil. *Wilson Bulletin* 11(13): 397–414.

Blanco, J. L. A. and Garcia, G. J. 1997. A study of habitat fragmentation in southeastern Brazil using remote sensing and geographic information systems (GIS). *Forest Ecology and Management* 98(1): 35–47.

Brooks, T., Tobias, J., and Balmford, A. 1999. Deforestation and bird extinctions in the Atlantic Forest. *Animal Conservation* 2: 211–222.

CABS (Center for Applied Biodiversity Science). 2000. *Designing sustainable landscapes: the Brazilian Atlantic Forest.* Washington, DC: Center for Applied Biodiversity and Instituto de Estudos Socio-ambientais do Sul da Bahia.

Cardoso da Silva, J. M. and Tabarelli, M. 2000. Tree species impoverishment and the future flora of the Atlantic Forest of Northeast Brazil. *Nature* 404: 72–74.

Chiarello, A. G. 1999. Effects of fragmentation of the Atlantic Forest mammal communities in southeastern Brazil. *Biological Conservation* 89: 71–82.

Crawshaw Jr., P. G. 1995. *Comparative ecology of ocelot (*Felis pardalis*) and jaguar (*Panthera onca*) in a protected subtropical forest in Brazil and Argentina.* Doctoral dissertation. University of Florida, Gainsville.

Cullen Jr., L., Bodmer, R. E., and Valladares, C. P. 2000. Effects of hunting in habitat fragments of the Atlantic forests, Brazil. *Biological Conservation* 95: 49–56.

Faria, D. and Baumgarten, J. 2000. *Bats and forest fragmentation in the Brazilian Atlantic Rainforest.* Society for Conservation Biology, June 2000 Abstract Database.

Faria, D., Pardini, R., and Lápis, R. R. 2000. *Comparative response of birds, bats and nonvolant small mammals facing habitat fragmentation in the Atlantic Forest.* Society for Conservation Biology, June 2000 Abstract Database.

Grelle, C. E. V., da Fonseca, G. A. B., Fonseca, M. T., and Costa, L. P. 1999. The question of scale in threat analysis: a case study with Brazilian mammals. *Animal Conservation* 2: 149–152.

IPE (Instituto de Pesquisas Ecologicas). 2000. *The ecological detectives.* Online: http://www.ipe.org.br/INGLES/detetives.htm.

Leite, M. R. P. and Boulhosa, R. L. P. 2002. Ecology and conservation of jaguar in the Atlantic coastal forest, Brazil. In: Medellín, R. A., Chetkewicz, C., Rabinowitz, A., Redford, K. H., Robinson, J. C., Sanderson, E., and Taber, A. (eds.). *Jaguares en el nuevo milenio. Una evaluación de su estado, detección de prioridades y recomendaciones para la conservación del Jaguar en América.* México, D.F.: Universidad Nacional Autónoma de México (UNAM)/Wildlife Conservation Society (WCS).

Machado, R. A., Bernarde, P. S., Morato, S. A. A., and Anjos, L. 1999. Análise comparada da riqueza de anuros entre duas áreas com diferentes estados de conservação no município de Londrina, Paraná, Brasil (Amphibia, Anura). *Revista Brasileira de Zoologia* 16 (4): 997–1004.

Marini, M. A. 2000. Efeitos de fragmentação florestal sobre as aves em Minas Gerais. In: dos Santos Alves, M. A., Cardoso da Silva, J. M., Van Sluys, M., de Godoy Bergallo, H., and Duarte de Rocha, C. F. (eds.). pp. 41–54. *A Ornitologia no Brasil.* Rio de Janeiro, Brasil: Pesquisa Atual e Perspectivas ed. Universidade do Estado do Rio de Janeiro (UERJ).

Marsden, S. J., Whiffin, M., and Galetti, M. 2001. Bird diversity and abundance in forest fragments and eucalyptus plantations around a Brazilian Atlantic Forest reserve. *Biodiversity and Conservation* 10: 737–751.

Myers, N., Mittermeier, R. A., Mittermeier, C. G., da Fonseca, G. A. B., and Kent, J. 2000. Biodiversity hotspots for conservation priorities. *Nature* 403: 853–858.

Ranta, P., Blom, T., Niemela, J., Joensuu, E., and Siitonen, M. 1998. The fragmented Atlantic rain forest of Brazil: size, shape and distribution of forest fragments. *Biodiversity and Conservation* 7: 385–403.

Projeto RestaUna (Remanescentes de Floresta da Região de Una). 2000. *Equipe técnica.* Universidade Estadual de Santa Cruz (UESC). Online: http://www.restauna.org.br/equipe.htm.

Silva, J. M. and Tabarelli, M. 2000. Tree species impoverishment and the future flora of the Atlantic Forest of northeast Brazil. *Nature* 404: 72–74.

Turner, I. M. and Corlett, R. T. 1996. The conservation value of small, isolated fragments of lowland tropical rain forest. *Trends in Ecology and Evolution* 11(8): 330–333.

Valladares-Padua, C., Padua, S. M., and Cullen Jr., L. 2002. Within and surrounding the Morro do Diabo State Park: biological value, conflicts, mitigation and sustainable development alternatives. *Environmental Science and Policy* 5: 69–78.

Viana, M. V., Tabanez, A. A. J., and Batista, J. L. F. 1997. Dynamics and restoration of forest fragments in the Brazilian Atlantic moist forest. In: Laurance, W. F. and Bierregaard Jr., R. O. (eds.). *Tropical forest remnants.* pp. 351–365. Chicago, London: University of Chicago Press.

Willis, E. O. 1979. The composition of avian communities in remanescent woodlots in southern Brazil. *Papeis Avulsos de Zoologia, São Paulo* 33: 1–2.

Chapter 32

Endangered Forests, Vanishing Peoples: Biocultural Diversity and Indigenous Knowledge

Thomas R. Jacobsen

Because indigenous peoples inhabit many of the world's highly biodiverse regions (McNeely 1995; Linares 1997; Posey 2001), cultural diversity and biological diversity are inextricably linked. Human societies have developed an intimate knowledge of how to use and manage the environments in which they live, and this knowledge has led to a variety of effective approaches to human, plant, and animal coexistence. Today, this knowledge, along with other aspects of the cultures that embody it, is greatly threatened. Globally, cultural diversity is disappearing faster than biological diversity (Linares 1997). As a result, the variety of approaches available to develop initiatives for sustainable resource use is decreasing, the possibility of developing imaginative new strategies is diminishing, and biodiversity is being further impoverished (McNeely et al. 1995; WWF 2000). If conservation strategies are to succeed in the long term, they must strive to include the knowledge of rural indigenous inhabitants in management plans and planning processes (Posey 1992; Ghimere and Pimbert 1996).

In the Atlantic Forests of Brazil, Paraguay, and Argentina, indigenous peoples have developed unique responses to their environments and should therefore play an important role in the conservation of the area (Hanazaki et al. 2000). Yet these societies are increasingly threatened by the loss of their forests to urbanization, commercial agriculture, and cattle ranching (Chapters 10, 19, and 26, this volume). Despite the establishment of indigenous reserves, the few surviving native societies in the region are being forced to relocate to small areas, where they are experiencing abject poverty and high rates of disease (Guzman 1993; Leitao 1994; Chapter 19, this volume). The conservation of the biological and cultural diversity of the Atlantic Forests depends greatly on the successful integration of conservation measures with indigenous livelihoods.

Historical Context

At the end of the last glacial period, approximately 12,000 years ago, humans began migrating to the continent from the north, probably via the Isthmus of Panama (Dean 1995). Evidence of human presence in the Atlantic Forest region of South America dates back to 11,000 years ago, when hunter-gatherers used stone implements (Dean 1995). Thus began a long period of interaction between humans and the natural environment, which fostered a wealth of indigenous knowledge of the Atlantic Forests and led to the development of land-use practices that, to varying degrees, have transformed much of South America's flora and fauna.

The first humans to migrate to South America lived in hunter-gatherer societies whose skillful use of hunting weapons and fire defined their relationship with the natural landscape. As they progressed south, hunters in the savanna regions used fire to drive large herbivorous prey. This practice cleared the plains of woody growth and replaced it with new shoots attractive to herbivores. Early societies that inhabited gallery forests along waterways also used fire in the surrounding primary forests. Here, fire acted to prevent the regeneration of plant species, thereby holding back the spread of the forest and maintaining the grasslands and savanna (Dean 1995).

Soon after the arrival of humans on the American continents, large mammal species rapidly went extinct, first in North America, then in South America. Contending theories propose that either climatic and geological changes or intensive hunting by humans caused these Pleistocene extinctions (Brown and Gibson 1983). As the early hunter-gatherer societies lost their primary source of protein, they migrated into lowland areas such as the Atlantic Forest, where they used forest resources more intensively and began to develop an in-depth knowledge of many plants. Some species provided a moderate amount of protein; the Brazilian pine (*Araucaria angustifolia*), for example, was deliberately spread by some peoples (Dean 1995). Faced with increased population density, and in search of better protein sources, societies in the Atlantic Forest region gradually shifted from plant gathering to plant cultivation. They eventually developed into agricultural societies based on swidden agriculture. At one time or another over thousands of years, these land-use practices and population increases may have reduced the Atlantic Forest to secondary growth (Dean 1995).

The plight of the Amerindian population of the Atlantic Forests, and the ecosystem as a whole, changed dramatically in the early sixteenth century, when Spanish and Portuguese mariners reached the coasts of Brazil and Argentina. The Spanish made their way inland via the Río de la Plata, and by 1537, they had encountered the native populations of modern-day Paraguay (Mooney 1910). Over the next few centuries, contact with the Europeans resulted in the drastic reduction of native populations and the territory they occupied, and native culture virtually collapsed (Patis 1992). In just 130 years, from 1511 to 1641, an estimated 2 million Indians in southeastern Brazil were killed by colonists, died from epidemic diseases, or were captured by slave traders (Mooney 1910). This figure represents 90 percent of the most populous indigenous group at the time, the

Guaraní. Many subgroups disappeared altogether. Most of the remaining indigenous peoples fled farther into the forest or joined Jesuit missions to escape the slave traders (Chapters 14 and 24, this volume). Those who remained independent and avoided contact with missionaries were known as the Cainguá, meaning "inhabitants of the forest" (HRAF 2001).

Cultural Attributes and Economic Base

The Amerindian societies present in the Atlantic Forest region today can be divided into two main groups. The Tupi inhabit the Brazilian coast and highlands to the north, and the Guaraní inhabit the lowlands in the southern portions of the Atlantic Forest, including Brazil, Paraguay, and Argentina. The term "Tupi-Guaraní," commonly used to identify all indigenous populations in the Atlantic Forest region, not only designates a general cultural pattern characteristic of these peoples but also refers to the common language family. A variety of dialects within the Tupi-Guaraní language are spoken, with variations in vocabulary, accent, and intonation. The Tupi-Guaraní can be broken down into many small local groups, or bands, many of whom are named after the areas they inhabit or for their chiefs (Martin and Beierle 2001).

Historically, the most significant economic unit for the Tupi-Guaraní was the extended family, comprising up to 100 people. These families lived in longhouses, *tapui,* and owned the surrounding agricultural fields, which they worked communally. Elder men ruled the *tapui,* and shamans held a great deal of influence in both the supernatural and secular realms (Martin and Beierle 2001). Today, Tupi-Guaraní societies are made up primarily of smaller, nuclear-family households, which form the core of trade and religious activities. Villages of the Tupi-Guaraní usually contain 10 to 25 households, with a community elder providing religious leadership (Reed 1997).

One of the key similarities among Tupi-Guaraní societies is the intensive cultivation of plants (Balée 1992). The Tupi-Guaraní Indians of northern Paraguay are credited with being the first to domesticate pineapple (*Ananas comosus*), evidenced by the fact that the Atlantic Forest region of southern Brazil and Paraguay today contains the largest diversity of the crop's wild relatives (Smith et. al. 1992). In addition, the cashew (*Anacardium occidentale*) was first domesticated by indigenous populations along the northeastern coast of Brazil long before the arrival of Europeans (Smith et. al. 1992). Other crops originally cultivated include manioc, bananas, maize, peanuts, capsicum peppers, pumpkins, and sweet potatoes. The Tupi-Guaraní had 28 named varieties of manioc alone (Dean 1995). Most of these crops are still cultivated today.

Most Tupi-Guaraní of the Atlantic Forest are primarily sedentary agriculturalists, practicing swidden agriculture, foraging, hunting, fishing, and—for some, such as the Aché of Paraguay—wage labor (HRAF 2001). Families generally garden small plots and shift them every 3 years, as part of a rotational cycle based on the long-fallow swidden system common throughout Latin America's lowland forests (Reed 1995). Curiously, some Tupi-Guaraní societies, such as the Aché of

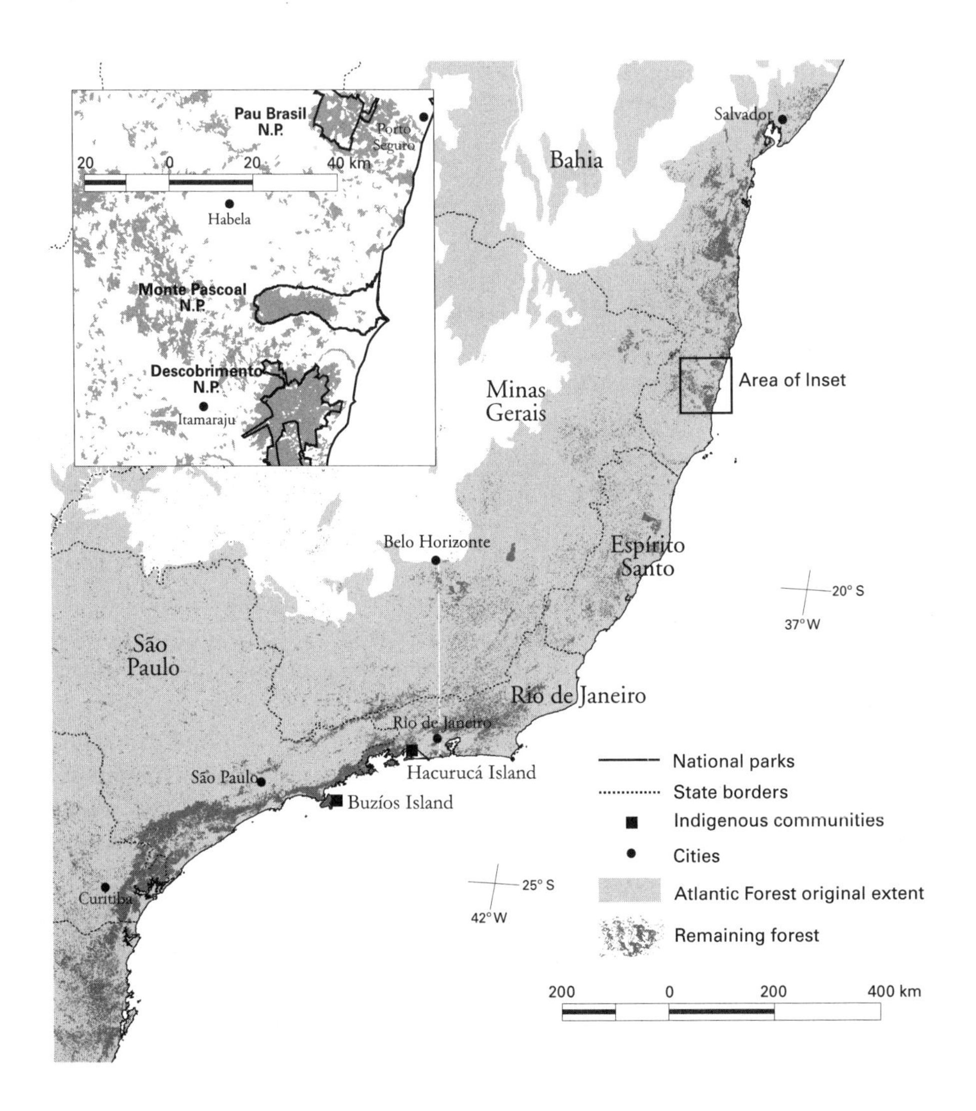

Forest cover data from Fundação S.O.S. Mata Atlântica (Paraná and Rio de Janeiro 2000; other states 1995). Protected areas data from Instituto Brasileiro do Meio Ambiente e dos Recursos Naturais Renováveis (IBAMA).

Figure 32.1. Location of Monte Pascoal National Park, where the Pataxos live, and Itacurucá and Buzios Islands, where Caiçara communities live.

Paraguay and the Héta of Paraná State, Brazil (now extinct), are primarily foraging societies that have "regressed" from agricultural societies. This transformation began at the time of European contact, when the onset of epidemic diseases, slave raids, and colonial warfare devastated settled village life and caused the loss of horticulture and semi-sedentarism. Some indigenous peoples began to pursue a more nomadic lifestyle not involved in the cultivation of plants (Balée 1992). It is interesting that these groups today maintain a strong reliance on plants such as the mucajá palm (*Acrocomia sclerocarpa*), which is a dominant plant in old agricultural fallows. To a certain degree, then, these hunter-gatherer societies still rely on agriculture (Bailey et al. 1989).

Some peoples, such as the Chiripá of Paraguay, integrate commercial extraction into systems of agroforestry, collecting commodities such as *yerba mate* (*Ilex paraguayensis*), used to make tea, as well as oils, furs, skins, wood for fence posts, and citrus oil for sale in the market economy. Instead of simply harvesting the leaves of *yerba mate,* they use techniques that not only preserve the existing trees but also encourage the growth of new ones (Reed 1995). This type of agroforestry allows the communities to participate in the market economy without harming the environment (Linares 1997).

Although the majority of indigenous societies in the Atlantic Forests are Amerindian, certain communities maintain a hybridized environmental knowledge fused from multiple cultural contexts. Such is the case among Caiçara communities of southeastern Brazil. These native inhabitants descend from Indians and Portuguese colonists and also preserve African influences (Begossi 1998). Over time, they have adapted to the coastal forests and maintain an extensive understanding of their environment. For example, among Caiçara communities living on Itacurucá Island in the state of Rio de Janeiro, 90 species of plants belonging to 40 families are used for a variety of purposes, including food, construction, handicrafts, and medicine (Figueiredo et al. 1993). Additionally, on nearby Búzios Island, approximately 61 plant species are used for food, 53 species for medicine, and 32 for construction and handicrafts (Begossi et al. 1993). The Caiçara communities also maintain specific knowledge about forms of manioc cultivation and practice taboos against harvesting some plants and hunting certain animals (Figure 32.1).

Current Populations

Today, the Tupi-Guaraní peoples of the Atlantic Forests of Brazil, Paraguay, and Argentina number approximately 134,000 individuals (Table 32.1).

About one-third of the 330,000 indigenous peoples of Brazil (FUNAI 2001) inhabit the Atlantic Forest region of Brazil, representing 0.06 percent of the entire national population (Table 32.1; IBGE 2001). They inhabit the coastal states of Brazil from Ceará in the north to Rio Grande do Sul in the south, as well as a part of the state of Mato Grosso do Sul, and can be broken into many subgroups.

In Paraguay, the population of all indigenous societies in the Atlantic Forests is approximately 20,000 (Table 32.1; DGEEC 2001), or 0.35 percent of Paraguay's total population (MNSU 2001). Two major ethnic groups are present in this

Table 32.1. Populations of indigenous people in Atlantic Forest hotspot.

Country	Total Population in Country	Indigenous People in Atlantic Forest	Percentage of Total Population
Brazil	172,594,833	110,000	0.06
Paraguay	5,743,117	20,000	0.35
Argentina	37,849,700	4,000	0.01
TOTAL	216,187,650	134,000	

region, the Aché and the Ava, both of whom speak dialects of Guaraní. The Ava are subdivided into three groups: Mbyá, Chiripá, and Pai Tavytera. Although these groups are politically separate, they are culturally almost identical (Hill, pers. comm. 2002). The Aché speak a dialect that is related to the dialects of Guaraní spoken by the other three groups, but their social organization and cultural systems are fairly different. The Mbyá inhabit the state of Guairá, scattered in small villages between Yutí to the south and San Joaquín to the north. The Chiripá have established their villages to the north of San Joaquín, and the Pai Tavytera are still farther north, near the Paraná River (Metraux 1948).

The Mbyá comprise all the indigenous peoples of the Atlantic Forests of Argentina. Inhabiting the province of Misiones, the Mbyá number approximately 4,000 (Table 32.1; Chapter 18, this volume) and represent 0.01 percent of the population of Argentina (World Gazetteer 2001). Collectively, they make up approximately 53 communities containing 720 families (Bertolini, pers. comm. 2002). Discrepancies that have been noted in the demographic data may result from the constant migrations of the Mbyá communities within the province of Misiones, where Guaraní villages are widely distributed (see Figure 18.1), and toward Brazil and Paraguay (Faúndez 2001).

Indigenous Rights and Conservation

Although the demographic, historical, and sociocultural characteristics of indigenous peoples of the Atlantic Forests have gained increased attention, their sovereignty over the forests and their integration into plans for the region's conservation are tenuous. As systems of production such as large-scale agriculture and ranching replace traditional resource use, indigenous peoples are confronted with forces out of their control and face situations to which they cannot adapt (Reed 1995). And even though indigenous groups, government agencies, and local and international conservation organizations all agree that indigenous peoples should have rights to use the forests, they have difficulty reaching consensus about whether those rights override conservation initiatives that seek to protect forests from any human influence (Reed 1995).

Some conservationists argue that biodiversity in tropical forests will not be depleted if indigenous populations live inside conservation areas. They believe that local populations should be integrated into management and decision-making processes and that traditional knowledge should be tapped as a source of insight into how to achieve sustained use of the natural resources (Schmink et al. 1992; Posey 1992; Colchester 2000; Schwartzman et al. 2000; Lizarralde 2001). Others, however, believe that these communities have a negative effect on biodiversity and that their goals do not include conservation (Chicchón 2000; Redford and Sanderson 2000; Terborgh 2000; Galetti 2001). In the Atlantic Forests, the relationships forged by local communities with their natural and political environments are highly diverse, varying among societies, over time, and among different locales.

Brazil

Brazil's newly created constitution of 1988 gave special emphasis to human rights. Article 67 ordered the demarcation of all indigenous territories in Brazil within 5 years. This process was under way in 1996, when President Cardoso signed Decree 1775, which gives anyone interested in occupying indigenous lands—including ranchers, loggers, and settlers—the right to challenge the demarcation process (Survival International 2001). As a result, the conflict over land rights throughout Brazil has escalated sharply, and more powerful forces regularly invade many indigenous territories, despite demarcation. As of 1997, approximately 50 percent of indigenous territories had been demarcated, and little progress has been made since (Brazilian Embassy 1996; SAIIC 2001).

The situation of indigenous lands in the Brazilian Atlantic Forest may sometimes lead to a crisis, as has happened with the Guaraní-Kaiowa of Mato do Sul. Over the last decade, more than 300 Indians have committed suicide out of despair and suffering. The Kaiowa have been expelled from their lands by cattle ranchers and resettled into recognized territories that are too small to provide subsistence. In the case of one village, 3,700 Kaiowa have been forced into an area of 36 km^2, making subsistence virtually impossible (SAIIC 2001; Survival International 2001).

Some indigenous peoples are defending their traditional lands and defining the patterns of resource use allowed within their borders. This is the case with the Pataxó of southern Bahia, who inhabit Monte Pascoal National Park (Figure 32.1). Ever since the park was created in 1961 to protect the Atlantic Forest, the Pataxó community has struggled to claim rights over resource use. Over the last 20 years, the Pataxó have engaged in the commercial extraction of timber, selling their products to local loggers (Rocha 1995). The Ministry of Environment and law-enforcement officials consider their occupation within the reserve's borders to be illegal. The Brazilian Indigenous National Fund (FUNAI) has been attempting to persuade the Pataxó to abandon their lands in return for consumer products (WRM 1999). With their population increasing almost 1.2 percent between 1965 and 1990, subsistence has become increasingly difficult for the Pataxó, and they live in poverty, dependent on assistance from the state government (Rocha 1995).

Paraguay

The 1992 Paraguay Constitution expands and raises to constitutional status the rights of indigenous peoples, strengthens the protection of their cultures, and officially identifies them as "peoples" (SDNP 2001). Despite the government's legal recognition of Paraguay's indigenous peoples, however, their well-being is threatened by continued encroachment on their lands and a lack of economic security (Haxton and Miller 1995).

One attempt to halt the process of rural transformation that is altering the natural and cultural landscape of Paraguay's Atlantic Forest was the creation of the Mbaracayú Nature Forest Reserve in eastern Paraguay (Chapter 28, this volume).

In 1992, the reserve was established to protect a large remnant of the Interior Atlantic Forest that is also the ancestral home of the Aché and Chiripá. To conserve this area effectively, management plans incorporated the Aché's traditional subsistence economy of hunting and gathering within the boundaries of the reserve, provided that only traditional weaponry, such as bows and arrows, is used (FMB 2001). Clearly, an understanding of Aché resource use was vital for successful conservation planning in Mbaracayú (Hill et al. 1997).

The law creating the reserve recognizes that the Aché have traditionally used the forest, and archaeological evidence points to the habitation of Paraguay's eastern forests for at least 10,000 years (Hill and Padwe 2000). In the 1970s, the Aché were forced to settle on government reservations. Since then, however, they have continued to trek into the forest to hunt, a practice they have pursued in Mbaracayú for at least the last century. Traditionally, the Aché also subsisted on honey, palm starch, and insect larvae. Today, swidden agriculture of manioc and maize complements their economic livelihood, as does the harvesting of fruits from the forest. Two Aché reservations, totaling approximately 500 people, are located near Mbaracayú (Hill et al. 1997). Analysis of hunting in the reserve by Aché communities suggests that this activity does not threaten populations of game animals (Hill and Padwe 2000).

The Chiripá system of agroforestry, notably the sustained extraction of *yerba mate,* has been integrated into the management scheme of Mbaracayú as a way to preserve the flora and fauna of the area (Reed 1995; TNC 2001). With assistance from the conservation community, the Chiripá have gained legal title to a strip of land between one of their settlements and the reserve, in which the sustainable extraction of *yerba* is taking place, thereby providing income to the community (TNC 2001).

Argentina

The Mbyá Guaraní of Misiones have two reserves in the province; both, however, have been continually invaded by settlers. As in most areas of the world, the indigenous people of Argentina are discriminated against even though the national constitution recognizes their political, social, economic, and cultural rights (Burke 1995). Large-scale commercial and private interests are developing lands that were demarcated as indigenous reservations, while the courts are delaying land disputes filed by native people (Burke 1995). The Mbyá still have a profound traditional knowledge of many of the resources of the Atlantic Forest (Chapter 18, this volume).

Conclusions

Since the European colonization of the region, the Atlantic Forests have undergone dramatic changes. As the world becomes increasingly globalized, the conservation community must look beyond state bureaucracies for help in defending protected areas of high biodiversity (Colchester 2000). The various groups of indigenous peoples in the Atlantic Forest region may or may not have conservation as a goal. In many cases, local peoples need to secure land and resource rights before they can be expected to conserve the biological resources upon which they depend (Linares 1997). Ultimately, creating alliances with forest-dwelling peoples, and recognizing them as crucial political actors in building an environmental constituency, is a far more effective way to achieve conservation than expelling them from their lands (Schwartzman et al. 2000). National governments have a responsibility to enforce mandated laws regarding the livelihood of indigenous peoples in Brazil, Paraguay, and Argentina. Recognition of the cultural needs and priority issues of indigenous peoples—such as land rights, education, and health care (Nations 2001)—must proceed hand in hand with conservation of the Atlantic Forests' biodiversity.

References

Bailey, R. C., Head, G., Jenike, M., Owen, G., Rechtman, R., and Zechenter, E. 1989. Hunting and gathering in tropical rainforest: is it possible? *American Anthropologist* 91: 59–82.

Balée, W. 1992. People of the fallow: a historical ecology of foraging in lowland South America. In: Redford, K. H. and Padoch, C. (eds.). *Conservation of neotropical forests: working from traditional resource use.* pp. 35–57. Columbia University Press: New York.

Begossi, A. 1998. Resilience and neo-traditional populations: the Caiçaras (Atlantic Forest) and Caboclos (Amazon, Brazil). In: Berkes, F. and Folke, C. (eds.). *Linking social and ecological systems: management practices and social mechanisms for building resilience.* pp. 129–157. Cambridge: Cambridge University Press.

Begossi, A., Leitão-Filho, H. F., and Richerson, P. J. 1993. Plant uses in a Brazilian coastal fishing community (Búzios Island). *J. Ethnobiology* 13(2): 233–256.

Brazilian Embassy. 1996. *Procedures for demarcation of indigenous lands in Brazil.* Online: http://www.brasilemb.org/indian_demarcation_procedures.shtml.

Brown, J. H. and Gibson, A. C. 1983. *Biogeography.* St. Louis: C. V. Mosby.

Burke, P. 1995. *Indigenous people in Argentina.* Centro de Documentación Mapuche. Online: http://linux.soc.uu.se/mapuche/indgen/Indarg01.html.

Chicchón, A. 2000. Conservation theory meets practice. *Conservation Biology* 14(5): 1368–1369.

Colchester, M. 2000. Self-determination or environmental determinism for indigenous peoples in tropical forest conservation. *Conservation Biology* 14(5): 1365–1367.

Dean, W. 1995. *With broadax and firebrand: the destruction of the Brazilian Atlantic Forest.* Berkeley, CA: University of California Press.

DGEEC (Dirección General de Estadística, Encuestas y Censos). 2001. Census 2000. Online: http://www.dgeec.gov.py/index.htm.

Faúndez, C. 2001. *El pueblo Mbyá Guaraní.* Online: http://www.geocities.com/RainForest/Andes/8976/mbya.htm.

Figueiredo, G. M., Leitão-Filho, H. F., and Begossi, A. 1993. Ethnobotany of Atlantic Forest coastal communities: Diversity of plant uses at Sepetiba Bay (SE Brazil). *Human Ecology* 25(2): 353–361.

FMB (Fundación Moisés Bertoni). 2001. *Institutional programs: Mbaracayú Forest Nature Reserve.* Online: http://www.mbertoni.org.py/ingles.htm.

FUNAI (Fundação Nacional do Indio). 2001. Online: http://www.funai.gov.br/funai.htm.

Galetti, M. 2001. Indians within conservation units: lessons from the Atlantic Forest. *Conservation Biology* 15(3): 798–799.

Ghimere, K. and Pimbert, M. P. 1996. *Social change and conservation.* London: Earthscan.

Guzman, D. 1993. *Argentina: Guaraní people vs. apartheid.* Online: http://nativenet.uthscsa.edu/archive/nl/9308/0259.html.

Hanazaki, N., Tamashiro, J. Y., Leitão-Filho, H. F., and Begossi, A. 2000. Diversity of plant uses in two Caiçara communities from the Atlantic Forest coast, Brazil. *Biodiversity and Conservation* 9: 597–615.

Haxton, M. L. and Miller, T. 1995. *Indigenous peoples in Paraguay.* Online: http://www.bsos.umd.edu/cidcm/mar/indpar.htm.

Hill, K. and Padwe, J. 2000. Sustainability of Aché hunting in the Mbaracayú Reserve, Paraguay. In: Robinson, J. G. and Bennett, E. L. (eds.). *Hunting for sustainability in tropical forests.* New York: Columbia University Press.

Hill, K., Padwe J., Bejyvagi, C., Bepurangi, A., Jakugi, F., Tykuarangi, R., and Tykuarangi, T. 1997. Impact of hunting on large vertebrates in the Mbaracayú Reserve, Paraguay. *Conservation Biology* 11(6): 1339–1353.

HRAF (Human Relations Area Files). 2001. *EHRAF collection of ethnography.* Published annually by Human Relations Area Files (HRAF), Yale University. Online: http://www.yale.edu/hraf/workbook.htm.

IBGE (Instituto Brasileiro de Geografia e Estatística). 2001. *Popclock, population for Brazil.* Online: http://www.ibge.gov.br.

Leitao, A. 1994. Indigenous peoples in Brazil, the Guaraní: a case for the UN. *Cultural Survival Quarterly* Spring 1994: 48–50.

Linares, O. F. 1997. Creating cultural diversity: tropical forests transformed. In: Raven, P. H. (ed.). *Nature and human society: the quest for a sustainable world.* pp. 420–434. Washington, DC: National Academy Press.

Lizarralde, M. 2001. Biodiversity and loss of indigenous languages and knowledge in South America. In: Maffi, L. (ed.). *On biocultural diversity: linking language, knowledge and the environment.* pp. 265–281. Washington, DC: Smithsonian Institution Press.

Martin, M. and Beierle, J. M. 2001. *Society-GUARANI.* : http://lucy.ukc.ac.uk/EthnoAtlas/Hmar/Cult_dir/Culture.7843.

McNeely, J. A., Gagdil, M., Leveque, C., Padoch, C., and Redford, K. 1995. Human influences on biodiversity. In: Heywood, V. H. (ed.). *Global biodiversity assessment.* pp. 711–821. Cambridge: Cambridge University Press.

Metraux, A. 1948. The Guaraní. In: Steward, Julian H. (ed.). *Handbook of South American Indians,* vol. 3. pp. 69–94. Washington, DC: Government Printing Office.

MNSU (Minnesota State University). 2001. *Population of Paraguay.* Online: http://emuseum.mnsu.edu/information/population/country.php?FILE=PA&NAME=Paraguay.

Mooney, J. 1910. Guaraní Indians. In: *The Catholic Encyclopedia, Volume VII.* Online: http://www.newadvent.org/cathen/07045a.htm.

Nations, J. D. 2001. Indigenous peoples and conservation: misguided myths in the Maya tropical forest. In: Maffi, L. (ed.). *On biocultural diversity: linking language, knowledge and the environment.* pp. 379–396. Washington, DC: Smithsonian Institution Press.

Patis, M. L. 1992. *Dossier Atlantic Rain Forest (Mata Atlântica).* São Paulo: Fundação SOS Mata Atlântica.

Posey, D. A. 1992. Interpreting and applying the "reality" of indigenous concepts: what is necessary to learn from the natives? In: Redford, K. H. and Padoch, C. (eds.). *Conservation of neotropical forests: working from traditional resource use.* pp. 35–57. New York: Columbia University Press.

Posey, D. A. 2001. Biological and cultural diversity. In: Maffi, L. (ed.). *On biocultural diversity: linking language, knowledge and the environment.* pp. 379–396. Washington, DC: Smithsonian Institution Press.

Redford, K. H. and Sanderson, S. E. 2000. Extracting humans from nature. *Conservation Biology* 14(5): 1362–1364.

Reed, R. 1995. *Prophets of agroforestry: Guaraní communities and commercial gathering.* Austin: University of Texas Press.

Reed, R. 1997. *Guide to the SM04 Guaraní file.* Online: http://ets.umdl.umich.edu/cgi/e/ehraf/hraf-idx?type=html&rgn=GUIDE&byte=467557953.

Rocha, C. F. D. 1995. Monte Pascoal National Park: indigenous inhabitants versus conservation units. In: Amend, S. and Amend, T. (eds.). *National parks without people? The South American experience.* pp. 147–158. Gland, Switzerland: IUCN (International Union for Conservation of Nature).

SAIIC (South and Meso American Indian Rights Center). 2001. *Indigenous rights in Brazil: stagnation to political impasse.* Online: http://saiic.nativeweb.org/brazil.html.

Schmink M., Redford, K. H., and Padoch, C. 1992. Traditional peoples and the biosphere: framing the issues and defining the terms. In: Redford, K. H. and Padoch, C. (eds.). *Conservation of neotropical forests: working from traditional resource use.* pp. 3–16. New York: Columbia University Press.

Schwartzman S., Nepstad, D., and Moreira, A. 2000. Arguing tropical forest conservation: people versus parks. *Conservation Biology* 14(5): 1370–1374.

SDNP (Sustainable Development Networking Programme). 2001. *Indigenous rights in the constitutions of countries in the Americas.* Online: http://www.sdnp.org.gy/apa/topic4.htm.

Smith, N. J. H., Williams, J. T., Plucknett, D. L., and Talbot, J. P. 1992. *Tropical forests and their crops.* Ithaca: Cornell University Press.

Survival International. 2001. *General information about the Guaraní.* http://www.survival-international.org/tc%20guarani.htm.

Terborgh, J. 2000. The fate of tropical forests: a matter of stewardship. *Conservation Biology* 14(5): 1358–1361.

TNC (The Nature Conservancy). 2001. *Paraguay program.* Online: http://nature.org/wherewework/southamerica/paraguay/index.html.

World Gazetteer. 2001. *Argentina.* Online: http://www.gazetteer.de/fr/fr_ar.htm.

WRM (World Rainforest Movement). 1999. Brazil: Pataxó recover traditional lands. *WRM. Bulletin* 26 (August). Online: http://www.wrm.org.uy/bulletin/26/Brazil.html.

WWF (World Wildlife Fund). 2000. *Environmental degradation aggravated by loss of traditional knowledge.* Online: http://www.panda.org/news_facts/newsroom/news.cfm?uNewsId=2169&uLangId=1.

Chapter 33

Unwanted Guests: The Invasion of Nonnative Species

Jamie K. Reaser, Carlos Galindo-Leal, and Silvia R. Ziller

South America has experienced three major waves of biotic invasion. Approximately 3 million years ago, the rise of the Isthmus of Panama enabled the influx of wildlife from North America during what has become known as the Great Faunal Interchange. Humans began to arrive in large numbers from Asia approximately 10,000 to 15,000 years ago, and Europeans within the last 500 years. All three waves changed the fauna and flora of South America dramatically. The first caused the decline of ungulates, large carnivorous marsupials, and flightless birds. The second contributed to the extinction of many large mammals. During the third wave, the Europeans converted large expanses of natural vegetation to plantation agriculture (Figure 33.1) and brought diseases to the indigenous populations (Crosby 1993; Dean 1996).

South America's fourth wave of invasion is currently under way: a global-scale reshuffling of biotas due to rapidly expanding international trade and travel and ongoing changes in land use and climate. The rate of invasion, its diversity, and the volume of invaders are historically unprecedented (Bright 1998; McNeely et al. 2001).

Sometimes new species are introduced intentionally, for agriculture, forestry, and ranching, or to stock fisheries (Rapoport 1992; Gurgel and Fernando 1994; Lukefahr et al. 1994). At other times, they are introduced accidentally. For example, wood packaging materials can transport nonnative insects, and tourists can transport microbes or seeds on their shoes (Bright 1998; McNeely et al. 2001).

Although most translocated species either do not survive or exist in populations that are small enough to manage (Bright 1998, 1999), about one out of every thousand thrives in its new environment (Williamson 1996). These "invasive alien species," as they are collectively known, reproduce, spread, and cause serious harm to the environment, economy, or human health.

Society pays a great price for invasive alien species—costs measured not just in currency but in unemployment, damaged goods and equipment, power failures, food and water shortages, environmental degradation, loss of biodiversity,

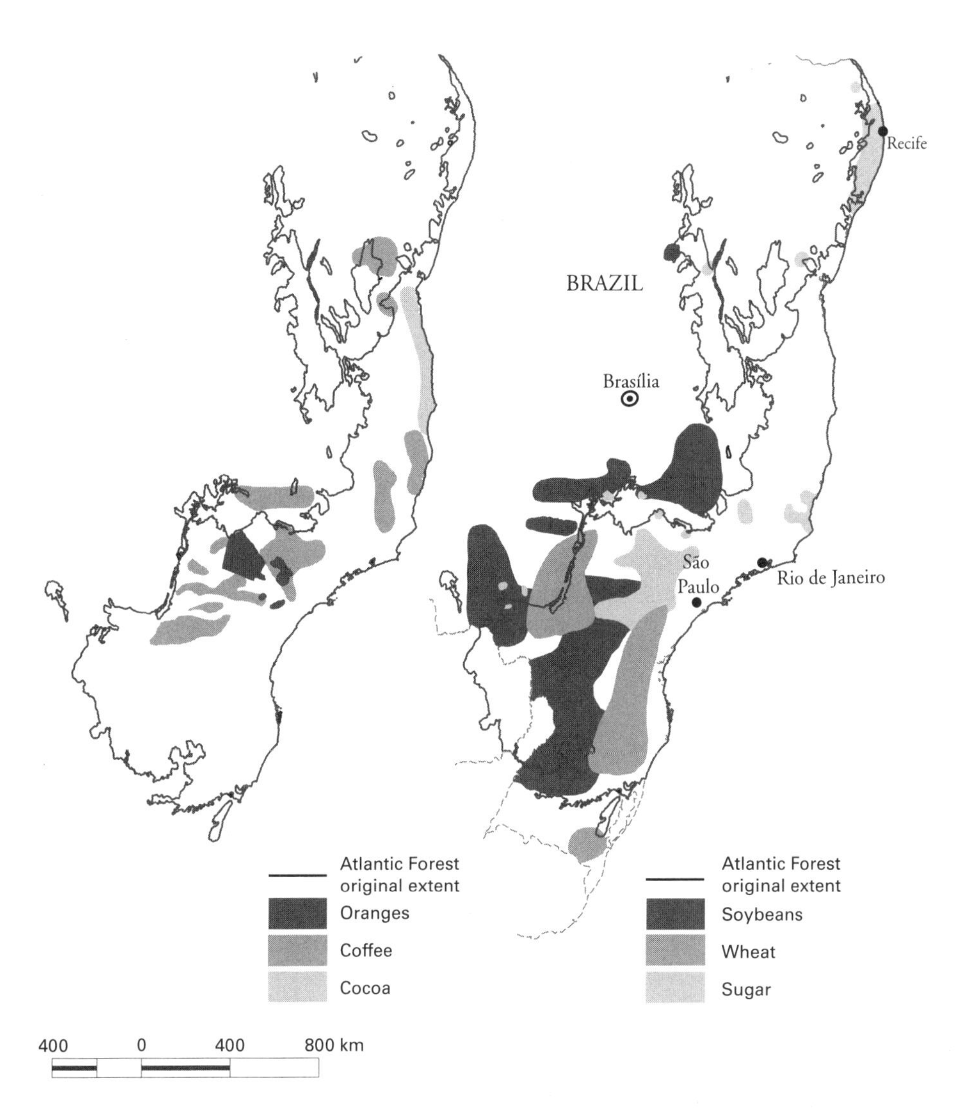

Figure 33.1. Distribution of main agricultural crops in the Atlantic Forest hotspot (NOAA/USDA).

increased rates and severity of natural disasters, disease epidemics, and lost lives (Bright 1998; McNeely et al. 2001). Invasive alien species often have synergistic and cascading impacts, influencing many aspects of environmental and human well-being over long periods of time.

Relatively little is known about which invasive alien species inhabit the Atlantic Forest, and their impacts have received even less study. Thus, we offer some broad-scale examples of the impacts of invasive alien species, as well as a few well-documented case studies from the Atlantic Forest.

Environmental Impacts

Globally, one of the most significant drivers of environmental change is invasive alien species (Rapoport 1992; McNeely 2001; McNeely et al. 2001). Invasive alien species now rank second to habitat conversion as a cause of species endangerment and extinction worldwide (Wilcove et al. 1998). Even the most well-protected natural areas are not immune to the invasion of nonnative species (GISP 2001).

Fish Introductions

Fish are introduced around the world for a wide variety of purposes—aquaculture, sport fishing, mosquito and weed control, or as ornamentals (Godinho 1996; Fontenelle and Wille 2001)—but such introductions have often resulted in both economic and ecological losses. The latter include displacement and extinction of native fish species through competition, predation, pathogens, or changes in habitat structure. Africa offers one of the most dramatic examples: In the 1950s, the Nile perch (*Lates niloticus*) was introduced into Lake Victoria (Uganda, Kenya, Tanzania) to increase the regional food supply. As a result of its highly predatory nature, coupled with overfishing and lake eutrophication, the diversity of endemic fishes (more than 300 cichlids) plummeted. About two-thirds of the species in the lake have gone extinct or are threatened with extinction (Ogutu-Ohwayo 1990; Kaufman 1992).

In Brazil, fish introductions were documented as early as the late 1800s and still continue (Godhino 1996; Orsi and Agostinho 1999; Alves et al. 1999; Fontenelle and Wille 2001). Some of the introductions occur through government programs that fail to assess and mitigate the consequences.

In northeastern Brazil, the Ministry of National Integration's Department of Action Against Droughts (DNOCS) has introduced 42 species of fish and crustaceans into approximately 100 reservoirs. Fourteen fish species and one shrimp have become established, yet stocking continues (Gurgel and Fernando 1994). In the state of Rio de Janeiro, 37 species of nonnative fish have been recorded (Fontenelle and Wille 2001). In several lakes in Minas Gerais, the number of native fish species was reduced by 50 percent (Godinho 1996) within 10 years of the introduction of Amazon tucunaré (*Cichla ocellaris*), cará-amazonas (*Astronotus ocellatus*), and piranhas (*Pygocentrus nattereri*). Recently, a North African catfish (*Clarias gariepinus*) was recorded in the Rio Paraopeba (São Francisco River basin), the Rio Grande (Paraná River basin), and the Rio Doce (Doce River basin) in Minas Gerais (Alves et al. 1999). This is a voracious and opportunistic feeder on insects, crabs, plankton, snails, and fish, but it will also take young birds, carrion, plants, and fruits.

During catastrophic flooding in the Paranapanema and Tibagi rivers in January 1997, almost 1.3 million fish comprising 10 nonnative species and 1 hybrid escaped from aquaculture facilities into neighboring rivers. The disaster is attributed to the illegal human occupation of the riverbanks and to the nonobservance of standard measures designed to prevent escape (Orsi and Agostinho 1999).

Africanized Bees

In 1956, honeybee queens (*Apis mellifera scutellata*) from South Africa and Tanzania were brought to Rio Claro, Brazil, to improve honey production (Kerr 1967). A year later, a few swarms escaped, established, and began expanding their range at a rate of 330–500 km/year, reaching Mexico in the 1980s and the United States in the 1990s (Merrill and Visscher 1995) (Figure 33.2). The African bees

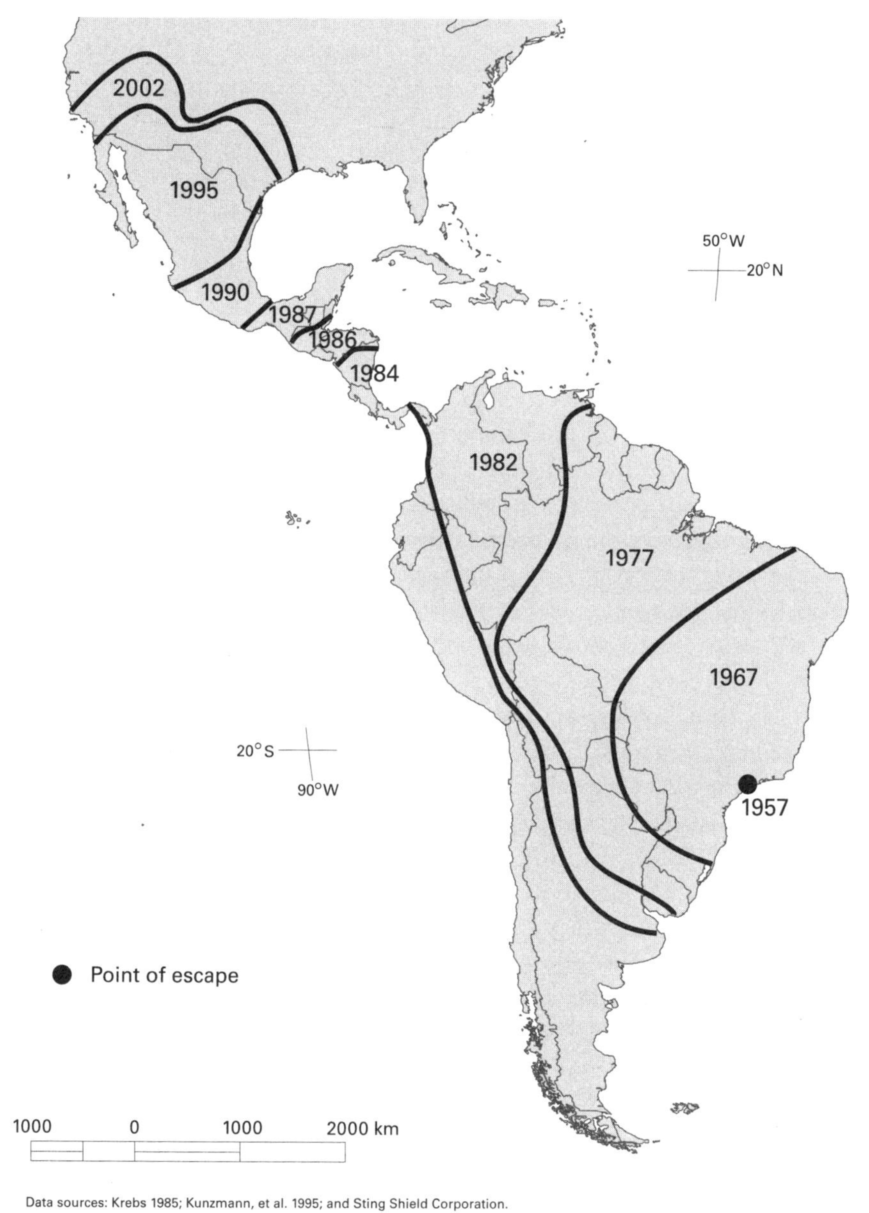

Figure 33.2. Expansion of African bees from Rio Claro in São Paulo, Brazil, to the United States during the last 45 years.

hybridized with European races (*Apis mellifera ligustica*), giving rise to extremely aggressive "Africanized bees."

The ecological consequences of the invasion, although poorly documented, suggest increased competition for floral resources and tree cavities. At Boraceia Biological Station in the state of São Paulo, Brazil, the most abundant flower visitor is the Africanized honeybee. A highly eusocial native bee community composed of 17 species of stingless bees (*Meliponini, Trigonini*) may be declining as result (Wilms et al. 1996). Africanized bees have greater dispersal abilities and a higher frequency of swarm formation, and they are less selective in their choice of nesting cavities than other bees (Schmidt and Thoenes 1990, 1995; Schmidt and Hurley 1995). As a result, they can form large densities—five to nine colonies per square kilometer in tropical regions (Ratnieks et al. 1991)—and compete for cavities with a wide variety of animals. Even macaws, which typically are aggressive birds, have been displaced from their nests (Pittman 1997).

Plantation Invasions of Sand Dune Habitats

Two nonnative species of pine (*Pinus elliottii* and *P. taeda*) are now common in southern and southeastern Brazil. They were introduced from North America in the 1960s and 1970s and extensively planted with the support of government incentives for increased forest production (Caruso 1990). Many experimental plots were subsequently abandoned, which enabled the trees to establish and spread. Sensitive habitats suffered losses to biodiversity as a result. On Santa Catarina Island's Mozambique Beach (southern coast of Brazil), for example, pines became established along 14 km of sand dune–covered coastline. Because of strong and constant winds, the pines grow low to the ground and spread sideways, outcompeting the once-diverse native plant species, including orchids and bromeliads, for light. This decline in biodiversity and scenic values has greatly reduced potential revenues from ecotourism (Ziller 2000).

Similar impacts have occurred with the introduction of casuarinas trees (*Casuarina equisetifolia*). Originally brought from Australia for dune stabilization, they were extensively planted along the coast of Santa Catarina State from the 1960s to the 1980s. As with the invasive pines, the ability of casuarinas to outcompete native dune vegetation for light has had devastating consequences for species diversity, as well as for scenic landscape values and the potential for ecotourism (Ziller, pers. obs. 2002). The impact of casuarinas has been especially great on Rosa and Ouvidor beaches in the municipality of Garopaba and the Mozambique Beach in Florianopolis. Because control efforts are lacking and strong prevailing winds from the ocean efficiently disperse the seeds, the plants are able to become established long distances from the point of initial introduction. The species is also widely invasive along the coast of Rio Grande do Sul State (Ziller, pers. obs. 2002).

Sand olive (*Dodonaea viscose;* "vassoura-vermelha"), originally from Australia and the West Indies, was also planted in coastal zones around the world for dune stabilization. In Brazil's Santa Catarina State, it dominates the initial phase of establishment of plants on inner sand dunes, shading out the native vegetation composed of herbs and low bushes. Strong ocean winds disperse seeds widely, so the species is also abundant in degraded forest areas on low hills along the coast, where sand olive outcompetes native species in the initial stage of development. On slopes, it is eventually outcompeted by taller native trees. Sand olive is also present in the central, northern, and western regions of Rio Grande do Sul State, and it is the most common species of shrub along roads in the state's southeastern region. It is particularly invasive in degraded subtropical forest areas and grasslands used for cattle ranching (Ziller, pers. obs. 2002).

Plant Invasions of Forested Habitats

Patient Lucy (*Impatiens walleriana;* "beijo," "maria-sem-vergonha") was introduced from the archipelago of Zanzibar, in eastern Africa, for ornamental purposes and is cultivated throughout Brazil. Seedlings can be purchased in most nurseries. Now established in the wild, its dense and colorful thickets are a tourist attraction along 20 km of a scenic road from Curitiba to the coast of Paraná State. The plant has invaded Marumbi State Park and the Atlantic Forest, spreading by seeds and stolons. Patient Lucy is shade tolerant and outcompetes native herbs, and it dominates the understory of Brazil pine (*Araucaria angustifolia*) forests throughout the three southern states of Brazil, especially on damp soils (Ziller, pers. obs. 2002).

The guava tree (*Psidium guajava*; "goiabeira"), originally from Mexico, has been planted so widely in Brazil that most people believe it is native. Its fruits are edible and are widely dispersed by birds. It is invasive on degraded lowland forest areas along the southern coast, in the interior southeastern range, and in the central depression in Rio Grande do Sul State. Although the species is a dominant pioneer tree, it tends to be shaded out by native trees in subsequent successional stages (Ziller, pers. obs. 2002).

Invasion of Wetland Habitats

White ginger (*Hedychium coronarium;* "lírio-do-brejo," "açucena") is native from the Himalayan region to China. Shade tolerant, it dominates large areas of herbaceous wetlands and the undergrowth of lowland forests on the coast of São Paulo, Paraná, Santa Catarina, and the north of Rio Grande do Sul. Although invasions are denser on wet soils, the species is also present in the undergrowth of forests on the low slopes along the coast, as well as along roads and railroad tracks crossing the coast range (Ziller, pers. obs. 2002).

Spreading by seed and stolons, white ginger suppresses native species in early stages of succession on the coastal plain. The invasions result in the loss of biodiversity and of plants traditionally used by local people for production of baskets, ropes, and other products. The plant is one of the most abundant species on the entire southern coast and was formerly used for paper production in the city of Morretes, near the coast of Paraná State (Ziller, pers. obs. 2002).

Economic Impacts

Invasive alien species such as those described above can take a heavy economic toll on governments, industries, and private citizens, although there are few quantitative studies that document the impact (Perrings et al. 2000). A recent study estimated that invasive alien species cost the United States more than $100 billion a year (Pimentel et al. 2000). Worldwide, the losses to agriculture have been estimated to be between $55 billion and nearly $248 billion annually (Bright 1999). But, controlling a single species can carry a price tag in the millions. For example, the Formosan termite (*Coptotermes formosanus*) costs an estimated $1 billion annually in property damage, repairs, and control measures in the southeastern United States (Suszkiw 1998).

Costs from invasive alien species are also incurred when specific commodities or transport systems are affected. Because trade disputes may arise over "pest risks" (Bright 1999), the spread of invasive alien species may result in restrictions on the importation of certain commodities. These include bans on (1) certain food products that might spread destructive pests or diseases that kill crops, livestock, or people; (2) commodities such as horticultural products, seeds, and pets that could escape into and damage the environment, or (3) certain types of shipping containers that could harbor pests potentially destructive to agriculture, forestry, fisheries, or the natural environment.

Little information exists on economic losses from introduced species in Brazil, but annual losses due to alien weeds, mites, and plant pathogens in croplands alone amount to $42.6 billion (Pimentel et al. 2001). In addition, introduced livestock and human diseases are enormously costly to Brazilian society. Weevils, witches broom, and signal grass are three examples of highly destructive invasives that have impacted both biodiversity and the economies of the Atlantic Forest.

Weevils versus Cotton

Cotton, originally from Central America, has been an important crop of Brazil since the arrival of the Portuguese. Herbaceous cotton (*Gossypium hirsutum* L. var. *latifolium* Hutch) was particularly productive in the states of Paraná, São Paulo, and Bahia, and arboreous cotton (*Gossypium hirsutum* L. variety *marie-gallant* Hutch) was productive in the northeast (Pernambuco). Together, they formed the basis of the Brazilian economy for nearly a century. In the last 15 years, however, cotton production has steadily declined. Brazil, once a great exporter, is now one of the biggest cotton importers in the world, at an annual cost of millions of dollars. The decline is due in large part to the invasion of the boll weevil (*Anthonomus grandis*), native to Mexico or Central America (Pierce 1998). The weevil probably entered the region as a "hitchhiker" on imported plant material. It was first detected in 1983 in the states of São Paulo and Paraíba, and dispersal throughout the major cotton-producing areas was phenomenally rapid. By 1991, every cotton-growing state in Brazil had been infested (Lukefahr et al. 1994).

Witch's Broom versus Cocoa

As recently as 1985, Brazil was the world's second largest cocoa (*Theobroma cacao*) producer, after Africa's Côte d'Ivoire. The balance has reversed. The Côte d'Ivoire now produces 43 percent of the world's supply, and Brazil's production has dropped from 21 percent of the world' supply in 1985–1986 to only 9 percent in 1989–1999 (Gray 2000). Production in Bahia tumbled from 378,000 tons in 1990–1991 to 118,700 tons in 1999–2000 (Gray 2000). Global competition was partly responsible, as was the spread of witch's broom, a disease caused by a fungus (*Crinipellis perniciosa*) that spread from Amazonia and is now present on 98 percent of the state's cocoa farms. The estimated economic losses to Brazil in the last 7 years range from $400 million to $1 billion and the loss of jobs from 50,000 to 250,000. During 2001, Brazil spent more than $61 million on cocoa imports (*Los Angeles Times* 2001).

Invasion of Signal Grass

Several species of grasses ("braquiária," "capim d'angola") in the genus *Brachiaria* (*B. mutica, B. decumbens, B. brizantha,* and *B. humidicola* are the most common species) were introduced from Africa as forage crops and have become invasive in pasture areas throughout Brazil. They are also used in revegetation programs associated with road construction. The plants dominate vast tracts of grassland, savanna, wetlands, and forest areas converted to agriculture. In the southern grasslands, native species contain higher proportions of protein and nutrients than signal grass, so invasions result in lower productivity of milk and beef (Fabio Rosa, pers. comm. 2001). On the Atlantic coast of Paraná State, these species often invade small streams, transforming conditions for aquatic fauna.

Health Impacts

Invasive alien species also can affect the health of humans, plants, and animals. Pathogens and parasites may themselves be invasive alien species or may be introduced by invasive vectors (Bright 1998). Recently, foot-and-mouth disease has become a significant concern in many regions of the world (Enserink 2001). Bubonic plague, spread by nonnative rats carrying infected fleas, is a well-known historic case. Cholera (*Vibrio cholerae*) and some of the microorganisms that can cause harmful algal blooms can be transported in and released from the ballast water carried by large ships (Wilson 1995).

One of the most predominant tropical diseases, schistosomiasis (also known as bilharzia) was introduced from Africa to South America during the slave trade (Desprès et al. 1993). An intestinal disease caused by blood flukes of the genus *Schistosoma,* it is the second most pervasive tropical disease in the world, after malaria. An estimated 150 million to 200 million people are infected worldwide (WHO 1998). Three local freshwater snails (*Biomphalaria glabrata, B. tenagophila,* and *B. straminea*) are mainly responsible for its transmission. The infection causes chronic urinary tract disease and often results in cirrhosis of the

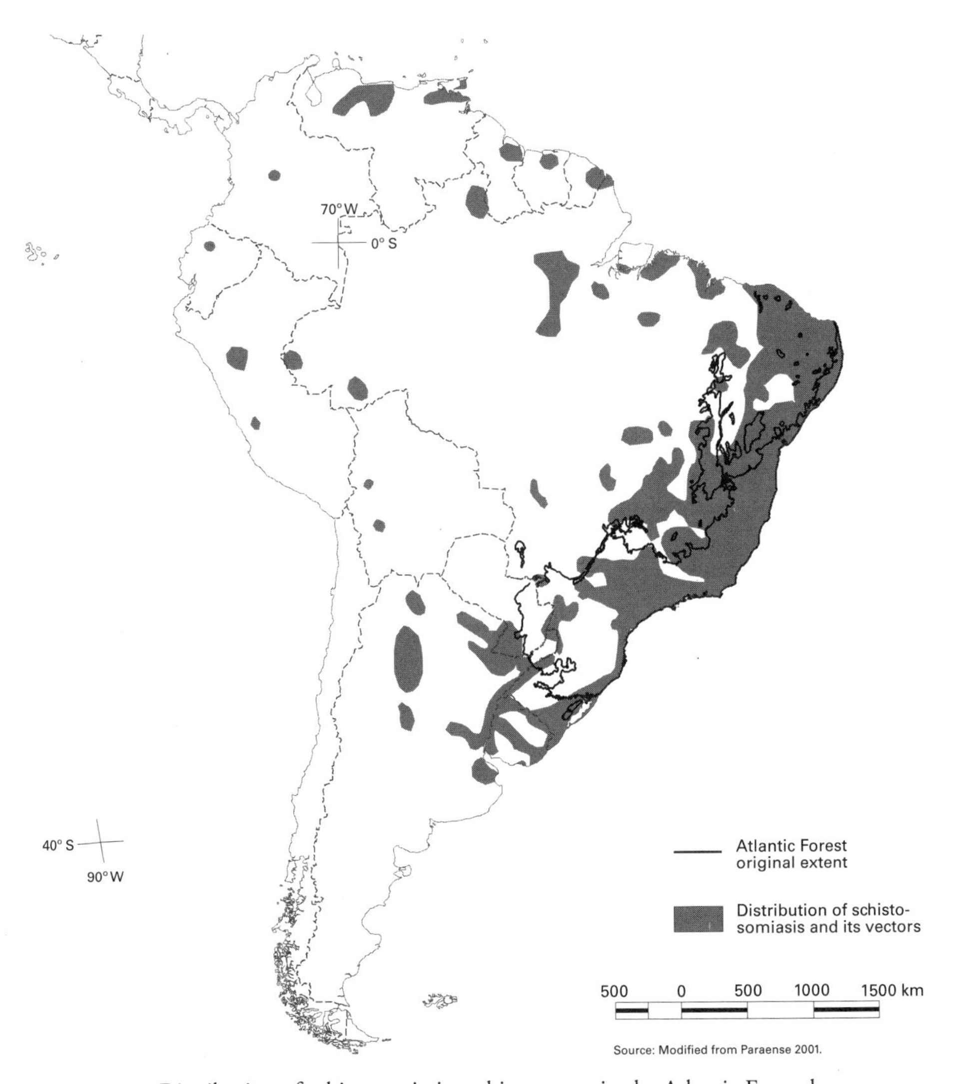

Figure 33.3. Distribution of schistosomiasis and its vectors in the Atlantic Forest hotspot.

liver and bladder cancer. It is most prevalent among children. Incidence is greatest in Brazil, with 3 million infected (WHO 1998), and distribution of the disease closely matches that of the Atlantic Forest and part of the *Cerrado* and *Caatinga* (Figure 33.3).

Population movements from northeastern Brazil have extended the area of infection to other regions (WHO 1998). The spread of schistosomiasis to northern Paraná is relatively recent. This previously forested region was settled in 1867 with emigrants from São Paulo and Minas Gerais. After the establishment of large-scale coffee plantations in the 1920s, new immigration followed from northeastern states, the seat of the oldest foci of schistosomiasis (Paraense 2001). Ecological changes due to dam construction (Chapter 37, this volume) have been responsible for the increase of the disease in many countries.

The Malaysian trumpet snail (*Melanoides tuberculata*), widely used in aquariums, was introduced for biological control of freshwater snails (*Biomphalaria* spp.) in several parts of the Atlantic Forest including Rio de Janeiro, São Paulo, and Minas Gerais (De Marco 1999; Giovanelli et al. 2001). These snails compete for food with and eat the eggs of the host snails. Unfortunately, the Malaysian trumpet snail is, in turn, an intermediate host for other parasites, especially a lung fluke (*Paragonimus* spp.) that can also affect humans. Paragonomiasis occurs through the ingestion of raw infected crustaceans (a second intermediate host), a practice that is spreading among fishermen and nearshore populations along the Paraná River and Rio Grande (São Paulo) in Brazil (Giovanelli et al. 2001).

Challenges for Prevention and Control

The prevention and control of invasive alien species presents many scientific, political, and ethical challenges (McNeely 2001). The process of invasion is often complex, resulting in considerable scientific uncertainty (Bright 1998; Bright 1999; Mooney and Hobbs 2000). Implementing effective prevention and control measures can be costly and can require new policy approaches, as well as significant advances in ecological knowledge and natural resource management (Shine et al. 2000; McNeely et al. 2001; Wittenberg and Cock 2001).

Because every country is an exporter and importer of goods and services, every country is also a facilitator and victim of the invasion of nonnative species. Furthermore, not only are invasive alien species moved—intentionally or inadvertently—but they can move themselves. Thus, once invasive alien species become established within one country, they pose a threat to an entire region, as well as to trading partners and every country along a trading pathway (McNeely et al. 2001; NISC 2001). As demands for international trade, tourism, and travel increase, minimizing the spread and impact of invasive alien species will become more challenging unless we adjust our values and develop a new ethic of responsibility (Hattigh 2001; Reaser 2001).

More than 40 international agreements and numerous codes of conduct directly address invasive alien species (Shine et al. 2000). Few countries, however, have developed well-coordinated national policies and programs to manage the problem. Neighboring countries and trading partners are often unaware of one another's policies and practices (NISC 2001).

Methods to limit the spread of invasive alien species can be controversial. Some animal rights groups oppose the eradication of invasive alien species, especially large mammals (Low 1999). Pesticide application, such as the use of DDT to control mosquitoes in malaria-infested regions, raises human health concerns (Bright 1998; Parker 2001). And some scientists and environmental groups worry that biological control agents (living organisms imported to control pests) pose risks greater than those of the already established alien species (Bright 1998; Strong and Pemberton 2000).

Regional Opportunities: Assessment and Strategic Planning

In closing, it is clear that no country will be able to succeed in addressing its domestic problems with invasive alien species unless it actively engages in international cooperation. Multisector partnerships need to be forged that include governments sharing borders and trade routes, intergovernmental organizations, industries (especially agriculture, forestry, and aquaculture), nongovernmental organizations, and donor agencies. In the case of the Atlantic Forest, it will be particularly important to encourage the participation of agencies that manage the region's numerous protected areas.

The evidence we present here is critical to effectively addressing the threat of invasive species in the Atlantic Forest. We recommend five priorities for immediate action:

1. Undertake a baseline assessment of invasive alien species and invasive alien species projects in the Atlantic Forest
2. Identify, rank, and monitor the vectors and pathways by which invasive alien species enter the Atlantic Forest
3. Contribute resultant data to the Inter-American Biodiversity Information Network (IABIN; http://www.iabin-us.org/) and the databases of the Inter-American Institute for Agricultural Cooperation (IICA)
4. Share this information with policymakers within the region, as well as through the Convention on Biological Diversity (CBD; http://www.biodiv.org) and International Plant Protection Convention (IPPC; http://www.fao.org/legal/treaties/004t-e.htm)
5. Develop and implement a plan for the early detection of and rapid response to the introduction of nonnative species in the Atlantic Forest. Park guards and ecotourism guides should be employed as members of "early detection and rapid response teams."

Acknowledgments

Grateful thanks to Jeff Waage, Penny Langhammer, Mario Luis Orsi, Dick Mack, Laurie Neville, and participants of the workshop "Prevention and Management of Invasive Alien Species: Forging Cooperation throughout South America."

References

Alves, C. B. M., Vono, V., and Vieira, F. 1999. Presence of the walking catfish *Clarias gariepinus* (Burchell) (Siluriformes, Clariidae) in Minas Gerais State hydrographic basins, Brazil. *Revista Brasileira de Zoologia* 16(1): 259–263.

Bright, C. 1998. *Life out of bounds: bioinvasion in a borderless world.* New York: W. W. Norton.

Bright, C. 1999. Invasive species: pathogens of globalization. *Foreign Policy* Fall 1999: 50–64.

Caruso, M. M. L. 1990. *O desmatamento da Ilha de Santa Catarina de 1500 aos dias atuais.* 2nd ed. Florianópolis, Brazil: Editora da UFSC.

Crosby, A. W. 1993. *Ecological imperialism, the biological expansion of Europe, 900–1900.* Cambridge: Cambridge Press.

Dean, W. 1996. *A ferro e fogo: A história e a devastação da Mata Brasileira.* São Paulo: Companhia das Letras.

De Marco Jr., P. 1999. Invasion by the introduced aquatic snail *Melanoides tuberculata* (Mueller, 1774) (Gastropoda: *Prosobranchia thiaridae*) of the Rio Doce State Park, Minas Gerais, Brazil. *Studies on Neotropical Fauna and Environment* 34(3): 186–189.

Desprès, L., Imbert-Establet, D., and Monnerot, M. 1993. *Molecular and Biochemical Parasitology* 60(2): 221–230.

Enserink, M. 2001. Barricading U.S. borders against a devastating disease. *Science* 291: 2298–2300.

Fontenelle, C. R. S. and Wille, L. N. R. 2001. Espécies de peixes introduzidas nos ecossistemas aquáticos continentais do estado do Rio de Janeiro, Brasil. *Comunicações do Museu de Ciências Techologia, Série Zoologia* 14(1): 43–59.

Giovanelli, A., Pinto Ayres Coelho da Silva, C. L., Medeiros, L., and Carvalho de Vasconcellos, M. 2001. The molluscicidal activity of the latex of *Euphorbia splendens* var. *hislopii* on *Melanoides tuberculata* (Thiaridae), a snail associated with habitats of *Biomphalaria glabrata* (Planorbidae). *Memorias do Instituto Oswaldo Cruz* 96(1): 123–125.

GISP (Global Invasive Species Programme). 2001. *Regional invasive alien species resource directory for South America.* Version 1. Global Invasive Species Programme, Stanford University, Stanford, California.

Godinho, A. L. 1996. *Peixes do Parque Estadual do Rio Doce.* Belo Horizonte, Brazil: Instituto Estadual de Florestas, Universidade Federal de Minas Gerais.

Gray, A. 2000. *The world cocoa market outlook.* LMC International Report. Yaoundé, Cameroon.

Gurgel, J. J. S. and Fernando, C. H. 1994. Fishes in semi-arid northeast Brazil with special reference to the role of tilapias. *Internationale Revue der gesamten Hydrobiologie.* Berlin 79(1): 77–94.

Hattinh, J. 2001. Human dimensions of invasive alien species in philosophical perspective: towards an ethic of conceptual responsibility. In McNeely, J. A. (ed.). *The great reshuffling: human dimensions of invasive species.* Cambridge: World Conservation Union (IUCN).

Kaufman, L. 1992. Catastrophic change in species-rich freshwater ecosystems: the lessons of Lake Victoria. *BioScience* 42(11): 846.

Kerr, W. E. 1967. The history of the introduction of African bees to Brazil. *South African Bee Journal* 39: 3–5.

Los Angeles Times. 2001. Brazil's cocoa blight points to global spread of farm pests. p. 15. 15 April 2001.

Low, T. 1999. *Feral future: the untold story of Australia's exotic invaders.* New York: Viking Press.

Lukefahr, M. J., Barbosa, S., and Braga, R. S. 1994. The introduction and spread of the boll weevil, *Anthonomus grandis* Boh. (Coleoptera) in Brazil. *Southwestern Entomologist* 19(4): 414–418.

McNeely, J. A., Mooney, H. A., Neville, L. E., Schei, P. J., and Waage, J. K. (eds.). 2001. *Global strategy on invasive alien species.* Cambridge: IUCN.

McNeely, J. (ed.). 2001. *The great reshuffling: human dimensions of invasive alien species.* Gland, Switzerland: IUCN.

Merrill, L. D. and Visscher, P. K. 1995. Africanized honey bees: A new challenge for fire managers. *Fire Management Notes* 55(4): 25–30.

Mooney, H. A. and Hobbs, R. J. (eds.). 2000. *Invasive species in a changing world.* Washington, DC: Island Press.

NISC (National Invasive Species Council). 2000. *Meeting the invasive species challenge: the U.S.'s first national invasive species management plan.* Washington, DC: National Invasive Species Council. Online: http://http://www.invasivespecies.gov/council/draft1002.pdf.

Ogutu-Ohwayo, R. 1990. The decline of the native fishes of lakes Victoria and Kyoga (East Africa) and the impact of introduced species, especially the Nile perch, *Lates niloticus,* and the Nile tilapia, *Oreochromis niloticus. Environmental Biology of Fishes* 27(2): 81–96.

Orsi, M. L., and Agostinho, A. A. 1999. Introdução de especies de peixes por escapes acidentais de tanques de cultivo em rios da Bacia do Rio Paraná, Brasil. *Revista Brasileira de Zoologia* 16(2): 557–560.

Paraense, L. 2001. The schistosome vectors in the Americas. *Memórias do Instituto Oswaldo Cruz.* 96(suppl.): 7–16.

Parker, V. 2001. Listening to the earth: a call for protection and restoration of habitats. In McNeely, J. A. (ed.). *The great reshuffling: human dimensions of invasive species.* pp: 43–54. Cambridge: IUCN.

Perrings, C., Williamson, M., and Dalmazzone, S. (eds.) 2000. *The economics of biological invasions.* Northampton, MA: Edward Elgar.

Pierce, J. B. 1998. *Cotton boll weevil biology.* Guide A-232. Agricultural Science Center at Artesia, New Mexico State University.

Pimentel, D., Loch, L., Zuniga, R., and Morrison, D. 2000. Environmental and economic costs of nonindigenous species in the United States. *BioScience* 50: 53–65.

Pimentel, D., McNair, S., Janecka, J., Wightman, J., Simmonds, C., O'Connell, C., Wong, E., Russel, L., Zern, J., Aquino, T., and Tsomondo, T. 2001. Economic and environmental threats of alien plant, animal, and microbe invasions. *Agriculture, Ecosystems & Environment* 84(1): 1–20.

Pittman, T. 1997. Latest update on the Hyacinthine Macaw project. *Parrot Society Magazine* Vol. XXXI. Online: http://www.bluemacaws.org/hyawild5.htm.

Rapoport, E. H. 1992. Las implicaciones ecológicas y económicas de la introducción de especies. *Ciencia & Ambiente* III: Jan/Jan.

Ratnieks, F. L. W., Piery, M. A., and Cuadriello, I. 1991. The natural nest and nest density of the Africanized honey bee (Hymenoptera, Apidae) near Tapachula, Chiapas, Mexico. *Canadian Entomologist* 123: 353–359.

Reaser, J. K. 2001. Invasive alien species prevention and control: the art and science of managing people. In McNeely, J. A. (ed.). *The great reshuffling: human dimensions of invasive species.* pp.: 89–104. Cambridge: IUCN.

Schmidt, J. O. and Hurley, R. 1995. Selection of nest site cavities by Africanized and European honey bees. *Apidologie* 26: 467–475.

Schmidt, J. O. and Thoenes, S. C. 1990. Honey bee (Hymenoptera: Apidae) preferences among artificial nest cavities. *Annals of the Entomological Society of America* 83: 271–274.

Schmidt, J. O. and Thoenes, S. C. 1995. Criteria for nest site selection in honey bees

(Hymenoptera: Apidae): Preferences between pheromone attractants and cavity shapes. *Physiology and Chemical Ecology* 21: 1130–1133.

Shine, C., Williams, N., and Gundling, L. 2000. *A guide to designing legal and institutional frameworks on alien invasive species.* Cambridge: IUCN.

Strong, D. R. and Pemberton, R. W. 2000. Biological control of invading species: risk and reform. *Science* 288: 1969–1970.

Suszkiw, A. R. S. 1998. The Formosan termite, a formidable foe. *Agricultural Research Magazine* (USDA, October): 1–9.

Wilcove, D., Rothstain, D., Dubow, J., Phillips, A., and Loses, E. 1998. Quantifying threats to imperiled species in the United States. *BioScience* 48: 607–615.

Williamson, M. 1996. *Biological invasions.* London: Champan & Hall.

Wilms, W., Imperatriz-Fonseca, V. L., and Engels, W. 1996. Resource partitioning between highly eusocial bees and possible impact of the introduced Africanized honey bee on native stingless bees in the Brazilian Atlantic rainforest. *Studies on Neotropical Fauna and Environment* 31(3–4): 137–151.

Wilson, M. E. 1995. Travel and the emergence of infectious diseases. *Emerging Infectious Diseases* 11(2). Online: http://www.cdc.gov/ncidod/eid/vol1no2/wilson.htm.

Wittenberg, R. and Cock, M. J. W. 2001. *Invasive alien species: a toolkit of best prevention and management practices.* Wallingford, Oxon, UK: CAB International.

WHO (World Health Organization). 1998. *Report of the WHO informal consultation on schistosomiasis control,* WHO/CDS/CPC/SIP/99.2. Online: http://www.who.int/ctd/schisto/99_2en.pdf.

Ziller, S. R. 2000. *A Estepe Gramíneo-Lenhosa no segundo planalto do Paraná: diagnóstico ambiental com enfoque à contaminação biológica.* Doctoral thesis. Curitiba: Universidade Federal do Paraná, Setor de Ciências Agrárias, Brazil.

Chapter 34

Harvesting and Conservation of Heart Palm

Sandra E. Chediack and Miguel Franco Baqueiro

It is ironic that one of the biologically richest areas of the world—the Atlantic Forest—houses some of the poorest human communities. The dependence of these communities on the natural resources of the Atlantic Forest ecosystems links the fate of people and ecosystems inextricably (Redford and Padoch 1992). The challenge for conservationists is to improve the livelihood of these communities without detriment to the resources. There are many approaches to this task, and they depend on the scale at which resources are managed. In this chapter, we use heart of palm (*Euterpe edulis,* or palmito) harvesting in Misiones, Argentina, as a case study of potential sustainable management in the Atlantic Forest.

The need for cash in Atlantic Forest communities places tremendous pressure on the available biological resources, and one of its most immediate consequences is deforestation. The forests of this region are being relentlessly transformed into silvicultural and agricultural land (Chapters 5, 15, and 25, this volume). From the point of view of biodiversity, the transformation to agricultural land is irreversible, at least in the immediate future, and avoiding it requires both social and political changes (Chapter 19, this volume). Sustainable management of the forest is therefore the only option that will satisfy both the communities' need for cash and the need to preserve biodiversity. Valuable timber species—*Cedrela fissilis,* Meliaceae, and *Cordia trichotoma,* Borraginaceae—and palmito are the resources most often extracted directly from the forest. Palm-dominated forests are known as *palmitales.* They form patches that are usually associated with certain topographic and soil conditions, as well as with a particular land-use history (Aguilar and Fuguet 1988). *Palmitales* are the most diverse forests in Argentina (Brown et al. 1993). Argentina houses the southernmost portion of the Atlantic Forest and of *palmitales;* their distribution is apparently limited by frost (Chediack, pers. obs. 1998).

Palmito

Palm heart (palmito) is the edible apical meristem (growing portion) and young folded leaves of palm trees (Balick 1984). Although various palm species have an edible heart, the quality of *Euterpe edulis* (Palmae) in the Atlantic Forest of

Argentina, Brazil, and Paraguay is judged superior to that of other species (Ferreira and Paschoalino 1988; Orlande et al. 1996). Overexploitation has decreased supply, however, and many companies have moved to the Amazon Basin to exploit related species (*E. oleracea* and *E. precatoria*) (Galetti and Chivers 1995). Perhaps as a result, commercialization has switched to cultivated varieties of *pejibaye, Bactris gasipaes* Kunth (Clement et al. 1996). Despite this change in the world market, the populations of *E. edulis* in the Atlantic Forest continue to be exploited, and their conservation status outside protected areas is uncertain. In the last decade, *E. edulis* has been declared threatened by extinction in Argentina (FVSA 1993) and vulnerable in Brazil (Galetti and Fernández 1998) and Paraguay (Molas 1989). The World Conservation Union (IUCN) has recommended that specific plans be developed for the protection of this species (IUCN 1996).

This palm has a single trunk with one terminal meristem. Cutting this meristem results in the death of the palm. Because the base of living leaves wraps around this portion of the stem, the upper part of the stem looks swollen. Within this green part of the stem, one finds the palm heart, "corazón," "cogollo," or palmito. The palm's crown of pinnate leaves spreads above this point. The fruits are violet and form racemes hanging below the green portion of the stem (Dimitri et al. 1974). The tallest palms reach a height of 18 m and a diameter at breast height (dbh) of 19 cm (Chediack, unpubl. data). The weight of the palm's heart correlates closely with its diameter (Reis et al. 1994). A palm with a 10-cm dbh produces approximately 350 g of palm heart (Nodari, Guerra 1988). This means that an adult palm produces just one marketable can of palm heart. Fruits mature synchronously and are consumed by a wide variety of animals, which may also aid dispersal (Placci et al. 1992; Galetti and Aleixo 1998).

Commercialization of Palmito

The most widespread model of commercialization is the sale of standing palms. The packing factory usually must pay for harvesting, transport, and processing. The property owner applies for a permit to harvest a certain amount of palmito. The plan must be backed by a forest inventory carried out by an approved forestry firm. The Ministry of Ecology reviews the proposed plan and issues a permit for extraction and transport. Often, these permits are forged; even if legitimate, they may be used in more than one area or on more than one occasion. There is no reliable information for Argentina, but in Brazil, Reis and Guerra (1999) estimated that the proportion of illegal commercialization of palm heart is 10 percent of the total. The owner must reforest the area using palmito seeds. The buyer pays approximately US$1 per palm heart. According to local peasants, the time between harvests is 8 years, but our own research on palm demography indicates that this may be too short a time for populations to recover fully.

Effects of Palm Harvesting on the Natural Forest

Harvesting large, mature palms has consequences for the future population dynamics of *E. edulis* and its accompanying species. Because these large individuals produce the majority of the seeds required for continuous replacement of the population, their death results in impoverishment of the seed bank. Seeds germinate within a few months after dispersal, and seedlings remain on the forest floor for several years before developing a visible stem. Since passive dispersal is limited to distances of a few meters, unless more seed arrives naturally from neighboring plots, the population tends toward extinction. For example, *E. edulis* forest patches in Iguazú National Park (Argentina) hold an average of 492 mature palms per hectare. After legal extraction, the remaining density is 35 mature (albeit younger) palms per hectare. Illegally harvested plots, however, do not contain a single remaining mature palm (Chediack, unpubl. data). Adding the likely decrease in genetic variability brought on by the practice of intensive harvesting, particularly in small forest patches, the demographic situation of these *palmitales* is rather poor.

The harvesting of palm heart also affects other species. Several species of frugivorous birds were forced to change their diet when the density of adult heart palms decreased in the Parque Estadual Intervales in the mountains of southern São Paulo, Brazil (Galetti and Aleixo 1998). Similarly, the incidence of the weevil *Rhynchophorus palmarum,* which feeds on the meristems of young palms and thereby kills them, increased substantially in harvested areas as compared to reserve areas in the Argentinean part of Iguazú (Gatti 1999). The abundance of trunks with succulent pith and young leaves lying on the forest floor apparently increases the availability of ovipositing sites, which then results in higher population densities of this predator.

Sustainability of Palm Heart Harvest

The sustainable harvest of a nontimber forest product is carried out in a limited forest area. This practice can be implemented indefinitely with an insignificant impact on the structure and dynamics of the population (Peters 1996). The observations discussed earlier in this chapter regarding current harvesting of *Euterpe edulis* and its consequences, however, suggest that sustainable harvest cannot be achieved in this species, at least in the short term. A modified version of sustainable harvest with a longer time frame, to allow for wide population fluctuations and longer times between harvests, is therefore required.

Population numbers aside, monitoring changes in genetic diversity may also be warranted. For example, to keep the demographic and genetic structure of the *palmitales* in the Atlantic Forest relatively stable, only palms with diameters larger than 9 cm should be harvested, and a minimum of 50 adults per hectare should be left standing to allow rapid regeneration (Reis and Guerra 1999). The Argentinean government, however, grants permission to harvest palmito when density is as low as 35 adult individuals per hectare (Chediack, unpubl. data).

Consequently, to increase the future value of exploited areas and to restore degraded ones, several researchers have conducted seed enrichment experiments. The best results in terms of seed germination, seedling survival, and cost have been achieved when ripe fruits (scarified or not) have been scattered annually or biannually (Bovi et al. 1988; Nodari et al. 1988b, Reis and Guerra 1999). Although the permits to harvest *palmitales* in Argentina specifically request that reforestation practices be implemented, these procedures are not always carried out. Even when they are, there is no monitoring of the fate of seedlings to guarantee that the practice has been successful. Studies conducted in Campinas, Brazil, using perennial crops such as banana yielded good results (Bovi et al. 1988). However, similar results are unlikely in the subtropical forests of Argentina, where winter frost kills a large number of seedlings. Winter frost may not occur every year, but if it coincides with the germination of naturally occurring or introduced seeds, the results can be disastrous (Chediack, unpubl. data).

Artificial Plantations and the Introduction of Hybrids

Due to the ease with which *E. edulis* and *E. oleracea* cross-pollinate when grown closely together (e.g., in a greenhouse), hybrids of these two species have been planted in Brazil (Aguiar 1988). These hybrids produce a high-quality palmito and have multiple stems. However, because the genetic and demographic consequences of introducing these hybrids into the southernmost populations of *E. edulis* have not been investigated, their introduction is not recommended. For example, if the hybrid turns out to be more sensitive to frost than *E. edulis* already is at the seedling stage, the consequences of gene flow from *E. oleracea* into the natural populations of *E. edulis* could be devastating.

The World Market

Unfortunately, there is no information on world trade specific to palmito (*E. edulis*). The available information groups together all species that provide heart of palm. Also, the available information on imports and exports of various countries is unreliable. For example, Weiss (1998) reports 24,100 metric tons globally exported in 1996, whereas Reis and Guerra (1999) report 4,853 metric tons for the same year. Part of the difference may result from the fact that some countries resell the product. The price per ton also varies according to author, but the difference is not as marked (Figure 34.1).

Brazil is the main producer of palmito, and 80 to 85 percent of its production is consumed domestically. Brazil is followed by Costa Rica and Ecuador. Paraguay has recently started production, and Argentina is its main buyer. Despite producing its own palmito, Argentina is not self-sufficient. Argentina's main providers are Brazil, Ecuador, and Paraguay (Weiss 1998).

Both demand and supply of palmito have increased in the last 10 years, and it is likely that supply will increase further as young plantations of *Bactris gasipaes* (*pejibaye*) mature in Ecuador, Peru, and Hawaii, among other countries. *E. edulis,* however, is preferred over *B. gasipaes.* Its ivory color makes it more appealing to

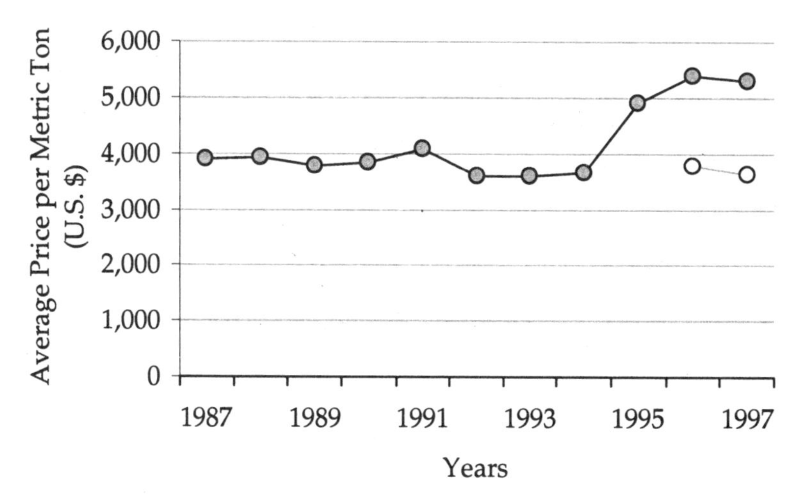

Figure 34. 1. Average price of palmito per metric ton from 1987 to 1997 (Reis and Guerra 1999, solid circles; Weiss 1998, empty circles).

the consumer than the yellowish color of *pejibaye.* More important, its fleshy tissue is softer and has a higher water content, and it lacks the fibers that characterize the *pejibaye.* In tests of palatability, *E. edulis* is consistently rated higher than *pejibaye* because of it sugar content and the presence of tannins (Ferreira and Paschoalino 1988; Clements et al. 1996). Consumers also prefer *E. edulis* because it is extracted from natural areas and, unlike palmito extracted from plantations, is deemed free of agrochemicals. Among the disadvantages of *E. edulis* are its natural variability (which means that there is little control over quality) and the ease with which the product oxidizes (Ferreira and Paschoalino 1988). The ease of oxidation forces the producer to sell *E. edulis* already processed, whereas *B. gasipaes* may be brought to market while still fresh. Furthermore, with little information available on present and predicted future stock and its distribution, it is difficult to make long-term production plans. Another disadvantage of *E. edulis* is that the maximum yields in Brazil are on the order of 17–312 kg/ha/year (Bovi et al. 1988), whereas *pejibaye* plantations produce a yield of 500 to 1,000 kg/ha/year (Clements et al. 1996). To compensate for this, the price per can of *E. edulis* heart of palm must be higher.

Conclusions

Palm heart is extracted together with other valuable timber species. A fair economic analysis would have to consider these other species in the equation. Overall, properly planned, conscientious exploitation of natural populations of *E. edulis* probably represents the best opportunity to preserve this species and its accompanying ecosystem. A system of management that aims at the sustainable use of these resources would be an ideal candidate for certification and would aid in protecting the unique communities of the Atlantic Forest. In a typical tragedy of the commons (use it or lose it), peasant communities prefer small but secure earnings to promises of riches to come. It is therefore necessary to change the current practice of mining forests to a practice of managing them (Putz et al. 2000). The social, cultural, and political barriers to bringing about this change represent the biggest challenge to conservation.

Acknowledgments

Thanks to J. Herrera, G. Gatti, S. Holz, and G. Placci, with whom we conducted research involving collaboration with the Laboratory of Population Dynamics of the Institute of Ecology, UNAM, Mexico; the University of Plymouth, United Kingdom; and the Vida Silvestre Argentina foundation; with financial support from the World Wildlife Fund (WWF) and Russell Train Education for Nature program, the Centro de Investigaciones Ecológicas Subtropicales, Parque Nacional Iguazú, and the Consejo Nacional de Investigaciones Científicas y Técnicas of Argentina. Further financial support has been provided by UNAM, Mexico. Thanks also to numerous field assistants.

References

Aguiar, C. J. S. 1988. Contribução da cultura do açaizeiro (*Euterpe oleracea* Mart.) no litoral Paulista. In: *Palmito, anais do primeiro encontro nacional de pesquisadores.* pp: 75–90. Curitiba, Brazil: EMBRAPA-CNF.

Aguilar, M. I. and Fuguet, M. F. 1988. *Palmito: descripción, distribución y diferentes manejos de* Euterpe edulis *Mart. Actas del VI Congreso Forestal Argentino.* Sgo. del Estero. 14–16.

Balick, M. J. 1984. Ethnobotany of palms in the neotropics. *Advances in Economic Botany* 1: 9–23.

Bovi, M. L. A., Godoy Jr., G., and Saes, L. A. 1988. Pesquisas os gêneros *Euterpe* e *Bactris* no Instituto Agronomico de Campinas. In: *Palmito, anais do primeiro encontro nacional de pesquisadores.* pp: 1–44. Curitiba, Brazil: EMBRAPA-CNF.

Brown, A. D., Placci, L. G., and Grau, H. R. 1993. Ecología y biodiversidad de las selvas subtropicales de Argentina. In: Goñi, F. and Catalén, L. (eds.). *Elementos de política ambiental H.* Cámara *de Diputados de Buenos Aires.* 215–222.

Clements, C. R., Manshardt, R. M., Cavaletto, C. G., DeFrank, J., Mood Jr., J., Nagai, N. Y., Fleming, K., and Zee, F. 1996. Pejibaye heart-of-palm in Hawaii: from introduction to market. In: Janick, J. (ed.), *Progress in new crops.* pp. 500–507. Arlington, VA: ASHS Press.

Dimitri, M. J., Hualde, I. R. V., Brizuela, C. A., and Fano, F. A. T. 1974. La flora arbórea del Parque Nacional Iguazú. *Anales de Parques Nacionales.* Tomo XII.

Ferreira, V. L. P and Paschoalino, J. E. 1988. Pesquisa sobre Palmito no Instituto de Tecnologia de Alimentos (ITAL). In: *Palmito, anais do primeiro encontro nacional de pesquisadores.* pp: 45–62. Curitiba, Brazil: EMBRAPA-CNF.

FVSA (Fundación Vida Silvestre Argentina). 1993. Situación ambiental de la Argentina, recomendaciones y prioridades de acción. *Boletín Técnico* 14: 70.

Galetti, M. and Chivers, D. J. 1995. Palm harvest threatens Brazil's best protected area of Atlantic Forest. *Oryx* 29(4): 225–226.

Galetti, M. and Fernandez, J. C. 1998. Palm heart harvesting in the Brazilian Atlantic

Forest: changes in industry structure and the illegal trade. *Journal of Applied Ecology* 35(2): 294–301.

Galetti, M. and A. Aleixo, A. 1998. Effects of palm heart harvesting on avian frugivores in the Atlantic rain forest of Brazil. *Journal of Applied Ecology* 35(2): 286–293.

Gatti, M. G. 1999. *El picudo de la palma,* Rhynchophorus palmarum *L.* (Coleoptera: Curculionidae), *en palmitales con y sin aprovechamiento forestal.* Tesis de licenciatura. Universidad Nacional de Córdoba, Argentina.

IUCN (World Conservation Union). 1996. *Palms, their conservation and sustainable utilization.* IUCN/SSC Palm specialist group, Johnson, D. (ed). Gland, Switzerland and Cambridge, UK: IUCN

Molas, P. J. 1989. El Palmito: una interesante alternativa de producción. *Rev. Forestal* 5(1): 27–30. Paraguay.

Nodari, R. O., Guerra, M. P., Reis, A., Reis, M. S., and Merizio, D. 1988. Eficiência de sistemas de implementação do palmiteiro em mata secundária. In: *Palmito, anais do primeiro encontro nacional de pesquisadores.* pp: 165–172. Curitiba, Brazil: EMBRAPA-CNF.

Nodari, R. O., Reis, M. S., Reis, A., and Guerra, M. P. 1988. Relaçaõ entre parámetros não destrutivos e o rendimento de palmito. Estudo preliminar. In: *Palmito, anais do primeiro encontro nacional de pesquisadores.* pp: 181–182. Curitiba, Brazil: EMBRAPA-CNF.

Orlande, T., Laarman, J., and Mortimer, J. 1996. Palmito sustainability and economics in Brazil's Atlantic coastal forest. *Forest Ecology and Management* 80: 257–265.

Peters, C. M. 1996. *Aprovechamiento sostenible de recursos no maderables en bosque Húmedo tropical: un manual ecológico.* Programa de Apoyo a la Biodiversidad. Serie General del Programa de Apoyo a la Biodiversidad No. 50. WWF, TNC, WRI and USAID.

Placci, L. G., Arditi, S. I., Giorgis, P. A., and Wutrich, A. A. 1992. Estructura del palmital e importancia de *Euterpe edulis* como especie clave en el Parque Nacional Iguazú, Argentina. *Yvyrareta* 3(3): 93–108.

Putz, F. E., Dykstra, D. P., and Heinrich, R. 2000. Why poor practices persist in the tropics. *Conservation Biology* 14(4): 951–956.

Redford, K. H., and Padoch, C. (eds.) 1992. *Conservation of neotropical forests, working from traditional resource use.* New York: Columbia University Press.

Reis, M. S. and Guerra, M. P. 1999. *Euterpe edulis* Martius (palmito). I Seminário Nacional "Recursos florestais da mata Atlántica," São Paulo, Brazil.

Reis, A., Reis, M. S., and Fantini, A. C. 1994. Manejo do palmiteiro (*Euterpe edulis*) em regime de rendimento sustentable. *Florianópolis,* UFSC: 78.

Weiss, K. D. 1998. *Un estudio del mercado mundial para el Pijuayo.* Winrock international: proyecto de desarrollo alternativo. USAID/Contradrogas. Lima, Peru.

Chapter 35

The Effects of Dams on Biodiversity in the Atlantic Forest

Colleen Fahey and Penny F. Langhammer

The construction of hundreds of dams in the Atlantic Forest over the past 50 years has dramatically affected biological and human communities in the region. Some of the largest dams in the world are found within this hotspot. The impacts are not unique to the Atlantic Forest region—60 percent of the world's large river basins are highly or moderately fragmented by dams and irrigation systems (WCD 2000). Although the number of extinctions directly caused by dams worldwide has not been determined, it is likely to be significant. Dams are considered to be the main reason for the endangerment and extinction of one-fifth of the world's freshwater fish species (McCully 2001).

Dams are built principally for hydropower, water storage, and flood control. Other benefits include irrigation and recreation. The array of negative consequences resulting from dam construction, however, suggests that the benefits may not be worth the costs.

Impacts on Ecosystems and Biodiversity

The impacts of dams on biodiversity occur at several ecological levels, and some consequences cannot be predicted or measured. Most obviously, the construction of dams and the subsequent filling of reservoirs inundates untold numbers of plants and animals. For organisms with distributions entirely within the area to be flooded, a dam can wipe out an entire species. Typically, biodiversity is affected through habitat loss; blocked migration; the introduction of exotic species; and alterations of the hydrology, water quality, morphology, and composition of river and riparian systems.

Habitat Loss

The construction of dams, and the subsequent flooding of large areas of land, results in significant habitat loss. Dams also isolate the remaining habitat of many terrestrial species, and the resulting smaller populations face greater risks of

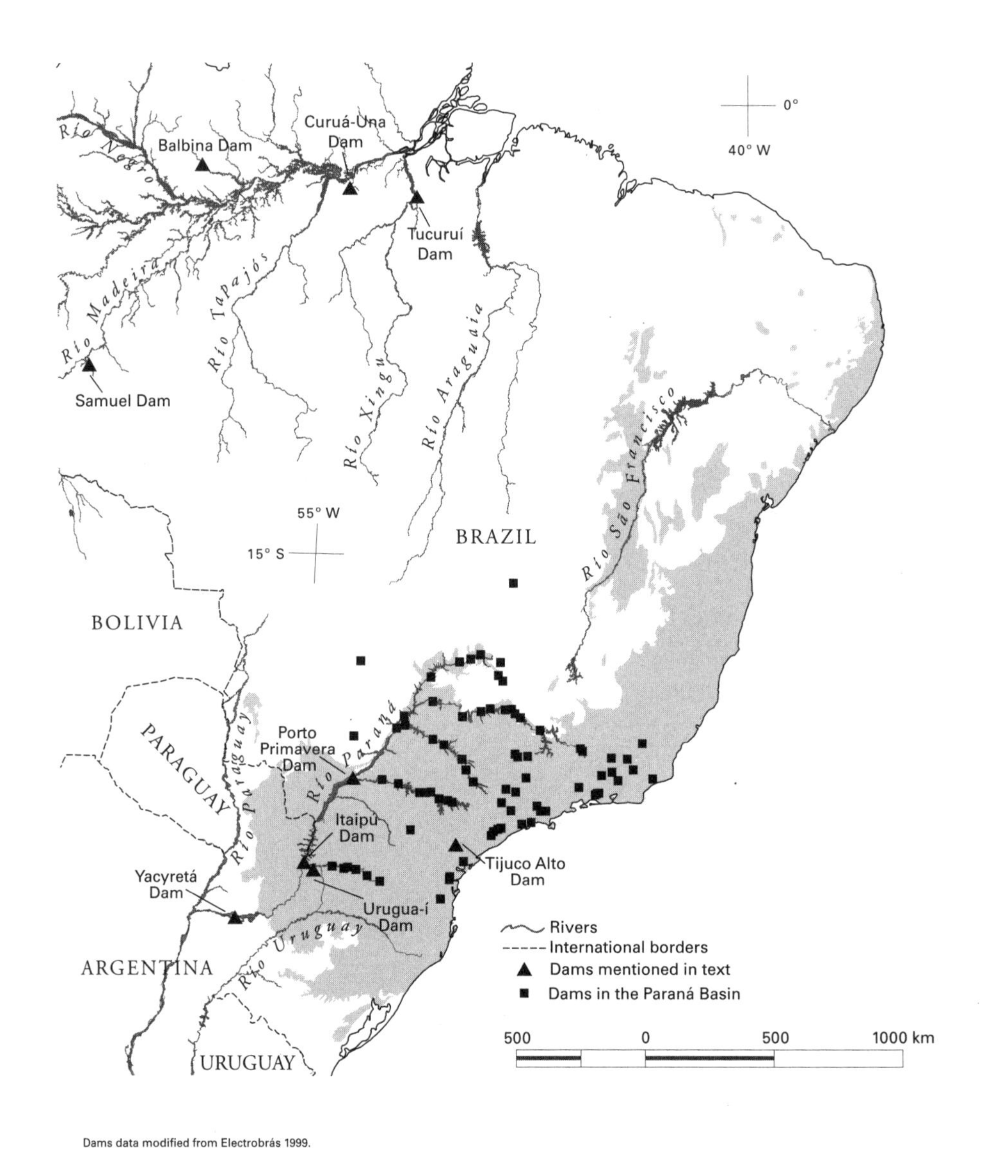

Figure 35.1. Location of major dams in the Atlantic Forest hotspot and other dams mentioned in the text.

extinction. Itaipú Dam, one of the world's largest hydroelectric power plants, has flooded at least 1,350 km^2 of Interior Atlantic Forest and associated ecosystems between Brazil and Paraguay (Figure 35.1) (McCully 2001). The reservoir of the recently completed Porto Primavera Dam on the Paraná River, straddling the Brazilian states of São Paulo and Mato Grosso do Sul, has flooded more than 2,200 km^2 of habitat (Valente 1997). This area supported important populations of the marsh deer (*Blastocerus dichotomus*), the broad-snouted caiman (*Caiman latirostris*), and the capybara (*Hydrochaeris hydrochaeris*) (Mourao and Campos 1995).

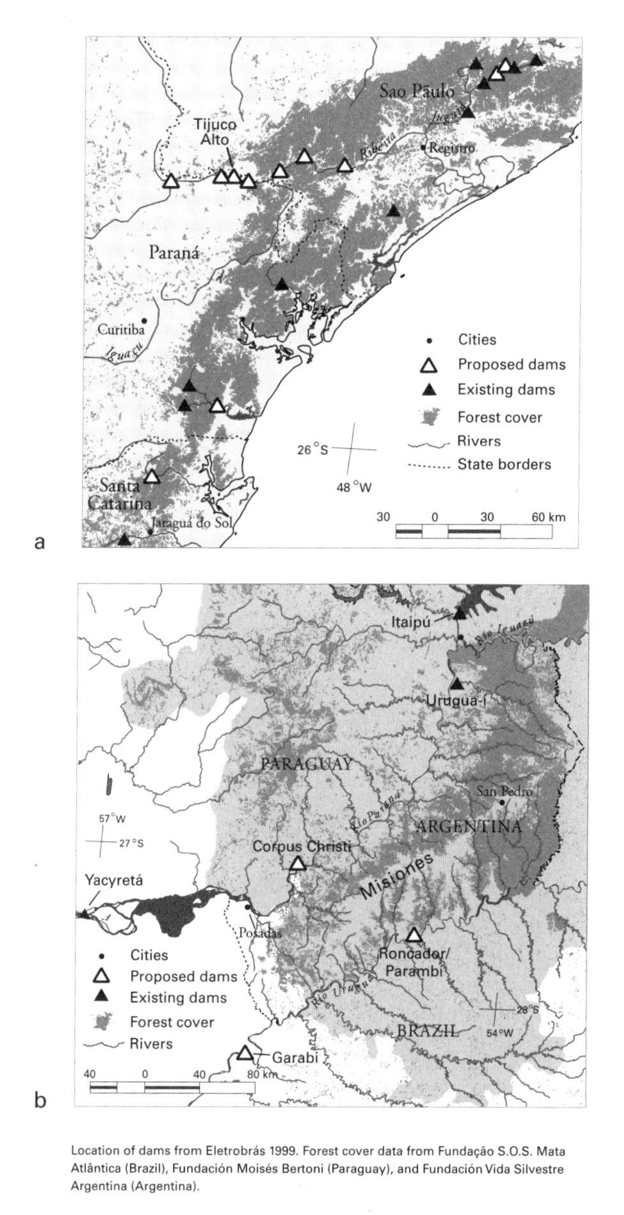

Location of dams from Eletrobrás 1999. Forest cover data from Fundação S.O.S. Mata Atlântica (Brazil), Fundación Moisés Bertoni (Paraguay), and Fundación Vida Silvestre Argentina (Argentina).

Figure 35.2. Proposed dams in the southeastern (a) and southwestern (b) regions of the Atlantic Forest hotspot.

Rivers, floodplains, and their associated habitats are some of the most diverse ecosystems on earth (Ward 1998, McCully 2000). Located on the Paraná River between Paraguay and Argentina, the Yacyretá Dam (Figures 35.1 and 35.2) has flooded more than 1,000 km^2 of riparian islands and forests and affected more than 500 vertebrate species (Chebez, pers. comm. 2002). In addition to riparian forest, savanna communities along riverbanks—characterized by grasses, small bushes, herbs, and occasionally small trees—have been completely eliminated by the flooding of reservoirs (Jiménez and Espinoza 2000).

Impeded Migration

Reservoirs inhibit the movements and migrations of both terrestrial and aquatic species. Terrestrial animals whose habitat has been isolated and fragmented by a dam may not be able to swim across the reservoir to other habitat patches. Dams impede the migration of fish and change species composition and the biotic structure of downstream communities by interrupting or stopping the regular flow patterns of rivers (Petts 1984; Walker 1985; Dudgeon 1992). Many fish species depend on regular flow patterns for migration, feeding, and reproduction. For example, fish ladders have been touted as the solution for restoring the migration patterns of many fish species, but these have met with limited success. Fish ladders, or elevators, installed at the Yacyretá Dam on the Paraná River, at a cost $30 million, were based on a model used for salmon and are inadequate for many species in the Paraná River that travel up and down the river (McCully 2001).

Invasive Exotic Species

Nonnative fish species are often introduced into reservoirs to enhance the recreational value of dams and to generate economically productive fisheries (Chapter 33, this volume). Habitats are dramatically modified by the construction of dams and the filling of reservoirs, and the new habitats are often favorable to exotic fish species as well as many other organisms that may be introduced either intentionally or accidentally (WCD 2000). The Malaysian trumpet snail (*Melanoides tuberculata*) has been introduced into many parts of the Brazilian Atlantic Forest to control the native snail species that carry the tropical disease schistosomiasis. This exotic species is now acting as vector of other diseases (Chapter 33, this volume). In addition to carrying diseases, nonnative plant and animal species may be predators or competitors of the native species or may modify ecosystems to such an extent that many native species can no longer survive (WCD 2000).

Altered Hydrology, Water Quality, Morphology, and Composition

The effects of dams on the physical environment include changes in the flood regime of rivers, interruption of nutrient and water flow, degradation of water quality, and alteration of the morphology of rivers and the surrounding landscape (Allan and Flecker 1993; Noss and Cooperrider 1994). The water quality in a dam's reservoir is often degraded by decomposing vegetation that was not removed prior to construction of the reservoir. Decomposition of submerged trees and aquatic weeds creates anoxic and acidic water conditions (Garzon 1984), often resulting in sizable fish kills. Dams can also change the water temperature both upstream and downstream, which can affect many species. Furthermore, the reduction in the amount of water flowing downstream can increase the salinity of water in river deltas to levels that are lethal to fish and other aquatic organisms. The life cycles of many aquatic species depend on pulses of water and/or nutrients delivered in an annual flood regime, which is typically modified by dam construction (WCD 2000; McCully 2001).

As a result of the changing ecological and environmental conditions that accompany dam construction, the composition of biological communities changes dramatically. Following the creation of the Yacyretá Dam, marked changes in the fish community of the Paraná River occurred. Detritivorous varieties increased, frugivorous fish decreased, and carnivorous fish sharply declined (Corcuera 1997; Bertonatti and Corcuera 2000).

Emission of Greenhouse Gases

Dam proponents argue that hydroelectric plants are "cleaner" than fossil fuel alternatives. However, reservoirs emit an enormous amount of carbon dioxide and methane, primarily as a result of decaying vegetation that is not removed (Fearnside 1995). In 1990 alone, the four largest dams in the Amazon region—Balbina, Curuá-Una, Samuel, and Tucuruí—emitted a quarter of a million tons of methane gas and 38 million tons of carbon dioxide (Fearnside 1995) (Figure 35.1). Reservoirs were previously thought to release most of their greenhouse gases immediately after closing, but recent studies have indicated that the amount decreases little over time (McCully 2001). Greenhouse gases are emitted from the surface of a reservoir and upon the release of water from a dam. The amount of methane released during discharge can be up to 13 times greater than that emitted from the surface of the reservoir (Fearnside 2001). In the tropics, where dams typically flood huge areas of forest, the impact on global warming can be much greater than if the same power were generated using fossil fuels (Fearnside 1995).

Vanishing Treasures

In the Atlantic Forest, the loss and alteration of habitat accompanying the development of dams have pushed several species to the brink of extinction. In early 1997, Brazilian ornithologists discovered a new species of bird that inhabits tall wetland grasses. The macuquinho-da-várzea, or tall-grass wetland tapaculo (*Scytalopus iraiensis*), was found in three locations along the Iraí River in eastern Paraná State (Bornschein et al. 1998). The discovery of the bird coincided with plans to close a dam on the Iraí River in 1998, which would have flooded the unique wetland habitat of the tapaculo. Protection of the species was argued on several grounds: small populations exist in only a few, extremely fragmented areas; little is known about its natural history; its habitat is continuously declining; and its habitat also harbors other threatened species, such as the sickle-winged nightjar (*Eleothreptus anomalus*). Despite a long campaign of protest by ornithologists and others, the dam was completed, and the area where the tapaculo was originally discovered is now under water (Pioli pers. comm., 2001). This recently discovered species is now listed as globally endangered, with its population numbers and habitat range "declining rapidly" (Hilton-Taylor 2000).

Four snail species endemic to the Paraná River basin—*Aylacostoma chloroticum, A. guaraniticum, A. stigmaticum,* and *A. cinculatum* (Thiaridae)—are extinct in the wild as a direct result of flooding of the habitat by the Yacyretá reservoir (Bertonatti 1999). The habitat of the *Aylacostoma* snails has been flooded by 37 of the 42 dams in the subbasin, including Yacyretá and Itaipú (Quintana and Mercado Laczko 1997). The murky, stagnant waters of the Yacyretá reservoir now cover what were once clear, oxygen-saturated rapids, the only remaining habitat range of the imperiled snail species. Before construction of the reservoir, the snail populations were abundant; after the reservoir was closed in 1993, their numbers declined drastically in just three years. By 1996, all specimens gathered during one collection were already dead (Quintana and Mercado Laczko 1997). Efforts are being made to cultivate the snails in aquaria, but because of their strict habitat needs and unusual methods of reproduction, this is an extremely difficult task.

The Brazilian merganser (*Mergus octosetaceus*) is another species that depends on rivers and streams with clear, running water. Native to the Atlantic Forest and the adjacent Cerrado, the Brazilian merganser is now critically endangered, primarily because of habitat loss, and the species continues to decline (Hilton-Taylor 2000). Tiny Atlantic Forest populations can be found on the Rio Tibagi in Paraná State, Brazil, and the Río Urugua-í in Misiones Province, Argentina (Chapter 30, this volume). The filling of the Urugua-í reservoir (Figures 35.1 and 35.2), which occurred between 1989 and 1991, had a major impact on the mergansers in Argentina. As its free-flowing habitat was converted into a lake, the population in this region declined. After the Urugua-í Dam was created, the merganser was observed only on the Uruzú stream, a tributary of the Urugua-í (Chapter 16, this volume).

Social Impacts

The social consequences of dams can be as dramatic as the environmental impacts. Most troubling is the forced resettlement of hundreds and sometimes thousands of people from their homes, which often accompanies dam projects. The secondary effects of dam construction, such as increased incidence of disease and damaged livelihoods, can severely disrupt social institutions. The public, sadly, does not have a good history of successfully opposing dam projects that negatively affect them.

Displacement and Resettlement

The Tucuruí Dam in the Amazon brought extremely high environmental and social costs (Fearnside 1999, 2001). More than 30,000 people were displaced from their homes (Magalhães 1990), and despite resettlement promises, 1,500 families remained homeless a year after the project was completed (Teixeira 1996). In the Atlantic Forest, the Yacyretá Dam has already displaced some 30,000 people, but the number could climb to 50,000 if the reservoir is filled to the projected 83 m above sea level (IRN 2001).

Resettlement often occurs, if at all, far from the river that once sustained the people's livelihood, affecting their culture and lifestyle and potentially causing them psychological distress. In all of Brazil, dams have forced about 1.5 million people out of their homes and their ways of life (IRN 1997). Years after dams are completed, families are still waiting for compensation, in the form of either land or money, from energy authorities.

Disease

The role of dams in the spread of disease is a social impact of dams that is often overlooked. Insects, snails, and other animals that are vectors of water-borne diseases and parasites thrive in habitats created by the construction of dams and irrigation systems (McCully 2001). The construction of dams is largely responsible for the spread of schistosomiasis into new regions and the rapid increase in the number of people infected (McCully 2001). This intestinal disease is caused by blood flukes of the genus *Schistosoma* transmitted by three species of freshwater snails (Chapter 33, this volume). The disease occurs in 70 countries and affects over 200 million people.

The increased standing water and other environmental changes that resulted from the Yacyretá Dam construction in the Atlantic Forest region promoted an increase in the incidence of dengue, hantavirus, leishmaniasis, and yellow fever, in addition to malaria and schistosomiasis.

Opposition and Public Will

Dam construction projects are periodically revived even if they have been strongly opposed in the past. In April 1996, a 90 percent plebiscite in Misiones, Argentina (representing 62.85 percent of registered voters), rejected the proposed Corpus Christi Dam regardless of its location along the Paraná River. However, the project was later reproposed and supported by local politicians, who showed an evident disregard of the public will (IRN 2001). In Curitiba, Brazil, people from 20 countries gathered on March 14, 1997, to approve the Declaration of Curitiba: Affirming the Right to Life and Livelihood of People Affected by Dams and to establish the International Day of Action against Dams and for Rivers, Water, and Life. The declaration summarizes the adverse impacts of dams that people around the world have suffered and demands an immediate moratorium on the construction of dams until the conditions of the declaration are met (McCully 2001).

Politics and Economics

Dams are often promoted as tools for economic development, and the political will for them can be immense. The costs of dam construction, which are typically underestimated, coupled with the environmental costs, which are rarely assessed accurately (if at all), make dams an inefficient vehicle for economic stimulus.

Cost and Time Overruns

Dams notoriously take longer to build and are more costly than estimated prior to construction. The World Bank attributes cost and time overruns to construction difficulties resulting from geologic conditions and problems with the resettlement of people living in the inundated area. Among World Bank–financed hydropower projects, resettlement costs have run an average of 54 percent higher than originally anticipated (McCully 2000). In addition, the area to be flooded by the reservoir is often grossly underestimated, either as a result of poor topographic data or even as a means of deliberately downplaying the environmental effects of a dam (Fearnside 1989). Hugely consumptive hydroelectric projects are known as "pharaonic works" in Brazil because, like the pyramids of Egypt, they require an entire society to build, yet they yield almost no economic benefits once completed (Fearnside 1989).

Yacyretá Dam is one of the largest hydroelectric projects in Latin America. Because of construction delays, cost overruns, mismanagement, and political corruption, the project is only about two-thirds complete, with the water level at 76 m above sea level. The project has become a notorious financial black hole and has even been called a "monument to corruption" by Argentine ex-President Carlos Menem (Treakle 1998). The projected cost of Yacyretá Dam was $2.3 billion in 1977; the latest estimate is that it will cost $11.5 billion (Table 35.1).

Originally projected in 1973 to cost $3.4 billion, Itaipú Dam ended up costing $20 billion. This is nearly what it would cost to protect the remaining habitat in the biodiversity hotspots adequately (Myers et al. 2000) and five times more than what it would cost to protect 4 million km^2 in three major tropical wilderness areas (Pimm et al. 2001).

Table 35.1. Estimated and actual costs and area flooded for four large dams (≥10-Mw installed capacity) in Brazil, Paraguay, and Argentina.

Name of Dam (Location)	Estimated Cost (U.S. $)	Actual Cost (U.S. $)	Estimated Area to be Flooded (km^2)	Actual Area Flooded (km^2)
Yacyretá (Argentina/ Paraguay)	2.3 billion	11.5 billion	Unknown	1,720
Itaipú (Brazil/Paraguay)	3.4 billion	20 billion	Unknown	1,350
Balbina (Amazon, Brazil)	383 million	750 million	1,240	3,147
Tucuruí (Amazon, Brazil)	2.9 billion; later upgraded to 5.4 billion	8.77 billion (without interest payments); total is over 10 billion	1,630	2,850

(Source: McCully 2000, World Commission on Dams 2000)

Impact Assessments and Mitigation Efforts

New hydroelectric projects are often planned to increase the power output or efficiency of nearby existing dams, resulting in a series of dams along a river. The 1986 Brazilian legislation requiring that environmental impact assessments (EIAs) be performed for every major development project refers only to the first project in a series along a river (Fearnside and Barbosa 1996). Little attention is paid to whether a new dam is necessary or whether other less destructive alternatives are available.

Impact assessments concluding that a project should be rejected are practically nonexistent. A biological survey, conducted independently in the area to be flooded by the Porto Primavera Dam, estimated the region to hold the second largest population of the marsh deer in Brazil, which is a globally threatened species (Pinder 1996). The dam project was completed in 1998. A major flaw in the EIA process is that the consulting firms that write the reports are often contracted by the organizations building the dams, thus biasing the results of the studies (Fearnside 1989).

Energy authorities have undertaken some efforts to reduce the impact of dams on the environment (Table 35.2), but these actions seem to be more for the sake of the authorities' public image than for the health of the environment. Mitigation activities are often highly publicized, along with claims that environmental conditions will be closely monitored (Fearnside 1989). When a dam is closed, fish are killed in numbers large enough to motivate embarrassed utility companies to close the dams unannounced, or even on holidays, so as to avoid public attention. Before the dedication ceremony of the Tucuruí Dam in the Amazon region, dead fish were removed by the truckload to improve the aesthetic appeal (Fearnside 1989).

Table 35.2. Mitigation activities promoted to counter some of the environmental impacts of dams, compared with their actual effects (Fearnside 1989, Cardenas 1986).

Mitigation Activity	Actual Effect
Faunal "salvage." Involves removing wildlife (a) to land outside area flooded by reservoir, or (b) to wildlife reserves.	(a) "Salvaged" animals are forced to compete with groups that are already present, resulting in population declines; (b) reserves may not be correct habitat type; quality of land is inferior to original habitat. Follow-up studies not conducted.
Construction of fish ladders to allow passage of fish through a dam.	Ladders are impractical for certain species or fish that are small, they are not widely implemented, and their efficacy is unknown. Some ladders in South America have been designed based on the ecological needs of salmon and do not meet the needs of local species.
Removal of vegetation before filling of reservoir to maintain water quality after dam is closed	Reservoir is quickly inundated with aquatic vegetation, resulting in anoxic conditions.

Promoting "Development"

Despite their environmental, social, and economic costs, dams continue to be built largely because political leaders view hydroelectric projects as a means of "developing" their countries, with little regard for their long-term consequences. Once a project is begun, it is almost never aborted. Money is continually poured into construction costs, and once plans are designed, the projects are seen as irreversible. Politicians are often proponents of dams because the massive amount of labor required to construct a dam generates thousands of jobs. Additional pressure to build dams is applied by the *barrageiros,* or dam builders, who are interested in ensuring their income. *Barrageiros* enjoy a high social status, comfortable living quarters provided at each project location, and schools that are separate from those of lower-class residents (Fearnside 1989; Fearnside and Barbosa 1996).

Looking to the Future

What can be done to address the social and environmental impacts of dams in the near-term future? The conservation community should critically evaluate, and lobby against where necessary, the scores of dam projects that have been proposed or are in the works. Developing and promoting "best practices" for dam projects is also extremely important, as dams will continue to be constructed for the foreseeable future.

Proposed Dams

A number of new hydroelectric projects have been proposed that will have serious impacts on the remaining fragments of the Atlantic Forest (Figure 35.2). Others are in the advanced planning or early construction stage. In particular, the Corpus Christi Dam on the Paraná River along the Argentina-Paraguay border would flood over 90 km^2 in Misiones Province; this project seems imminent despite widespread public opposition. The Garabí Dam on the Uruguay River, along the border of Argentina and Brazil, threatens to flood at least 300 km^2 in southern Misiones. The Roncador Dam, also on the Uruguay River, is another threat to biodiversity in the region (Chapter 19, this volume). Although Argentina is experiencing an increased demand for electricity, the greatest draw on the hydropower resources of Misiones is likely to come from Brazil, which is experiencing severe electricity shortages.

In Brazil, new dams have been proposed throughout the Atlantic Forest region. Several of these would be in or adjacent to some of the largest remaining forest fragments in the hotspot, primarily within the states of São Paulo and Rio de Janeiro (Figure 35.2). Among the most worrisome are dams planned for the Ribeira de Iguape Valley, including the Tijuco Alto Dam in São Paulo. In March 2001, families from this region protested the decision by the Instituto Brasileiro do Meio Ambiente e dos Recursos Naturais Renováveis (IBAMA) to grant Tijuco Alto an environmental license (ENS 2001). They joined more than 1,500 Brazilians in a larger protest at the Ministry of Energy on the International Day of Action against Dams. Social and environmental organizations continue to oppose construction of the Iguape Valley dams.

Best Practices

Lessons learned about the impacts of existing dams on biodiversity and human communities are critical for assessing the potential impacts of proposed dams. In the future, every effort should be made to avoid highly sensitive areas, such as sizable forest fragments in the biodiversity hotspots or the last remaining habitats of endangered species and communities. Assessing existing dams is also necessary, because in many cases they can be managed in ways that reduce environmental impacts. In other cases, decommissioning may be warranted and practical, given the tremendous cost of dam maintenance. The best strategy, from a proactive perspective, is to pursue alternative energy sources—such as wind, solar, and hydrogen fuel cells—and to improve the efficiency of current energy production and consumption through technological advances (McCully 2001).

Although dams have been responsible for tremendous environmental and social impacts around the world, they have also provided electricity, flood control, and irrigation services. It would not be realistic to urge the destruction of all existing dams and the abolishment of all planned dams. Rather, dam construction should proceed only when and where the dams will provide true economic benefits that exceed those of alternative energy sources, when the people affected are involved in the decision-making process, and only after environmental impact assessments based on good science are conducted and reported. Inevitable damage to property and the environment must be adequately mitigated and fairly compensated.

The World Commission on Dams issued a landmark report following a 2-year evaluation of the impacts of dams globally (WCD 2000). The commission, born out of a 1997 IUCN-World Bank–sponsored workshop, was composed of numerous stakeholders, including representatives from the dam industry and from the antidam movement. The report presents a decision-making framework that can serve as a guide for evaluating existing and future dam projects. It provides criteria and standards for considering the social, environmental, technical, economic, and financial effects of dams.

References

Allan, J. D. and Flecker, A. S. 1993. Biodiversity conservation in running waters: identifying the major factors that threaten destruction of riverine species and ecosystems. *BioScience* 43(1): 32–43.

Bertonatti, C. 1999. Caracoles de Apipé: ¡Viven!. *Revista Vida Silvestre* 65: 16–20. Buenos Aires: Fundación Vida Silvestre Argentina (FVSA).

Bertonatti, C. and Corcuera, J. 2000. *Situación ambiental Argentina 2000.* 440 pp. Buenos Aires: FVSA.

Bornschein, M. R., Reinert, B. L., and Pichorim, M. 1998. Descrição, ecología e conservação de um novo *Scytalopus* (Rhinocryptidae) do sul do Brasil, com comentários sobre a morfología da família. *Ararajuba* 6(1): 3–36.

Corcuera, J. 1997. Cuenca del Plata, una integración incompleta. *Revista Vida Silvestre* 54: 6–15. Buenos Aires: FVSA.

Dudgeon, D. 1992. Endangered ecosystems: a review of the conservation status of tropical Asian rivers. *Hydrobiologia* 248: 167–191.

ENS (Environment News Service). March 14, 2001. *Dam protesters occupy Brazil's Ministry of Energy.* Online: http://forests.org/archive/brazil/damprocc.htm.

Fearnside, P. M. 1989. Brazil's Balbina Dam: environment versus the legacy of the Pharaohs in Amazonia. *Environmental Management* 13(4): 401–423.

Fearnside, P. M. 1995. Hydroelectric dams in the Brazilian Amazon as sources of "greenhouse" gases. *Environmental Conservation* 22(1): 7–19.

Fearnside, P. M. 1999. Social impacts of Brazil's Tucuruí Dam. *Environmental Management* 24(4): 483–495.

Fearnside, P. M. 2001. Environmental impacts of Brazil's Tucuruí Dam: unlearned lessons for hydroelectric development in Amazonia. *Environmental Management* 27(3): 377–396.

Fearnside, P. M. and Barbosa, R. I. 1996. The Cotingo Dam as a test of Brazil's system for evaluating proposed developments in Amazonia. *Environmental Management* 20(5): 631–648.

Garzon, C. E. 1984. *Water quality in hydroelectric projects: considerations for planning in tropical forest regions.* World Bank Technical Paper No. 20. Washington, DC: World Bank.

Hilton-Taylor, C. (compiler) 2000. *2000 IUCN red list of threatened species.* Gland, Switzerland: World Conservation Union (IUCN).

IRN (International Rivers Network). 1997. *First International Conference of Dam-Affected People begins today.* March 11. Online: http://www.irn.org/programs/hidrovia/pr970311.html.

IRN (International Rivers Network). 2001. *Mobilization in Misiones against the Corpus Dam project and rejecting proposals to raise the level of Yacyretá reservoir.* March 14. Online: http://www.irn.org/dayofaction/2001/010315.yacyretadoa.html.

Jiménez, B. and Espinoza, C. 2000. *Manual de Plantas útiles de la Reserva Priva YPETÎ.* Asunción, Paraguay: Fundación Moisés Bertoni (FMB).

Magalhães, S. B. 1990. Tucuruí: a relocation policy in context. In: Santos, L. A. O. and de Andrade, L. M. M. (eds.). *Hydroelectric dams on Brazil's Xingu River and indigenous peoples.* Cultural Survival Report 30. Cambridge, MA: Cultural Survival.

McCully, P. 2000. *Expensive and dirty hydro: why dams are uneconomic and not part of the solution to global warming.* International Rivers Network. Paper prepared for the Bratislava Hearing on Dams. Online: http://www.nextcity.com/ProbeInternational/worldbank/paper0001.html.

McCully, P. 2001. *Silenced rivers: the ecology and politics of large dams.* London: Zed Books.

Mourao, G. and Campos, Z. 1995. Survey of broad-snouted caiman (*Caiman latirostris*), marsh deer (*Blastocerus dichotomus*) and capybara (*Hydrochaeris hydrochaeris*) in the area to be inundated by Porto Primavera Dam, Brazil. *Biological Conservation* 73: 27–31.

Myers, N., Mittermeier, R. A., Mittermeier, C. G., da Fonseca, G. A. B., and Kent, J. 2000. Biodiversity hotspots for conservation priorities. *Nature* 403: 853–858.

Petts, G. E. 1984. *Impounded rivers.* Chichester: Wiley.

Pimm, S. L., Ayres, M., Balmford, A., Branch, G., Brandon, K., Brooks, T., Bustamante, R., Costanza, R., Cowling, R., Curran, L. M., Dobson, A., Farber, S., da Fonseca, G. A. B., Gascon, C., Kitching, R., McNeely, J., Lovejoy, T., Mittermeier, R. A., Myers, N., Patz, J. A., Raffle, B., Rapport, D., Raven, P., Roberts, C., Rodriguez, J. P., Rylands, A. B., Tucker, C., Safina, C., Samper, C., Stiassny, M. L. J., Supriatna, J., Wall, D. H., and Wilcove, D. 2001. Can we defy nature's end? *Science* 293: 2207–2208.

Pinder, L. 1996. Marsh deer (*Blastocerus dichotomus*) population estimate in the Paraná River, Brazil. *Biological Conservation* 75: 87–91.

Quintana, M. G. and Mercado Laczko, A. C. 1997. Biodiversidad en peligro: caracoles en los rápidos en Yacyretá. *Ciencia Hoy* 7(41): 22–31.

Teixeira, M. G. C. 1996. *Energy policy in Latin America.* Aldershot, UK: Ashgate Publishing.

Treakle, K. 1998. *Accountability at the World Bank: what does it take? Lessons from the Yacyretá Hydroelectric Project, Argentina/Paraguay.* Paper prepared for the Bank Information Center. Online: http://www.bicusa.org/publications/yacyreta.htm.

Valente, R. 1997. Obra de 17 años causa desastre ambiental. *Folha de São Paulo* (August 4): 12–13.

Walker, K. F. 1985. A review of the ecological effects of river regulation in Australia. *Hydrobiologia* 125: 111–129.

Ward, J. V. 1998. Riverine landscapes: biodiversity patterns, disturbance regimes, and aquatic conservation. *Biological Conservation* 83(3): 269–278.

WCD (World Commission on Dams). 2000. *Dams and development: a new framework for decision-making.* London: Earthscan Publications, Ltd.

Chapter 36

Populating the Environment: Human Growth, Density, and Migration in the Atlantic Forest

Thomas R. Jacobsen

As the global human population grows beyond 6 billion, demand for food, fuel, fresh water, and other natural resources continues to increase. Approximately 95 percent of this growth, estimated at over 80 million people per year, is occurring in developing countries (Engelman et al. 1997), where many households live in conditions of poverty and scarcity and low incomes predominate. Projections indicate that in 2025, approximately 98 percent of global population growth will take place in developing countries (UNFPA 1999). In these regions, expanding human populations clearly create greater competition for increasingly limited resources (ICRW 1999). As a result, more natural areas are modified, and the earth's biological diversity is seriously impacted.

However, the link between human populations and biodiversity loss is based not solely on population density and growth but also on several underlying causes (Cincotta and Engelman 2000; Stedman-Edwards 2000) rooted in patterns and levels of consumption, human settlement, and natural resource use. Demographic factors such as high fertility rates, urbanization, and migration often contribute to the fragmentation, isolation, and reduction of natural habitats witnessed in areas of high biodiversity (ICRW 1999). In many instances, population growth exacerbates poverty and migration, leading to the dissolution of social mechanisms responsible for adequate natural resource management (McNeely et al. 1995).

In 1995, over 1.1 billion people, almost 20 percent of the global population, were living in the biodiversity hotspots. In comparison to the world's population growth rate between 1995 and 2000 (1.3 percent per year), population increase in the hotspots is estimated at 1.8 percent per year, higher than that of developing countries (1.6 percent per year) and almost 40 percent higher than the world's overall population growth rate (Cincotta et al. 2000). In the Atlantic Forest of Brazil, Paraguay, and Argentina, population increase is occurring at a rate of 1.8 percent per year. With approximately 122,100,000 people, this hotspot contains an average of 82 people/km^2 (considering an area of 1,482,700 km^2). The most densely populated region is in the municipality of São João de Meriti, Rio de Janeiro, where density reaches approximately 12,842 people/km^2 (IBGE 2000). The highest increase in growth has been occurring in the Brazilian municipality of Canoinhas (Santa Catarina), where population grew 30 percent per year from 1991 to 2000.

Population Distribution

The Atlantic Forest directly impacts the lives of urban, rural, coastal, and indigenous communities. Aside from the forest's value as a historical and cultural landmark, its ecosystems regulate sources of water flow, maintain soil composition and fertility, control climate, and support the slopes of mountains, all of which are crucial to human welfare (ISA 1999). Because of different settlement histories and land-use policies, the distribution of people does not occur evenly in the Atlantic Forest. European colonization of South America first took place in this region, and today the natural environment has been greatly altered (MMA 1999).

Brazil is the fifth most populous country in the world, and approximately 118 million people, or 68 percent of its population (IBGE 2000), are concentrated in the Atlantic Forest region. Due to differences in defining the Atlantic Forest boundaries, population estimates may vary slightly. For example, Hirota (Chapter 6, this volume) identifies 108,000,000 inhabitants in the Brazilian Atlantic Forest. During the early and mid-twentieth century, Brazil's growth rate steadily increased, peaking in the 1960s at 3.1 percent. Over the last several decades, however, Brazil has undergone a sharp decrease in population growth rate; its current rate is 1.5 percent per year. This trend represents a significant demographic transformation for the country (PRCDC 1999). In the Brazilian Atlantic Forest, the increase in growth rate from 1996 to 2000 was 1.8 percent per year, higher than the national average.

Two of the world's 15 largest urban areas (Figure 36.1) are found in the Atlantic Forest. The megacities of Rio de Janeiro and São Paulo house Brazil's largest industrial and timber production centers and generate 80 percent of the gross domestic product of Brazil (CEPF 2001).

The São Paulo metropolitan area extends over 1,500 km^2 and is the second largest city in Latin America. It represents a core area of the country, concentrating financial capital and information control, including technology, science, and computer hardware and software (Rossini and Calió 2002). Population has increased rapidly over time in all the states housing the Brazilian Atlantic Forest, but most notably in the heavily populated regions of Rio de Janeiro and São Paulo (Figure 36.2). Recent census data indicate, however, that growth trends in these two cities are decreasing (Matos and Baeninger 2001).

IBGE (2000) estimates population density in the Atlantic Forests of Brazil at 74.5 people/km^2. Although Brazil contains the most populated areas in the three countries housing the Atlantic Forest, there is not a direct relationship between population and habitat loss. For example, areas where population density

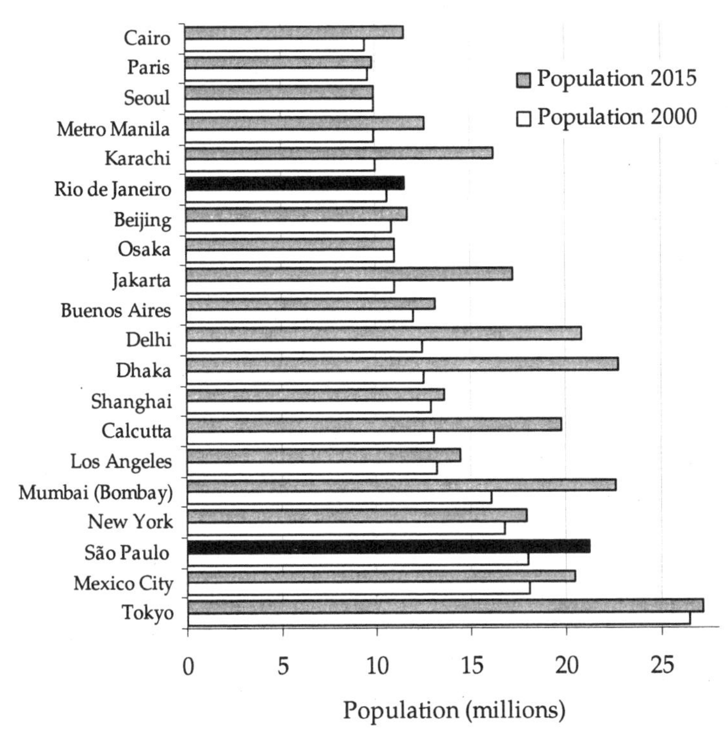

Figure 36.1. Two of the world's 20 largest cities are in the Atlantic Forest hotspot (UNFPA 1999).

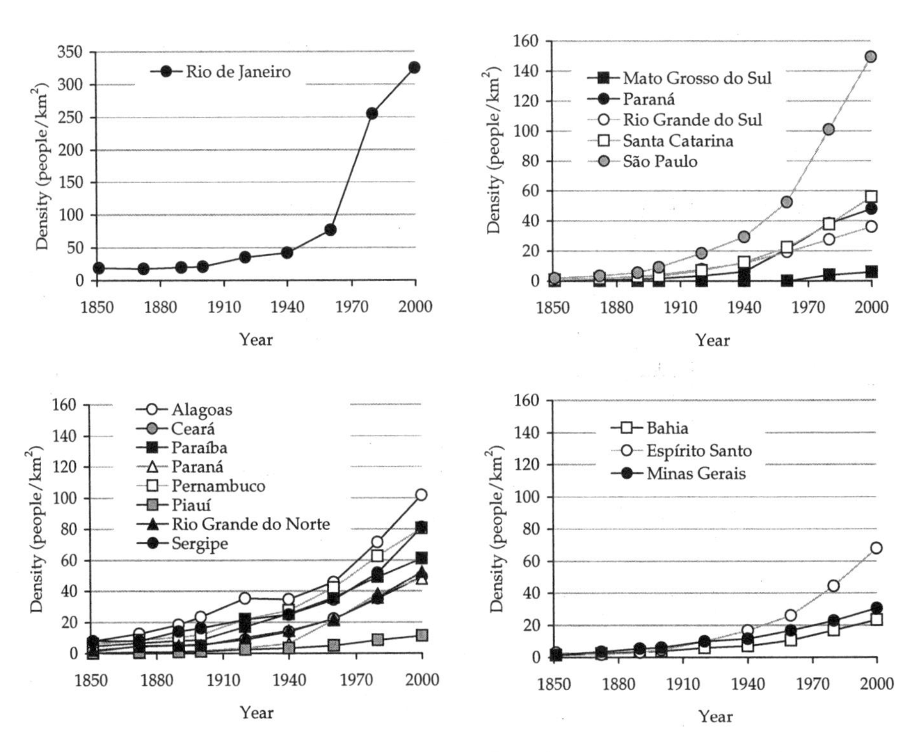

Figure 36.2. Population trends in states across the Atlantic Forest in Brazil from 1850 to 2000.

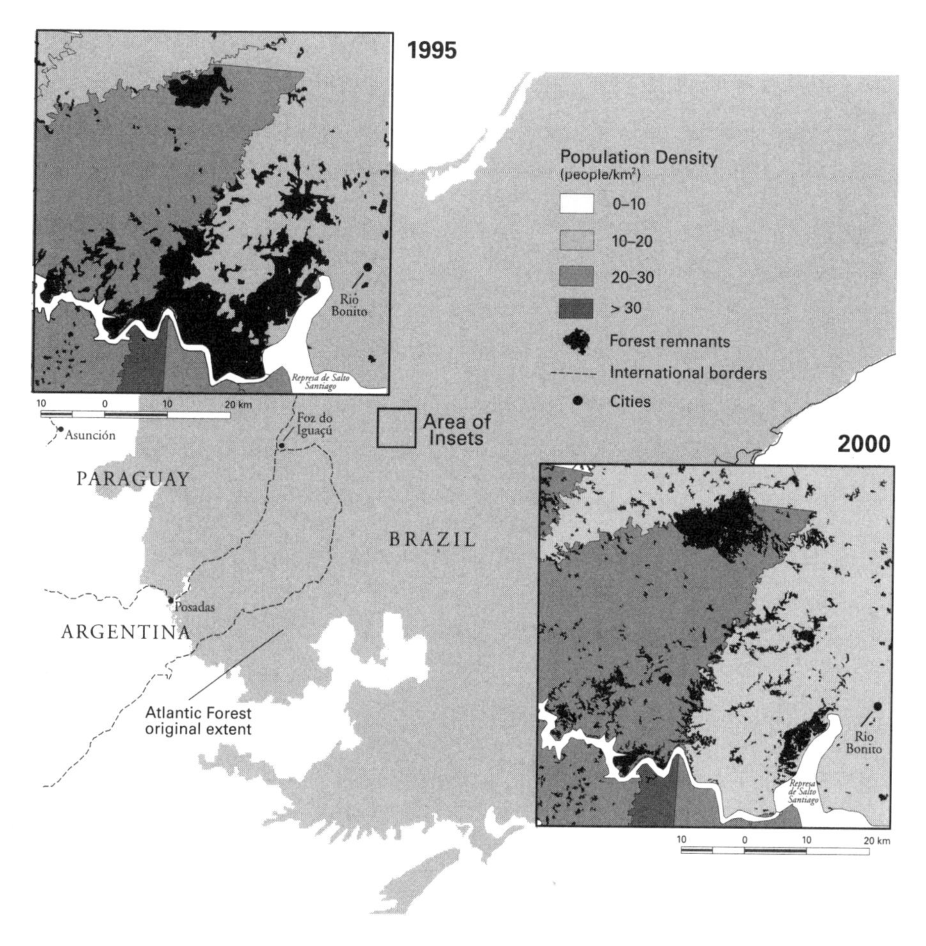

Population density data from Instituto Brasileiro de Geografia e Estatística (IBGE).

Figure 36.3. In only five years (1995–2000), a huge area was deforested in Rio Bonito do Iguaçu, Brazil, despite a nearly stable human population.

has remained steady over the last decade have experienced severe deforestation (Figure 36.3).

Conversely, areas of dense forest are found in close proximity to highly urbanized regions such as São Paulo and Rio de Janeiro. Although the mountainous slopes prevent human settlement in these forests (Chapter 11, this volume), it is clear that the density of human populations does not always represent a significant threat to biodiversity in the region. Often the greatest pressures come from the exploitation of forests by large-scale timber, farming, ranching, and mining interests.

Nonetheless, as the most urbanized and densely populated region in Brazil, the Atlantic Forest displays a highly unstable balance between humans and nature (Hogan 2001). Urbanization has led to the degradation of water sources and groundwater, to contaminated soils, and to untreated sewer emissions. The region's already fragile ecosystems are overburdened by excessive waste, and the high

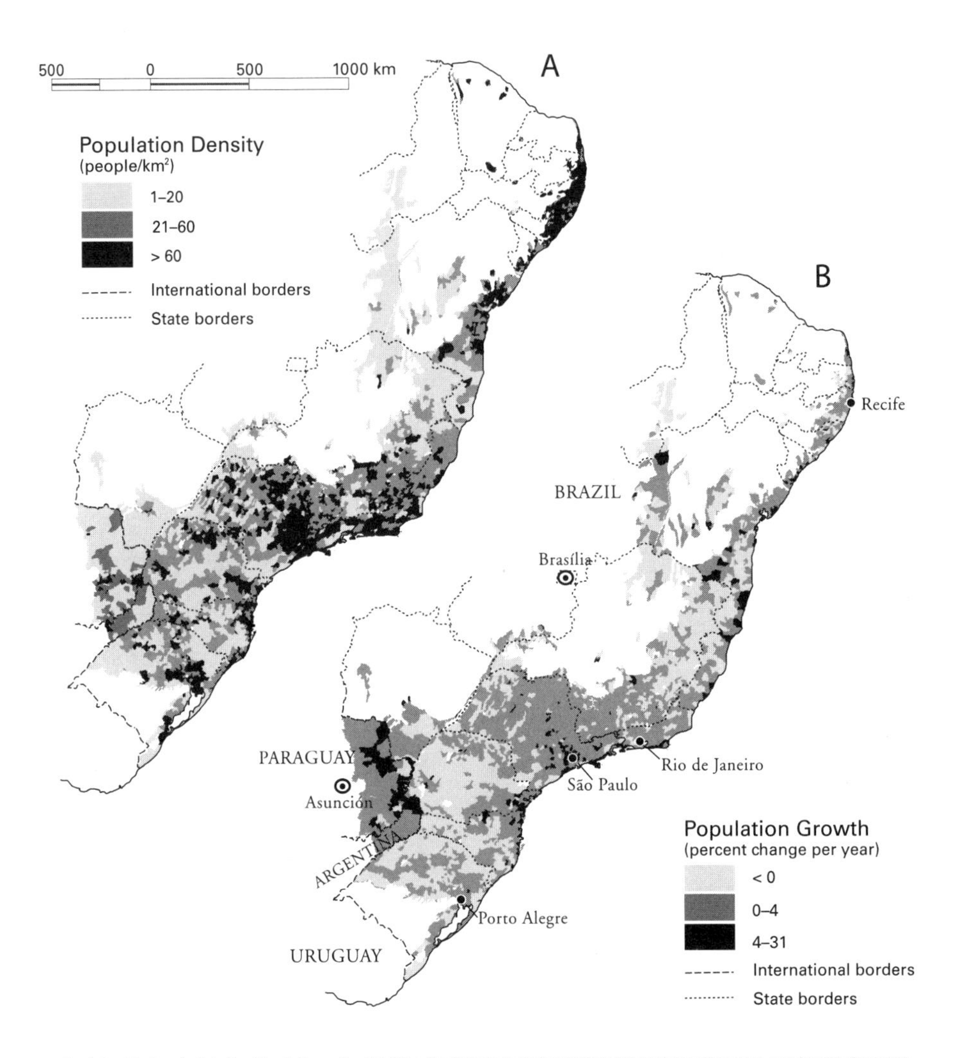

Population data from Instituto Brasileiro de Geografia e Estatística (Brazil), Instituto Nacional de Estadística y Censos (Argentina), and Dirección General de Estadística, Encuestas y Censos (Paraguay).

Figure 36.4. Human population density (a) and population growth (b) in the Atlantic Forest hotspot.

level of pollution production has caused acid rain to fall over the remaining forest fragments. Urban expansion is also threatening the integrity of *restinga* ecosystems in the region (CEPF 2001) and has led to increased consumerism and large-scale tourist development, all of which represent population-related threats to remaining forests in the region. In fact, such pressures are felt on a global scale. Around the world, 146 major cities are located in or next to a hotspot (Cincotta and Engelman 2000). Projected rates of global urbanization demonstrate that by 2030, almost 5 billion people—or 61 percent of the world's inhabitants—will live in cities, with rates expected to be higher in Latin America (UNFPA 1999). Although population density is high in Brazil (Figure 36.4), growth rates are more moderate. Areas of high growth are not extensive and are localized to specific municipalities.

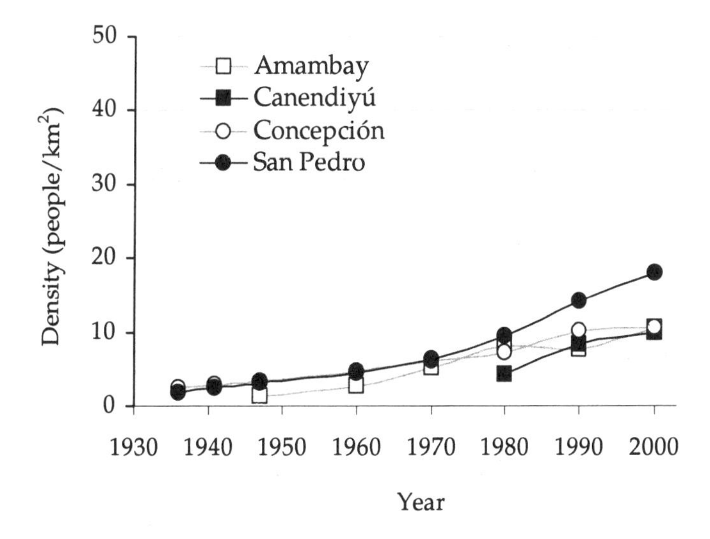

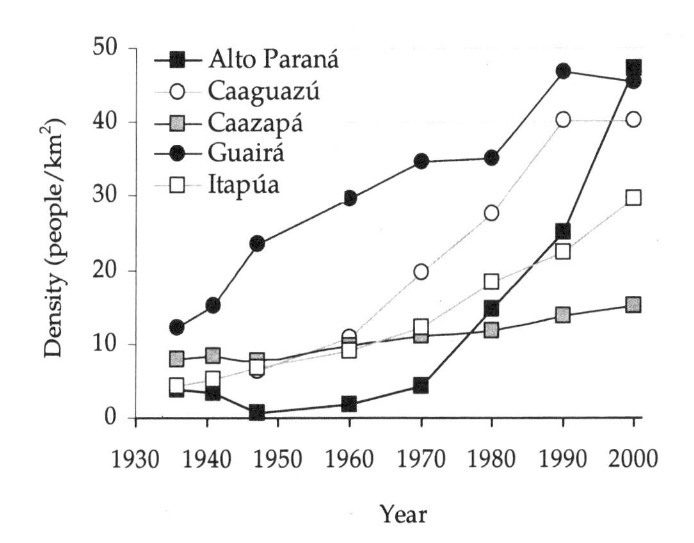

Figure 36.5. Trends in population density in the Interior Atlantic Forest of Paraguay from 1935 to 2000.

However, high human population growth rates occur on a widespread scale throughout the Atlantic Forests of Paraguay and Argentina (Figures 36.5 and 36.6). Over recent decades, urbanization in Paraguay has increased rapidly, contributing to the growth of cities that now house over half the population (Chapters 24 and 26, this volume). The total human population in the Atlantic Forest provinces of Paraguay is approximately 2,822,300 (DGEEC 2000). Although the forests cover only part of the provinces of Itapuã, Caazapã, San Pedro, and Concepción, total population numbers for these provinces were used in the calculation. Population density has historically increased steadily in this region (Figure 36.5) and is today calculated to be approximately 25 people/km^2 (DGEEC

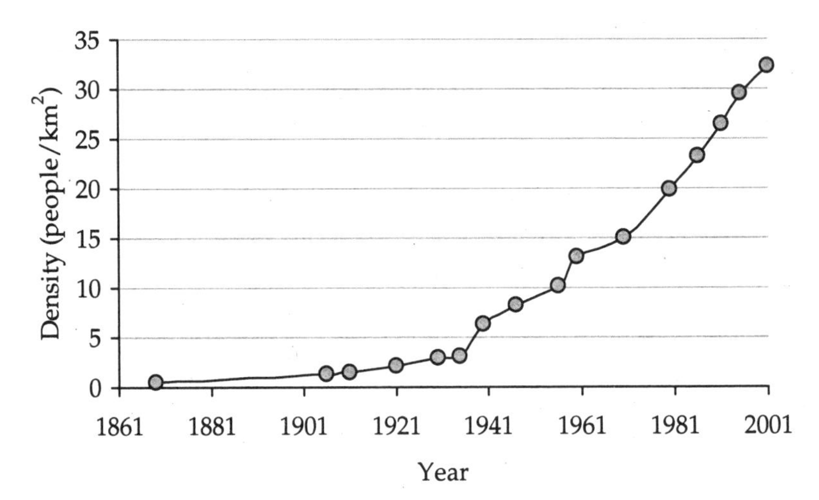

Figure 36.6. Changes in population density in the Interior Atlantic Forest of Misiones, Argentina, from 1865 to 2001.

2000). Population increase in these states from 1992 to 2000 is estimated at 3.7 percent per year (the highest is 14 percent in Hohenau, Itapuã).

In Argentina, the province of Misiones displays one of the highest human population densities in the country at 32.3 people/km^2 (INDEC 2001). With a total population of approximately 961,000 (INDEC 2001), this region is characterized by extreme poverty and rapid population growth (Chapter 19, this volume). Over the last half-century, population numbers in the province have increased significantly (Figure 36.6). From 1991 to 2001, population increased by an average of 2.9 percent per year (INDEC 2001), with the highest growth (8.9 percent) occurring in the municipality of General Belgrano in the northeast.

Migrations

In the Atlantic Forest hotspot, as in the rest of the world, the distribution of human populations is strongly affected by migration patterns, particularly from rural to urban areas. Underlying these pressures are various factors—such as social inequality, poverty, and the expansion of commercial production—that drive human migratory patterns and significantly threaten biodiversity. Urbanization is one of the greatest concerns caused by migration, as exemplified in the São Paulo metropolitan area, where human impact on the Atlantic Forests is high.

Over the last half-century, a large-scale rural exodus has been occurring in Brazil, leading to a rural population decrease in both relative and absolute terms (Sawyer and Rigotti 2001) that is directly impacting the environment (Rossini and Calió 2002). Combined with population increases, disparities in income and poor living standards have fueled migration from rural to urban environments and migration from poor regions to those with better employment opportunities and social infrastructure. Since 1975, Brazil has experienced an absolute decline in rural population, as a result of people moving to the cities (McNeely et al. 1995).

At the same time, Brazil is witnessing the migration of human populations toward forested, frontier areas. Between 1950 and 2000, approximately 50 million people in Brazil migrated to rural areas, representing one of the largest movements of its kind in the world (Sawyer and Rigotti 2001). Migrations to rural areas in Brazil have coincided with the expansion of agriculture and cattle ranching as well as the increase in mining. In search of better lands, migrants have also invaded indigenous territories. Overall, migration patterns in Brazil have led to the occupation of new areas and the loss of biological diversity (MMA 1999).

Globally, the makeup of national populations will become increasingly affected by migration not only within but also between countries (UNFPA 1999). This trend is exemplified in the Atlantic Forests of Argentina and Paraguay, where migrants from neighboring countries contribute significantly to population dynamics. Based on census information in Paraguay from 1992, Brazilians make up 56.8 percent of total migrants into the country, and 25.8 percent come from Argentina (DGEEC 1992; Chapter 26, this volume). Factors influencing these patterns include road construction as well as the establishment of a bridge between Paraguay and Argentina.

The Argentine province of Misiones is also witnessing much internal and external migration, especially in the northeastern part where migrants continue to arrive from southern Brazil. This contributes to the high growth rates exemplified in Misiones (Figure 36.6). In search of land, migrants ultimately reduce forest cover through slash-and-burn agriculture. Overall, however, a rural exodus toward urban areas is occurring in Misiones (Chapter 19, this volume).

Conclusion

The complexity of the relationships between humans and the environment makes it difficult to establish clear indicators of biodiversity loss based on human population variables. In the Atlantic Forests, where areas of low population density occur in highly deforested areas, it is evident that the major threats to biodiversity do not stem from human population growth or density, but rather from large-scale logging, mining, and cattle-ranching interests. Overall, statistics on population density and growth rates do not demonstrate the patterns driving population distribution, nor do they illuminate the cultural and technological contexts that shape levels and patterns of resource use underlying biodiversity loss. Often, as is the case with national and international migration patterns seen in the Atlantic Forests, the location of population growth, as well as the levels of consumption and patterns of development, are as important for understanding biodiversity loss as the absolute numbers (Stedman-Edwards 2000).

Trends in human population throughout the hotspots demonstrate a high risk that the loss of biological diversity will continue in these regions as human-dominated ecosystems expand (Cincotta and Engelman 2000). Globally, as in the Atlantic Forest hotspot, human fertility rates are decreasing, resulting in a decline in population growth. Although this represents a positive sign for biodiversity conservation, Atlantic Forest remnants still decrease in size every year. As more humans are becoming dependent on a finite amount of resources in this region, urbanization and migration patterns are significantly affecting biodiversity and are important factors to be taken into account in developing strategies for conservation throughout the Atlantic Forest.

References

CEPF (Critical Ecosystem Partnership Fund). 2001. *Critical ecosystem partnership fund. Ecosystem profile, Atlantic forest biodiversity hotspot, Brazil.* Washington, DC: Conservation International.

Cincotta, R. P. and Engelman, R. 2000. *Nature's place: human population and the future of biological diversity.* Washington, DC: Population Action International.

Cincotta, R. P., Wisnewski J., and Engelman, R. 2000. Human population in the biodiversity hotspots. *Nature* 404: 990–992.

DGEEC (Direccion General de Estadistica, Encuestas y Censos). 1992. *Direccion General de Estadistica, Encuestas y Censos, Paraguay.* Online: http://www.dgeec.gov.py.

Engelman, R., Cincotta, R. P., Dye, B., Gardner-Outlaw, T., and Wisnewski, J. 1997. *People in the balance: population and natural resources at the turn of the millennium.* Washington, DC: Population Action International.

Hogan, D. J. 2001. *Demographic dynamics and environmental change in Brazil.* Presented at the XXIV General Population Conference, International Union for the Scientific Study of Population, Salvador, Bahia, Brazil, August 18–24.

IBGE (Instituto Brasileiro de Geografia et Estatística). 2000. Online: http://www.ibge.gov.br.

ICRW (International Center for Research on Women). 1999. *Population, consumption, and environment linkages in mangrove ecosystem in the Gulf of Fonseca, Information Bulletin, March 1999.* Online: http://www.icrw.org/mangroves.htm.

INDEC (Instituto Nacional de Estadistica y Censos). 2001. *Población Censo 1991 y Censo 2001, total del país por provincia.* Online: http://www.indec.mecon.ar/default.htm.

ISA (Instituto Socioambiental). 1999. *Atlantic rainforest program.* Online: http://www.socioambiental.org/website/english/sociamb/programas/mata_atlantica.html.

Matos, R. and Baeninger, R. 2001. *Migration and urbanization in Brazil: process of spatial concentration and deconcentration and the recent debate.* Presented at the XXIV General Population Conference, International Union for the Scientific Study of Population, Salvador, Bahia, Brazil, August 18–24.

McNeely, J. A., Gagdil, M., Leveque C., Padoch C., and Redford, K. 1995. Human influences on biodiversity. In: Heywood, V. H. (ed.) *Global biodiversity assessment.* pp. 711–821. Cambridge: Cambridge University Press.

MMA (Ministério do Meio Ambiente). 1999. *First national report for the Convention on Biological Diversity.* Brazil: Ministry of Environment.

PRCDC (Population Resource Center). 1999. *A demographic profile of Brazil.* Online: http://www.prcdc.org/summaries/brazil/brazil.html.

Rossini, E. R. and Calió, S. A. 2002. *Women, migration, environment and rural development policy in Brazil.* Food and Agriculture Organization (FAO). Online: http://www.fao.org/DOCREP/x0210e/x0210e00.htm.

Sawyer, D. and Rigotti, J. I. R. 2001. *Migration and spatial distribution of rural population in Brazil, 1950–2050.* Presented at the XXIV General Population Conference, International Union for the Scientific Study of Population, Salvador, Bahia, Brazil, August 18–24.

Stedman-Edwards, P. 2000. A framework for analysing biodiversity loss. In: Wood, A., Stedman-Edwards, P., and Mang, J. (eds.). *The root causes of biodiversity loss.* London: World Wildlife Fund (WWF).

UNFPA (United Nations Population Fund). 1999. *State of world population 1999.* Online: http://www.unfpa.org/swp/1999/thestate.htm.

Chapter 37

Mercosur and the Atlantic Forest: An Environmental Regulatory Framework

María Leichner

Origins of the Regulatory Framework

The need to protect the environment is widely recognized today, and ensuring this protection is a matter of law and regulation. In many cases, we know how human activity will impact ecological systems. In others, we can surmise what the effects will be and take steps to prevent them, using various regulatory mechanisms and instruments. Still, we are not doing everything possible. To avoid irreversible damage to plant and animal populations and to protect biological diversity, we must recognize the limits on the regenerative capacity of ecosystems (Leichner 2000a). The value of a healthy environment can only be calculated by internalizing the costs of environmental losses. Such internalization is one indicator of environmental protection, but it is not sufficient. This chapter underscores the need for effective environmental policy instruments supported by a solid institutional structure.

International trade policies and trade agreements have environmental repercussions. Thus, regional trading blocs such as the Southern Common Market (Mercosur) cannot ignore environmental issues (Bertucci 1996). Mercosur originated when the Treaty of Asunción was signed in 1991. The treaty sought to reverse an economic situation—characterized by stifling foreign debt, lack of investment, and stagnation of international trade—that led to what has been called "the lost decade" in Latin America (Mariño Fages 1999).

The four southern Latin American countries of Argentina, Brazil, Paraguay, and Uruguay were the charter members of Mercosur. Subsequently, Bolivia and Chile were incorporated as associate members. The histories of the Mercosur countries have been shaped by a misconception of what constitutes progress. This misconception is based on development models and styles characteristic of a civilization that is ecologically unsound, socially unjust, and economically untenable (Coria et al. 1997).

The group was formed to ensure free circulation of goods, services, and factors of production through a common external tariff and trade policy. The agreement establishing Mercosur also calls for states to coordinate their macroeconomic policies and harmonize their legislation. After an initial stage, Mercosur evolved into a customs union in 1994 and is still striving to become, in keeping with its name, a true common market (Leichner 2000b).

Mercosur law should be viewed as an ongoing process of decision making (Boggiano 1998). Mercosur could potentially develop environmental law dynamically, with the help of the various community law institutions that have made headway toward sustainable development. This chapter explores whether Mercosur provides an adequate regulatory framework to support efforts to conserve the Atlantic Forest.

The integration process has brought to the fore a multitude of issues that were obscured behind formal and informal trade barriers. At the same time, while the trade agreement was being negotiated at the regional level, environmental issues were receiving increased attention on the international front, culminating in the Earth Summit (United Nations Conference on Environment and Development), held in Rio de Janeiro in 1992. In addition, multilateral environmental agreements linking trade with the environment have proliferated. Even more important, environmental analysis in the context of trade agreements has become more effective.

Neither international trade nor globalization are ends in themselves. Rather, they are the means for achieving a truly equitable multilateral trading system. Free trade can yield tremendous benefits for the environment and for sustainable development (Brañes et al. 2000). Unquestionably, the ultimate aim of regional integration is to improve the well-being of the region's citizens. To achieve this goal, a structure for ensuring sustainable development is essential.

Economic interdependence has had an impact on the treatment of environmental issues. Growing economic competition at the international level has turned local environmental issues into global concerns (Esty 2001). Within Mercosur, the member countries have relatively similar social, political, and economic features. The region is characterized by large disparities in gross domestic product (GDP) growth rates, continuous increases in income inequality, and a credibility gap in relation to economic policies and government structures. These characteristics are the ultimate cause of the region's environmental degradation, which is increasing at an alarming rate (Leichner 2000a).

Environmental Legal Instruments

Among the milestones in the history of environmental policy within Mercosur, two stand out. The 1992 Canela Declaration has led to significant progress in internalizing environmental costs into production processes. The Specialized Meetings on the Environment (known by their Spanish and Portuguese acronym, REMA) follow the European Union (EU) model of establishing environmental protection directives through the Single European Act (1986). The REMAs have approved basic directives on the sustainable management of renewable natural resources (Directive No. 4), the use of clean technologies (Directive No. 6), and the reduction of environmental deterioration resulting from productive processes (Directive No. 7).

In 2001, after a lengthy period of negotiation and debate, Mercosur approved the Framework Agreement on the Environment, marking the first concrete step toward an adequate regulatory framework for the region. The agreement provides the broad outlines of a regional framework for regulatory policy on sustainable development and could at least serve as a catalyst for change. However, it needs to be strengthened by an effective regional environmental policy that incorporates the values embodied in the environmental protocol.

Regulations for Protection through Mercosur

The Mercosur member countries are also members of the World Trade Organization (WTO). Compatibility of regional trade agreements with the multilateral trading system is a fundamental issue in the WTO. The main requirement is that the regional agreement must facilitate trade among the member territories without creating trade barriers for other WTO members that are not parties to the regional pact. This requirement is critical because Article XXIV of the 1994 General Agreement on Tariffs and Trade (GATT) establishes that if a free-trade zone or customs union is created, duties and other barriers to trade should be reduced or removed.

The coverage of multilateral agreements and provisions is currently much greater than was the case when Mercosur was in its infancy. The influence of such accords will probably continue to expand, in terms of both the number of signatory countries and the commitments assumed. Similarly, the number and complexity of the issues and subject matter under discussion have grown exponentially.

The need to regulate environmental policy is reinforced by multilateral environmental agreements that incorporate both the traditional negotiating format of the WTO (rounds of negotiation and consensus building among all parties) and the effectiveness of traditional basic instruments of the trading system (such as the principles of most-favored nation, nondiscrimination, and national treatment). Regional agreements appear to be a feasible and reasonable alternative—without discounting the value of multilateral agreements—for negotiating some issues for which it is difficult to achieve multilateral agreements.

The Structure of the Multilateral Trading System: Progress in Environmental Policy Making?

The WTO promotes well-being by establishing rules and principles for an open and nondiscriminatory multilateral trading system. Multilateral Environmental Agreements (MEAs) also contribute to general well-being by establishing rules, principles, institutions, and mechanisms for environmental protection. In these regards, the two systems support each other. MEAs will remain an effective means for international negotiation of policies on environment and trade. In turn, trade measures are an attractive aspect of these agreements when they are directly related to the management and conservation of environmental resources.

The Mercosur countries are parties to most of the main multilateral environmental agreements (PNUMA 2000). The provisions contained in regional instruments have been incorporated into existing regulatory frameworks—which, unfortunately, have not led to any substantial modifications in institutional structures. At most, these frameworks have facilitated the creation of administrative units within existing institutions. Economic instruments have not generally been used to implement regional MEAs. This is a clear need in the region.

With the establishment of a multilateral trading system, represented by the WTO, the first steps were taken toward effective interaction between the trade and environmental sectors. In the recent Doha Ministerial Declaration (2001), the WTO acknowledged the need to adopt sustainable trading systems, expand access to markets for environmental products, and recommend the execution of environmental sustainability studies.

Mercosur now faces three complex and demanding negotiating processes, in addition to any others that may occur with individual countries. The main one will take place within the sphere of the WTO, which is the principal forum for trade negotiations aimed at opening markets for goods and services under stable and predictable conditions. The other two are the negotiations for the Free Trade Area of the Americas (FTAA) and those with the European Union. In all three cases, it will be essential to have a strategic approach shared by all stakeholders and to maintain open lines of communication during the negotiation process.

This process of simultaneous trade negotiations, however, entails more than strategic and organizational definition. Mercosur must deal with challenges of consolidation and deepening if it is to take its place as a bona fide participant in such negotiations.

The EU-Mercosur Interregional Framework Cooperation Agreement (Article 17) calls for interregional cooperation in the areas of environmental protection and rational use of natural resources, with the aim of achieving sustainable development. This framework agreement has translated into concrete activities: the exchange of information and know-how; training and education regarding the environment; technical assistance; the implementation of joint research projects; and, where appropriate, institutional assistance.

The sanitary and phytosanitary restrictions that the European Union places on access to its markets are not only the largest obstacles to trade but also the most relevant from the standpoint of negotiations and the linkage of environment and trade.

The Mercosur countries have ratified most of the multilateral environmental agreements. Although the impact of these agreements on environmental protection is difficult to assess, they have succeeded in raising awareness of environmental issues among decision makers and the general public in the region. This increased awareness, however, has seldom led to environmental issues being prioritized on political agendas or resource allocation being prioritized in national budgets.

The existing institutional and legal framework is adequate to respond to the need for conservation and sustainable use of biodiversity in the region. All that is required is better enforcement of legal instruments—especially some MEAs, such as the Convention on International Trade in Endangered Species of Wild Fauna and Flora (CITES)—and ratification of the principal conventions by more countries (CITES, the Ramsar Wetlands Convention) and implementation of their additional protocols, especially the Protocol on Protected Areas (Cabrera 2001). Equally necessary is coordination of the countries' positions on global agreements (such as those on climate change and biodiversity and the Montreal Protocol) and on internalization of environmental costs of production processes.

The Convention on Biological Diversity in the Atlantic Forest

The Convention on Biological Diversity (CBD), a direct outcome of the Earth Summit, has brought about progress in the institutional sphere (formation of national biodiversity commissions and focal points), in policy and regulatory frameworks (national biodiversity action plans and strategies), and in knowledge and assessment (country reports, inventories). It has even led to the negotiation of a protocol on the risks of modern biotechnology (the Cartagena Protocol on Biosafety).

The first task in applying the CBD to the Atlantic Forest is to analyze the myriad causes of biodiversity loss: unsustainable consumption of forest resources, economic systems that have failed to acknowledge the value of the environment, inequity in land ownership and in distribution of the benefits of conservation and the use of biological resources, gaps in knowledge about sustainable management, use of inappropriate economic incentives, and lack of a coherent regional regulatory framework for conservation and regional political cohesion (see several chapters in this book). Considering the close linkage of these factors with the economic situation, it is not surprising that the region is experiencing rapid deterioration of its biodiversity.

Habitat loss due to deforestation, changes in land use, and overexploitation of resources—coupled with the introduction of exotic species—are among the factors that account for the deterioration of the Atlantic Forest. The issue of biodiversity began to acquire new policy implications in 1992, when the signing of the CBD launched a movement to stem the loss of biodiversity. The new focus is founded on a much broader concept: the need for sustainable use of natural resources and for fair and equitable distribution of the benefits deriving from their use. This broad concept includes other issues, such as biotechnology, biosafety, intellectual property rights, protection of knowledge and innovations, and improvement of local communities' relationship with the forest.

Some concrete progress has been made in multilateral cooperation through the national biodiversity strategies, among other elements that facilitate conservation of biological resources. The numbers of national, state, and private protected areas have increased, although this increase has not necessarily resulted in complete protection of ecosystems.

Multidimensional Bioethics within Mercosur

Three levels in the ecological structure have been identified: ecosystems and communities, habitats and species, and genomes and genes (Leff and Bastida 2001). These levels are closely linked by interactions and interdependences that maintain them over time and space. However, the CBD multilateral environmental agreement does not stress the importance of the interdependence among these levels or even between these levels and the cultural level. Under the CBD, each level is treated separately, without regard for the interactions with other biological or physical units. In the case of genes and genomes, for example, the CBD permits the introduction of genetic material from other organisms into the DNA of a plant, which may render its offspring infertile. Such drastic interference with the ecological and evolutionary processes of plant species violates basic principles of bioethics (Leff and Bastida 2001).

The CBD's failure to recognize the interdependence of the three levels of ecological structure makes the Atlantic Forest and other megadiverse regions vulnerable. Under the current criteria, genes, species, and ecosystems are viewed as components disconnected from one another, and they may be isolated, manipulated, managed, and conserved, in situ or ex situ. To separate them conceptually and materially is to act without regard for minimum bioethical standards.

Multidimensional bioethics seeks to remedy the CBD's failure to acknowledge these connections. It aims to avoid any confusion about what direction harmonization of policies on environment and trade should take. It should also be noted that the CBD makes no mention of the importance of a healthy and balanced physical environment.

Conclusions

Given the present regulatory framework, our challenge is to work toward greater cohesion that will embrace the environmental mechanisms and tools that currently exist in the global sphere, while finding our own answers to sustainability within Mercosur. To achieve these objectives, we should seek to expand the current legal framework, especially by strengthening the Framework Agreement on the Environment and by implementing its basic directives. We should also resume the work, with the necessary updates and revisions, undertaken by Subgroup 6, one of several technical working groups within Mercosur (which evolved from the REMAs) to try to breathe life into the environmental protocol.

A regional strategy on biodiversity for the Mercosur countries should be developed within the existing institutional framework. The working group formed to devise the strategy should include trade and environmental experts from the region. This work should serve as the basis for effectively implementing protection of the Atlantic Forest and all its priority conservation areas. A unified regional system of ecological easements should be implemented, and flexible tools for land tenure should be created, with economic incentives that encourage conservation.

Mercosur should become a supranational organization. If it does, the provisions emanating from its decision-making bodies will be espoused by the member countries automatically and true regional strategies can thus be developed.

Perhaps the 2002 "Rio+10" Summit in South Africa will spawn genuine efforts to achieve sustainable development. Perhaps it will yield forceful decisions that will be implemented by governments and civil society alike with a view to creating a new order, in which respect for the environment will be paramount and only one kind of development will be contemplated—development for future generations.

References

Bertucci, R., Cunha, E., Cunha, T., Devia, L., Figueiras, M., Ruiz Díaz Labrano, R., and Vidal Perera, R. 1996. *Mercosur y medio ambiente.* Buenos Aires: Ediciones Ciudad Argentina.

Boggiano, A. 1998. *Derecho internacional publico y privado y derecho del Mercosur. Ius Inter Iura.* Buenos Aires: Ediciones La Ley S.A.

Brañes, R., Caillaux, J., González, M., and Silva, C. D. 2000. *Medio ambiente y libre comercio en América Latina: los desafíos del libre comercio para América Latina desde la perspectiva del área del Libre Comercio de las Américas (ALCA).* México: Programa de las Naciones Unidas para el Desarrollo) (PNUD) and Asociación Latinoamericana de Derecho Ambiental (ALDA).

Cabrera, J. 2001. *El impacto de las conferencias de Río y Estocolmo sobre el derecho y las políticas ambientales en ALC.* Paper prepared for the Development Observatory of the University of Costa Rica, San José.

Coria, S., Devia, L., and Gaudino, E. 1997. *Integración, desarrollo sustentable y medio ambiente.* Buenos Aires: Ediciones Ciudad Argentina. (Cuadernos de integración 1).

Esty, D. 2001. *El reto ambiental de la Organización Mundial de Comercio. Sugerencias para una reconciliación.* Barcelona: Editorial Gedisa S.A.

Guajardo, C. A. 1999. *Comercio internacional y globalización.* Mendoza, Argentina: Ediciones Jurídicas Cuyo.

Leff, E. and Bastida, M. (eds.). 2001. *Comercio, ambiente y desarrollo sustentable: perspectivas de América Latina y el Caribe.* Serie Foros y Debates Ambientales, núm. 2. México: Programa de las Naciones Unidas para el Medio Ambiente (PNUMA)-Universidad Nacional Autónoma de México (UNAM)-Mexican Council for Sustainable Development (COMEDES)-International Institute for Sustainable Development (IISD).

Leichner, M. 2000a. *Environmentalism and the new logic of business.* Mercosur Economic Summit 2000. World Economic Forum, Rio de Janeiro. Online: http://www.weforum.org/site/knowledgenavigator.nsf/Content/Environmentalism%20and%20the%20New%20Logic%20of%20Business?open&country_id=®ion_id=801002.

Leichner, M. 2000b. *Mercosur: su dimensión ambiental. Comercio y prioridades políticas de inversión.* Washington, DC: World Wildlife Foundation (WWF).

Mariño Fages, R. J. 1999. *La supranacionalidad en los procesos de integración regional.* Mario A. Viera, ed. Corrientes, Argentina: Editorial Mave.

PNUMA (Programa de las Naciones Unidas para el Medio Ambiente), Oficina Regional para América Latina y el Caribe. 2000. *GEO América Latina y el Caribe: Perspectiva del Medio Ambiente 2000.* San José: Observatorio del Desarrollo, Universidad de Costa Rica.

Chapter 38

A Challenge for Conservation: Atlantic Forest Protected Areas

Alexandra-Valeria Lairana

The Atlantic Forest contains a large number of endemic and threatened species (Chapters 8 and 30, this volume). The in situ safeguarding of some of these species in reserves, parks, and other protected areas constitutes a strategy of enormous importance for the conservation of the last forest remnants. This chapter summarizes available information about the number, size, and overall extent of the protected areas within the Atlantic Forest hotspot.

A protected area is defined as "an area of land and/or sea especially dedicated to the protection and maintenance of biological diversity, and of natural and associated cultural resources, and managed through legal or other effective means" (IUCN 1994; Green and Paine 1997). Protected areas are established to

- promote scientific research,
- protect wilderness,
- preserve species and genetic diversity,
- maintain environmental services,
- protect specific natural and cultural features,
- promote tourism and recreation,
- educate,
- encourage sustainable use of resources from natural ecosystems, and
- maintain cultural and traditional attributes (IUCN 1994).

History of the Establishment of Protected Areas

Throughout the history of Argentina, Brazil, and Paraguay, many environmental movements and organizations have proposed and designated protected areas (Chapters 4, 14, and 24, this volume). Such areas are not considered officially protected, however, until they have been legally recognized, registered, and accepted by governments and local communities.

With the exception of a small area in São Paulo (Parque Estadual da Cidade), established in 1898, all protected areas in the Atlantic Forest were created during

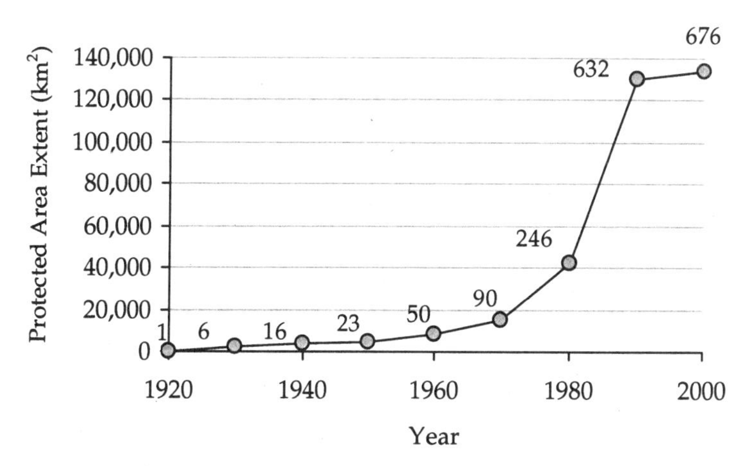

Figure 38.1. Protected areas designated in the Atlantic Forest hotspot since 1920. Numbers of protected areas are displayed above each circle.

the twentieth century. The earliest protected areas were Iguazú National Park (Argentina), established in 1934; Itatiaia National Park (Brazil), established in 1937; and Iguaçu National Park (Brazil), established in 1939 (IUCN 1998). The creation of new protected areas in the Atlantic Forest has increased significantly over time (Figure 38.1). Since the beginning of the twentieth century, 676 protected areas—totaling 134,000 km^2—have been created in the Atlantic Forest and adjacent coastal areas.

Analysis Constraints

Exactly how much land is really being protected in the Atlantic Forest is difficult to determine, a fact that constrains any analysis of the status of its protected areas. Brazil, Argentina, and Paraguay each use different classifications for their national systems of conservation units (Table 38.1). Moreover, within each country there are a variety of administrative systems (national, federal, state, provincial, departmental, municipal), and their functions and responsibilities often overlap. Additional complications arise from the fact that some protected areas are still being defined and delimited. In other cases, the boundaries have not yet been fixed (e.g., Yarará Municipal Nature Park and the Ictica Caraguatay and Corpus nature reserves in Misiones, Argentina) (APN 2002). Other areas have overlapping boundaries, both within and between territories. For example, some conservation units lie within others, as in the case of the Yabotí Biosphere Reserve in Misiones, Argentina, which encompasses the Moconá and Esmeralda provincial parks, the Papel Misionero Natural Cultural Reserve, and the Guaraní Experimental Reserve (Chapter 21, this volume). Overlapping also occurs between protected areas; for example, the Curumbataí-Botucatu-Tejuca Environmental Protection Area and the Piracicaba-Juqueri-Mirim Environmental Protection Area in the state of São Paulo, Brazil, partially overlap (SEA 2001).

Determining the true extent of protected areas is further hindered by the fact

Table 38.1. Protected area categories used in the countries of the Atlantic Forest hotspot.

Brazil	Argentina	Paraguay
State Environmental Protection Area	Experimental Area	Scientific Monument**
Environmental Protection Area*	Natural Monument	National Park**
State of Relevant Ecological Interest Area	National Natural Monument	Biological Refuge**
Relevant Ecological Interest Area*	Protected Landscape	Biological Reserve**
State Ecological Station	Municipal Park	Biosphere Reserve**
Ecological Station*	National Park	Ecological Reserve
State Forest	Municipal Natural Park	National Reserve**
National Forest	Provincial Park	Natural Reserve
State Ecological Park	Wildlife Refuge	Private Natural Reserve
State Park	Private Wildlife Refuge	Ecological Easement
State Tourist Park	Natural Cultural Reserve	Managed Resources Reserve
State Forest Park	Biosphere Reserve	
National Park*	Multiple Use Reserve	
State Biological and Archaeological Reserve	Wildlife Reserve	
State Biological Reserve	Forest Reserve	
Biological Reserve*	National Reserve	
State Ecological Reserve	Strict Natural Reserve	
Ecological Reserve*	Fish Natural Reserve	
Extractive Reserve*	Municipal Natural Reserve	
State Forest Reserve	Private Reserve	
Private Natural Heritage Reserve	Provincial Reserve	
*Classification IBAMA	All within the records of APN	**Classification de SINASIP

that some areas come under more than one jurisdiction or are comanaged by different entities (e.g., Itaipú Binational Reserve, Paraguay/Brazil). Similarly, some protected areas, such as the Cerro Corá National Park (Paraguay) and the Yabotí Biosphere Reserve (Argentina), comprise both public and private lands. The dissolution and recategorization of former protected areas poses an additional problem for analysis when official records have not been updated—as is the case, for instance, with the General Belgrano Forest Reserve in Misiones, Argentina, which is now known as Guardaparque (Park Ranger) Horacio Foerster Forest Reserve (Chapter 21, this volume). Moreover, some protected areas are currently functioning as such without official recognition (e.g., Roberto Cametti Provincial Park in Misiones, Argentina).

For these various reasons, protected areas that are effectively fulfilling their functions must be distinguished from those that are not, since many are no more than "paper parks" (Bertonatti and Corcuera 2000; Chapters 21 and 28, this volume). In other words, the official declaration that an area is protected, though necessary, is not sufficient to ensure conservation in the long term.

Total Area under Protection

The information used to estimate the Atlantic Forest's total area under protection was extracted from lists prepared by L. P. Pinto of Conservation International Brazil (CI Brazil), Silva and Casteleti (Chapter 5, this volume), Giraudo et al. (Chapter 21, this volume), and Yanosky and Cabrera (Chapter 28, this volume). These lists were developed using similar but not identical criteria. Nevertheless, the criteria are similar enough to allow for a reasonably accurate classification of protected areas and an estimate of the total surface area under protection (Table 38.2).

At the national level, official lists of protected areas are maintained by the Brazilian Institute for the Environment and Renewable Natural Resources (IBAMA), the Argentine National Parks Administration (APN), and the Paraguayan National System of Protected Wilderness Areas (SINASIP). At the international level, the 1997 United Nations List of Protected Areas (IUCN 1998) includes a selected group of protected areas located in the Atlantic Forest.

These lists, however, suffer from three kinds of limitations: (1) insufficient current, official information; (2) duplicated information (overlapping areas); and (3) insufficient standardized information. For the Brazilian protected areas, the IBAMA list mentions only the federal protected areas; no state or privately owned areas are included. Silva and Casteleti (Chapter 5, this volume) include only Brazilian areas under comprehensive protection (i.e., ecological and biological stations; national, state, and forest parks; and biological, ecological, and forest reserves). Environmental protection areas, areas of special ecological interest, national forests, and extraction reserves are thus excluded. The list compiled by CI Brazil (unpubl.) provides information on protected areas according to their jurisdiction (federal/state), ownership (private/public), and location (continental, coastal, transition area)—although it does not indicate areas that overlap. Finally, because Argentina's official list of protected areas (APN 2002) is not up to date, information on Argentina from other sources was used (Chapter 21, this volume).

Preliminary analysis indicates that 676 protected areas (public and private) exist in the Atlantic Forest hotspot, encompassing a total area of approximately 134,000 km^2. Brazil has the largest number of protected areas (577 units), with a combined area of approximately 125,300 km^2. The Brazilian list includes protected areas that are located in coastal and transition areas, as well as those in the Planalto (Plateau) and Campos do Sul regions, the *Araucaria* pine forests, and the Restinga and Atlantic Forest. Argentina has 60 protected areas (approximately 4,600 km^2), and Paraguay has 39 (around 4,000 km^2). As I demonstrate below,

Table 38.2. Categories of protected areas according to IUCN (IUCN 1994).

Category	Name	Main Objective	Description
Ia	Strict Nature Reserve	Science	Area of land and/or sea possessing some outstanding or representative ecosystems, geological or physiological features and/or species, available primarily for scientific research and/or environmental monitoring.
Ib	Wilderness Area	Wilderness protection	Large area of unmodified or slightly modified land and/or sea, retaining its natural character and influence, without permanent or significant habitation, which is protected and managed so as to preserve its natural condition.
II	National Park	Ecosystem protection and recreation	Natural area of land and/or sea, designated to (a) protect the ecological integrity of one or more ecosystems for present and future generations; (b) exclude exploitation or occupation inimical to the purposes of designation of the area; and (c) provide a foundation for spiritual, scientific, educational, recreational, and visitor opportunities, all of which must be environmentally and culturally compatible.
III	Natural Monument	Conservation of specific natural features	Area containing one or more specific natural or natural/cultural features, which is of outstanding or unique value because of its inherent rarity, representative or aesthetic qualities, or cultural significance.
IV	Habitat/Species Management Area	Conservation through management intervention	Area of land and/or sea subject to active intervention for management purposes so as to ensure the maintenance of habitats and/or to meet the requirements of specific species.
V	Protected Landscape/ Seascape	Landscape/ seascape conservation and recreation	Area of land, with coast and sea as appropriate, where the interaction of people and nature over time has produced an area of distinct character with significant aesthetic, ecological and/or cultural value, and often with high biological diversity. Safeguarding the integrity of this traditional interaction is vital to the protection, maintenance, and evolution of such an area.

Table 38.2. Continued

Category	Name	Main Objective	Description
VI	Managed Resource Protected Area	Sustainable use of natural ecosystems	Area containing predominantly unmodified natural systems, managed to ensure long-term protection and maintenance of biological diversity, while providing at the same time a sustainable flow of natural products and services to meet community needs.

however, analyses of the degree to which the Atlantic Forest is being protected depend on the criteria used to include protected areas (Figures 38.6a and 38.6b).

Number and Size

Although the number of protected areas in the Atlantic Forest is relatively high (676 units), the total surface area under protection is of greater importance. Protected areas should be as large as possible, not only to maximize the integrity of the species, habitats, and ecosystems they contain, but also to minimize the risk of species extinction and to maximize the representation of ecological communities and species therein (Green and Paine 1997). Internationally, areas larger than 10 km^2, or 1 km^2 in the case of wholly protected islands, are considered more likely to achieve conservation objectives (IUCN 1994). Areas that simply meet this minimum standard, however, are still extremely small areas (Chapter 31, this volume).

For this analysis, protected areas are divided into four size categories: (1) up to 10 km^2, (2) 10–100 km^2, (3) 100–1,000 km^2, and (4) larger than 1,000 km^2. In the Atlantic Forest, many "small" protected areas and only a few "large" ones exist (Figure 38.2). Fifty-nine percent (395 units) have an area of less than 10 km^2.

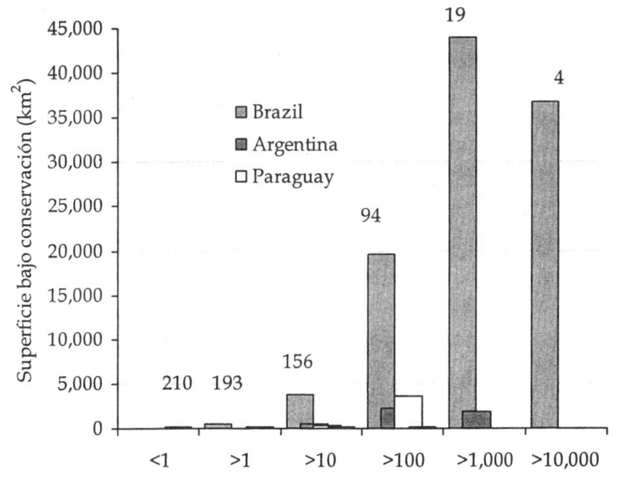

Figure 38.2. Number and extent of protected areas in the Atlantic Forest hotspot classified by size.

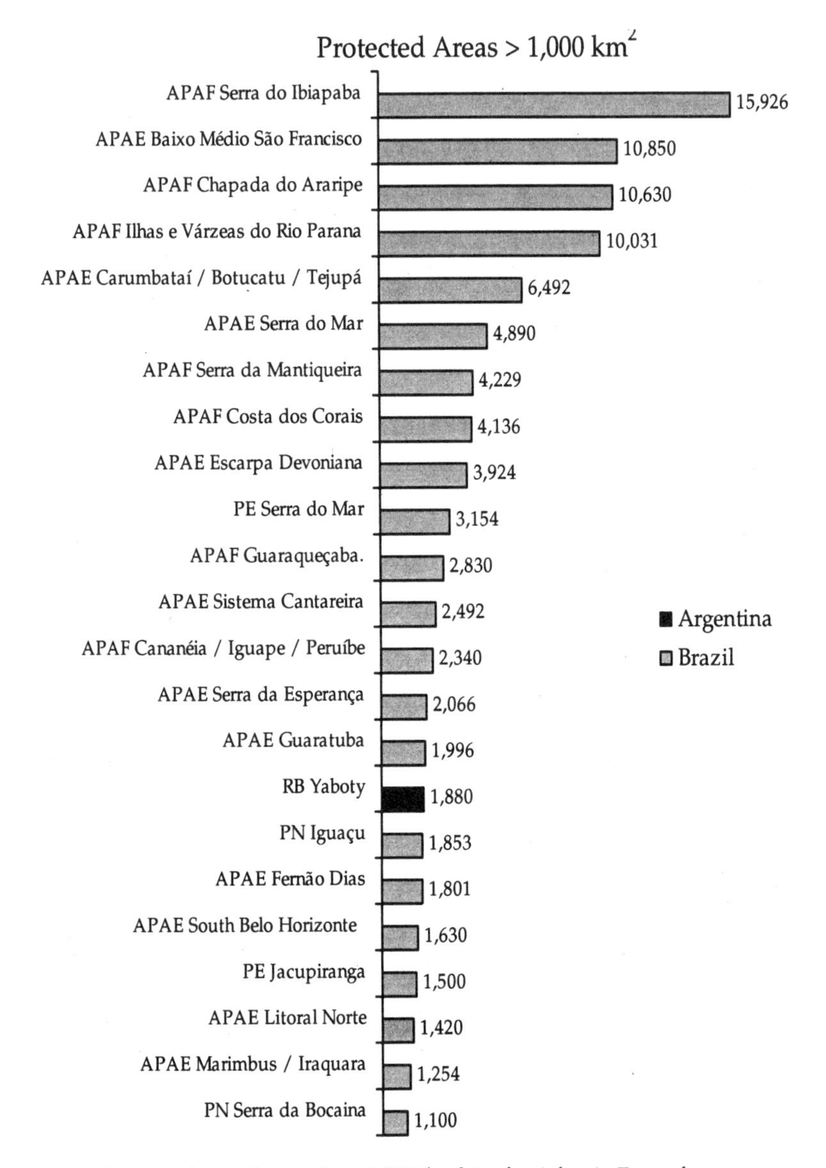

Figure 38.3. Protected areas larger than 1,000 km² in the Atlantic Forest hotspot.

The remaining 41 percent (273 units) have an area of 10 km² or larger. For five protected areas in Argentina and three in Paraguay, size data are unavailable.

The fourth size category (larger than 1,000 km²) includes 23 units (3 percent) (Figure 38.3) with an approximate total area of 98,900 km². Virtually all of the protected areas in this group are Brazilian, the sole exception being the Yabotí Biosphere Reserve in Misiones, Argentina. The most common type of protected area (20 out of 23) within this group is the "environmental protection area" (*área de proteção ambiental*, APA), a Brazilian designation. Within this category, the Serra Ibiapaba APA (15,926 km²) is the largest protected area in the Atlantic Forest (Figure 38.3). Unfortunately, APAs lack implementation and are strongly

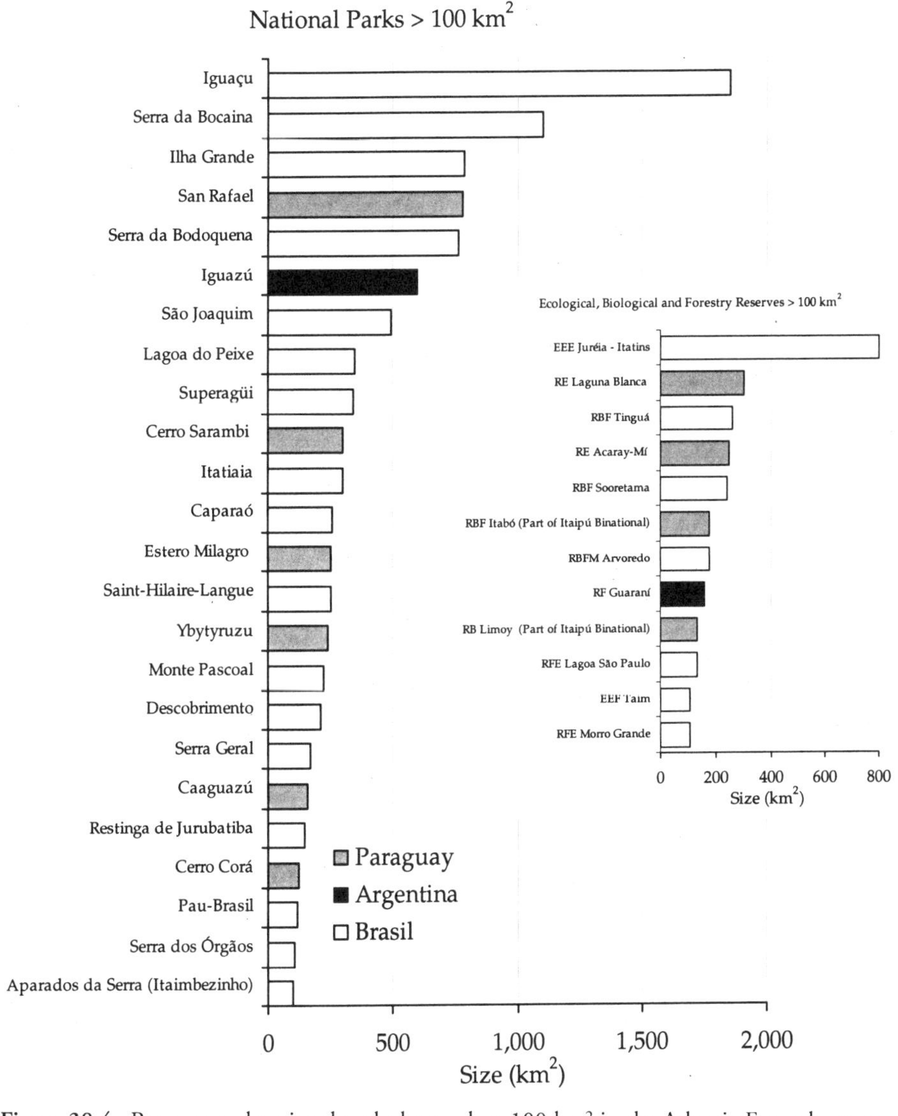

Figure 38.4. Reserves and national parks larger than 100 km^2 in the Atlantic Forest hotspot.

questioned by Brazilian environmentalists. The category of areas larger than 1,000 km^2 also includes two Brazilian national parks: Iguaçu (1,850 km^2) and Serra da Bocaina (1,100 km^2). No protected areas in the Atlantic Forest of Paraguay fall into this category. The largest protected area in that country is the privately owned Serranías de San Rafael National Park, which is 780 km^2.

The third size category (100–1,000 km^2) comprises 117 units (17.5 percent). Of these, 25 are environmental protection areas (APAs), 17 are national parks, and 15 are state parks. The remaining 60 units have a variety of designations (Figure 38.4).

A total of 202 protected areas (30 percent) have areas of less than 1 km^2. Brazil has the largest number of such areas: 179 units, of which 141 are "private natural heritage reserves" (*reserva particular do patrimônio natural*). Argentina and Paraguay have 21 and 2 units, respectively, in this category.

Private and Public Protected Areas

Governments have a fundamental responsibility to establish and maintain national systems of protected areas, which are considered important components of national strategies for conservation and sustainable development (IUCN 1994). In practice, however, the responsibility for managing individual protected areas often ends up resting with central or regional governments, nongovernmental organizations, the private sector, or the local community. Accordingly, the protected areas in the Atlantic Forest may be private, public, or a combination of both. Publicly owned lands generally fall into the following categories: national (Argentina and Paraguay), federal or state (Brazil), provincial (Argentina), or municipal (Argentina and Brazil). Privately held areas include private natural heritage reserves (Brazil); private reserves (Argentina); private natural reserves, managed resource reserves, and ecological easements (Paraguay). Mixed public-private areas are generally co-managed by the local government and a nongovernmental organization.

Although the registered number of public protected areas (56 percent) is slightly larger than that of private areas (44 percent), the privately owned areas under protection total only 7,000 km^2, whereas the publicly owned areas amount to 126,800 km^2. Brazil has the largest area under public protection (124,000 km^2), and Paraguay has the largest amount of privately owned protected land (Figure 38.5). In the Misiones Forest in Argentina, approximately 540 km^2 are

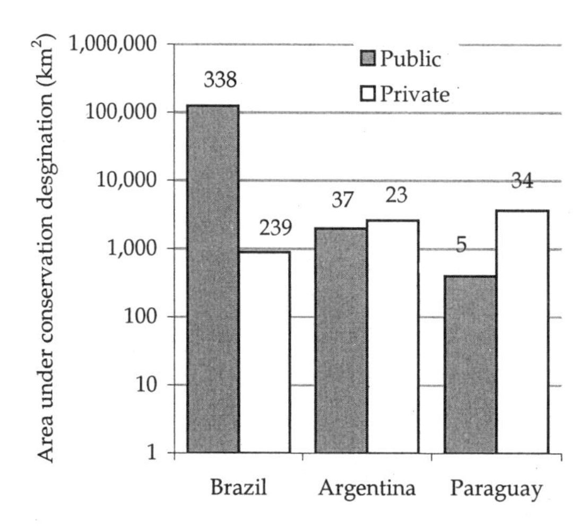

Figure 38.5. Private and public protected areas in the Atlantic Forest hotspot. The number of protected areas is shown on top each bar.

privately managed. Fundación Vida Silvestre Argentina (Argentine Wildlife Foundation, FVSA) comanages some of the private protected areas in the province (e.g., the Papel Misionero Natural Cultural Reserve). In Paraguay, some 3,700 km^2, which include fragments of the Interior Atlantic Forest, are privately protected (e.g., the Mbaracayú Natural Forest Reserve, 644 km^2). In Brazil, approximately 890 km^2 of protected lands (239 private natural heritage reserves) are privately held. Most of these do not exceed 1 km^2 in size.

IUCN Protected Area Management Categories

The designations for Atlantic Forest protected areas vary because each country has its own classification system (Table 38.1). With a view to standardizing the categorization of protected areas at the global level, the World Commission on Protected Areas (WCPA) of the World Conservation Union (IUCN) has established six protected area management categories (Table 38.2). Although activities that occur in protected areas fulfill more than one objective, their categorization under the IUCN system is based on a primary management objective, and at least three-fourths of the protected area should be managed for that primary purpose (IUCN 1994). It is important to clarify that the names governments assign to protected areas do not define their category in accordance with the IUCN system. For example, the fact that a protected area is called a national park does not necessarily mean that it falls under IUCN category II.

In the case of the Atlantic Forest hotspot, only 40 percent (274 units) of all the areas included are classified according to the IUCN management categories. This value is equal to approximately 59,800 km^2 of protected areas (public and private, coastal and continental).

Those categories in the IUCN system that are classified as comprehensive protection (categories Ia, Ib, II, and III) apply to areas that are managed strictly for conservation purposes and do not allow for permanent human settlements (IUCN 1994). Of the 59,800 km^2 classified under the IUCN categories, 25,600 km^2 (43 percent) are under comprehensive protection (categories Ia, II, and III), while 34,200 km^2 (57 percent) have resource management objectives (categories IV, V, and VI; Table 38.3). Of all the areas classified under some IUCN management

Table 38.3. Number and extent of protected areas under different IUCN categories in the Atlantic Forest.

Class	Category	Units	Extent (km^2)	>10 km^2	>100 km^2	Percentage of Total Land Area under Protection
Ia	Strict Nature Reserve	85	3,700	45	10	6.15
II	National Park	109	22,000	70	40	36.57
III	Natural Monument	1	1.9			0.00
IV	Habitat/Species Management Area	22	1,050	13		1.74
V	Protected Landscape/Seascape	39	33,000	36	25	54.85
VI	Managed Resource Protected Area	18	407	5	3	0.67

category, 25,600 km^2 (115 coastal and inland units) are under comprehensive protection and measure more than 10 km^2. It is important to emphasize that the protected areas in this group afford greater possibilities for conserving the Atlantic Forest, owing to their size and the fact that they are strict protection areas.

International Designations

Earlier versions of the IUCN protected area management categories included "Category IX: Biosphere Reserve" and "Category X: World Heritage Site" (Green and Paine 1997). The most recent United Nations list of protected areas (IUCN 1998), however, excludes these designations and eliminates categories VII, VIII, IX, and X. Within the Atlantic Forest, about 293,000 km^2 have been designated as biosphere reserves, 58,000 km^2 as world heritage sites, and approximately 700 km^2 as Ramsar wetland sites.

Biosphere Reserves

In 1993, MAB-UNESCO (Man and the Biosphere Program, United Nations Environmental, Scientific and Cultural Organization) recognized approximately 290,000 km^2 in 14 Brazilian states as biosphere reserves (*reserva da biosfera da Mata Atlântica*) (Bertolini 1999). In 1995, 2,363 km^2 in Misiones Province, Argentina, were designated as the Yabotí Biosphere Reserve (MAB-UNESCO 2001). In 2000, the Mbaracayú Natural Forest Reserve (644 km^2) was recognized in Paraguay (MAB-UNESCO 2001). Unfortunately, many biosphere reserves exist only on paper, particularly when they are exaggeratedly large.

World Heritage Sites

Among the 167 sites inscribed on the World Heritage List (UNESCO 2001), five are in the Atlantic Forest, with four in Brazil—Iguaçu National Park (1,852 km^2, 1986), Serra das Capibaras National Park (1,000 km^2, 1991), the Discovery Coast Atlantic Forest Reserves (which include 8 protected areas with a total area of 11,200 km^2, 1999) and the Atlantic Forest Southeast Reserves (29 protected areas with a combined area of 47,000 km^2, 1999)—and one in Argentina: Iguazú National Park (599 km^2, 1984) (UNESCO 2001).

Ramsar Sites

The Ramsar Convention on Wetlands ("Ramsar Convention") was signed in the city of Ramsar, Iran, in 1971 and came into force in 1975 (Ramsar 2000). Currently, 127 nations are parties to the convention, and 1,085 sites have been designated for inclusion on the List of Wetlands of International Importance ("Ramsar List") (Ramsar 2000). In 1993, more than 450,000 km^2 of marine ecosystems on the Atlantic coast were added to the Ramsar List. Lagoa do Peixe National Park (344 km^2) and Estero Milagro National Park (250 km^2) are the only Atlantic Forest wetlands included on the list (Ramsar 2000).

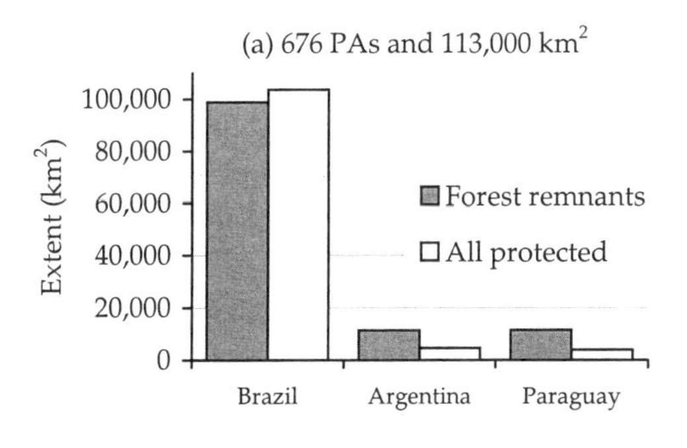

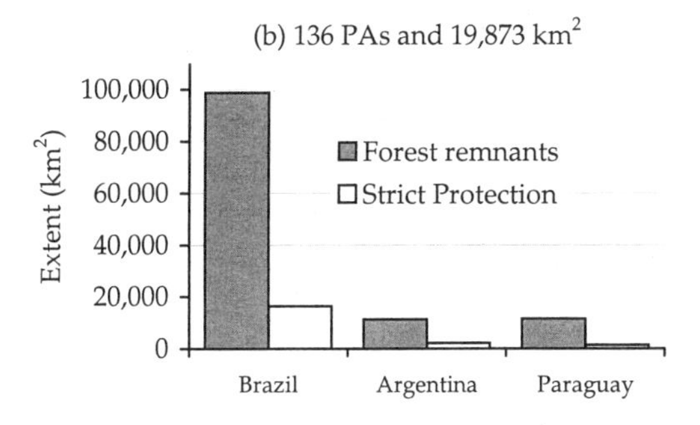

Figure 38.6. Forest remnants in the Atlantic Forest and protected area extent. (a) All continental, coastal, and transitional areas; (b) only strictly protected areas.

Forest Remnants and Protected Areas

Seventeen Brazilian states, nine Paraguayan departments, and one Argentine province together contain around 121,600 km^2 of remnants from original forests in the Atlantic Forest hotspot (Chapters 5, 6, 21, and 28, this volume). The largest percentage is in Brazil (98,800 km^2). Argentina has an estimated 11,300 km^2, and Paraguay has around 11,500 km^2. Excluding protected areas located in coastal regions (9,550 km^2) and transition areas (11,000 km^2), approximately 113,000 km^2 (93 percent) of the forestland in the Atlantic Forest is under some form of legal protection (CI Brazil, unpubl. data; Chapters 15 and 28, this volume; Figure 38.6a). However, this is a gross overestimation of protection because many of the largest areas (APAs) contain degraded areas and even urban centers. If just strictly protected areas are considered, then only around 19,873 km^2 are currently under protection (Figure 38.6b), which represents approximately 16 percent of the remaining forested area.

Conclusions

To evaluate the protection status of the Atlantic Forest, researchers must know both the extent of the remaining ecosystems and the area occupied by protected areas that effectively fulfill their objectives. Analysis of these two factors indicates that only a fraction (less than 20 percent) of the area under protection is being strictly protected. This appears to be the case regardless of which lists are consulted or which methodology is used to classify the information. The situation may look even worse if protected areas are evaluated according to their quality of management and their actual state, since we can assume that ineffective management of protected areas is detrimental to the conservation of Atlantic Forest remnants.

Ideally, protected areas should be large (preferably larger than 100 km^2) so that they are better able to ensure the continuity of viable populations and biological, ecological, and evolutionary processes. In addition, protected area management practices should differentiate between strict (comprehensive) protection and resource management. Protected areas should also be officially recognized under international classifications and designations and should be managed by governments and local communities that have the capacity to assume responsibility for conservation of the forest remnants. Ideally, of course, all protected areas would conform to these specifications. The reality, however, is much different: many officially declared areas either are not fulfilling the objectives for which they were created or are not operational because they were never fully implemented, thus jeopardizing efforts to conserve the last remnants of the Atlantic Forest.

Acknowledgments

Thanks to Carlos Galindo-Leal, Monica Fonseca, and Luiz Paulo Pinto for their comments, suggestions, and updated information.

References

APN (Administración de Parques Nacionales). 2002. Buenos Aires, Argentina. Online: www.medioambiente.gov.ar/sian/apn.

Bertolini, M. 1999. *Plan de Manejo del Parque Provincial Salto Encantado del Valle del Cuña-Pirú.* Ministerio de Ecología y Recursos Naturales Renovables. Gobierno de la Provincia de Misiones, Argentina.

Bertonatti, C. and Corcuera, J. 2000. *Situación ambiental argentina 2000.* Buenos Aires: Fundación Vida Silvestre Argentina (FVSA).

Green, M. and Paine, J. 1997. *Summary of protected areas recorded in the WCMC protected areas database.* Cambridge: World Conservation Monitoring Centre.

IBAMA (Instituto Brasileiro do Meio Ambiente e dos Recursos Naturais). Online: http://www2.ibama.gov.br/unidades.

IUCN (World Conservation Union). 1994. *Guidelines for protected area management categories.* Part II, The management categories. Gland, Switzerland, and Cambridge, UK: IUCN.

IUCN. 1998. *1997 United Nations list of protected areas.* Prepared by United Nations Environment Program (UNEP) World Conservation Monitoring Center (WCMC) and World Commission on Protected Areas (WCPA). Gland, Switzerland and Cambridge, UK: IUCN. Online: http://www.unep-wcmc.org/protected_areas/UN_list/index.htm.

MAB-UNESCO (Man and the Biosphere Program, United Nations Environmental, Scientific and Cultural Organization). 2001. *World network of biosphere reserves: 411 reserves in 94 countries.* Paris: Secretariat, Division of Earth Sciences (SC/ECO), UNESCO. Online: http://www.unesco.org/mab/wnbr.htm.

Ramsar 2000. Ramsar Convention on Wetlands. Online: www.ramsar.org.

SEA (Secretaria do Estado do Meio Ambiente). 2001. *Piracicaba/Juqueri Mirim área 1.* São Paulo, Brazil. Online: http://www.ambiente.sp.gov.br/apas/mapas_apas/piracicaba_1.htm.

UNESCO (United Nations Environmental, Scientific and Cultural Organization). 2001. *World Heritage List.* Online: http://www.unesco.org/whc/heritage.htm.

PART VI

Conclusion

Chapter 39

Outlook for the Atlantic Forest

Carlos Galindo-Leal, Ibsen de Gusmão Câmara, and Philippa J. Benson

The increasing likelihood of unprecedented extinctions driven by human activities—the biodiversity crisis—is a global phenomenon, but some regions are more threatened than others. The Atlantic Forest of South America is one of these regions. The impending threats to a broad range of species and ecosystems in the Atlantic Forest represent a dangerous trend that is becoming more common in many other regions around the world. The lessons we learn in the Atlantic Forest about how to assess and reverse these trends are key to building effective conservation efforts elsewhere.

Scientists, artists, and poets have long recognized the Atlantic Forest's rare beauty and the diversity of its animals, plants, landscapes, waterways, and indigenous cultures. Despite this recognition, planned and unplanned development has promoted the rapid expansion of agricultural industries and infrastructures that have devastated the forest and intensified social and economic inequities throughout the region.

No single book could fully analyze all aspects of the conservation problems and their potential solutions in a region as biologically, socially, and politically complex as the Atlantic Forest, which spans three countries and covers 1,350,000 km^2. Nonetheless, the authors of this volume have described the status of biodiversity throughout the hotspot, laid out the most important socioeconomic factors driving the loss of biodiversity, and discussed the capacity of people and institutions to address the challenges most crucial to conserving biodiversity.

The authors have examined the most significant concerns about biodiversity loss: identifying species in most urgent need of protection, preserving the remaining forested landscapes, and building the capacity of institutions with the greatest potential to promote conservation goals. We have tried to present a balanced perspective by bringing together the views of a broad range of seasoned experts, from prominent scientists to leaders of governmental and nongovernmental organizations. Their views, and the goals of the organizations for which they work, complement one another in many ways.

Plentiful evidence supports our belief that immediate, concerted efforts must be made to halt any further loss of biodiversity in the Atlantic Forest. Every remaining forest must be conserved, every remaining forest patch must be managed and protected, and every area must be supported to restore and maintain normal ecological and evolutionary processes. Without such focused and vigorous action, natural resources, cultural heritages, and wonders of the world will be lost forever.

Throughout this book, the authors have outlined a range of strategies to address the tremendous conservation challenges we face. These strategies include improving baseline information, shoring up environmental laws and government policies, and developing the capacity of institutions that support conservation. We can translate these strategies into specific actions: if baseline information is needed, we can compile data; if laws are weak, we can lobby to change them. All those committed to conservation must remember that these strategies and actions, however important, will not address the deepest root causes of the biodiversity crisis in the Atlantic Forest, nor those of other hotspots. The origins of the biodiversity crisis lie deep in cultural and economic attitudes that exhibit themselves in the daily behaviors of individuals, families, communities, institutions, and governments. We must focus on these historical attitudes as the seminal components of the problem of biodiversity loss and as essential ingredients in the solutions we must create together.

Where Do We Go from Here: Strategies and Actions

As the authors of this book have pointed out, improving baseline information is a huge and complex endeavor, even though many hundreds of species have already been identified as needing immediate protection to avoid extinction. Despite wide recognition of the extraordinary array of rare and precious species and ecosystems in the Atlantic Forest, there is still a dearth of basic information about the biological systems it harbors. New species are still being discovered throughout the Atlantic Forest; this is just one of many indications that there is much more to learn from the region's magnificent flora and fauna. We must assess the actual status of many species and their habitats, as well as the threats facing them, so that we can develop additional targeted conservation and protection actions. It is particularly important to evaluate species with restricted ranges because their populations are fragile and the opportunities to protect and conserve them are limited.

We must focus increased attention on developing appropriate scales for conservation actions. These should range from threatened and endemic species selected by their geographic distribution to priority sites determined by the occurrence of natural vegetation and viable populations. Conservation actions should also be chosen to preserve and promote ecological and evolutionary processes within defined ecosystems. We have underscored the specific need to strengthen efforts in areas where land protection is already in place, as well as in those that contain environmental services vital for local communities and in those where impacts and threats are particularly concentrated. The conservation community has already targeted many sites for priority conservation action throughout the

Atlantic Forest hotspot, some of which are integrated in biodiversity corridors. By strengthening existing protected areas, creating new ones, and providing incentives that increase the extent of interconnected natural landscapes, biodiversity corridors maintain and restore ecosystem connectivity. In doing so, they protect and enhance the services they deliver.

We have also pointed out the urgent need to develop human resources as a core component of consolidating conservation capacity in the Atlantic Forest. The people trained to address conservation challenges represent every discipline related to the task of saving the Atlantic Forest. Civil servants, politicians, researchers, park guardians, and advocates in nongovernmental agencies and in the general population need to be able to discuss conservation problems and solutions from informed perspectives. In addition, of course, sheer numbers matter, and the more people who understand the exigent need to move ahead rapidly with conservation actions, the more quickly and successfully conservation plans can be put in place.

Environmental policies and regulations in the three countries of the Atlantic Forest region need to move from their customary marginal position to the forefront of their national agendas. Although some laws have improved in recent years, comprehensive environmental policies are often lacking. Legal norms may be insufficient and enforcement inadequate. As laws and policies in Brazil, Argentina, and Paraguay reflect a stronger commitment to achieving concrete conservation outcomes, it will become easier to coordinate activities among government agencies, both within and between nations. Funding is another aspect of coordinated policy and law. Surprisingly, many people—including politicians—do not consistently link the health of the environment with the health of human populations nor with a healthy economy. As a consequence, funding for conservation projects often diminishes when there is competition from other initiatives that seem to have more immediate benefits to human or economic health. The truth is that conservation actions support the health of human and economic systems, particularly over the long term. Nonetheless, because limits on funding will always be an issue, conservation planners should strive to make their projects as self-sufficient as possible.

Finally, we have stressed the need to monitor robust indicators of biodiversity status so that we can determine if our actions are actually making a difference. Monitoring the dynamics of policies, land use, population increase, social conditions, and economies can also help us forecast and prepare for future events. Although some indicators now used are effective, others need to be improved. We also need to develop new and better indicators. Assessing indicators and communicating them to one another will allow us to track progress and make timely adjustments of conservation investments.

Attending to Root Causes

In addition to developing specific strategies and actions, conservationists must remain attentive to the underlying dynamics of biodiversity loss. The basic causes of biodiversity destruction are insidious and surprisingly easy to overlook.

First, many government policies and international loans encourage or include subsidies specifically designed to create financial conditions that promote further unrestrained development. For example, many subsidies and loans support programs that encourage the export of commodity products; promote the creation of monoculture forest plantations; and provide monies for large infrastructure projects such as dams, power plants, and highways. Few of these subsidies and loans carry any conditions that effectively mitigate environmental impacts or contribute to the preservation of biodiversity within the programs they support. Human populations are economic as well as biological entities, and the national and international subsidies that promote growth without strategic long-term planning for conservation of species and ecosystems are key forces of biodiversity destruction.

Second, as in other hotspots around the planet, human populations in the Atlantic Forest are large, and they continue to increase. As populations expand and develop, their demands put huge pressures on natural resources. Such pressures range from increased levels of subsistence hunting and gathering by underprivileged indigenous communities to rapid and poorly planned development of large dams systems to provide water and energy for urban dwellers. Unchecked in their demands, expanding human populations are a significant threat to biodiversity.

Finally, the economies that are driving the destruction of the Atlantic Forest today began in colonial times, when the wanton extraction and consumption of natural resources was the favored model of development. As populations and their demands have increased, this model has grown out of control, fed by free trade and globalization. The model persists today and eats away at the fabric that clothes human society: the ecosystems that provide clean air, clean water, and fertile soil. The desire for material wealth and comfort leads directly to uncontrolled destruction of the earth's resources and underlies many of the proximate factors that are demolishing biodiversity. The colonial development model has thoroughly transformed the extraordinary biome of the Atlantic Forest and reduced it to its present impoverished state.

A Resilient Forest

Despite these challenges, there are reasons to be hopeful for the future of the Atlantic Forest. Unlike the case in many other hotspots, much is known about where conservation actions are most needed in areas at risk throughout the Atlantic Forest in Brazil, Argentina, and Paraguay. Many encouraging developments reflect an authentic commitment to conservation goals by local, regional, and national leaders throughout the region. Perhaps most important, conservation science continues to document the amazing resiliency of nature. Species, ecosystems, waterways, and soils can be ravaged to within a breath of annihilation; yet given one chance, however slim, they persist and then flourish once more.

About the State of the Hotspots Series

The State of the Hotspots series documents the status and future prospects of biodiversity in the world's hotspots. The biodiversity hotspot concept comes from British ecologist Norman Myers. In 1988, he defined the idea to help conservationists determine what areas are the most important for preserving species. Biodiversity hotspots are regions that harbor a great diversity of endemic species yet have been significantly impacted and altered by human activities. The 25 hotpots identified around the world together contain 60 percent of the world's plant and animal species in less than 2 percent of the earth's land surface. The hotspot concept provides the scientific underpinnings for targeting research and conservation strategies. In this series, experts from a variety of disciplines will present and analyze current data on ecology, economics, policy, and conservation planning in each biodiversity hotspot. The combined expertise in each volume represents an international effort by a broad range of researchers to collect, synthesize, and present information critical to protecting and restoring these unique regions.

About the Contributors

Alexandre Pires Aguiar (Ph.D., Ohio State University, USA), Postdoctoral fellow at the Museum of Zoology of the University of São Paulo, Brazil.

Vanesa Arzamendia (Ph.D. candidate, Instituto Nacional de Limnología, Argentina), Researcher, National University of Litoral, Santa Fe, Argentina.

Diego Baldo, Researcher, Department of Genetics of the National University of Misiones, Argentina.

Manuel J. Belgrano, Researcher, Instituto de Botánica Darwinion, Argentina.

Philippa J. Benson (Ph.D., Carnegie Mellon University), is the Managing Editor at the Center for Applied Biodiversity Science, Conservation International, USA.

María Paula Bertolini, Researcher working with the Ministry of Ecology, Natural Resources and Tourism of Misiones Province, Argentina.

Mario S. Di Bitetti (Ph.D., New York State University, USA), Postdoctoral fellow, National University of Tucumán, Argentina.

Thomas Brooks (Ph.D., University of Tennessee, USA), Director of the Conservation Synthesis Department at Center for Applied Biodiversity Science at Conservation International, USA.

Elizabeth Cabrera, Head of Environmental Education at Guyra Paraguay and former member of the Fundación Moisés Bertoni, Paraguay.

José Maria Cardoso da Silva (Ph.D., University of Copenhagen, Denmark), Director for Amazonia at Conservation International–Brazil.

José Luis Cartes (M.Sc., Universidad de Córdoba, Argentina), Director of Conservation Area in Guyra Paraguay, Paraguay.

Jose Carlos Carvalho, Forestry Engineer, graduate of Federal Rural University of Rio de Janeiro. Executive Secretary, Brazilian Ministry of the Environment (1999–2000), Minister of State for the Environment and Sustainable Development, State of Minas Gerais.

Miguel Castelino, former park warden of the National Park Iguazú, Argentina.

Juan Carlos Chebez, Regional Office Director for Northeast Argentina, National Park Administration in Iguazú National Park, Argentina.

Sandra Emilia Chediack (Ph.D. candidate, National Autonomous University of Mexico), Researcher specializing in demography and sustainable use of forest resources in Misiones, Argentina.

Juan Pablo Cinto (M.Sc. candidate, University of Misiones, Argentina), General Director of Ecology in the Province of Misiones, Argentina.

Robert Clay (Ph.D., University of Cambridge, UK), Director of Research and Programs at Guyra Paraguay, Paraguay.

Juan Francisco Facetti (Ph.D., Foundation Universitaire Luxembourgeoise, Belgium), Professor, National University of Asunción, Paraguay, and former Minister of the National Environmental Authority of Paraguay.

Colleen Fahey (M.Sc. candidate, Duke University, USA), Research intern, Center for Applied Biodiversity Science at Conservation International, USA.

Frank Fragano, Principal Technical Advisor for the Paraguayan Wildlands Protection Initiative funded by the UNDP-Global Environmental Facility in Paraguay.

Miguel Franco Baqueiro (Ph.D., University of Wales, UK), Senior Lecturer in Ecology at the Department of Biological Sciences, University of Plymouth, UK.

Carlos Eduardo Frickmann Young (Ph.D., University College London, UK), Associate Professor, Economics Institute, Federal University at Rio de Janeiro, Brazil.

Carlos Galindo-Leal (M.Sc., Ph.D., University of British Columbia, Canada), Senior Director for State of the Hotspots Program, Center for Applied Biodiversity Science at Conservation International, USA.

Adriano Garcia Chiarello (Ph.D., University of Cambridge, UK), Professor of Vertebrate Zoology of the Catholic University in Minas Gerais, Brazil.

Alejandro R. Giraudo (Ph.D., Córdoba University, Argentina), Researcher at the

National Council of Technical and Scientific Research specializing in the Atlantic Forest in Argentina.

Ibsen de Gusmão Câmara, Admiral, NGO advisor, and member of the National Environmental Council in Brazil.

Norma Hilgert (National University of Córdoba, Argentina), Researcher, National Ministry of Science and Technology, Argentina.

Márcia Makiko Hirota (M.Sc., Catholic Pontifícia University of Campinas, Brazil), Project Director for Fundação SOS Mata Atlântica in São Paulo, Brazil, and coordinator of the atlas of forest remnants of Mata Atlântica.

Silvia Cristina Holz (Ph.D. candidate, National University of Buenos Aires, Argentina), Researcher, Fundación Vida Silvestre Argentina and World Wildlife Fund in Misiones, Argentina.

Silvio Jablonski (Ph.D., Federal University of Rio de Janeiro, Brazil), Professor, Department of Oceanography, Rio de Janeiro State University, Brazil.

Thomas R. Jacobsen (M.A., University of Kent, Canterbury, UK), Researcher, Center for Applied Biodiversity Science at Conservation International, USA.

Maria Cecília Kierulff (Ph.D., University of Cambridge, UK), Coordinator of the Central Corridor in Conservation International–Brazil and former coordinator of the Golden Lion Tamarin Translocation project in Rio de Janeiro, Brazil.

Devra Kleiman (Ph.D., University College London, UK), Former Assistant Director for Research and Head of the Department of Zoological Research, National Zoological Park, Washington, DC, USA.

Ernesto R. Krauczuk, Researcher, specializing in inventories and biodiversity studies in the Ministry of Ecology in Misiones, Argentina.

Alexandra-Valeria Lairana Ramírez (M.Sc., International University of Andalucía, Spain), Biologist from Mayor University of San Andrés (UMSA), La Paz, Bolivia.

Penny F. Langhammer (M.Sc., Duke University, USA), Researcher, Center for Applied Biodiversity Science at Conservation International, USA.

María Leichner Reynal (Ph.D., Law National University of Litoral, Argentina), executive director and cofounder of Fundación Ecos, Uruguay.

Daniel Ligier, Researcher in the Experimental Agriculture Station of the Instituto Nacional de Tecnología Agropecuaria, Corrientes, Argentina.

Sérgio Lucena Mendes (M.Sc., Federal University of Brasilia; Ph.D., Campinas State University, Brazil), Professor, Federal University of Espírito Santo, and founder and Director of the Institute for Atlantic Forest Research in Brazil.

Ana María Macedo Sienra, Coordinator of Private Initiatives Conservation Program of Moisés Bertoni Foundation and Paraguayan Representative of the Trinational Green Corridor Initiative.

Carlos Henrique Madeiros Casteleti, Researcher, Federal University of Pernambuco, Brazil.

Denise Marçal Rambaldi, Executive Director of Golden Lion Tamarin Association, Rio de Janeiro, Brazil.

Amalia M. Miquelarena, Professor of Icthiology, National University of La Plata, Argentina.

Eloina Neri de Matos (M.Sc., Centro Tropical de Investigación y Enseñanza, Costa Rica), Researcher, Instituto de Estudos Sócio-Ambientais do Sul da Bahia in Brazil.

Silvio Olivieri, Vice President of Conservation Knowledge Department, Center for Applied Biodiversity Science at Conservation International, USA.

Ulyses F. J. Pardiñas (Ph.D., National University of La Plata, Argentina), researcher, National Council of Technical and Scientific Research (CONICET), Argentina.

Luiz Paulo Pinto (M.Sc., Federal University of Minas Gerais, Brazil), director of the Mata Atlântica Program at Conservation International Brazil.

Guillermo Placci (Ph.D., National University of La Plata, Argentina), Director of the regional office of Fundación Vida Silvestre Argentina and World Wildlife Fund and administrator of the Wildlife Reserve Urugua-í in Misiones, Argentina.

Hernán Povedano (Ph.D. candidate, National University of La Plata, Argentina), Researcher specializing in sustainable development and regional planning projects in Misiones, Argentina.

Jamie K. Reaser (Ph.D., Stanford University, USA), Assistant Director for International Policy, Science, and Cooperation for the U.S. National Invasive Species Council and member of the Global Invasive Species Programme, USA.

Luis Alberto Rey, President of the Directory of the National Park Administration of Argentina, former Minister of Ecology and Natural Renewable Resources of Misiones Province, and former Minister of Agriculture and Production.

Cláudia Maria Rocha Costa, Project supervisor for Biodiversitas Foundation, Brazil.

Anthony B. Rylands (Ph.D., University of Cambridge, UK), Senior Director for Conservation Biology, Center for Applied Biodiversity Science at Conservation International, USA.

Angela Sánchez, Professor of Education, Escuela Normal de Villarrica, Paraguay, and former instructor at the first aboriginal school in the Mbyá community in Misiones, Argentina.

Marcelo Tabarelli (Ph.D., University of São Paulo, Brazil), Professor, Federal University of Pernambuco, and Director of the Environmental Research Center of the Northeast, Brazil.

M. Cecília Wey de Brito (Ph.D. candidate, University of São Paulo, Brazil), Coordinator of the Alliance for the Conservation of Atlantic Forest, SOS Mata Atlântica Foundation and Conservation International Brazil.

Alberto Yanosky (M.Sc., Ph.D., National University of La Plata, Argentina), Chief Executive Officer at Guyra Paraguay and former Deputy Director of Fundación Moisés Bertoni in Paraguay.

Silvia Renate Ziller (M.Sc., Ph.D., Federal University of Paraná), president of the Horus Institute for Environmental Conservation and Development in southern Brazil.

Index